Lecture Notes in Economics and Mathematical Systems

For information about Vols. 1–183, please contact your bookseller or Springer-Verlag

Vol. 184: R. E. Burkard and U. Derigs, Assignment and Matching Problems: Solution Methods with FORTRAN-Programs. VIII, 148 pages. 1980.

Vol. 185: C. C. von Weizsäcker, Barriers to Entry. VI, 220 pages. 1980.

Vol. 186: Ch.-L. Hwang and K. Yoon, Multiple Attribute Decision Making – Methods and Applications. A State-of-the-Art-Survey. XI, 259 pages. 1981.

Vol. 187: W. Hock, K. Schittkowski, Test Examples for Nonlinear Programming Codes. V. 178 pages. 1981.

Vol. 188: D. Bös, Economic Theory of Public Enterprise. VII, 142 pages. 1981.

Vol. 189: A. P. Lüthi, Messung wirtschaftlicher Ungleichheit. IX, 287 pages. 1981.

Vol. 190: J. N. Morse, Organizations: Multiple Agents with Multiple Criteria. Proceedings, 1980. VI, 509 pages. 1981.

Vol. 191: H. R. Sneessens, Theory and Estimation of Macroeconomic Rationing Models. VII, 138 pages. 1981.

Vol. 192: H. J. Bierens: Robust Methods and Asymptotic Theory in Nonlinear Econometrics. IX, 198 pages. 1981.

Vol. 193: J. K. Sengupta, Optimal Decisions under Uncertainty. VII, 156 pages. 1981.

Vol. 194: R. W. Shephard, Cost and Production Functions. XI, 104 pages. 1981.

Vol. 195: H. W. Ursprung, Die elementare Katastrophentheorie. Eine Darstellung aus der Sicht der Ökonomie. VII, 332 pages. 1982.

Vol. 196: M. Nermuth, Information Structures in Economics. VIII, 236 pages. 1982.

Vol. 197: Integer Programming and Related Areas. A Classified Bibliography. 1978 – 1981. Edited by R. von Randow. XIV, 338 pages. 1982.

Vol. 198: P. Zweifel, Ein ökonomisches Modell des Arztverhaltens. XIX, 392 Seiten. 1982.

Vol. 199: Evaluating Mathematical Programming Techniques. Proceedings, 1981. Edited by J.M. Mulvey. XI, 379 pages. 1982.

Vol. 200: The Resource Sector in an Open Economy. Edited by H. Siebert. IX, 161 pages. 1984.

Vol. 201: P. M. C. de Boer, Price Effects in Input-Output-Relations: A Theoretical and Empirical Study for the Netherlands 1949–1967. X, 140 pages. 1982.

Vol. 202: U. Witt, J. Perske, SMS – A Program Package for Simulation and Gaming of Stochastic Market Processes and Learning Behavior. VII, 266 pages. 1982.

Vol. 203: Compilation of Input-Output Tables. Proceedings, 1981. Edited by J. V. Skolka. VII, 307 pages. 1982.

Vol. 204: K. C. Mosler, Entscheidungsregeln bei Risiko: Multivariate stochastische Dominanz. VII, 172 Seiten. 1982.

Vol. 205: R. Ramanathan, Introduction to the Theory of Economic Growth. IX, 347 pages. 1982.

Vol. 206: M. H. Karwan, V. Lotfi, J. Telgen, and S. Zionts, Redundancy in Mathematical Programming. VII, 286 pages. 1983.

Vol. 207: Y. Fujimori, Modern Analysis of Value Theory. X, 165 pages. 1982.

Vol. 208: Econometric Decision Models. Proceedings, 1981. Edited by J. Gruber. VI, 364 pages. 1983.

Vol. 209: Essays and Surveys on Multiple Criteria Decision Making. Proceedings, 1982. Edited by P. Hansen. VII, 441 pages. 1983.

Vol. 210: Technology, Organization and Economic Structure. Edited by R. Sato and M. J. Beckmann. VIII, 195 pages. 1983.

Vol. 211: P. van den Heuvel, The Stability of a Macroeconomic System with Quantity Constraints. VII, 169 pages. 1983.

Vol. 212: R. Sato and T. Nôno, Invariance Principles and the Structure of Technology. V, 94 pages. 1983.

Vol. 213: Aspiration Levels in Bargaining and Economic Decision Making. Proceedings, 1982. Edited by R. Tietz. VIII, 406 pages. 1983.

Vol. 214: M. Faber, H. Niemes und G. Stephan, Entropie, Umweltschutz und Rohstoffverbrauch. IX, 181 Seiten. 1983.

Vol. 215: Semi-Infinite Programming and Applications. Proceedings, 1981. Edited by A. V. Fiacco and K. O. Kortanek. XI, 322 pages. 1983.

Vol. 216: H. H. Müller, Fiscal Policies in a General Equilibrium Model with Persistent Unemployment. VI, 92 pages. 1983.

Vol. 217: Ch. Grootaert, The Relation Between Final Demand and Income Distribution. XIV, 105 pages. 1983.

Vol. 218: P. van Loon, A Dynamic Theory of the Firm: Production, Finance and Investment. VII, 191 pages. 1983.

Vol. 219: E. van Damme, Refinements of the Nash Equilibrium Concept. VI, 151 pages. 1983.

Vol. 220: M. Aoki, Notes on Economic Time Series Analysis: System Theoretic Perspectives. IX, 249 pages. 1983.

Vol. 221: S. Nakamura, An Inter-Industry Translog Model of Prices and Technical Change for the West German Economy. XIV, 290 pages. 1984.

Vol. 222: P. Meier, Energy Systems Analysis for Developing Countries. VI, 344 pages. 1984.

Vol. 223: W. Trockel, Market Demand. VIII, 205 pages. 1984.

Vol. 224: M. Kiy, Ein disaggregiertes Prognosesystem für die Bundesrepublik Deutschland. XVIII, 276 Seiten. 1984.

Vol. 225: T. R. von Ungern-Sternberg, Zur Analyse von Märkten mit unvollständiger Nachfragerinformation. IX, 125 Seiten. 1984

Vol. 226: Selected Topics in Operations Research and Mathematical Economics. Proceedings, 1983. Edited by G. Hammer and D. Pallaschke. IX, 478 pages. 1984.

Vol. 227: Risk and Capital. Proceedings, 1983. Edited by G. Bamberg and K. Spremann. VII, 306 pages. 1984.

Vol. 228: Nonlinear Models of Fluctuating Growth. Proceedings, 1983. Edited by R. M. Goodwin, M. Krüger and A. Vercelli. XVII, 277 pages. 1984.

Vol. 229: Interactive Decision Analysis. Proceedings, 1983. Edited by M. Grauer and A. P. Wierzbicki. VIII, 269 pages. 1984.

Vol. 230: Macro-Economic Planning with Conflicting Goals. Proceedings, 1982. Edited by M. Despontin, P. Nijkamp and J. Spronk. VI, 297 pages. 1984.

Vol. 231: G. F. Newell, The M/M/∞ Service System with Ranked Servers in Heavy Traffic. XI, 126 pages. 1984.

Vol. 232: L. Bauwens, Bayesian Full Information Analysis of Simultaneous Equation Models Using Integration by Monte Carlo. VI, 114 pages. 1984.

Vol. 233: G. Wagenhals, The World Copper Market. XI, 190 pages. 1984.

Vol. 234: B. C. Eaves, A Course in Triangulations for Solving Equations with Deformations. III, 302 pages. 1984.

Vol. 235: Stochastic Models in Reliability Theory. Proceedings, 1984. Edited by S. Osaki and Y. Hatoyama. VII, 212 pages. 1984.

continuation on page 613

Lecture Notes in Economics and Mathematical Systems

Managing Editors: M. Beckmann and W. Krelle

335

A.G. Lockett G. Islei (Eds.)

Improving Decision Making in Organisations

Proceedings of the Eighth International Conference on Multiple Criteria Decision Making Held at Manchester Business School, University of Manchester, UK, August 21st–26th, 1988

Springer-Verlag
Berlin Heidelberg New York London Paris Tokyo Hong Kong

Managing Editors

Prof. Dr. M. Beckmann
Brown University
Providence, RI 02912, USA

Prof. Dr. W. Krelle
Institut für Gesellschafts- und Wirtschaftswissenschaften
der Universität Bonn
Adenauerallee 24–42, D-5300 Bonn, FRG

Editors

Professor A. G. Lockett
Dr. G. Islei
Manchester Business School
Booth St. West
Manchester, M15 6PB, United Kingdom

ISBN 978-3-540-51795-5 ISBN 978-3-642-49298-3 (eBook)
DOI 10.1007/978-3-642-49298-3

2142/3140-543210 – Printed on acid-free paper

PREFACE

MCDM has been an active research area for over 20 years and the previous conferences clearly showed a tremendous growth of interest. A variety of successful applications and recent developments of interactive computer software to support decision making confirm a sustained progress. We therefore decided to make our theme "Improving Decision Making in Organisations". The aim was to take stock of the impact of multicriteria concepts in organisations and to involve management practitioners from a wide range of backgrounds. To this end the conference was organised round five broad themes and papers were solicited on the following topics:-

- Psychology - how do individuals in practice use and relate to the methodologies.
- Organisation - how do our models fit into the decision making framework of real organisations.
- Application - how have the models been used in practice and what is the users view.
- Methodology - what are the new areas in model development.
- Related Areas - is there complementary work eg. Expert Systems which may be attempting to solve very similar problems.

The call for papers produced an overwhelming response of over 100 papers. They were from a variety of disciplines and applications, and we decided to devote approximately one day to each of the five areas. We are very impressed by the results which can be seen in this edited proceedings.

The papers are too numerous to discuss individually and it would be improper to highlight just a few. However, some impressions remain from the event which we think are worth noting. Many techniques have passed the "age of consent" and are used in real life situations by interested decision makers. Although technical problems still confront us the emphasis has now shifted to process issues and implementation. How and why do people use these methods, and how can they be adopted to every day decision making in the management environment? We should be encouraged by the conference but at the same time appreciate the research that needs to be done.

Such an edited text, however, cannot convey the interaction and integration of the live event. Over 100 people from 27 different countries took part - a mini United Nations. For five days we listened, talked, and discussed in both formal and informal sessions. We hope we played some part in creating a broader based MCDM network which will continue long after the event.

Manchester
United Kingdom 1989

Geoff Lockett
Gerd Islei

The papers have had to receive some degree of selection and the ones chosen reflect the original intention behind the theme. Some papers presented at this conference are not included in the proceedings. Their abstracts can be found at the end of the volume.

Acknowledgement

We would like to thank the organisations who sponsored our event. Without them the international character of the conference would have been greatly diminished. We also wish to express our gratitude to the staff at Manchester Business School for caring for our creature comforts.

TABLE OF CONTENTS

PART I - Psychology

PART II - Organisation

PART III - Applications

APPENDICES

ACQUISITIONS, MERGERS AND HABITUAL DOMAINS ANALYSIS

Deepak K. Datta
Assistant Professor of
Strategic Management
School of Business
University of Kansas
Lawrence, KS 66044

P.L. Yu
Carl A. Scupin Distinguished
Professor
School of Business
University of Kansas
Lawrence, KS 66044

Abstract

Acquisitions and mergers, which have been important mechanisms in corporate growth and diversification, involve the integration and assimilation of two previously independent organizations with distinct habitual domains (HD). Therefore, along with the choice of the right acquisition candidate, effective integration of the two habitual domains is key to the success or failure of an acquisition. This paper proposes a framework for the systematic analysis of a firm's habitual domain and its impact on acquisitions and mergers, with the objective of enhancing the probabilities of success and minimizing the chances of failure.

I. INTRODUCTION

Organizations in the 1980s have been operating in an environment which has become increasingly volatile and competitive. In order to survive, grow and be profitable, they are being forced to look beyond their current operations and diversify into related and unrelated areas of business. The transition can be effected in a number of ways (e.g., joint ventures, internal development and acquisitions and mergers) -- however, acquisitions remain as the most commonly used mode [22]. Billions of dollars are invested each year by U.S. corporations in such ventures -- in 1986, for example, there were more than 4000 mergers involving a total dollar value of more than $190 billion. A similar trend has been observed in the U.K., where the number of acquisitions increased from 940 in 1985 to 1320 in 1986 [4]. However, acquisitions have also been characterized by a high failure rate -- 40-50% according to statistics compiled by W.T. Grimm & Co. [11]. Thus, the effective management of the acquisition process remains a task of high priority in the minds of many corporate executives.

Acquisitions have figured quite prominently both in the empirical literature, especially in the areas of finance [14], industrial economics and strategic management [5,18,26], and also in the more prescriptive, practitioner-oriented literature [15,19,29]. A variety of frameworks have also been suggested by various authors [2,3,13,28] providing a number of alternative approaches towards improving the chances of success. However, past efforts have generally been characterized by a rather narrow focus and have, consequently, been of limited help to managers involved in the making of effective acquisition decisions.

On the other hand, scholars and researchers in management science have been developing various optimization techniques under the rubric of multicriteria decision making [27,31,34].

Unfortunately, the merging of two entities is generaly a complex, dynamic process and decision situations that are encountered do not easily lend themselves to the traditional mathematical and static approaches in multiple criteria decision making (MCDM). In this context, concepts in second order games and habitual domains (HD) analysis [31,32,33], which seeks the dynamic restructuring of MCDM systems, provide at least a partial remedy.

In this paper we suggest an application of habitual domains analysis towards a more balanced and effective analysis of acquisitions and mergers. First, we introduce the concept of organizational habitual domains. Then, we provide a framework (IFFS) which allows the systematic analysis of acquisition opportunities and problems. Finally, we discuss how habitual domains analysis can help facilitate negotiations and post-acquisition implementation processes in acquisitions and mergers.

2.HABITUAL DOMAINS

The concept of habitual domains can best be illustrated using the following example:

> A retiring Chairman was faced with the difficult choice of selecting a heir from two prospective finalists, invited them to his ranch. Providing them with two identical horses (one white and the other black), he posed an unique and challenging problem -- "whoever's horse is slower is completing the course will be declared the winner."
>
> The puzzled candidates were taken aback. The problem did not fit in with their habitual ways of thinking -- the emphasis and evaluation criteria in a horse race, after all, is generally "quickness" and speed. Finally, A getting out of his habitual ways of thinking, jumped on B's horse and sped to the finish line. When B realized what was happening, it was too late!

The above example provides an illustration of the fact that sticking closely to one's existing habitual domain might not be particularly conducive to the finding of a good solution in a complex problem . Indeed many non-trivial decision problems are solved by being able to "jump out" of the constraints of the current habitual domain. Many real life situations pose a broad array of "non-trivial" and "challenging" problems and decisions which typically cannot be analyzed or solved by traditional mathematical or optimal control theory. Second order games and habitual domains analysis, in such cases, can be particularly helpful [32,33].

Broadly defined, habitual domains (HD) are a "collection of ideas and actions that can potentially be activated at point in time 't'. Depending on our emphasis, we may focus on the dimension related to "information processing" (e.g., goal setting, state valuation, memory, charge structures, attention, charge release methods, external information inputs, physiological conditions etc.) or focus on the dimension related to "individual events" (decisions problems, situations etc.). The elements of habitual domains can be further decomposed into potential domains, actual domains, activation possibility and reachable domains. A detailed description of the above and on how habitual domains can reach their "steady states" most of the time can be found in [31,32]. As habitual domains in individuals reach a steady state, he or she will manifest a steady pattern of behavior and/or decisions.

2.1 Organizational Habitual Domains

Organizations, as living entities, also have their own habitual domain. While individual habitual domains can be described using the two broad sets of factors described above, organizational habitual domains are more complex and a third dimension, namely, 'organizational units', is required for its definition (Figure 1). The way organizational members perceive, evaluate and react to various events often depend on the functional area of the members. Thus, the habitual domains of those in marketing might be very different from those in production or in finance. However, it has been observed that organizations, over a period of time, develop habitual ways in which they process information available to them. They habitually emphasize certain goals and objectives, functions and functional areas (e.g., while in some organizations sales and marketing may be emphasized, in others it may be the production function) and habitually adopt certain corporate and business strategies. Quite often, it may be a reflection of the habitual domain of the top management or the dominant coalition [25]. Environmental and organizational changes, however, might bring about changes in the composition of the dominant coalition and might result in a shift in habitual domains.

The enduring nature of the organizational habitual domains has been partially discussed in the literature on corporate cultures [23] For example, the regional Bell operating companies (RBOCs or baby Bells) formed as a result of the divestiture forced on AT&T, show a very high degree of similarity in terms of habitual domains as evident from an examination of their diversification patterns. Moreover, the diversification attempts have been triggered by similar "charge structures" -- e.g., excess cash and a desire to move into areas which provide the momentum for future growth.

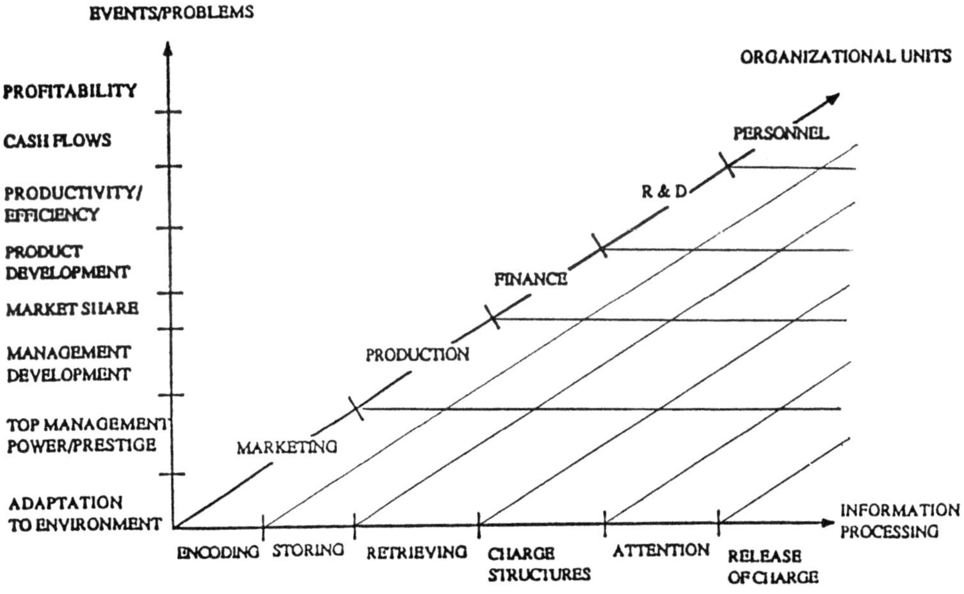

Figure 1: ORGANIZATIONAL HABITUAL DOMAINS

3. ACQUISITION DECISIONS AND HABITUAL DOMAINS

Why do organizations undertake acquisitions? As previously mentioned, organizations and organizational members have certain goals they seek to achieve through mergers and acquisitions [20]. In other words, acquisitions are an effort by organizations to move from the current habitual domain (HD^0) towards the ideal or desired habitual domain (HD^*). The gap ($HD^0 \backslash HD^*$) creates a "charge" (e.g., in terms of excess cash, poor profitability, lack of growth and organizational vulnerability to being acquired) which can be 'released' in the form of acquisition attempts. The dimensions related to "information processing" in organizational habitual domains may be studies according to the model in Figure 2.

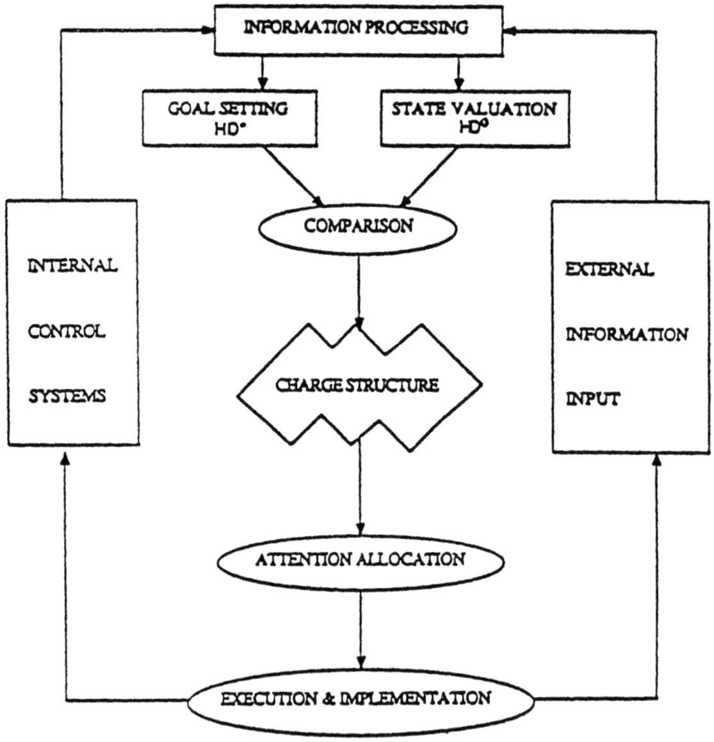

Figure 2: ORGANIZATIONAL DECISION MECHANISMS

3.1 Identifying Current Habitual Domains and Defining the Desired Habitual Domain

A corporation must begin with a process of self-evaluation -- one of identifying its current habitual domain (HD^0). The evaluation would necessarily include an evaluation of the external environment within which the firm is currently operating, and also of its internal environment (strengths and weaknesses). While industry attractiveness can be measured along the five broad dimensions of (1) market factors, (2) competitive factors, (3) financial factors, (4) technological factors, and (5) socio-political factors, a sixth dimension, namely, "organization/management" is essential for the evaluation of firm strengths. Each dimension, in turn, has several sub-factors along which industry attractiveness and business strengths can be initially measured and then consolidated (an example of the various subfactors associated with "competitive" dimension of industry attractiveness is provided in Appendix I). Also, the existing fit in terms of various organizational elements (shared values, structure, systems, management style, strategy and skills) needs to be carefully evaluated. Understanding the current state, while often overlooked, is extremely important in the identification of the desired state and the subsequent development of appropriate acquisition strategies.

Along with an identification of the current habitual domain (HD^0), the organization must also identify the desired habitual domain (HD^*) which it and its members seek to achieve. It could be in terms of positioning the organization in more attractive industries in the future (with high profitability and growth potential), developing organizational strengths along certain key success factors, increased organizational size and management power and prestige, or any other goal (stated or unstated) which the organization and its key members seek to achieve. Identification of the desired habitual domain (HD^*) can be done along factors similar to those used in the evaluation of the current habitual domain.

Evaluating HD^0 and identifying the desired HD^* allows a corporation to assess the 'gap' between where it currently 'is' and where 'it wants to be'. The nature and the extent of this gap, in turn, acts as a 'trigger' mechanism for the development of appropriate strategies which would allow the organization to make an effective transition from HD^0 to HD^*. Often this is achieved via acquisitions and mergers, which, over the years have increased in popularity. However, as previously mentioned, they have also been plagued by high failure rates. While the causes of such failure are many, failed acquisitions are often characterized by an inability on the part of the organization to set clear direction (define the desired HD^*), a lack of proper evaluation of acquisition candidates, or, an inability on the part of the acquiring firm management to effectively assimilate the two organizations once the acquisition is consummated. In the following paragraphs we present an analytical framework (IFFS) which should help organizations make the right choices and avoid the pitfalls so commonly associated with acquisitions.

3.2 Candidate Selection using the IFFS Framework

Selection of a candidate, the acquisition of which is expected to move the organization towards the desired habitual domain generally starts with an initial pool of possible candidates. The selection process, then, is essentially one of narrowing down the choices to a much smaller, ordered set of alternatives using an appropriate criteria set. Once this is accomplished and the candidate chosen, the organization can move to the next step, namely, negotiations, which, if successfully completed, will result in the acquisition being consummated.

The candidate selection stage typically involves two major steps, (1) identifying an initial set of candidates, and (2) defining the set of criteria to be used in the selection.

(1) Identifying an initial set:

Various informational inputs, both from 'inside' and 'outside' the organization (e.g., market information, informal contacts of top management, management consultants, investment bankers or even computerized databases) serve to generate a starting list of candidates. Obviously, thorough analyses of all such candidates is neither feasible nor desirable from the viewpoint of time and cost. Consequently, there is a need for 'cursory screening' -- a process which serves to narrow the pool to a more manageable "initial set" by eliminating the obviously unacceptable or dominated ones. This process relies extensively on managerial judgement and intuition and is influenced largely by elements of the organization's HD. However, caution must be exercised by management to avoid being constrained by their current habitual domain and, thereby overlook the not-so-obvious opportunities. In fact, ideally, effort should be made so that the process is guided not by current habitual domains but by the habitual domain that the organization desires to achieve.

(2) Defining a criteria set for analysis and evaluation - the IFFS framework

Perhaps the most crucial task in the selection phase involves the definition of the multiple criteria and sub-criteria along which the alternatives need to be analyzed in narrowing the field towards the 'final choice' of candidate. It means considering not only 'industry attractiveness' and the 'firm strengths' of the target firm along the dimensions suggested earlier, but also the extent of 'interorganizational fit' or compatibility between the two organization [7] and the extent of synergistic benefits that the acquisition is likely to provide [19] (Figure 3).

Thus, there are four sets of criteria that need to be considered in a complete, unbiased analysis of acquisition opportunities. These form the basic dimensions of HD^* for effective acquisition decisions.

(i) **Industry Attractiveness:** The current and the potential (or future) attractiveness of the industry of the candidate's business is certainly of prime concern to the acquiring firm in making a choice. Such attractiveness, which can be defined by the extent to which a particular acquisition helps close the gap between $HD^* \backslash HD^0$, can be measured along each dimension using the various criteria and subcriteria mentioned earlier (e.g., the degree of rivalry in the industry, the threat of new entry, the bargaining power of buyers and suppliers and the threat of substitutes for the 'competitive' dimension). Once evaluated using a scoring mechanism, (e.g., on a Likert type 1-5 scale with '1' and '5' signifying 'very low' and 'very high' attractiveness respectively) the criteria scores can then be consolidated to arrive at a composite score for industry attractiveness. The evaluation and the scoring can be done either jointly or individually by members of the task force based on a consideration of all possible sources of data and information.

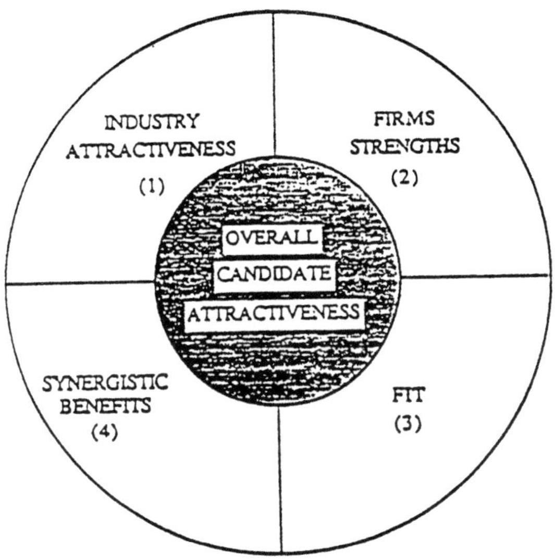

Figure 3: IFFS ANALYSIS

(ii) **Firm Strengths**: This refers to the set of criteria required to assess the position of a candidate firm within its industry in terms of its relative strengths and weaknesses. As stated in section 3.1, the multiple criteria can be grouped into six broad categories. The assessment is done along the subfactors and then consolidated using a system of weights and scores. A good acquisition candidate is one which has the strengths and competences necessary for the acquiring firm to move towards the desired habitual domain (HD^*).

(iii) **Fit**: For an acquisition to be successfully assimilated, management of the acquiring firm will also have to assess the extent of fit between its own organization and the potential candidate. Lack of fit, which can be along the various organizational elements discussed earlier, can have particularly undesirable consequences, especially from the perspective of post-acquisition integration and assimilation. For example, a lack of fit between the management styles of the two firms or differences in organizational systems can make it extremely difficult to integrate or assimilate the acquired entity towards the achievement of organizational objectives [7]. This was evidenced in the acquisition of EDS by General Motors where the two management had very different styles of operation making it virtually impossible for them to work together. A lack of fit can have a particularly severe impact on acquisition performance in related acquisitions, primarily because of lost synergistic benefits. Moreover, a lack of fit, has also been associated with post-acquisition turnover among key personnel in the acquired firm -- often jeopardizing the very success of an acquistition [12].

(iv) **Synergistic Benefits**: Another important set of criteria in the evaluation of a candidate in a related acquisition is the extent of available synergistic benefits. Merging or integration of activities following an acquisition can result in more efficient operations -- benefits which are generally referred to as synergies or economies. These benefits accrue from the integration of operations and activities in areas such as production, warehousing, transportation/distribution and R&D, due to resultant cost savings in the form of economies of scale and scope. Additional cost savings can be achieved through improved utilization of managerial skills. Also, effective sharing of complementary resources and skills (parts of the habitual domain of the involved organization) often results in further benefits and attendant competitive advantages. It provides the acquiring firm with an increased set of strengths and fewer weaknesses and helps it move towards the desired HD^*. While an evaluation of potential benefits undoubtedly forms an important input in the selection decision one must, however, not forget the costs (both tangible and intangible) that are associated with their realization [6,8,12]. Problems and difficulties faced in combining two previously independent firms (e.g., conflicts in amalgamating firms with differing habitual domains) given rise to costs which need to be assessed. The potential benefits should then be adjusted downwards to more accurately reflect the net achieveable synergistic benefits.

3.3 Overall Candidate Attractiveness Score and Candidate Choice

Once the appropriate set of criteria have been developed and candidates in the initial set evaluated along each criterion and sub-criterion in the decision tree, the scores for each can then be collapsed into a single consolidated score for each of the IFFS dimensions using relevant weights and the backward integration process [21,32]. In evaluating the candidates, the absence of available objective data in many instances might necessitate judgmental evaluation. To reduce the element of bias and increase the validity of the evaluation, a quasi-delphi approach using multiple evaluators, internal and external, is recommended. Particularly helpful would be the use of outside experts (industry experts or consultants) who are not constrained by their own and the acquiring firms' habitual domains and are, therefore, better able to provide an independent and unbiased evaluation.

Provided a candidate's score exceeds the minimum threshold level along each of the dimensions of industry attractiveness, firm strengths and fit, an overall attractiveness score for the candidate can be computed. The score can then be used to (i) screen out undesirable candidates and drop them from further consideration, and (ii) establish a rank order among desirable candidates, which should enable the organization to move towards the next step, namely, negotiations.

4. NEGOTIATIONS

The noted Chinese scholar and strategist, Sun-Tzu, in his classical book "Principles of War" once said,

"Knowing yourself and knowing your enemy, will allow you to survive every battle"

While the target or candidate firm is not necessarily an 'enemy', 'knowing' the acquired firm (i.e. knowing its habitual domain) provides the acquiring firm management with a significant edge in planning its negotiations strategy. Along with issues of industry attractiveness, firm strengths, fit and synergies, questions to which answers need to be sought include, "Who are the key players in the candidate firm?" "What are their interests and habitual domains?" "What is the acquiring firm's current level of understanding of their habitual domain?" "Are there common goals which the acquiring firm management can emphasize during negotiations and thus obtain the target firm's cooperation?" "Can the acquiring firm change the rules of the game (the time and the habitual constraints) in a way which enables the outcomes to be in its favor?"

While it is generally acknowledged that negotiations in acquisitions is as much an art as a science [9], a careful analysis of the candidate firm and its habitual domain makes an analytical approach feasible. Moreover, proper planning, in addition to defusing the hostility factor, also enables participants in the negotiations process to focus on objective rather than emotional factors. It will help put together a constructive acquisitions package which would enable the negotiations to proceed smoothly to closing. Successful negotiations and closing, in turn, allows the acquiring firm to proceed to the next stage -- namely, implementing the acquisition and assimilating the two organizations.

5. IMPLEMENTATION AND ASSIMILATION

Unfortunately, the emphasis in acquisitions and mergers among managers is often more on "making the deal" than on "making the deal work". Many promising acquisitions do not achieve their potential because of the acquiring firm's inability to manage the acquisition after it was consummated [16,17,24. It is here, perhaps more than in any other stage, that an organization should make a concerted effort at understanding the habitual domains of the other firm to avoid dysfunctional conflicts and devise ways by which they reinforce one another. Failure to understand the implications of differences in habitual domains -- (e.g., in terms of motives, goals, strategies and styles of operation) might manifest itself in post-acquisition conflicts which can seriously jeopardize an acquisition. Also, all too common is the 'conquering army syndrome' - an attempt on the part of the acquiring firm management to undermine the habitual domain of the acquired firm and impose its own. Often such action is followed by disagreements and frustation leading to an exodus of talented people in the acquired firm -- probably the very people who had made the firm an attractive candidate in the first place. The acquisition of Fairchild Camera and Instruments Ltd. by Schlumberger provides an interesting example of this phenomenon. Schlumberger imposed its own management team and systems on Fairchild immediately following the acquisition, resulting in most of the key engineering people leaving the organization. As a result, product development withered and profits dissapeared [1].

A careful analysis, evaluation and understanding of the habitual domains of the acquired company also helps the acquiring firm management make the right post-acquisition decisions. One such decision involves the extent of autonomy that should be given to the acquired firm -- a decision which depends both on the type of acquisition (related or unrelated to the company's core business) and the observed differences in the habitual domains of the two companies. In related acquisitions, where autonomy may be dysfunctional (because it hinders the realization of operating synergies), knowledge of the acquired firms' habitual domain helps in the coordination of activities and in the installation of appropriate control systems to effectively monitor the integration process. Moreover, a better understanding of the interests of different organizational members is useful in the resolution of conflicts that might arise and in channeling their energies towards the achievment of overall goals.

Differences in habitual domains need to be carefully monitored because they tend to generate considerable stress, anxiety and uncertainty among organizational members - especially those of the acquired firm. Here again, by expanding its own habitual domain to understand that of the acquired firm and its members, the acquiring firm management can take appropriate actions and decisions which helps both in reducing undesirable stress and anxiety but also helps maintain the level of motivation and commitment necessary to make the acquisition a success.

6. CONCLUSION

Acquisitions and mergers have become an integral part of the business world. Yet study after study have shown that a large number of them end up as dismal failures. The resultant costs to stockholders and other stakeholders are enormous. In this paper we have illustrated how habitual domains analysis can play a crucial role in the making of effective acquisition decisions and help organizations enhance the probabilities of achieving success in the risky business of acquisitions and mergers.

7. REFERENCES

[1] Arnold, J.D., "Saving Corporate Marriages: Five Cases" *Mergers and Acquisitions*, Winter 1983, 53-58.

[2] Bermer, R.J. and M.R. Wade, "the Planned Approach to Acquisitions" in S.S. Lee and R.D. Coleman (eds.) *Handbook of Mergers, Acquisitions and Buyouts,* Prentice Hall, Englewood Cliffs, NJ 1981.

[3] Brockhaus, W.L., "A Model for Sucess in Mergers and Acquisitions" *S.A.M. Advanced Management Journal*, Winter 1975, 40-69.

[4] Coley, S, Reinton, S. and J. Welch, "The high mortality rate in plying unfamiliar turf", *Mergers and Acquisitions*, 21, 1, 1987, 56-59.

[5] Datta, D.K., *Assimilation of Acquired Organizations*, Unpublished Doctoral Dissertation, University of Pittsburgh, 1986.

[6] Davis, M.S., "Two plus two doesn't equal five", *Fortune*, December 9, 1985, 171-79.

[7] Davis R.L., "Compatibility in Organizational Marriages", *Harvard Business Review*, 46, July-August 1968.

[8] Diven D.L., "Organizational Planning: Neglected Factor in Merger and Acquisitions Strategy", *Managerial Planning*, July-August 1984, 4-12.

[9] Fray, L.L., Gaylin, D.H. and J.W. Down, "Successful Acquisition Planning", *Journal of Business Strategy*, 5, 1, Summer 1984, 46-55.

[10] Gill, J. and I. Foulder, "Manging a merger: The Acqusition and the Aftermath", *Personnel Management*, 10, 1978, 14-18.

[11] W.T. Grimm & Co., *Mergerstat Review*, Chicago, 1986.

[12] Hayes, R.H., "The Human Side of Acquisitions", *Management Review*, 68, 1979, 44-66.

[13] Howell R.A., "Plan to Integrate Your Acquisitions", *Harvard Business Review*, November-December 1970, 66-76.

[14] Jensen, M.C. and R. Ruback, "The market for corporate returns", *Journal of Financial Economics*, April 1983.

[15] Kierulf, H.E., "Finding the best acquisition candidates", *Harvard Business Review*, January-February 1981, 66-67.

[16] Kitching, J., "Why do mergers miscarry?", *Harvard Business Review*, 45, Nov-Dec 1967, 84-101.

[17] Leighton C.M. and G.R. Tod, "After the Acquisition: The Continuing Challenge", *Harvard Business Review*, March-April 1969, 90-102.

[18] Lubatkin, M., "Mergers and the performance of the acquiring firm", *Academy of Management Review*, 8, 2, 1983, 218-225.

[19] Parsons A.J.,"The Hidden Value: Key to Successful Acquisitions", *Business Horizons*, March-April 1984, 30-37.

[20] Reilly R.F., "Planning for an Acquisition Strategy", *Managerial Planning*, March-April 1982, 36-42.

[21] Saaty, T.L., *The Analytic Hierarchy Process*, MsGraw-Hill, New York, N.Y., 1980.

[22] Salter, M.S. and W.A. Weinhold, *Diversification through Acquisitions: Strategies for creating economic value*, The Free Press, New York, 1979.

[23] Schwartz H. and S.M. Davis, "Matching Corporate Culture and Business Strategy", *Organizational Dynamics*, Summer 1981, 30-48.

[24] Seed III, A.H., "Why do corporate marriages fail", *Financial Executive*, December 1974, 56-62.

[25] Simon, H.A., "On the Concept of Organizational Goals", *Administrative Science Quarterly*, Vol 9, 1964, 1-22.

[26] Singh, H., *Corporate Acquisitions and Economic Performance*, Unpublished Doctoral Dissertation, The University of Michigan, ann Arbor, 1984.

[27] Steuer, R.E., *Multiple Criteria Optimization*, Wiley, New York, 1985.

[28] Temple J.E., T.G. Veal Jr., W.K. Smith, "Strategic Growth through Merger and Acquisition", *Financial Executive*, May 1981, 23-26.

[29] Walter, J.R. and D.M. VanAlstyne, "Mergers and Acquisitions: A How-to perspective on avoiding the pitfalls", *Management Review*, April 1984, 20-26.

[30] Yu, P.L., "Second-Order Game Problem: Decision Dynamics in Gaming Phenomenon", *Journal of Optimization Theory and Applications*, 27, 1, January 1979, 147-166.

[31] Yu, P.L., *Multiple Criteria Decision Making: Concepts, techniques and Extensions*, Plenum, New York, N.Y. 1985.

[32] Yu, P.L., "Behavior mechanism in decision making", in Y.Y. Haines and V. Chankong (eds.), *Decision Making with Multiple Objectives*, Springer-Verlag, New York, N.Y. 1985, 1-22.

[33]. Yu, P.L., "Second Order games and Habitual Domain analysis", in J.R. Avula, G. Leitmann, C.D. Mote and E.Y. Rodin (eds.), *Mathematical Modelling in Science and Technology*, Pergamon, Elmsford, N.Y., 1987, 7-12.

[34] Zelany, M., *Multiple Criteria Decision making*, McGraw-Hill, New York, N.Y., 1982.

APPENDIX I

COMPETITIVE FACTORS :

- Industry concentration :
  Current concentration index
  Projected change in concentration.

- Entry barrier/exit barrier:
  Capital requirements
  Product differentiation
  Economies of scale
  Distribution channels
  Brand identity
  Switching costs
  Access to raw material.

- Buyer power :
  Number of buyers
  Switching costs
  Dependence on industry.

- Supplier power:
  Number of suppliers
  Dependence on industry
  Switching cost.

- Threats of substitutes:
  Price/performance tradoff.

- Overseas competitors:
  Number of major foreign players
  Cost factors (factor costs differences)
  Technology availability in other countries.

- Rivalry among competitors:
  Number of competitors
  Industry capacity vs. demand
  Diversity of competition
  Degree of product differentiation.

AN EXPERIMENTAL STUDY OF THE EFFECTIVENESS OF COMPUTER-PROGRAMMED DECISION SUPPORT

Daniëlle Timmermans and Charles Vlek, Department of Psychology
and Laurie Hendrickx, Traffic Research Center
University of Groningen

INTRODUCTION

Currently there exists a growing interest in computer-programmed decision support. A wide variety of programs have been and are being designed to improve the quality of decision making. One category of decision support programs is based on models from multi-attribute (expected) utility theory. These programs embody procedural rather than substantive criteria for making good decisions.

The justification of decision aids lies in the fact that man has a "bounded rationality" (Simon, 1957, 1976); often (s)he constructs a simplified model of the world and acts accordingly. People are limited, sequential, non-linear and often non-compensatory information processors. Limitations on short-term memory make it difficult to integrate large quantities of information. Therefore one tends to reduce the cognitive requirements by simplificing the decision process (e.g., by using an elimination-by-aspects strategy). This tendency will be more pronounced as the decision problem increases in complexity, and it can lead to a suboptimal decision process (Payne,1976).

However, there is not yet much empirical support for the purported effectiveness of decision aids. Proponents of formal decision analysis usually advance logical (e.g. consistency) rather than empirical arguments in its support, reflecting the theory's axiomatic foundations and deductive mathematical structure. A good decision would be reached when it is made according to the model. However, it is not at all evident that the user or decision maker will accept the advice which was calculated using the model or that the method aided the decision maker in an other way. In this paper we will discuss an experiment conducted in order to test a decision aid based on multi-attribute utility (MAU) theory. Before doing so we will briefly present the MAU-model and some other empirical studies.

THE MULTIATTRIBUTE UTILITY MODEL

The multi-attribute utility or MAU-model is a normative model which has its roots in mathematical and economic theories of what good decisions are (Keeny and Raiffa,

1976). In order to make a choice among alternatives, the choice problem is decomposed. An alternative is evaluated on all criteria and the importance or weight of each criterion is indicated. The overall utility of an alternative is the weighted sum of the utilities of the decomposed parts (i.e. the several criteria). The alternative with the highest overall utility is the preferred one. In the MAU-model it is assumed that preferences are consistent and independent of each other. The decomposition principle implies that there will be less systematic and random error when judgments are concerned with simple, unidimensional components of a decision problem. Central concepts are rationality and optimality. The rationality of models from decision analysis such as the multi-attribute utility model is primarily instrumental: a decision is rational if it maximizes the decision maker's (expected) utility under the subscription of specific assumptions and a limited time horizon. It does not aid the user in evaluating the optimality of the criteria themselves.

EMPIRICAL STUDIES ON MAU DECISION SUPPORT

Studies about the effectiveness of decision support based on the MAU model are scarce (see Rohrmann,1986; Pitz and Sachs,1984; Schoemaker,1981). There are even fewer studies on computerised versions. Humphreys and McFadden (1980) published the results of an experiment with an interactive computer-based decision aid called MAUD (MultiAttribute Utility Decomposition). Their findings indicate that MAUD led to reduced goal confusion and increased the awareness of the structure of the importances of the various attributes. They concluded (rather tentatively) that MAUD made people more capable of deciding in general, rather then merely solving the immediate decision problem.

John, Von Winterfeldt and Edwards (1983) compared multi-attribute utility analysis of personal decision problems with a human analyst versus an analysis with a computerised MAU-model. Among other things, they tried to establish to what extent the satisfaction of the user is related to the formal methods and procedures of the theory in contrast to other factors such as personal interaction. John et al.'s subjects found the analyst session more agreeable than the session with the computer programme. Agreement with and acceptance of the analyst and MAUD results were the same.

Bronner and De Hoog (1983) investigated the use of a decision-aiding technique for personal decision making based on MAUD using a wide variety of subjects regarding age and educational level. The aim of their study was to find out whether the aid could be effectively used by people with no specific training in decision analysis. They concluded that this was the case, although the interface still needed improvements. Satisfaction with the programme's advice occurred in 86% of the cases.

THE EXPERIMENT

The aim of the experiment was to test the effectiveness of a multi-attribute method in solving different kinds of choice problems. We designed the experiment in order to answer four questions: (1) How effective is decision support based on a MAU-model?; (2) What is the effect of the computerisation of the model?; (3) How does a formalised method compare to the unaided decision process?; (4) What is the effect of task complexity on the decision processes?

Subjects went through a standard MAU-procedure which in one condition was run on a personal computer, and in the second condition by a trained interviewer. In the third condition subjects went through an unaided decision process. For the computer-supported condition (CS) we selected the Dutch programme MIDAS (de Hoog, 1983) which is based on MAUD, an interactive MAU programme designed by Humphreys and McFadden (1980). MIDAS is a 'stand alone' programme, so there is no need for an analyst to assist the user in analysing his or her decision. In the interviewer supported condition (IS) each subject was led through the procedure by a trained interviewer. In the unaided decision (UD) procedure subjects were asked to make a decision in their own way.

In each condition subjects were asked to solve four well-defined choice problems differing in content and formal characteristics. The formal characteristics were an operationalisation of "task complexity", which has been proven to influence decision processes (Payne, 1976; Paquette and Kida, 1988). By giving the subjects construed problems and not using "real" problems, we were able to vary the characteristics of these problems systematically. The drawback of this construction is that the choice problems may not be "real" enough for the subjects. However, the subjects generally considered the problems as realistic (mean judgment 4.2 on a 5-point scale) and thought they had enough information to make a choice (mean 3.2 on a 5-point scale).

METHOD

Subjects: Forty-eight students (25 female and 23 male) at the University of Groningen participated in the computer-supported condition (CS), 24 subjects (15 female and 9 male) in the interviewer-supported condition (IS) and 24 subjects (15 female and 9 male) in the unaided condition (UD). The computer experience of subjects (Ss) in the CS condition varied from "none" to "a lot" with an average of 2.6 on a 5-point scale. They found the instructions for using the computer "very clear" (mean 4.7) and had few difficulties in handling the computer (mean 3.4 on a 5 point-scale).

Problems: We construed 4 problems sampled from 4 different domains: choosing a medical treatment for an affliction of the ankles, choosing a job, buying a house; and selecting a suitable applicant for a teaching job. A second independent variable was task complexity, operationalised on 3 dimensions: the number of available alternatives, the homogeneity of the option set and the degree of uncertainty about the value of the attributes (only between subgroups of the subjects in the computer-supported condition). A greater amount of information (number of alternatives and number of attributes) is considered to make decisions more difficult (Huber,1980; Abelson and Levi,1985). There were either 2 or 5 choice alternatives. The option set was made more or less heterogeneous by enlarging or reducing the range of the values of attributes (e.g., number of years of experience of the job candidates) and by increasing the differences between the alternatives (e.g., jobs only in the Netherlands or also abroad) . In this way we tried to increase the number of possible aspects for comparison of the alternatives. The more heterogeneous the option set is, the greater the variation of attribute values. This would make a decision easier, because it is less difficult to differentiate among the alternatives (Van Raaij, 1977). Furthermore, the uncertainty about the value of the attributes (the third factor in the CS-condition) complicated the task (Meyer, 1981). This last manipulation was omitted in the second and third condition because it turned out to have only little effect. Decision problems specified according to the relevant condition were presented on paper, with a general description of each problem followed by a short description of the various choice alternatives.

Design: Computer-supported (48 Ss) versus interviewer-supported (24 Ss) MAU-analysis versus unaided decision process (24 Ss) constituted the main between-subjects variable in this study. In addition, half of the computer-supported group worked on decision problems with uncertainty about the value of one attribute; the other half received choice alternatives with certain positions on all attributes, as did the 24 Ss in the interviewer condition and the unaided condition. There were three within-subjects variables: problem domain, problem complexity (a 2x2 combination of number, and homogeneity versus heterogeneity of choice alternatives), and order of presentation of choice problems. These within-Ss variables were systematically balanced across Ss according to a Graeco-Latin square design. Each subject was asked to solve 4 problems consecutively, one from each domain.

Procedure: In the CS condition subjects were told the aim of the study and were asked to go to an experimental room and to read the first choice problem. After they finished reading they filled in a questionnaire and were asked how difficult they thought the problem was and if they already had formed a preference. If the problem consisted of 5 choice alternatives, Ss were also asked if they wanted to reject any alternatives rightaway. After this they worked through the MIDAS-programme. This programme aided the subject in: (a) formulating utility attributes, (b) scoring the various alternatives on the attributes, (c) indicating their "ideal position" on each attribute, and (d) assigning importance weights to the attributes. After the Ss typed in the required information, the programme calculated the overall utility rank order of the choice alternatives. The Ss then filled in another questionnaire containing questions like how difficult they thought the procedure had been, how they evaluated the procedure, and related questions. They were also asked to indicate their final preference, which was not necessarily the same alternative as had been calculated by the programme. In the IS condition, the formal procedure was the same except that it was the interviewer who assisted the subject and calculated the overall preference rank ordering. In the UD condition subjects were asked to make a decsion while thinking aloud. They also filled in a questionnaire after they had made a decision.

Criteria of effectiveness: Traditionally, for the evaluation of the effectiveness of decisions, there are two kinds of criteria: outcome and process criteria. These can both be subjective (e.g.,"satisfaction") or objective (e.g.,"best choice", "coherence"). Most real life decisions cannot be evaluated by (immediate) outcome. Procedural quality criteria have recently been debated on several occasions. A procedure is said to be more effective if it takes e.g. less time, is more thorough and consistent (Edwards, Kiss, Majone and Toda, 1984; Rohrmann,1986).

In the present experiment the following objective measures were used: (a) time spent on analysing the decision; (b) number of attributes generated during the process; (c) the convergence of the intuitive and the MAUA-predicted preference ordering (see also Humphreys and McFadden, 1980). Low convergence might indicate either that the decision maker was not consistent in his or her evaluations or that his or her preferences are not covered by the model. High convergence might indicate that the (initial or final) intuitive preference ordering and the MAU-produced ordering are based on the same cognitive structure of the decision problem. High convergence between the initial intuitive preference and the calculated preference might indicate that decision support is less useful. Increased convergence between intuitive and MAU-preference before compared with convergence after the decision procedure could be interpreted as an indication of the effect of the method.

The subjective or questionnaire measures were: (d) satisfaction of the user with the procedure and satisfaction with the resulting choice; (e) experienced difficulty of the decision procedure; and (f) acceptance (difficulty and agreeableness) of the method (CS and IS condition). These questionnaire items were measured on a 5-point scale, 5 being the highest score. In the following section we limit ourselves to the effects of the decision procedure (CS, IS or UD) and the complexity of the problem as operationalised in the number of choice alternatives. For lack of space we also omit the results of the analysis of the think-aloud protocols from the UD condition.

RESULTS

(a): Time spent on the decision process: The amount of time spent while going through the process varied greatly, from 5 minutes to 56 minutes with an average of 22 minutes in the CS condition, of 19 minutes in the IS condition. In the UD condition the average was 14 minutes (range: 5 to 34 minutes). When the subjects had to make a choice out of 5 alternatives, the time needed for the process was longer than for the choice problem having 2 alternatives (9 min. for CS and 12 min. for IS). There was also a substantive reduction in time over the 4 sessions (from 31 min. to 20 min. in CS and IS). This could partly be explained by the fact that subjects needed less and less time in reading or listening to the instructions.

TABLE 1: AVERAGE SCORES ON EFFECTIVENESS CRITERIA IN THREE CONDITIONS

	COMPUTER SUPPORTED (192 cases)	INTERVIEWER SUPPORTED (96 cases)	UNAIDED PROCESS (96 cases)
(a) TIME NEEDED	28 m.	32 m.	14 m.
(b) NR.ATTRIBUTES	4.7	4.9 *	4.3 *
(c) CONVERGENCE CALCUL.=FINAL PREFERENCE			
1.if init.= calc. (N = 78, 50)	97%	98%	
2.if init.≠ calc. (N = 26, 8)	46%	43%	
3.if no init.pref (N = 88, 37)	81%	71%	
TOTAL	83% (N = 160)	81% (N = 78)	
(d) SATISFACTION			
with choice	3.8	4.2	4.1
with process	3.2 **	3.4	4.0 **
(e) EXPERIENCED DIFFICULTY	2.7	2.9	2.8
(f) ACCEPTABILITY OF METHOD			
easy	4.0	3.2	
agreeable	3.6	3.3	

Note: The convergence of preference (initial and final) is only calculated in the CS and IS condition. The percentages refer to the number of cases (from each group (1,2 or 3) with a total of N cases) for which the calculated preference agree with the final preference. The subjective criteria (d,e and f) are measured on a 5-point scale with 5 as the highest (best) score. * $p<0.01$; ** $p<0.001$

(b): Number of attributes: Subjects generated more utility attributes in the supported MAU conditions (4.7 in CS and 4.9 in IS versus 4.3 in UD; $F=2.4$; $p<0.01$). In the UD condition there was an effect of problem complexity: more attributes were generated in problems with 5 alternatives than with 2 alternatives (4.8 versus 3.9; $F=3.6$; $p<0.001$).

(c): Convergence of preferences: In 54% (CS) and 61% (IS) of the cases the subjects formulated an initial preference (directly after reading the decision problem).

Subjects more often indicated an initial preference when the problem had 2 alternatives than when it had 5 alternatives (66% versus 42% in CS; Chi-square: 11.0; p<.001; 75% versus 48% in IS; Chi-square: 7.4; p<0.01).

In roughly 80% of the cases when subjects had an initial (I) preference, the calculated (C) preference agreed with the initial (I) preference (I=C in 75% for CS and in 86% of the cases for IS respectively). Not surprisingly they did not change their minds when they formulated their final (F) preference (C=F in 97% for CS and in 98% for IS of the cases respectively). In the cases when subjects had an initial preference which disagreed with the calculated preference (C≠F in 25% for CS and in 14% for IS of the cases respectively), in nearly half of these cases (46% for CS and 43% for IS) subjects changed their mind and opted for the calculated preference. In the cases when subjects did not have an initial preference (46% for CS and 39% for IS of the cases respectively), most subjects agreed with the calculated preference (C=F in 81% for CS and in 71% for IS of the cases respectively).

With regard to the effects of the various experimental variables the only significant difference occurred for the number of available alternatives. Decision problems with only 2 alternatives resulted in greater convergence between initial and calculated preference (Chi-square: 9.2; p<.003) and between calculated and final preference (Chi-square: 9.8; p<.002); see table 1 for results. A higher convergence suggests that the calculated preference and the final preference are based on the same cognitive representation of the problem.

(d) Satisfaction: Subjects were most satisfied with the decision process in the UD condition, i.e., when they did not get any support, (mean score 4.0 for UD versus 3.2 for CS and 3.4 for IS; F=8.0; p<0.001). There was no significant difference concerning the satisfaction with the final choice made. However, there was a different evaluation of the final choice in different situations. When the problem had only 2 alternatives Ss judged the advice to be more optimal (4.0 vs. 3.7; Chi-square: 7.9; p<.01) than when the problem had 5 options. When subjects had an initial preference, they evaluated the outcome more positively (4.1 vs. 3.5; F=4.6; p<0.001) than when no initial preference was expressed. When the calculated preference agreed with their final preference, they also judged the advice to be more optimal (4.3 vs. 2.8; Chi-square: 67.1; p<.000). Thus, in the cases where the decision process was easier (2 alternatives to choose from, an initial preference available or final preference in agreement with calculated preference) the advice was evaluated to be more optimal.

(e) Experienced difficulty: There were no significant differences in experienced difficulty of the decision procedure. With regard to both MAU-methods (CS and IS), some parts were considered to be more difficult than others. Assigning importance weights was evaluated as the most difficult part of the method. It has been demonstrated (e.g., Dawes and Corrigan, 1974, and Einhorn and Hogarth, 1975) that giving all attribute dimensions in the preference structure the same weight will often leave the overall preference order unaffected. In order to estimate the influence of differential weighting on the calculated preference, we calculated the subject's final preference with equal weights. In 83% for CS and 85% for IS, of the cases there was no change in preference order of the first and the second preferred alternative when equal weights were assigned to the attribute dimensions. This may be caused by the robustness of linear models. An other explanation is that subjects implicitly take into account these differences in importance of the attributes while evaluating the alternatives with respect to the attributes. However, the effects of assigning equal or differential weights are different under different conditions. When it was more difficult to make a choice, assigning weights had an effect in almost twice as many cases as in the simpler condition (see table 2). A problem is considered to be more difficult when there are more alternatives to choose from (i.e., 5 versus 2 alternative), or when subjects had no initial preference or when their initial preference did not agree with the calculated preference. A possible explanation is that in the latter, easier condition the weights are accounted for during the scoring of the alternatives.

TABLE 2: CHANGES IN MOST PREFERRED ALTERNATIVE WHEN EQUAL WEIGHTS WERE USED INSTEAD OF DIFFERENTIAL WEIGHTS

	CS	IS
Overall mean:	17%	15%
5 alternatives	21%	19%
2 alternatives	12%	10%
no initial preference	23%	14%
initial preference	12%	15%
init.pref.≠calc.pref.	27%	29%
init.pref.=calc.pref.	14%	12%

Note: Change in most preferred alternative is given in percentage of the total number of cases.

(f) Acceptance of the method: The subjects evaluated the MAU-method to be moderately insightful (mean score 3.8 on a 5-point scale for CS and IS). They generally indicated that the method was useful in supporting their decision process (3.4) and evaluated the method as reasonably agreeable, easy and trustworthy (3.5, 3.6, and 3.3 respectively). They evaluated the calculated choice as optimal (4.0), but indicated that they were not always willing to accept the advice if it differed from their own preference or they could not understand how the advice was calculated. In the CS and IS conditions, subjects liked the method because it helped them order the positive and negative aspects of their decision, gave more insight into the problem and made them think in more specific and explicit terms. However, they considered the method only as supporting and did not see the advice as binding. Many subjects believed that it was difficult to include the more intuitive and emotional aspects of a decision and saw this as a disadvantage. Some subjects also objected to the weighting of the attributes. They found it too difficult or not relevant, and that it distorted the final outcome. None of the subjects in the IS condition preferred a procedure without interviewer, and some of the subjects in the CS condition said that they would prefer an interviewer instead of a computer programme.

DISCUSSION AND CONCLUSIONS

In conclusion, the answers to our four beginning questions can be formulated as follows:

(1) How effective is formal decision support?

Regarding time needed, the supported decision procedure seems to be less efficient. It took subjects almost twice as much time to make up their minds. However, subjects used more information (considered more attributes) in the supported conditions. This can be taken as an indication of a more thorough decision process. Furthermore, going through the method made subjects change their minds in some cases and helped them to make up their minds in others. This can be considered as an effect of the MAU-method on the preferences of the subjects. The satisfaction with the final choice did not differ between the three conditions, but subjects indicated that the unaided decision process was their favourite. The supported process was not considered to be more difficult. Assigning differential weights was considered to be the most difficult part of the method and sometimes considered as being irrelevant. This seems to be supported by the finding that in more than 80% of the cases the use of equal weights would not have changed the identity of the most preferred alternative. This questions the usefulness of including differential weighting in multiattribute utility analysis. The present findings indicate that people are generally quite satisfied with the MAU-method and evaluate the advice as reasonable to good. Subjects indicated that they were not willing to accept the MAUA result if they did not understand how this had been calculated. A much expressed objection to the method was that it is not possible to include the less rational aspects of a decision which nevertheless may be very important.

(2) What is the effect of the computerisation of the method?

Comparing the results of the computer-supported and the interviewer-supported conditions, we found only few differences. This suggests that the computerisation of the method hardly made any difference. There were small and nonsignificant differences regarding the judged easiness of the method (more "easy" for CS) and the evaluation of the result (more "optimal" for IS). Time spent in making the decision and number of attributes generated were hardly different.

(3) How do formalized methods compare to the unaided process?

The unaided decision procedure took less time and was evaluated more positively. The evaluation of the final choice and the reported difficulty of the decision process did not differ from the CS and IS conditions. However, subjects used fewer attributes in the unaided decision process compared to the supported procedures. This may be seen as an indication of a less thorough decision.

(4) What is the effect of task complexity on decision processes?

The effects of decision support seems to be dependent on the experienced difficulty of the problem. The complexity of the problem was operationalised by varying the number of choice alternatives. Having no initial preference, and disagreement between initial and calculated preference can both be considered as an indication of the experienced difficulty of the problem. When there were 5 alternatives to choose from, or when there was no initial preference or when the calculated preference disagreed with the initial preference, assigning differential weights more often made a difference for the final preferred option. This suggests that the usefulness of differential weighting is related to the difficulty of the decision process. The more difficult the decision problem is, the more useful differential weighting seems to be. An explanation is that in the simpler decision problems weights could more easily be accounted for during the scoring of options on attributes. Furthermore, convergence of the initial and final preference with the calculated preference was higher for the simpler problems. When convergence is higher, the final preference more often may be based on a cognitive structure of the problem which is similar to the model structure. This suggests that the simpler choice problems are solved in a more MAU-like fashion (and support would be less useful), but also that the MAU-method is not completely adequate for supporting the more difficult decision problems. The final preference in complexer choice problems more often disagreed with the calculated preference and might be more often based on a different cognitive representation of the problem than the model structure. When the decision procedure was easier, the final choice was evaluated as more optimal and the process took less time.

In general, the results of our study suggest that:

(a) In the opinion of the subjects the method is especially useful for increasing insight and for ordering the pros and cons of choice alternatives, but the calculated preference is not seen as binding.

(b) Simple decision problems (e.g., 2-option problems with a limited number of attributes) do not require explicit decision support; the convergence of initial and calculated preference is already high.

(c) The resolution of more complex problems may profit from explicit decision support; the method had more effect when the problem was considered to be more difficult.

(d) However, the MAU decision support cannot be considered to be completely successful; convergence of the calculated and final preference was higher in the simpler problems and the final choice was considered to be more optimal than for the more complex problems.

The effectiveness of the MAU decision support seems to be dependent on the experienced difficulty of the decision problem. Therefore, in order to provide effective decision support it would be useful if this is tailored to the state of mind (knowledge and or experience with decision problem) of the user and the type of problem. This would seem to require a basic taxonomy of decision problems as well as more insight into the user's state of mind. Also, gaining more insight into the nature of the unaided decision processes regarding different kinds of problems seems to be an important starting point for building more effective decision support.

REFERENCES:

Abelson,R.P. and A.Levi (1985): Decision Making and Decision Theory. In:G.Lindzey and E.Aronson(eds.) The Handbook of Social Psychology,3rd ed., Random House, New York

Bronner F. and R. de Hoog (1983): Non-expert use of a computerized decision aid. In:P.Humphreys,O.Svenson and A.Vari(eds): Analysing and Aiding Decision Processes, Amsterdam, North-Holland Publishing Company.

Dawes.R.M. and B.Corrigan(1974): Linear models in decision making.Psych.Bul.81,95-106

Edwards W., I.Kiss, G.Majone, M.Toda (1984): What constitutes a good decision.Acta Psychologica,56, 5-27

Einhorn,H.J. and R.M.Hogarth (1975): Unit weighing schemes for decision making. Org. Beh. and Human Perf. 13,171-192

Huber,O (1980): The influence of some task variables on cognitive operations in an information-processing decision model, Acta Psychologica 45, 187-196

Humphreys,P and W.McFadden (1980): Experiences with MAUD: Aiding decision structuring versus bootstrapping the decision maker. Acta Psychologica 45 ,51-69

John,R., D.von Winterfeldt and W.Edwards (1983): The quality and user acceptance of multiattribute utility analysis performed by computer and analyst. In:P.Humphreys, et al.(eds):Analysing and Aiding Decision Processes,Amsterdam,North-Hol.Publ.Co.

Keeny, R.L. and H.Raiffa (1976): Decisions with multiple objectives: preferences and value trade-offs. New York, Wiley

Klayman,J (1982): Simulations of six decision strategies: Comparisons of search patterns, processing characteristics, and response to task complexity. Working paper, nr. 100, University of Chicago

Meyer,P. (1981): A model of multi-attribute judgments under attribute uncertainty and informational constraint, Journal of Marketing Research, Vol.XVIII, 428-441

Paquette,L. and T.Kida (1988): The effect of decision strategy and task complexity on decision performance. Organ. Beh. and Human Dec. Proc., 41, 128-142

Payne,J.W.(1976): Task complexity and contingent processing in decision making: an information search and protocol analysis. Org. Beh. and Human Perf.,16,366-387

Pitz,G.F. and N.J Sachs (1984): Judgment and Decision: Theory and Application, Ann.Rev.Psychol. 35:139-163

Raaij,W.F.van (1977): Informatieverwerking en keuzegedrag van consumenten, Jaarboek van de Nederlandse Vereniging van Marktonderzoek, 217-233

Rohrmann,B.(1986): Evaluating the usefulness of Decision Aids: a methodological perspective. Paper presented at the 10th SPUDM conference, Helsinki, Finland

Schoemaker,P.J.,1982: Behavioral issues in multiattribute utility modelling and decision analysis. In:J.N.Morse(ed):Organizations: Multiple Agents with Multiple Criteria, Springer-Verlag, Heidelberg

Simon, H. (1957, 1976):Administrative behavior: a study of decision-making processes in administrative organizations. New York, Free Press

DECISION RATIONALITY AND HABITUAL DOMAIN ANALYSIS

Po-Lung Yu# Shu-De Huang† Dazhi Zhang#

#: The School of Business, The University of Kansas, Lawrence, Kansas 66045-2003, USA

†: Dept. of Management Science, Jiaozuo Mining Institute, Henan, People's Republic of China

ABSTRACT: Rationality is a reflection of habitual domain. However it can strongly influence our judgement and decisions. This paper outlines: (i) habitual domain theory, (ii) rationality in terms of habitual domains, (iii) rationalities in different academic disciplines and (iv) effective processes for solving nontrivial problems.

1. INTRODUCTION

There are a number of methods in multiple criteria decision making (MCDM) (for instance, see [3,5,11,14,16,19] and quotes therein), representing a number of decision rationalities. There are also a number of decision rationalities described in various academic disciplines including economics, psychology, organizational behavior, philosophy and decision science (for instance, see [1,2,6,9,10,20] and quotes therein). Different rationalities require different information inputs to reach the final decision. For instance, for value function maximization, one needs to assess the preference information to construct value function and then to find the alternative which maximizes the value function over all alternatives; for compromise solutions one needs to identify the ideal point or target point and the appropriate regret (distance) function, then find the alternative which minimizes the regret over all alternatives; for Pareto solutions, all one needs is the assumption that more is better for each criterion. Obviously, different decision rationalities will produce different suggestions for the final solution.

It has been recognized that over time people sooner or later develop a fairly steady set of ways of thinking, judging and responding to various events and problems. This leads to the conception of habitual domains (HD) (see [15,16]). People do not always need sophisticated mathematical methods to reach a decision. Indeed, if a sophisticate method is beyond the comprehension of the HD of a decision maker (DM), it may be regarded as irrational and therefore rejected by the DM, even though he may politely excuse himself. This raises three serious questions: (1) What decision rationality can be more easily accepted by the DM? (2) should we expand our own decision rationality to make it comprehensible to the HD of the DM? or should we expand the DM's HD to make the decision rationality acceptable to him? or both? and (3) as a consultant or decision aid, how do we sharpen our skills to help the DM reach a final decision?

This article tries to answer the above questions. In addition, we should formally introduce the concept of rationality from DM's point of view. Specifically, in section 2 we shall briefly sketch the concept of HD analysis. Then we discuss the concept of rationality in terms of habitual domain in section 3. In section 4 we shall sketch (decision) rationalities scattered in various academic disciplines. In section 5 we will introduce effective processes for solving non-trivial problems by the habitual domain theory. Conclusion and further research will be sketched in section 6.

2. BASIC ELEMENTS IN HABITUAL DOMAIN THEORY

Much progress has been made in habitual domain theory [15-18] since its advent about a decade ago. Basically, habitual domain theory offers a dynamic model of decision / behavior processes based upon eight hypotheses of human behaviors [15-18]. These include

1) the **circuit pattern hypothesis**, which models the functions of the brain - the information processing center of human beings - similar to those of a computer. More specifically, each unit of thought, concept or idea is encoded as a circuit pattern and the circuit patterns will be reinforced when the corresponding ideas are repeated. Furthermore, the stronger the circuit pattern, the more easily the corresponding thoughts are retrieved in the thinking, rationalizing, and the decision making processes;

2) the **unlimited capacity hypothesis**, which assumes that each normal brain is capable of encoding and storing all thoughts and concepts that one ever attends to;

3) the **most efficient restructuring hypothesis**, which states that the encoded ideas are organized systematically as data bases for efficient retrieving and are continuously restructured so that the relevant ones can be most efficiently retrieved to release charges;

4) the **analogy/association hypothesis**, which states that when confronted with a new event, the brain first investigates its features and attributes to establish a relationship, an analogy, and/or association, of the new event to those already known, then the whole of the past knowledge is automatically brought to bear on the interpretation and understanding of the new event;

5) the **goal-setting and state evaluation hypothesis**, which states that each one of us has a set of goal functions and for each goal function we have an ideal state or equilibrium point to reach and maintain (goal setting). We continuously monitor, consciously or subconsciously, where we are relative to the ideal state or equilibrium point (state valuation). Goal setting and state valuation are dynamic and interactive and subject to the influence of self-suggestion, physiological forces, memory, and information processing capacity.
6) the **charge structure and attention hypothesis**, which assumes that each event is related to a set of goal functions. When there is unfavorable deviation of the perceived value from the ideal, the goal function will produce various level of charge. At any point of time, our attention will be paid to the event which has most influence on the totality of the charges.
7) the **discharge hypothesis**, which states that one selects the action which yields the lowest remaining charge (the remaining charge is the resistance to the total discharge, and this is called the *least resistance principle*); and
8) the **information inputs hypothesis**, which states that humans have innate needs of gathering external information and unless attention is paid to them, external information inputs may not be processed.

The main ideas in this model are (a) each individual is endowed with an internal information-processing capacity, (b) a number of state variables are used to describe human physiological conditions, social situations, and self-suggested goals, and a charge structure will be produced when the state value is significantly different from its ideal goal value, and (c) the purpose of the attention is to release charge in the most efficient way. These involve actions, acquiring external information, self-suggestion, and rationalization.

As a consequence of these eight hypotheses, it can be shown that humans sooner or later develop a fairly stable set of thinking, judging and reacting to various events and problems. See [15, Chapter 9 of 16]. This observation can be captured by the concept of habitual domain.

The habitual domain at time t, denoted by HD_t, is defined to be the collection of ideas and actions that can be activated at time t. The concept is further refined as to contain the **actual domain**, denoted by AD_t, which is the set of ideas/actions that are actually activated at time t; the **potential domain** PD_t, which is the set of ideas/actions that can be potentially activated; and the **reachable domain** $R(I_t, O_t)$, defined as the set of ideas /actions that can be reached from the initial set of ideas I_t through the set of operators O_t (operators means thinking methods or processes).

In [15,18] (especially Chapter 9 of [16]) the expansion of habitual domain, the interaction of different habitual domains, and the applications of habitual domains on various decision problems were extensively discussed; the explanation of some observations in social psychology was given by employing habitual domain analysis; and the existence of stable habitual domains was given and proved.

Let $P_t(I,\Sigma)$ $(0 \le P_t(I,\Sigma) \le 1)$ be the propensity that idea I will be activated when Σ is presented at time t. The α-core of Σ at time t is defined by

$$C_t(\alpha,\Sigma) = \{ I \mid P_t(I,\Sigma) \ge \alpha \} .$$

which is the collection of ideas/operators that have propensity larger than α to be activated. Note that $C_t(\alpha,\Sigma)$ is like the α-cut of the "fuzzy set" with $P_t(I,\Sigma)$ as the membership function.

3. RATIONALITY IN TERMS OF HABITUAL DOMAINS

To answer the first question raised in the introduction, let us formally introduce the concept of rationality by applying habitual domain analysis.

A thinking process, decision or event is *rational to a particular person* if it can be comprehended and accepted by his/her habitual domain.

The *rationality of person B from person A's point of view* can be defined as follows:

Let HD_a and HD_b be the habitual domains of persons A and B respectively. Define

$$R_a(b) = \mu(HD_a \cap HD_b)/\mu(HD_b)$$

to be the relative measure for the degree to which person A regards person B as rational, where μ is a properly defined measure on HD's. Thus person A thinks person B is totally rational if HD_a contains HD_b.

One of the key ideas of this definition is that the rationality of person B depends also on the other persons - the evaluators; a person's rationality is a function of the evaluators.

It should be noted that the concept of rationality also changes from time to time; what is rational today may not be as rational tomorrow. For example, having sex before getting married is considered totally irrational ten years ago but not too irrational these days in China. The bounded rationality theory is very time dependent because as new decision technologies are uncovered and new information-processing systems developed, the bounds of bounded rationality are pushed back [9]. Therefore, we should study rationality from the dynamic point of view: treating the concept of rationality by applying process approaches. These suggest us to add the time dimension to the measurement of rationality, i.e., the degree to which that person B is rational in person A's eyes (at time t) may be of the form

$$R_a(b,t) = \mu(HD_a(t-\Delta) \cap HD_b(t-\Delta))/\mu(HD_b(t-\Delta))$$

where $HD_a(t-\Delta)$ and $HD_b(t-\Delta)$ are the habitual domains at time t-Δ, where Δ is a small time interval. Therefore, person B's rationality is a function of both the evaluator and the time.

If HD_a and HD_b are the habitual domains of group A and group B respectively, then $R_a(b,t)$ may be regarded as a measure for the degree to which group B is rational from group A's point of view.

Other concepts which are closely related to that of rationality is **comprehensibility** and **observability**.

Let c(HD) be the set of all thinking processes, decisions or events that can be comprehended by HD, or simply called as *comprehensible Domain* of HD; and let the observable part of HD be denoted by o(HD), which for simplicity is called the *observable domain* of HD. Let $c(HD_i)$ and $o(HD_i)$ respectively be the comprehensible and observable domains of person i's HD. Then the larger the value of $\mu(c(HD_a) \cap o(HD_b))/\mu(o(HD_b))$, the more person A can comprehend person B. Thus the *comprehensibility* of person B from person A's point of view can be defined as

$$C_a(b) = \mu(c(HD_a) \cap o(HD_b)) \,/\, \mu(o(HD_b)).$$

4. RATIONALITIES IN DIFFERENT ACADEMIC DISCIPLINES

In this section, we sketch some well-known rationalities of organization behavior, economics and philosophy and show their underlying habitual thinking.

1. Organizational Behavior. In the field of micro-organizational behavior, a number of theories have been developed with each of them stressing a special kind of decision rationality and some particular aspects of human behavior. For example, the goal setting theory states that an individual's behavior is directed and governed by goals; the expectancy theory, on the other hand, argues that one acts in order to maximize his/her subjective expected utilities.

In his book "Organizations and Organization Theory", Pfeffer [9] offered a categorization of the organization theories from two perspectives - perspective on action and level of analysis. The perspective on action includes *prospective rationality* (rationalize first and take an action later), *retrospective rationality* (act first and rationalize the action later), and emergent process (people, problems and decisions emerge together and interact one another); the level of analysis consists of individualist as well as organizationist.

The habitual domain theory encompasses the category of individualist prospective rationality, since it can be inferred from the theory that an individual's action is taken in order to reduce his/her charge to the lowest level. The scope of the habitual domain theory, however, is much broader than that of some theories in the same category, for example, goal theory and needs theory, since both goal setting and needs are only part of the determinants of the charge structure.

On the other hand, habitual domain theory also encompasses retrospective rationality as it is capable of explaining most theories of different rationalities. For example, when a problem is repetitive, truly satisfactory solutions are usually readily available for discharge. As the circuit pattern of solving the problem is so strong that it is not necessary for one to rationalize the problem before taking the action. This is especially so when the decision problem involves only with low stakes. Obviously, these ideas can also be applied to explain why most problem solving in mature organizations is done without recourse to the rethinking of the basic premises which lead to a solution.

It also worth mentioning that according to the discharge hypothesis, *active search for solutions or avoidance justification* can be activated when the decision problem is of high stakes and/or involves great uncertainty. Of course, this depends on whether the decision maker has confidence in finding a truly satisfactory solution in due time. Therefore, one adopts a strategy according to what the real situation is instead of simply rationalizing prospectively or retrospectively.

Finally, habitual domain analysis is based on the dynamic process of human behavior resulting from the continuous interaction of external environment with internal information processes. Therefore, it also encompasses the emergent process. Furthermore, as organization can be regarded as a living entity it also has a habitual domain. Habitual domain theory can thus also cover the organization behavior (see [4]).

2. Economics. In economics, the rational expectation is usually defined to be the one which utilizes all the available information to the actual process in determining the interested variable [1]. In microeconomic theory, *rational behavior* is referred to the one which maximizes the expected utility after considering all available information. The concept of "rational" was criticized by questioning the plausibility of assuming that when forming expectations the typical individual is sufficiently sensible and capable to use all the available information. It is also questionable that a typical person would weigh up the costs and benefits of any activity and carry out that activity up to the point where the marginal costs equal the marginal benefits [1]. To partially overcome these problems, the notions of "satisficing" as well as *bounded rationality* has been proposed with some of their versions emphasizing their effects on producing the use of rules, standard operating procedures, and routinized, habitual patterns of acting [9].

The above economics concepts are closely related to the concept of habitual domain since the habitual domain theory stresses the overall "satisficing" (by releasing charge) which can be inferred from the hypothesis of goal setting and state valuation as well as that of discharge. In addition, the implausibility of the assumption of utilizing all the available information can also be seen from the eighth hypothesis, that is, the hypothesis of information inputs. The problem however, is what does "satisficing" mean, since this concept varies from person to person and from time to time. Note that if one is totally ingrained with the doctrine of maximizing expected utility, he or she feels extremely uncomfortable not using it (because of high level of charge for not using it). From his or her point of view, maximizing expected utility is totally rational no matter what other people may say.

3. Philosophy. In philosophy, seven concepts of rationality were distinguished [2]. These include 1) conceptual (minimizing fuzziness); 2) logical (striving for consistency); 3) methodological (questioning and justifying); 4) epistemological (caring for empirical support); 5) ontological (adopting a consistent world view); 6) valuational (striving for attainable and worthwhile goals); and 7) practical (adopting means likely to help attain the goals in view). The rationalities that mathematical theorists pursue are conceptual, logical and methodological ones. Scientists, on the other hand, are supposed to abide by methodological, epistemological, and ontological rationalities; while managers emphasize valuational and practical rationalities [2].

It was called *semirational* if some, not all, of the seven rationalities were satisfied, and *full rationality* if all were satisfied [2]. But whether a rationality was satisfied or to what extent it was satisfied heavily depends on how large the habitual domains of both the individual (agent) and the evaluators are. For example, a "logic" or "psychologic" which is fairly consistent with one person's opinion may not be consistent for the other one. "I enjoy taking this job because it is really challenging" and "I'm not going to take this job since it's too challenging" are commonly heard in everyday life. Therefore, one's full rationality may be regarded as totally irrational by someone else.

Usually, there are two aspects in the concept of rationality: 1) given what the agent believes, is a certain action "reasonable"?; and 2) are the beliefs on which the agents are acting reasonable? [7]. The first question which is related to the logic or psychologic of reasoning (habitual domain of thinking process or operators) and the second one which is related to the preassumptions (habitual domain of existing ideas) together take us into the fact/value controversy and the role of science in ethics. Just like the situations mentioned above, here the problem however, is what does "reasonable" mean, since it is quite clear that the concept of reasonability varies from culture to culture, nation to nation, group to group, and person to person. For example, greeting a stranger on a street is generally considered "reasonable" in the United States but "unreasonable" in China.

5 EFFECTIVE PROCESSES FOR SOLVING NONTRIVIAL PROBLEMS

It is suggested in habitual domain theory that associated with each problem Σ there is a set of $HD^*(\Sigma)$ for effective solution, i.e., $HD^*(\Sigma)$ is the collection of ideas/actions needed to solve problem Σ. When the decision maker thinks he/she has already mastered the HD*, he/she feels comfortable to make the decision or he/she has reached the spontaneity to deal with the problem. Related to the perceived HD* and how quickly HD* is activated to solve the problem is the concept of decision cycle, which is defined to be the time needed from identifying the problem to making the decision [17]. Obviously, the decision cycle involves the process of acquiring and mastering HD*; and the quality of decision is a function of HD* and environmental factors as well as the compatibility among them.

According to whether $HD^*(\Sigma)$ is known, fuzzily known, or unknown, the decision problem is classified as a *routine problem*, a *fuzzy problem* or a *challenging problem* in habitual domain theory [17].

In order to help attacking these problems, habitual domain theory proposed the following learning process [17]:

Implanting $\rightarrow$ Nurturing $\rightarrow$ Habituating

which should be implemented to every problem solving. Here implanting means seeking a way (or through learning) to have an event presented to the decision maker so that $P_{t'}(I,\Sigma) > 0$ ($t'>t$) for an idea I which does not belong to

HD_t, the current habitual domain. The nurturing, on the other hand, means practicing and rehearsing the usage of I, so that it can be strongly imprinted in our memory and $P_{t'}(I,\Sigma)$ can become larger and larger as t increases. Habituating, however, means having the new idea I become an element in the core of habitual domain through repeatedly nurturing and practicing.

For routine problems, $PD_t(\Sigma)$ may not contain $HD^*(\Sigma)$ (i.e.,the difference $HD^*(\Sigma)\backslash PD_t(\Sigma)$ may be nonempty) for most people, especially new workers. Thus it is necessary to acquire new elements in $HD^*(\Sigma)\backslash PD_t(\Sigma)$. The goal is to make $HD^*(\Sigma)$ part of $C_t(\alpha, \Sigma)$ with $\alpha \rightarrow 1$. For fuzzy problems, one should start at the core with a very low value; while for challenging problems, there is no effective training which can guarantee the solving since $HD^*(\Sigma)$ is unknown, it should therefore be the first step to seek ways to expand one's habitual domain.

Most theories and techniques in operations research are designed to solve routine problems; fuzzy mathematics can be applicable in attacking some fuzzy problems [12] & [13]; while suggestion theory, which is still been worked on in the field of habitual domain theory, is being expected to be powerful in finding solutions for challenging problems.

The above process of implanting, nurturing and habituating are crucial in expanding the decision rationality, the HD's of our own and the DM. To be effective in solving nontrivial decision problems, expansion of decision rationality and/or HD's involved is essential (refer to question 2 raised in the introduction). The same process also served as an effective means to answer the third question raised in the introduction.

6. CONCLUSION

We have briefly sketched the concept of habitual domain analysis and the concept of decision rationality in terms of habitual domain. Effective processes for solving nontrivial problems are also offered. A number of research problems remain to be explored. For instance, how do we refine the concepts of comprehensibility and observability so that more operational measures can be defined for applications? In organizational HD, group units play important roles (coalition, cooperation, and competition), how do we effectively study them? How does uncertainty relate to routine, fuzzy and challenging problems? How does it relate to HD analysis? We invite the interested reader to join us to explore them.

REFERENCES

[1] Attfield C.L. et al.(1985) Rational expectations in Macroeconomics Basil Blackwell Inc. New York 15-18, 21-23

[2] Bunge, M. (1987) Seven desiderata for rationality, in Rationality: the critical view Agassi and Jarvie (eds.) Martinus Nijhoff Publishers 5-15

[3] Chankong,V. and Y.Y. Haimes (1983) Multiobjective Decision Making: theory and methodology, New York, North-Holland

[4] Datta, D. and P.L. Yu (1988) "Acquisition, Merger, and Habitual Domain Analysis", working paper, School of Business, University of Kansas, Lawrence, Kansas

[5] Goicoechea, A., D. R. Hansen, and L. Duckstein (1982) Multiobjective Decision Analysis with Engineering and Business Application New York John Wiley & Sons

[6] Hogarth, R. (1987) Judgment and Choice, John Wiley & Sons

[7] Holmstrom, N. (1986) Rationality and Moral/Political Decisions in Rationality in thought and action Tamny and Irani (eds.) Greenwood press 61-72

[8] Mackenzie, K.D. (1978) Organizational Structures AHM Publishing Corporation Illinois 48-83

[9] Pfeffer, J. (1982) Organizations and Organization Theory Pitman Publishing Inc., Boston 5-18, 6-7

[10] Simon, H.A. (1957) "A behavioral Model of Rational Choice" in H.A.Simon, Models of Man. Wiley New York

[11] Steuer, R.E. (1985), Multiple Criteria Optimization, Wiley, New York

[12] Wang, P. and D. Zhang (1985) Fuzzy Decision Making - Theory and Applications (209 pages in Chinese), Beijing Normal University, Beijing

[13] Wang, P. and D. Zhang (1986), An exploratory study on mathematical form of Ideology, Applied Mathematics - A Journal of Chinese Universities, 1(1), 85-95

[14] White, D.J. (1982), Optimality and Efficiency, Wiley, New York

[15] Yu, P.L.(1980) Behavior bases and habitual domains of human decision/behavior-concepts and applications, in Multiple Decision Making Theory and Application, G. Fandel and T. Gal (eds.) Springer-Verlag, New York 511-539

[16] Yu, P.L. (1985) Multiple-Criteria Decision Making: Concepts, Techniques and Extensions Plenum Press, New York 291-293, 287-289

[17] Yu, P.L. (1988) Behavior Mechanism and Career Management Lecture notes at KU

[18] Yu, P.L. and S.D. Huang (1987) Knowing People and Making Strategic Decisions (in Chinese) Coal Press, Beijing, China

[19] Zeleny, M.(1982), Multiple Criteria Decision Making, McGraw-Hill, New York

[20] Zimbardo P.G. and Ruch , F.L. (1975) Psychology and Life Foresman & Company

A HOLISTIC APPROACH IN MULTICRITERIA DECISION AID

Manuel A.C.C. Matos and Vladimiro Miranda
FEUP-DEEC / INESC (norte) - Largo de Mompilher - 4000 Porto - PORTUGAL
Tel: 351.2.321006; Telex: 23023 INESC P; Fax: 351.2.318692; Ean: v_miranda@inescn.ctt.pt

Abstract

In this paper, the multicriteria decision problem is approached from a holistic point of vue, rather than searching for a preferred solution on the basis of an alternative-after-alternative checking procedure. The techniques described were conceived for multi-attribute problems with large sets of alternatives or for multiobjective problems where a sampling procedure can be used, leading to a large set of representative efficient solutions. The psychology and behaviour of the DM are taken into account in order to justify a holistic approach to the dialogue between the DM and the problem model.

The idea behind the new method is to replace the set of efficient solutions or alternatives by a small number of macro-solutions (MS) whose characteristics can be clearly perceived, compared and evaluated by the Decision Maker (DM). The procedure adopted to define the MS is to consider each one as a fuzzy cluster . The fuzzy clusters are determined, in the Anchor Point model developed, starting by having as prototypes the points that correspond to the individual extreme values of the attributes. The membership function values of each solution to every MS may be approximated in interaction with the DM without explicitly examining the alternatives.

The basis of a DM/DAS communication protocol will be described allowing to translate in terms of fuzzy (or hard) sets the linguistic declarations stated by the DM. This protocol also allows the system to present to the DM the macro-alternatives under consideration in a linguistic label form. A set of graduating modifiers, acting on the MS linguistic labels, will provide composite labels associated to subsets of the MS. The dialogue, search and reduction process can therefore be directed by the DM with linguistic declarations, close to his/her sensibility (along with, if desired, quantified indices).

Key words: multiple criteria decision aid; fuzzy sets; fuzzy clustering; natural language interfacing

1. INTRODUCTION

The scientific community has been aware of the fact that the main problem in the field of decision aid may not be a problem of modelling but of communication. In fact, the efforts in the recent years have been directed towards reducing in some way the relative importance of the role of the analyst in the decision process, leaving place for the decision maker to fully perform his/her task (which is to take decisions).

We feel that many of the proposed approaches in multi-criteria decision aid still carry the seal of the analyst with them. By this we mean that in order to be able to work with those decision aid tools, the DM must reach a certain degree of understanding both in the mathematical procedures behind the system and in the kind of behaviour he is supposed to exhibit before the analyst. The analyst acts in fact as a two way communication channel between the DM and the data and information relating to the problem under consideration. Being so, he must analyses the problem, interpret the results, pass the information to the DM, interpret his/her answers and commands, interact again with the problem (in fact, with its model) and so on. Of course, there is noise added in the communication process by this analyst link.

Being the role of the analyst so important, and being the methods usually proposed conceived by analysts, we reached the conclusion that the DM psychology, behaviour and point of vue had not yet deserved full consideration.

In order to reach a decision, it is now obvious that the DM must undertake a learning period during which he codes and recodes the characteristics of the problem he is dealing with, until he forms in his mind what we may call "the picture of the whole thing". This general image creation may depend on the cognitive style of the subject, and in some of them that are particularly bad performers that "general picture" will never occur; but in the present case we assume that the DM is a trained subject with some expertise and that his decision making responsibilities are associated with an ability to understand complex problems in a global way.

With this in mind, we have developed the following:

a)a mathematical model that provides a means of "organizing" the set of efficient solutions;

b)a translating process that allows communication in both ways and permits the mathematical tools to be used at the Decision Maker's (DM) will, by means of simple and close-to-natural language controls.

The mathematical model to be described below is appropriate to problems where the number of alternatives produced may be very large, making difficult the subsequent decision phase. It is adequate to any decision problem for which a preliminary solution generating method has been used; many multiple attribute decision formulations are certainly to be included in this class of problems, but multiobjective problems (and among them linear multiobjective formulations) can be dealt with as well. In the latter case, we do not propose to discuss thoroughly the process of generating discrete solutions; in any case we are led to the recognition that many continuous formulations actually relate to discrete problems for which integer models would be far difficult to manage.

The translating process is treated as a means to provide a communication protocol between the DM and the Decision Aid System (DM/DAS protocol) under the form of a close-to-natural language, eliminating the visible role of the analyst. This feature has the potential to open a way into the generalization of the use of MCDA methods, by making them easy and natural to work with.

We would like to emphasize that our approach tends to split the decision aid process into two general phases. First, an alternative generating method is supposed to be run; it will always depend strongly on the analyst and is supposed to be supported by a large computer if necessary. Afterwards, a learning and decision process will take place where the analyst role will tend to be minor and the grand stage will be free for the DM to move on at will; and this second phase can well be supported even by microcomputers worked by the DM himself, enhancing his feeling of mastering the problem.

2. JUSTIFYING A HOLISTIC APPROACH

Psychology and information theory bounded together in a cognitive approach have worked under a "human information processing" paradigm and since long have well established some limits on the human capacity of processing information, related to the Short Term Memory. In a well known article about "the magical number seven, plus or minus two", Miller (1956) first described the basis of what would become an important approach in the coming years. In short, the span of absolute judgement and the span of immediate memory impose severe limitations on the amount of information that the human being is able to receive, process and remember.

In the case of absolute judgement, the channel capacity of the human is directly related to the number of bits of transmitted information. If the input information is made in terms of only one variable, then a ceiling around 2.6 bits of channel capacity (roughly 6.5 categories) appears in untrained subjects. When multivariate stimuli are present, that capacity is increased; but while adding more variables increases the total capacity, the accuracy for any particular variable is decreased. This simply means that people are less accurate if they must judge more than one attribute simultaneously. This conclusion, shared by common sense, is of obvious interest to a decision aid environment and introduces the problem of strategies.

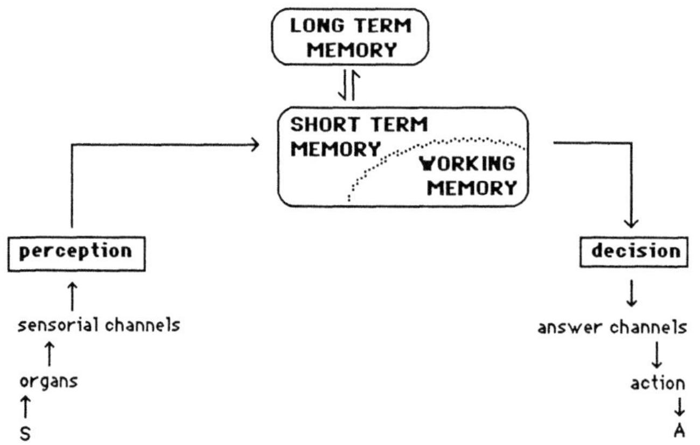

Fig 1 - Central role of the Short Term Memory in human information processing

The human information processing paradigm admits that any cognitive performance implies the use of programs to deal with the input data received through sensorial channels or recovered from storage (Long Term Memory and Semantic Memory) and produces an output. We have therefore the following steps: input, coding, storage, recovering and answer . The central role of the Short Term Memory in this process, as illustrated in fig. 1, is evident. There we can identify all high levels of information processing such as relations, inferences, comparisons, justifications (Sternberg, 1977).

It is therefore important to consider the span of immediate memory, which has a different nature from the absolute judgement process considered above. In fact, this span is quite independent of the number of bits of information processed, but is very directly related to the number of items or *chunks* of information involved (with a maximum around 7). This independence of the amount of information is obtained by the intensive use of the ability of *recoding* . The process goes like this: the input information is perceived in a code that contains many chunks with few bits per chunk; the operator recodes the input into another code that contains fewer chunks with more bits per chunk. In this way, following what could be called a labeling strategy, the operator can deal with more information while keeping trace of a small number of items. Take, for instance, the way people memorize telephone numbers, by sequences of two or three.

It has been demonstrated that the consistency of a learning period is crucial to the success of the recoding procedure, and this conclusion is also important to any decision aid approach. It must be added that a trial and error possibility in the decision aid system may enhance the capacity for the DM to build the most appropriate way of recoding the input information.

These spans and the limited capacity of the Quick Access Memory impose severe upper bounds on the amount of information one can handle at a time. Being such a poor link in a communication system, man had to put in action ways to abstract the essential features of the problems and translate them into alternative forms and simpler rules. And these abstraction capacities seem to be directly related to a biological condition, in the sense that some processes are preferred to others because the brain is like a hardware that is more adapted to some ways of working due to its built-in structure.

For example, the human ability to handle self-embedded recursion is very limited. However, in everyday situations people are able to resume what they were doing after having been interrupted. The thing is that the environment helps people to remember the re-entry point after executing a "sub-routine", overruling the cognitive limitations. This has a strong implication in the conception of a decision aid system: it must allow back and forth excursions of the DM through the various stages of the information processed, so that he may perform local operations and resume his previous reasonings; and coming back to a point where he had interrupted his line of thinking, he must be faced with the same situation in order to be helped in remembering the state of the problem at the interrupting time.

One last word for the analyst. Being a source of noise in the communication system, his role must tend to be minor or erased during the decision process. We feel that he is of primarily importance during the modeling phase, and even then there is noise introduced in the process, as some modeling options have hidden behind them decisions of some sort. But during the decision phase, he must in the ideal case be restricted to a function of assisting the DM in learning how to use the decision aid system.

3. THE ANCHOR POINT MODEL

In this chapter we will present one of the two general models that we have derived for grouping the alternatives into a small number of clusters. The objective is to provide the DM, right from the start, with global information. The number of clusters should not be greater than six or seven, and in all cases the DM must be able to control this number, so that no more information than what he can handle at a time should be presented to him.

We will take as departing point a limited (although large) set $\mathbb{Z}$ of alternatives, generated by any method. These alternatives are supposed to be non dominated, either in a traditional crisp sense or in any kind of fuzzy sense. Without loss of generality we will suppose a maximizing problem in all objectives. In order to deal with such a large set, the methodology defines or identifies a restricted number of Macro-Solutions (MS) in which the relevant information is condensed. Each MS $\mathbf{M}_i$ is considered as a fuzzy set of alternatives

$$\mathbf{M}_i = \{(\mathbf{z}, u(\mathbf{z},\mathbf{M}_i)) \mid \mathbf{z} \in \mathbb{Z}\}$$

where u: $\mathbf{z} \rightarrow [0,1]$ is the membership of $\mathbf{z}$ in $\mathbf{M}_i$.

The definition of the MS requires the assessment of the membership functions u_{ik} for all the solutions $\mathbf{z}_k$ referring to each $\mathbf{M}_i$. The alternative set may be then represented in several ways, such as by a set **S** of representative points of all the MS, to be defined later.The intersection of two MS can be given by the Zadeh operator

$$u(\mathbf{z}, \mathbf{M}_i \cap \mathbf{M}_j) = \min \{ u(\mathbf{z}, \mathbf{M}_i); \; u(\mathbf{z}, \mathbf{M}_j) \}$$

but other operators may be considered.

In the following paragraphs one technique to deal with the MS (in order to reduce the alternative set according to the DM's preferences) will be briefly sketched. Details of the procedures are found in Matos (1988) ; other techniques, namely a Successive Amplification Method, are mentioned in Matos and Miranda (1988).

Anchor Points

This model requires the pre-definition of a representative set **S** of the MS, using a number of anchor points in the attribute space, which may correspond to the individual maxima of the objective functions. Other anchors can be added such as the ideal solution in the sense of Zeleny (1982), if its meaning can be well established from the DM point of vue. Fuzzy membership functions u_{ik} relating any alternative $\mathbf{z}_k$ to each anchor point $\mathbf{S}_i$ are defined as distance-type measures, taking into account the maximum and minimum value of each objective for normalization purposes, but a solution is not penalized for being better than the anchor point in a particular attribute component.

The fuzzy membership functions $u(\mathbf{z}, \mathbf{M}_i)$ are defined as follows:

* for each attribute (or component) k,

$$u_k (\mathbf{z}, \mathbf{M}_i) = \min \{ 1 ; (\mathbf{z}_k - \mathbf{z}_k^{min}).(\mathbf{s}_{ik} - \mathbf{z}_k^{min})^{-1} \}$$

* a weight is defined to normalize the importance of the scale amplitude of each attribute:

$$w_{ik} = [(\mathbf{s}_{ik} - \mathbf{z}_k^{min}).(\mathbf{z}_k^{Max} - \mathbf{z}_k^{min})^{-1}]$$

* the membership function values are then obtained by

$$u (\mathbf{z}, \mathbf{M}_i) = (\Sigma_k w_k [u_k (\mathbf{z}, \mathbf{M}_i)]^r)^{1/r} . (\Sigma_k w_k)^{-1/r}$$

where $r \in [1, \infty[$ is an external parameter.

By this procedure we have evaluated "how distant" or "how different" is an alternative from any anchor point, measuring this by the value of the membership function. Each alternative is therefore associated with a set of membership values relating it to all anchor points.

An iterative procedure may then be followed to support a DM in the search for the preferred solution. It is based on determining the anchor points, assessing membership functions of every alternative to every MS, constructing α-level sets for every MS and calculating their mutual intersections, in order to be able to initiate a dialogue with the DM until a unique set is reached, from which the preferred solution may be chosen. We must stress that the intersections should be performed cautiously; for instance, the use of the "min" operator seems to be inadequate because it lacks trade-off properties that are implicit (or explicit) in the human comparison of alternatives.

4. BASIS FOR A DM/DAS PROTOCOL

One of the main features of the model presented above is the possibility to label the MS, (built on the anchor points) with linguistic descriptors related to the objectives under consideration. This has some similarity with the "close to the ideal" label used by Zeleny (1982) for the compromise set. Other obvious and typical labels could be "cheap solution" or "safe alternative" in problems with cost functions or safety concerns. Compound solutions originated by the intersection of MS or crisp α-level sets may also be labeled, using the natural language operator "and".

This labeling characteristic appears as very important since it can provide an easy way for the DM to *recode* the information he is getting and therefore keep control of the vast amount of information he must deal with.

It is also possible to establish some linguistic attributes to the membership functions by a suitable construction of their semantic representation. This is done by means of a nominal scale of graduating descriptors such as "most", "very", "medium", etc. (which in fact are fuzzy sets of membership function values), also referred to as intensities in the literature. The graduating descriptors are grammatically used as modifiers of the original labels adopted to identify the MS so that every solution **z** may be identified by a compound natural label arising from the relations between **z** and the MS.

In order to accomplish this, it is only necessary to define the possibility distributions associated to each modifier. This can be done *a priori* or in interaction with the DM, following a methodology suggested by Freksa (1982) for each objective; it goes without saying that one of the convenient shapes for those distributions, for operational purposes, is the trapezoidal one.

DM $\rightarrow$ DAS communication: interpretative

The communication in the sense DM $\rightarrow$ DAS is used to translate in terms of fuzzy (or hard) sets the linguistic descriptors used by the DM.

Consider the set $\mathbb{Z}$ and c macrosolutions $\mathbf{M}_i$, each one labeled. Admit that by some process the "direct" membership functions $u(\mathbf{z}, \mathbf{M}_i)$ have been assessed for all $\mathbf{z} \in \mathbb{Z}$. Take finally a set of graduating modifiers **G** defined by their possibility distributions such as in figure 1.

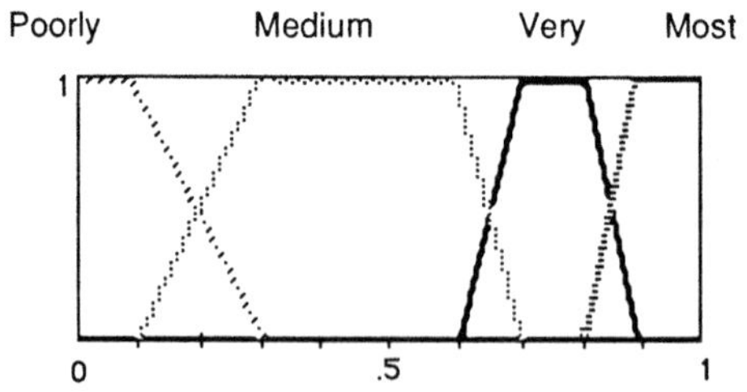

Figure 2 - Graduating descriptor possibility distributions (in this example, only four modifiers are presented, but the actual number proposed is seven - again, the magical number...)

The reunion will be required often to translate expressions such as "at least $G_j \cdot M_i$" , which has its formal equivalent in $\bigcup_{k\geq j} G_k \cdot M_i$. The use of the **max** operator would lead to coherency conflicts as it would generate aggregate non-convex membership functions. Instead, we use the cumulative formulation

$$d(\mathbf{z}, \mathbf{D}_{ji} \wedge \mathbf{D}_{pi}) = d_{ji} + d_{pi}$$

which requires that $\sum g_j(u)=1$ for all u. From a point of view, this way of defining the membership functions g_j can be seen as a fuzzy partition of the interval [0, 1].

DAS → DM communication: classifier

In this case the purpose is to present to the DM the labels of the alternatives under consideration. These labels should be related to the direct membership functions $u(\mathbf{z}, \mathbf{M})$, which are calculated in association with the MS and filtered by the graduating descriptors.

Take a general set $\mathbf{X}$ of elements x where fuzzy subsets $\mathbf{F}_i$ are defined such as

$$\mathbf{F}_i = \{ (x, u(x, \mathbf{F}_i)) \mid x \in \mathbf{X} \}$$

When one uses a fuzzy designation associated with a subset $\mathbf{F}_i$, it is at once known the possibility of an element $x \in \mathbf{X}$ being labeled in that way, through the value of the membership function. It is easily concluded that each element has associated with it a set of ordered pairs

$$\{ (\mathbf{F}_i, u(x, \mathbf{F}_i)) \mid \mathbf{F}_i \in \mathbf{X} \} .$$

If $\mathbf{F}_i$ is a designation, one can consider the set of all the designations $\mathbf{D}(\mathbf{X})$ of the fuzzy subsets of $\mathbf{X}$ or, if preferred, the class $\mathbf{D}(\mathbf{X})$ of the fuzzy subsets of $\mathbf{X}$. We have reached therefore a fuzzy subset of designations of x as

$$\{ (\mathbf{F}_i, u(x, \mathbf{F}_i)) \mid \mathbf{F}_i \in \mathbf{D}(\mathbf{X}) \} .$$

This subset forms a classifying system for an element x in relation with a specific property or attribute. Coming back to the decision aid problem, any alternative $\mathbf{z}$ of $\mathbb{Z}$ can receive a fuzzy classification $\mathbf{C}_i$ in respect to a macrosolution $\mathbf{M}_i$ by

$$\mathbf{C}_i(\mathbf{z}) = \{ (\mathbf{G}_j, d_j(\mathbf{z})) \mid d_j(\mathbf{z}) = g_j(u(\mathbf{z}, \mathbf{M}_i)) \}$$

Finally a label can be associated to $\mathbf{z}$ if desired using the maximum rule.

Then, an elementary statement $\mathbf{D}_{ji}$ produced by the DM will be of the form

$$\mathbf{D}_{ji} = \mathbf{G}_j \cdot \mathbf{M}_i$$

corresponding to a fuzzy set of solutions

$$\mathbf{D}_{ji} = \{ (\mathbf{z}, d_{ji}(\mathbf{z})) \mid \mathbf{z} \in \mathbb{Z} \quad \text{and} \quad d_{ji}(\mathbf{z}) = g_j(u(\mathbf{z}, \mathbf{M}_i)) \}$$

The following example illustrates the technique:

$$\mathbf{M}_i = \{ (\mathbf{z}_1, 0.9); (\mathbf{z}_2, 0.8); (\mathbf{z}_3, 0.7); (\mathbf{z}_4, 0.65); (z_5, 0.6) \} \quad - \quad \textbf{label: safe}$$

$\mathbf{G}_j$: **descriptor: very** (see figure 2)

$$\mathbf{D}_{ji} = \mathbf{G}_j \cdot \mathbf{M}_i = \{ (\mathbf{z}_2, 1); (\mathbf{z}_3, 1); (\mathbf{z}_4, 0.5) \} \quad - \quad \textbf{composite label: very safe}$$

If necessary, the α-level crisp core $\mathbf{N}^{\alpha}$ of $\mathbf{D}_{ji}$ may be used; this set is extracted without the construction of $\mathbf{D}_{ji}$, assuming a convexity condition on $g_j(u)$, by:

$$u^- = \min_{(u)} \{g_j(u) \geq \alpha \}$$
$$u^+ = \max_{(u)} \{g_j(u) \geq \alpha \}$$

and

$$\mathbf{N}^{\alpha}[\mathbf{D}_{ji}] = \{ \mathbf{z} \mid \mathbf{z} \in \mathbb{Z} \text{ and } u^- \leq u(\mathbf{z}, \mathbf{M}_i) \leq u^+ \}$$

In the previous example, for $\alpha = 0.7$, one would get $u^- = 6.7$, $u^+ = 8.3$ and $\mathbf{N}^{0.7} = \{\mathbf{z}_2, \mathbf{z}_3\}$.

More elaborate designations require the use of fuzzy set reunion and intersection rules. Taking two elementary designations $\mathbf{D}_{ji}$ and $\mathbf{D}_{qp}$ one gets

$$d(\mathbf{z}, \mathbf{D}_{ji} \wedge \mathbf{D}_{qp}) = \min \{d_{ji}, d_{qp}\} \quad , \quad i \neq p$$

$$d(\mathbf{z}, \mathbf{D}_{ji} \vee \mathbf{D}_{qp}) = \max \{d_{ji}, d_{qp}\} \quad , \quad i \neq p$$

The generalization of the intersection rule is quite straightforward, depending only on the satisfaction of the condition

$$\mathbf{G}_j \cap \mathbf{G}_q = \varnothing \qquad \text{for } |j-q| > 1$$

which means that the intersection of two modifiers is empty in all cases except if they are adjacent.

For example, admit the graduating descriptors in figure 2 and Macrosolution $\mathbf{M}_2$ label: **SAFE.** Take solution $\mathbf{z}_2$: u ($\mathbf{z}_2$,$\mathbf{M}_2$)= 0.8 . Then the classification: **VERY SAFE** will emerge from:

$$\mathbf{C}_2(\mathbf{z}_2) = \{ (\text{POORLY}, 0) ; (\text{MEDIUM}, 0); (\text{VERY}, 1); (\text{MOST}, 0) \}$$

5. QUALITATIVE DOMINANCE

Taking advantage of the α-level classifying properties of the system developed, one can also adopt another strategy in order to greatly reduce the number of alternatives that deserve separate consideration, if the concept of qualitative dominance is introduced. One solution is said to qualitatively dominate another if in each attribute it is classified at least in the same rank or preference degree and in at least one attribute this classification is strictly of a higher degree. The practical effect of this concept will turn clear with an example, extracted from a problem with three objective functions presented by Hwang and Masud (1979, pag. 111).

In Table I, 20 efficient alternatives are represented by their membership values to each Macrosolution, the anchor points having been defined as the best value in each objective. Classifying these solutions by means of the graduating descriptors depicted in figure 2 (assuming that they were built in interaction with the DM), one reaches Table II. It becomes then obvious that only five solutions are retained as qualitatively non dominated. We are therefore in the presence of a powerful filtering device, that can be adjusted by compounding more or less the graduating descriptors, and that can be used to highlight in a way some of the main trends that may be detected in the alternative set.

The important is to notice that these trends are revealed as a consequence of the qualitative nature of the graduating descriptors, whose definition was previously built in interaction with the DM. Besides, the flexibility of the model allows the range of these modifiers to be adjusted by the DM at any time (if, for instance, the DM reaches the conclusion that, after all, he had been too strict in defining what he felt "most" would correspond to).

The interpretation that can be given to this filtering process is that the mathematical models, by the very fact of their numerical nature, create a number of mathematically efficient solutions that may not have a psychological correspondence; in other words, numerically close solutions are certainly in many cases indifferent in respect to one another but nevertheless qualitative classes can well be distinguished by humans. Take the famous example of the cup of coffee: small differences in sugar content are not distinguishable, still graduating levels can be well defined and make sense to the human such as "little", "medium" or "very sweet".

A theoretical consequence to be explored is the fact that the objections to the transitivity of preference relations based on that famous paradox may well be overcome when a qualitative filter built from DM declarations is imposed on the set of alternatives. A "well behaved" system of preferences should probably be based on qualitative classes rather than on objects (alternatives) themselves.

TABLE I

	M_1	M_2	M_3
1	.632	.260	.934
2	.589	.395	.993
3	.707	.426	.792
4	.516	.062	1.00
5	.609	.462	.970
6	.907	.679	.658
7	.847	.749	.595
8	.761	.843	.452
9	.861	.646	.697
10	.911	.704	.619
11	.938	.677	.630
12	.852	.920	.617
13	.938	.667	.629
14	1.00	.688	.531
15	.673	.962	.226
16	.630	1.00	.207
17	.763	.949	.210
18	.965	.673	.589
19	.759	.956	.208
20	.752	.961	.207

TABLE II

	A_1	A_2	A_3	
1	m/V	P/m	Mt	*
2	m	m	Mt	*
3	V	m	V	
4	m	P	Mt	*
5	m/V	m	Mt	
6	Mt	V/m	V/m	
7	V/Mt	V	m	*
8	V	V/Mt	m	*
9	Mt/V	m/V	V/m	*
10	Mt	V	m/V	
11	Mt	V/m	m/V	*
12	Mt/V	Mt	m/V	
13	Mt	V/m	m/V	*
14	Mt	V/m	m	*
15	V/m	Mt	m/P	*
16	m/V	Mt	m/P	*
17	V	Mt	m/P	*
18	Mt	V/m	m	*
19	V	Mt	m/P	*
20	V	Mt	m/P	*

TABLES I and II - 20 efficient solutions from Hwang and Masud (1979). Table I includes the membership function values of each solution to every Macrosolution. Table II shows the qualitative classification to each Attribute as a result of the composition of Table I values with Fig. 2 graduating descriptors. All solutions marked with * are qualitatively dominated. The main trends are best represented by solutions 3, 5, 6, 10 and 12. Notice the elimination of all anchor points (4, 14 and 16). DESCRIPTORS: P - poorly; m - medium; V - very; Mt - most.

6. CONCLUSION

The merits of the approach referred to above are divided between the analytical algorithms developed and the close-to-natural interface conceived. Some important features of the proposed procedure are:

a) A clear cut separation between the alternative generating phase and the learning and decision phase is adopted, clarifying the roles of the analyst and the DM;

b) An overall insight of the alternatives is provided at the beginning of the process.

c) The DM has never to consider, at each step, more alternatives than he/she is prepared to deal with.

d) Except for the final decision, it is always possible to take multiple decisions.

e) All decisions are choices among alternatives or groups of alternatives - the DM never needs to provide (or decide on) weights, trade-offs or aspiration levels.

The last two points deserve emphasizing. Choices among alternatives discard the necessity of the modelling of any kind of utility function with its preference transitivity drawbacks. In fact, even if one admits that some kind of utility could be assigned to the DM at some stage, no one is prepared to state that it will remain unchanged till the end of the process. The Anchor Point method, by allowing a progressive reduction of the alternative set with easy backtracking, helps the DM to learn about the problem he is dealing with; therefore, any kind of utility notion is likely to change during the process.

On the other hand, the multiple choice possibility gives free hand to the subjective use of concepts such as preference, outranking or indifference without (at this stage) the necessity of modelling them. These concepts seem to be best handled if qualitatively modeled instead of quantitatively, specially when indifference between alternatives may be transformed into a positive preference relation between classes of alternatives. One way of achieving this is certainly by means of linguistic labels and graduating modifiers: they introduce the qualitative aspect in the problem model and provide the DM a means to recode the information extracted from the problem in a psychologically natural way, taking in account the limits on his capacity for processing information at a time.

REFERENCES

1. **Miller, G.A.**,"The magical number seven, plus or minus two: some limits on our capacity for processing information", in Psychological Review, vol. 63, n.2, Mar 1956.

2. **Sternberg, R.J.** - "Intelligence, information processing and analogical reasoning: the componential analysis of human abilities", Hillsdale, N.J., Erlbaum, 1977.

3. **Bezdek,J.C.** - "Pattern recognition with fuzzy objective function algorithms", Plenum Press, 1981.

4. **Matos, M.A.C.C.** - "Ajuda à decisão multicritério: novas contribuições", PhD. thesis, Faculdade de Engenharia da Universidade do Porto, Departamento de Engenharia Electrotécnica e de Computadores, 1988.

5. **Matos, M.A.C.C; Miranda, V.** - "A holistic approach and a close to natural interface in a decision aid system", 27th meeting of the Europ. Work. Group on MCDA, Faculté Polytechniquede Mons, Belgium, Mar 1988

5. **Zeleny** - "Multiple criteria decision making", McGraw Hill, 1982.

6. **Freksa, C.** - "Linguistic description of human judgements in expert systems and in the 'soft sciences", in "Approximate reasoning in decision analysis", M.M. Gupta & E Sanchez edt., North-Holland, 1982.

7. **Hwang, C.L.; Masud, A.S.M.** - "Multiple Objective Decision Making - Methods and Applications", Springer-Verlag, New York, 1979

ACKNOWLEDGEMENT

The authors would like to thank Dr. Leandro Almeida, from the Faculdade de Psicologia da Universidade do Porto, for the discussions on the human information processing theory.

A COMPARISON OF MULTI-ATTRIBUTE UTILITY GENERATION METHODS

Arthur Money Ph.D.
Henley Management College
Henley-on-Thames
UNITED KINGDOM

Peter Naude M.Sc.
University of Cape Town
Rondebosch 7700
REPUBLIC OF SOUTH AFRICA

Trevor Wegner Ph.D.
University of Cape Town
Rondebosch 7700
REPUBLIC OF SOUTH AFRICA.

INTRODUCTION :

Utility based techniques have been applied in recent years to a variety of business related areas for quantifying respondent perceptions and/or preferences relating to a specific object/product. The methods of Conjoint Analysis, Trade-Off Analysis, the Self-Explicated Rating approach, Cognitive Mapping and various hybrid models are amongst the most frequently advocated utility generation approaches (Green and Rao 1971; Neslin 1981; Green, Goldberg and Montemayer 1981; Holbrook 1981; Cattin and Wittink 1982; Green and Wind 1984; Goldberg, Green and Wind, 1984; Jaccard, Brinberg and Ackerman 1986, Conrath, Montazemi, Higgins 1987). A relatively new approach which has emerged is that of the Analytical Hierarchy Process (Saaty and Wind 1980).

The presence of an array of utility generation techniques has raised a research problem which has not yet been adequately addressed in the literature. This concerns the degree of agreement between these various utility generation measures. Research which has appeared on this issue has been conducted by, inter alia, Neslin (1981), Shoemaker and Waid (1982), Jaccard, Brinberg, Ackerman (1986), Lockett and Stratford (1987) and Money, Tromp and Wegner (1988). The results of convergent validity have however not been very encouraging.

This paper seeks to supplement this research area with a focus on the **convergence of derived utilities** between the Analytical Hierarchy Process (AHP) and the Conjoint Measurement approach. This comparison was proposed by Wind and Saaty (1980 : 657). Notwithstanding, the only reported comparative study involving the AHP has been conducted by Lockett and Stratford (1987). However, even these researchers call for further research. Convergent validity will be examined from the standpoint of similarity in ranked response profiles as reflected by the resultant utilities of attributes across the different methods.

A **secondary objective** is to examine the **ease of data capture** for each of the methods involved. To this end, two new data capturing instruments for the Conjoint method are examined in relation to the standard approaches (ie. Full Factorial design and Fractional Factorial design) as a means of providing a data capturing instrument of comparable reliability but which is cognitively less demanding on respondents and simpler to apply.

METHODOLOGY :

(i) The Problem Setting :

A common frame of reference was used to capture perceptions data from 33 respondents. Respondents were asked to reveal their preferences for various attributes associated with a particular motor vehicle. The primary attributes related to Price, Specific Features and Type of Guarantee offered. Each primary attribute in turn is associated with three sub-attributes resulting in 9 attribute utilities to be generated. The objective is to uncover the utility a respondent associates with each attribute in a likely purchase decision situation.

(ii) Utility Generation Methods :

Utilities are generated using five methods. The first is the **AHP approach** (Method 1) which employs eigenvector analysis in the generation of utilities. The next two methods use the **Conjoint measurement** technique as applied to the **Full Factorial** design (Method 2) and **Fractional Factorial** design (Method 3) respectively.

The remaining two methods are variants of the Conjoint and the AHP approaches. Both methods adopt a two stage approach to the generation of attribute utilities. The first of these variant methods, referred to as the **AHP/Conjoint Hybrid** approach (Method 4), uses the AHP approach to data capture. However, the pairwise trade-offs are performed on attribute bundles (as used in the Fractional Conjoint design), instead of the individual attributes. The AHP-generated utilities for the bundles are then ranked to provide the input to a Conjoint analysis from which individual attribute utilities are derived.

The second variant, referred to as the **Conjoint Pairwise Frequency** approach (Method 5), uses a modified AHP approach to data capture. Only the first component of the AHP pairwise trade-off data collection method is used, namely the indication of the preferred attribute bundle. These expressed preferences are used to produce initially a frequency count of preferred attribute bundles. Thereafter the bundles are ranked on the basis of the frequencies and subjected to a Conjoint analysis from which individual attribute utilities are again generated.

The modified data capturing instruments of these latter two utility generation methods is an attempt to combine the cognitively less demanding approach of pairwise trade-offs of the AHP with the inclusion of bundles which provide a more realistic representation of the problem (Goldberg, Green and Wind, 1984).

ANALYSIS.

Initially, the attribute utilities for each respondent within each method were standardised to corresponding utilities within a zero-one interval and a sum equal to one. The subsequent comparative analysis of profiles across Utility Generating (UG) methods is conducted firstly on an **individual** respondent basis and then on an **aggregated** level.

For **each respondent**, the derived attribute utilities within each method were ranked from 'least preferred' (rank = 1) to 'most preferred' (rank = 9). These ranked response profiles were then compared across UG methods to establish any measures of similarity. Kendall's Coefficient of Concordance (K-W) which measures the degree of association among several rankings of N objects/attributes was computed for each respondent.

From the distribution of K-W values in Figure 1, it is clear that, with the exception of 6 respondents, on an **individual level**, there is a moderate to high degree of similarity of ranked response profiles across the five UG methods. Over 80 percent of all respondents generated consistent profiles across methods.

To establish if this observation is valid across all respondents, it is necessary to aggregate response profiles for each UG method. Averaging across all respondents however, assumes homogenuity of response profiles. If this is not the case, then the averaging process will distort the overall response profile for each method. To examine whether the assumption of homogenuity of response profile across respondents for each method is true, respondents with significant K-W coefficients (K-W > 0.500) were clustered on the basis of their averaged standardised response profiles across methods.

FIGURE 1 : DISTRIBUTION OF CONCORDANCE COEFFICIENTS

```
                                               ___________
                                   ___________I           I
                                  I           I           I
                                  I           I           I___________
N = 33                            I           I           I           I
                  ________________I           I           I           I
                 I      I         I           I           I           I
          _______I      I         I           I           I           I
         I   2   I  4   I    4    I     8     I     9     I     6     I
         I_______I______I_________I___________I___________I___________I_

     K-W < .4   .4     .5       .549        .649        .749        1.00

Association     Poor   I      Moderate      I      High
                       I p<    I p <        I      p <
Significance           I 0.01  I 0.005      I      0.001
```

The complete linkage clustering process identified three separate clusters of respondents. To establish if the response profiles between clusters are significantly different, Kendall's Concordance coefficients were computed between the ranked average profile for each cluster. A comparison of the ranked profile of the overall sample with the clusters collectively is also included. (Refer Table 1.)

On an individual cluster basis, each cluster's response profile to the overall [27 respondents] ranked profile is examined using Spearman Correlation coefficients. The results appear in Table 2.

TABLE 1 : RANKED AVERAGE PROFILES PER RESPONDENT CLUSTER

ATTRIBUTE	ALL GROUPS (N = 27)	CLUSTER 1 (N_1 = 10)	CLUSTER 2 (N_2 = 9)	CLUSTER 3 (N_3 = 8)
U_1	7	9	7	7
U_2	5.5	7	2	6
U_3	3	1.5	5	1
U_4	4	5	6	2
U_5	2	3	1	4
U_6	9	6	8	9
U_7	8	8	9	8
U_8	5.5	4	4	5
U_9	1	1.5	3	3
K-W (Between Clusters 1,2 and 3 only)			0.7361	$p<0.001$ (df = 8)
K-W (Between Clusters and Overall Sample)			0.7781	$p<0.001$ (df = 8)

TABLE 2 : SPEARMAN RANK CORRELATIONS BETWEEN RESPONDENT CLUSTERS AND OVERALL SAMPLE.

Overall / Cluster 1	0.8167	$p < 0.010$
Overall / Cluster 2	0.7542	$p < 0.025$
Overall / Cluster 3	0.8625	$p < 0.005$

As seen from Tables 1 and 2, while separate clusters emerged from the analysis, their average ranked response profiles do not appears to be significantly different. There is a moderately high degree of intra-cluster association (K-W = 0.7361) and a higher degree of association between the separate clusters and the overall sample (K-W = 0.7781). The average ranked response profile across all respondents can therefore be assumed to be representative of each cluster's profile. This permits further analysis to be conducted on an aggregated level using the overall sample rather than the separate clusters.

The mean utilities across all 27 respondents is therefore produced as a basis for further examining the convergence of these UG methods on an **aggregated level.** Their values for each attribute derived from each of the five methods, together with a ranking of the attributes within each method is given in Table 3.

TABLE 3 : AVERAGED ATTRIBUTE UTILITIES AND RELATED RANKINGS

1 = Least Preferred Attribute : 9 = Most Preferred Attribute

	Method 1		Method 2		Method 3		Method 4		Method 5	
	Ave	R	Ave	R	Ave	R	Ave	R	Ave	R
U_1	0.254	8	0.183	9	0.134	7	0.131	7	0.134	7
U_2	0.073	6	0.123	6	0.120	6	0.105	5	0.091	3
U_3	0.038	2	0.027	1	0.080	2	0.094	3.5	0.105	5
U_4	0.054	4	0.096	4	0.109	5	0.094	3.5	0.097	4
U_5	0.052	3	0.064	3	0.076	1	0.081	2	0.086	2
U_6	0.182	7	0.173	8	0.149	9	0.160	9	0.151	9
U_7	0.258	9	0.162	7	0.143	8	0.155	8	0.147	8
U_8	0.060	5	0.109	5	0.108	4	0.108	6	0.109	6
U_9	0.029	1	0.063	2	0.082	3	0.072	1	0.079	1

Firstly, two Correlation Matrices are computed based upon the interval scaled utilities and the rankings of these averaged utilities as given in Table 3. The Pearson product moment correlation is computed from the interval scaled utilities, while the Spearman's rank correlation coefficient is computed from their rankings. The results are shown in Table 4 where the upper diagonal elements are Pearson product moment correlations, and the lower diagonal elements are Spearman rank correlations.

TABLE 4 : CORRELATIONS ACROSS THE FIVE METHODS

	Method 1 (AHP)	Method 2 (FULL)	Method 3 (FRAC)	Method 4 (HYBRID)	Method 5 (PAIR/W)
Method 1	0.0000	0.8753*	0.8375*	0.8731*	0.8817*
Method 2	0.9333*	0.0000	0.9517*	0.8581*	0.7897
Method 3	0.8667*	0.8833*	0.0000	0.9315*	0.8534*
Method 4	0.9042*	0.8625*	0.8792*	0.0000	0.9704*
Method 5	0.7833	0.7167	0.7667	0.9458*	0.0000

* Significant at $p < 0.005$ ($N = 9$) ($r_{(.005)} = 0.8330$)

Average Spearman Correlation Across all Pairs = 0.8542*

Average Pearsons Correlation Across all Pairs = 0.8822*

Kendall's Coefficient of Concordance = 0.882 ($p<0.001$/df=8)

A reasonably consistent pattern emerges between the two sets of correlations. Highly significant correlations are found between almost all pairs of Utility Generation Methods. The only exception is the correlation between the Full Conjoint method and the Pairwise Frequency method where the association is moderately lower.

These associations can be examined graphically through the use of **cluster analysis** which will identify any natural groupings of methods. Using the single linkage method of clustering on the Pearson correlation coefficients from Table 4, no distinct clusters of methods appear. From the dendrogram given in Figure 2, it can be seen that a single cluster emerges at a similarity level (Pearson correlation) which is still highly significant (r = 0.8817).

FIGURE 2 : DENDROGRAM FOR CLUSTERING OF METHODS

```
                           METHOD
Distance        1       2       3       4       5

  0.9704        I       I       I       I_______I
  0.9517        I       I_______I           I
  0.9315        I           I_______________I
  0.8817        I_____________________I
```

To establish a measure of overall similarity of attribute utility profiles generated by each of the five methods, Kendall's Coefficient of Concordance is computed using the rankings of Table 3. The value of 0.882 is statistically significant (p<0.001). This implies that essentially consistent rankings are being generated across the alternative Utility Generation Methods.

This perceived similarity across all five methods is further tested using Friedman's Two Way Analysis of Variance for related samples. The null hypothesis tested is that the rankings generated by the five methods are drawn from the same population. Based on the rankings in Table 3, the Chi Squared (calculated) statistic for the Friedman Test is 2.200 (df = 4), implying that the Null hypothesis cannot be rejected (p > 0.70).

In addition, overall measures of association, using Kendall's Concordance statistic, for each separate **Respondent cluster** as identified in Table 1, were also computed to establish if any improvement in association between the different methods resulted. These K-W coefficients are shown in Table 5.

TABLE 5 : KENDALL'S COEFFICIENTS OF CONCORDANCE BETWEEN METHODS WITHIN EACH CLUSTER OF RESPONDENTS.

Overall (Entire Sample N = 27)	0.882
Cluster 1 (N_1 = 10)	0.859
Cluster 2 (N_2 = 9)	0.814
Cluster 3 (N_3 = 8)	0.854

From Table 5, the degree of homogenuity of ranked profiles is only marginally lower than for the sample as a whole. This supports the earlier conclusion to aggregate over the entire sample rather than to perform analysis at the aggregated level on separate clusters.

DISCUSSION :

Five Utility Generation methods were examined for convergence validity. The Full and Fractional Conjoint methods were selected as they represented the "traditional" methods of utility generation. The AHP is a relatively new approach for which comparisons of results have not been produced. The Hybrid and the Pairwise approaches were proposed as methods which combines the realism of trading off bundles of attributes which is a Conjoint feature, with the ease of pairwise trade-offs of data capture of the AHP - and so avoids the cumbersome data capturing procedures of ranking which is cognitively demanding.

The results are encouraging. On an individual level, more than 80 percent of respondents generated significantly similar ranked response profiles across the five Utility Generation methods. On an aggregated level, highly significant ranked response profiles across the five methods were identified within three separately identified clusters of respondents and for the overall sample.

Notwithstanding the high overall consistency measures, moderate differences are observable between the aggregated ranked response profiles of each method. While the top three attributes identified from each method are similar, their ordering is not completely identical (refer to Table 3). A possible explanation may be found in an examination of the stress factors and consistency indices of the data capturing procedures. Closer attention needs to be paid to the removal of inconsistent responses in future studies of this nature.

This paper has however contributed in the following way. Firstly, it has provided evidence of the degree of association between the AHP method and the Conjoint approaches as was suggested by Wind and Saaty (1980). Secondly, it has proposed cognitively less demanding data capturing approaches based on the AHP methodology, but includes the realism of attribute bundles of the Conjoint methods. Thirdly, the overall significant relationships between the AHP and the Conjoint set of methods implies that these methods could conceivably be used interchangibly with consistent outcomes. However, before advocating complete compatibility, it is recommended that the results obtained in this research study need to be replicated in order to validate these conclusions beyond reasonable doubt.

ACKNOWLEDGEMENT : The authors acknowledge the support provided by the HSRC in the preparation of this paper.

BIBLIOGRAPHY

Cattin P, Wittink D R (1982) : Commercial Use of Conjoint Analysis, Journal of Marketing, Vol. 46 :44-53.

Conrath D W, Montazemi A R, Higgins C A (1987) : Evaluating Information in Ill-Structured Decision Environments. Journal of Operational Research Society ; Vol. 38 (5) :- 375-385.

Goldberg S M, Green P E, Wind Y (1984) : Conjoint Analysis for Price Premiums for Hotel Amenities, Journal of Business, Vol. 57 (1) : S111-S132.

Green P E, Goldberg S M, Montemayer M (1981) : A Hybrid Utility Estimation Model for Conjoint Analysis. Journal of Marketing. Vol. 45 : 33-41.

Green P E, Rao V R (1971) : Conjoint Measurement for Quantifying Judgemental Data, Journal of Marketing Research, Vol. VIII : 355-363.

Holbrook M B (1981) : Integrating Compositional and Decompositional Analysis to Represent the Intervening Role of Perceptions in Evaluative Judgements, Journal of Marketing Research, Vol. XVIII : 13-28.

Jaccard J, Brinberg D, Ackerman L J (1986) : Assessing Attribute Importance : A Comparison of Six Methods. Journal of Consumer Research. Vol. 12 : 463 - 468.

Lockett G, Stratford M (1987) : Ranking of Research Projects : Experiments with Two Methods, OMEGA Vol. 15(5) : 395-400.

Money A, Tromp, D, Wegner T : (1988) : The Quantification of DSS Benefits within the Context of Value Analysis, MIS Quarterly. Vol. 12(2) : 223-236.

Neslin S A (1981) : Linking Product Features to Perceptions : Self-Stated Versus Statistically Revealed Importance Weights, Journal of Marketing Research, Vol. XVIII : 80-86.

Saaty T L (1980) : The Analytic Hierarchy Process. McGraw-Hill.

Schoemaker P J H, Waid C C : (1982) : An Experimental Comparison of Different Approaches to Determining Weights in Additive Utility Models. Management Science : Vol. 28 : 182-196.

Wind Y, Saaty T L (1980) : Marketing Applications of the Analytic Hierarchy Process. Management Science. Vol. 26 (7) : 641 - 658.

INTERACTIVE OPTIMIZATION: AN OPTIMAL SCALING AND STEP SIZING APPROACH FOR THE METHOD OF ABSTRACT FORCES

Roy J. Clinton
Management and Marketing Department
Northeast Louisiana University
Monroe, Louisiana 71209/USA

Marvin D. Troutt
Management Department
Southern Illinois University
Carbondale, Illinois 62901/USA

Introduction

The method of abstract forces (MAF), suggested by Troutt and Hemming (1985), for interactive multiple criteria optimization was modified to handle heterogeneous criteria. The modified method of abstract forces (MMAF) was tested by comparing solutions derived with it against solutions found with the more traditional "what-if" method. Both methods aid the decision maker (DM) in solving problems that involve numerous alternatives and conflicting criteria. A unique aspect of the MMAF is its capability of reverse problem solving (RPS). In RPS the DM works backward to determine the inputs that would result in the desired solution. In constrast a what-if method (WIFM) tells the DM what the results are after the DM varies the inputs.

Most multiple criteria optimization (MCO) methods have major limitations. Their lack of practicality in real world problem solving, is perhaps one of their most serious limitations. Many methods involve confusing exercises, take excessive time to reach a solution, and often diverge before eventually converging on the optimal solution. Consequently they are not very user friendly.

Multiple Criteria Problems

Many MCO problems consist of criteria which are measured in different units. Units of measurement may be dollars, number of employees, number of items, weight, days of availability, etc. In MCO the DM must chose an **x** (the decision) from a feasible set **X**, which will optimize several criteria functions, $f_j(\mathbf{x})$. We assume that the DM wishes to optimize each criterion and that his preferences constitute a value function, $U(f_j(\mathbf{x}))$, which is not known explicitly (Keeney and Raiffa, 1976).

What-if Models

Decision makers have ready access to numerous computer programs for "what-if" analysis. Solutions are found in an interactive process where the user changes the inputs and observes the new outputs (see figure 1). If output values are not satisfactory the user continues to change the inputs until outputs are satisfactory. The resulting search is very time consuming since the number of possible inputs is enormous. If the relationships among criteria and constraints are very complex the user may have difficulty in selecting ideal values for the inputs. Some selections may not be feasible, given the constraints. Also, changing inputs in attempts to improve one criterion may cause other criteria to deteriorate drastically. These are some of the reasons why the MMAF may be an improvement over what-if and goal seeking methods.

Figure 1. Logic of a What-if Session

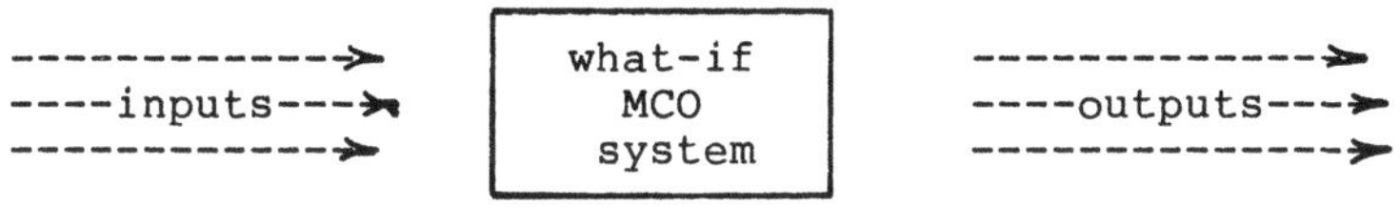

Newly proposed MCO methods should at least outperform existing what-if methods. However, in what-if an experienced user has the flexibility to rapidly ascend to the best solution, since he can promptly select ideal values, if known, for the inputs.

The Modified Method of Abstract Forces

The MMAF belongs to the family of direction finding methods which search the solution space for the optimal solution. A fundamental task in these methods is finding the search direction of greatest improvement in the user's utility function, such as the gradient of the utility function. This can be determined from marginal rates of substitution (MRS), e.g. trade-off information. If we assume that the objective function, $U(f_j(x))$, can be approximated by the weighted linear function

$$w_1 f_1(x)+,\dots,+w_k f_k(x) \qquad [1]$$

then at any solution point, x, the best direction of improvement of each criterion function, $f_j(x)$, at some current solution, x^o, is its gradient (Starr and Zeleny, 1977),

$$\nabla f_j(x^o) = \frac{\partial f_j}{\partial x_1^o}, \ldots, \frac{\partial f_j}{\partial x_n^o} . \qquad [2]$$

Therefore the best direction of improvement of the linear approximation for $U(f_j(x))$ is the weighted combination of the individual gradients of each criterion, e.g.

$$\nabla U(f_j(x)) = \sum_{j=1}^{k} w_j^o \nabla f_j(x^o) . \qquad [3]$$

In the popular method of Geoffrion et al (1972), the weights, w_j, assessed at x^o are approximated by the MRS between each $f_j(x)$ and an arbitrary reference criterion. In the Frank-Wolfe context the best direction of improvement is the direction from the current point, x^o, to an extreme point, x^e (found by solving [4] for x^e),

$$\text{Maximize} \sum_{j=1}^{k} w_j^o \nabla f_j(x^o) \cdot x^e . \qquad [4]$$

Once the direction is found we must also decide where on the path from x^o to x^e we should move to find the best solution on that path. One approach is to divide the path into a limited number of equal segments and compare the solutions at the division points. (Another approach will be described later). The user then selects the best solution and attempts to improve it. This search procedure reiterates until no further improvement can be found.

Similar to other MCO methods the MMAF involves conversion of the MCO problem, with assumed criterion weights, into an equivalent vector maximization problem. An algorithm solves the vector maximization problem using forces of effort, F_j, supplied by the DM, to improve each criterion's value. A central concern is to provide a simple method for soliciting direction of improvement information. Accordingly it lacks the limitations of ideal point knowledge and is parsimonious regarding user inputs. The user is not faced with the perplexing task of assigning weights to the various criteria, adjusting those weights as the environment changes, making complex decision choices, or answering numerous ranking questions. Instead, the MMAF presents good solutions and asks the user to choose the best, and to specify desired improvements in the chosen solution (see figure 2). It assumes that, at any point x, it is desirable to improve each criterion. When presented with a set of criterion values for some alternative solution the user applies different amounts of force, which are perceived as appropriately needed, to improve each criterion's value. The system then calculates the feasible inputs that are needed to give the desired outputs.

This intuitively appealing notion of applying efforts of force relative to the outcome desired, e.g. the more effort applied the more improvement gained, is preferred by the typical user.

Figure 2. Logic of a MMAF Session

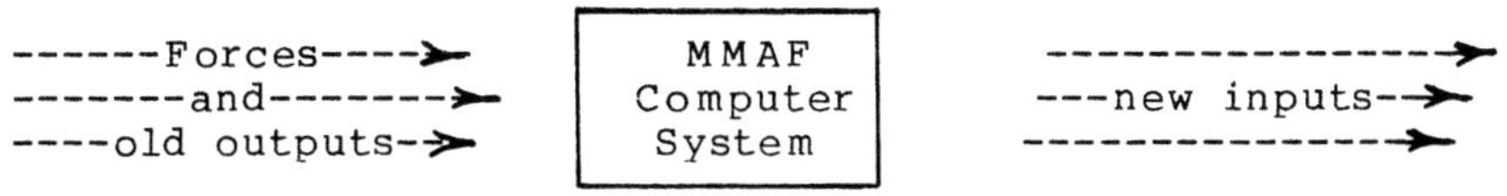

A major hypothesis is that the forces estimate, up to a positive scalar constant, the gradient direction of U in criterion space. That is

$$\nabla U(f_j) = kF_j \qquad [5]$$

is a good estimate for

$$\nabla U(f_j) = \partial U/\partial f_j, \qquad [6]$$

where k is some constant and the F_j are forces of improvement. This hypothesis was substantiated by Troutt and Hemming (1985).

The Need for Scaling Heterogeneous Criteria

Without scaling this method is virtually useless because it, like many other MCO methods, experiences a distortion due to a "magnitude of impact" problem. For example, if one criterion is workforce level and another criterion is inventory, measured in number of items on hand, there is considerable difference in the magnitudes in units of measurement. Relatively small changes in inventory levels of inexpensive items will have a significant impact in the determination of the LP coefficient vector, whereas large changes in workforce levels will have little impact. This diversity of impacts gives erroneous results in the LP maximization problem.

This is a serious problem since it appears that users have a natural expectation that equal forces produce equal benefits. We therefore seek to equalize the impact that each criterion has on the LP coefficient vector (used to find the decision space extreme point) by using a constant scaling algorithm. The algorithm provides values for an adjusting constant, A_j, for each product (force of improvement and the respective criterion gradient). The application of a force of improvement, F_j, by the user will then increase the impact of the respective criterion, relative to the amount of force applied, without undue influence from the criterion's unit of measurement.

At any point in decision space the impact that a given criterion has on the gradient of the value function may be represented by the size of the criterion gradient. This size may be measured by the norm of the criterion's gradient. To equalize the impacts of each criterion gradient, ∇C_j, at some current solution, we find values for adjusting constants, A_j, such that

$$A_1 \|\nabla C_1\| = A_2 \|\nabla C_2\| = \ldots A_n \|\nabla C_n\| = 1. \qquad [7]$$

After scaling the terms in the LP coefficient vector are found by

$$\frac{\partial V}{\partial x_i} = \sum F_j A_j \frac{\partial C_j}{\partial x_i} \quad . \qquad [8]$$

Given some starting solution, the A_j can be determined. These A_j, along with the user supplied F_j, are used to find the next solution, which yields still better values of A_j. (An interesting robust property is that as we progress from a good solution to the optimal solution, the values of A_j do not change appreciably from one iteration to the next, and stay constant for linear criteria).

A Variable Step Size Procedure

Since the display of an excessive number of alternatives both confuses the DM and slows the convergence process, a limited number of alternatives are displayed to the DM. Finding points on the search path which yield the best solutions is a perplexing problem. A procedure that uses the sum of the percent changes (SPC) in each criterion between iterations provides a basis for determining variable interval search lengths. This procedure is based on the assumption that as the DM's compromise solution approaches the optimal solution the relative changes in criteria values at each iteration diminish. By showing data for three successive iterations, h1, h2, and h3, figures 3 and 4 demonstrate that the SPC diminishes from a value of 228 to a value of 26.5 as the current solution approaches the optimal.

Figure 3: % Changes in Criteria Values (Iterations 1 and 2)

	Criterion Values				
h	C1	C2	C3	C4	C5
1	7500	10	150	20,000	4
2	5000	20	120	15,000	2
change Δ	2500	10	30	5,000	2
%change Δ	33	100	20	25	50
SPC	228				

Figure 4: % Changes in Criteria Values (Iterations 2 and 3)

h	C1	C2	C3	C4	C5
2	5000	20	120	15,000	2
3	5500	21	114	14,850	2.2
change Δ	500	1	6	150	.2
%change Δ	10	.5	5	1	10
SPC	26.5				

(Criterion Values: C1–C5)

In our step size determination procedure the ratio of the last iteration's sum of percent change (SPC_{h-1}) to the current value (SPC_h) is used to determine ρ. New alternative solutions are thus a function of both ρ and the extreme point generated. Assume we are using a MCO method where the old solution and four new alternatives are displayed to the DM. The equations for calculating A_i (e.g. the three intermediate solutions that lie between the old solution, A_0, and the extreme point solution, A_4) are:

$$A_1 = A_0 + .25(A_4 - A_0)\cdot\rho \quad [9]$$

$$A_2 = A_0 + .50(A_4 - A_0)\cdot\rho \quad [10]$$

$$A_3 = A_0 + .75(A_4 - A_0)\cdot\rho \quad [11]$$

where: $\rho = SPC_{h-1}/SPC_h$ but is never > 1 and A_i = alternative solution at the i-th interval. Whenever the process is not converging ρ will have a value of 1 and step sizes will be equal. Conversely, whenever the solutions converge, the criteria values converge and ρ becomes smaller and smaller. Hence, the intervals should decrease so that new alternatives (e.g. intermediate solutions A_1, A_2, A_3) are closer to the last solution, A_0. Relative changes in step sizes (intervals) along the search path for different values of ρ are demonstrated in figure 5. Note that as ρ diminishes new alternative solutions are in the neighborhood of the last solution.

Although at each iteration we search in a new direction, we now search closer to the current solution. Skeptics may argue that this step size determination procedure will result in exclusion from consideration of all solutions except local optimum. Note that although the first step is small the remaining steps are progressively bigger. If the optimum is distant from the current solution this procedure will still identify its existence. (At each iteration the solution represented by A_3 is still relatively distant from A_0). If somehow missed in previous iterations, the A_3 solutions should detect the presence of a hidden global optimum.

Figure 5: Variable Step Sizes and Their Relation to ρ.

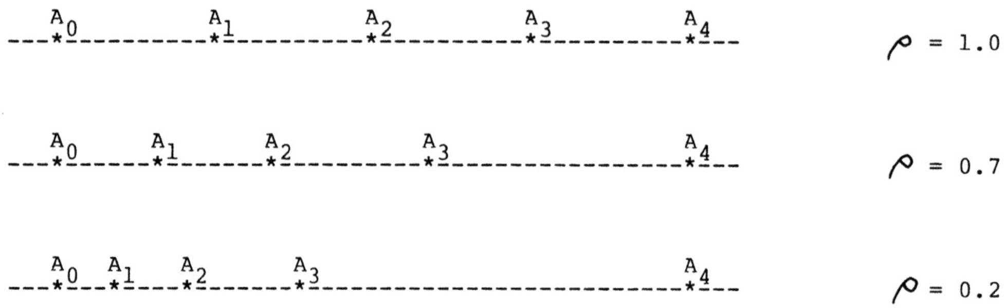

An Empirical Test

The MMAF was tested in an artificial decision making setting, e.g. a production planning problem, by comparing its results against results achieved with a WIFM. Decision variables were monthly production amounts, using regular time and overtime, in an environment of fluctuating demands. Criteria were workforce smoothing index (WFSI), inventory smoothing index (ISI), overtime costs (OT), back order costs (BO), holding costs (HOLD), and hiring/firing costs (HF). The values of the set of decision variables for the current solution were used to determine the size (impact) of each criterion's gradient. For each vector the mean absolute impact (MAI) was computed. The MAIs were normalized by dividing by the MAI of a base criterion. At a sample iteration the normalized values, e.g. adjusting constants, for the five criteria were 15, 0.004, 1, 1.1, and 0.67 respectively. The range in values for these adjusting constants is indicative of the severe disparity in criteria impacts.

Research Design

Students, 30 undergraduates and 20 graduates, comprised the four group convenience sample. Each subject used both the WIFM and the MMAF and solved different production planning problems in two consecutive exercises. To establish a reference point for an analysis each subject began with the same "start-up" plan. The research design, figure 6, controlled for order of presentation bias (by reversing the order between groups) and for learning effect (by using a different problem in the second exercise.)

Figure 6: Research Design

Subject Group	Sequence of Exercises				Attributes and Outcomes
	First Exercise		Second Exercise		
	Method	Problem	Method	Problem	
1	WIFM	1	MMAF	2	
2	MMAF	1	WIFM	2	
3	WIFM	2	MMAF	1	
4	MMAF	2	WIFM	1	

Findings

Attributes (number of trials and time required) and outcomes (solutions) were recorded and statistically analyzed. A summary of statistics for the criteria are given in tables 1 and 2. Both the number of trial solutions and the time required to reach an acceptable solution were significantly reduced when using the MMAF.

TABLE 1 Mean Values for Each Criterion, Problem 1

Method	Iteration	WFSI	ISI	OT	BO	HOLD	HF	TOTAL
	start-up	875	700	16000	0	12000	26500	54700
WIFM	trial 1	767	965	14723	9623	22660	17933	64939
WIFM	trial 6	305	377	7903	885	11992	11966	32746
MMAF	trial 1	124	220	8636	0	8480	8783	25899
MMAF	trial 6	106	253	1677	36	6396	9668	17776

TABLE 2 Mean Values for Each Criterion, Problem 2

Method	Iteration	WFSI	ISI	OT	BO	HOLD	HF	TOTAL
	start-up	2333	157	19250	360	26783	79500	125893
WIFM	trial 1	1939	2086	18461	7423	68366	41181	135432
WIFM	trial 6	341	1144	20450	5455	54002	22741	102649
MMAF	trial 1	256	118	24643	275	17839	27213	69970
MMAF	trial 6	143	105	26097	0	1424	19180	59401

Although results were achieved in half the time required by the WIFM the MMAF yielded much better solutions. Analysis of the data shows that the values of the criteria rapidly converge to an optimal solution (best mix of inputs and criteria values) when using the MMAF. Also there was less deviation in criterion values from one trial to the next. In the WIFM the performance criteria did not improve during initial iterations, and only improved slowly in later trials. When asked which method would provide the best plan and which method would be easier for the typical decision maker to use most subjects preferred the MMAF.

In problem 2 the relationships among the various criteria were somewhat unique in that a good mix of criteria values consists of high levels on one criterion, overtime costs, in order to achieve significantly lower levels on several other criteria. Using the WIFM subjects did not recognize this possibility and kept values on that criterion so low that a good mix did not occur. In the MMAF, however, subjects quickly found a good mix of criteria values.

Summary

In the application tested the MMAF was shown to be far superior to the traditional what-if interactive MCO system. The solutions were not only better, but were achieved with considerably fewer trials and in significantly less time than with the WIFM.

References

Dyer, J.S. (1973). A time-sharing computer program for the solution of the multiple criteria problem. Management Science, 19(12), 1379-83.

Frank, M., and Wolfe, P. (1956). An algorithm for quadratic programming. Naval Research Logistics Quarterly, 3(1&2), 95-110.

Geoffrion, A.M., Dyer, J.S., and Feinberg, A. (1972). An interactive approach for multi-criterion optimization with an application to the operation of an academic department. Management Science, 19, 357-368.

Keeney, R.L., and Raiffa, H. (1976). Decisions with Multiple Objectives: Preference and Value Trade Offs. New York: Wiley.

Korhonen, P. (1986). A hierarchical interactive method for ranking alternatives with multiple qualitative criteria. European Journal of Operational Research, 24, 265-276.

Starr, M.K., and Zeleny, M. (1977). Multiple Criteria Decision Making. North-Holland Publishing, Amsterdam.

Troutt, M.D., and Hemming, T. (1984). A method of intensities for interactive multicriteria optimization, Proceedings, National AIDS, Toronto, November.

Troutt, M.D., and Hemming, T. (1985). A method of forces for interactive optimization. (submitted for publication).

Zionts, S., and Wallenius, J. (1983). An interactive multiple objective linear programming method for a class of underlying nonlinear utility functions. Management Science, 29 (5), 519-29.

Zionts, S. (1981). A multiple criteria method for choosing among discrete alternatives. European Journal of Operational Research, 7, 143-147.

A Multidimensional Scaling Model for Qualitative Pair Wise Comparisons

Willem A Schreüder

Department of Mathematics, West Virginia University,
Morgantown, WV 26506, USA

and

Esbeth van Dyk

Mathematics and Physics Building, University of Sussex,
Falmer, Brighton, BN1 9QH, UK

A multidimensional scaling model, QualScal, is developed for the case where the decision maker is only prepared to make qualitative comparisons for pairs of policies, stating whether he is indifferent between two policies or which one he prefers. Quantitative distance information is derived using the successive neighbour method. The scaling is done by a modified version of the Kruskal nonmetric multidimensional scaling method. In this a novel procedure for scaling disjoint subsets, i.e. subsets containing no inter-subset indifferences, has been incorporated. The model is evaluated by using numerically valued policies generated with a random variate to introduce a measure of inconsistency and then comparing the results with the original numerical values. Two applications to published data are discussed.

1. INTRODUCTION

This research was prompted by Rivett's earlier work in this field [7,8]. We consider the situation where a decision maker is faced with a set of policies (decision alternatives) containing no obvious best one, and his aim is to find a rank ordering for them. We develop a multidimensional scaling (MDS) model, QualScal, for the case where the decision maker (DM) cannot or does not wish to specify the criteria according to which he evaluates the policies. We therefore merely ask him to make a qualitative comparison for each possible pair of policies, stating whether he is indifferent between them or which one he prefers. During the comparison he may use criteria consciously or subconsciously, but he is not required to do so explicitly. The term indifference is used to indicate equal desirability. Thus, the indifferences of the DM do not necessarily mean that the outcome of policies are similar.

This indifference data is used as input to a custom MDS routine, which produces a map in which policies are represented by points (nodes) and indifferences by links (edges). From the map we obtain the rank ordering of the policy set for this DM.

In the next section nonmetric MDS is explained briefly. In section 3 we discuss the nature of the input data supplied by the DM and the successive neighbour method which is used to derive additional quantitative information from this input. This is followed by an explanation of the procedure for the scaling of disjoint subsets in section 4. Section 5 addresses the significance of the preference information obtained from the DM. The method is evaluated in section 6 by applying it to numerically valued policies for which the correct rank ordering is known. In section 7 further evaluation is done by applying the method to two examples from the literature. Finally, some indications of future work are given in the last section.

2. NONMETRIC MULTIDIMENSIONAL SCALING

MDS is a class of techniques for constructing a configuration of n points in low-dimensional space such that the interpoint distances match in some sense the proximities of n objects. The algorithm for nonmetric scaling (NMDS) was originally developed by Kruskal [5,6]. He required the proximities to be monotonically related to the distances in the configuration.

We shall work with dissimilarities instead of proximities, since we are interested in an ascending monotone relationship between the Euclidean distances in the configuration and the dissimilarities among our n objects, here policies. The matrix of dissimilarities constitutes the input to the algorithm and the primary output consists of a spatial map, in order to make the results visually more comprehensible.

The scaling is done by using a steepest descent method, in an effort to minimize the stress S which is a measure of how well the configuration represents the data. Stress is defined as

$$S = \sqrt{\frac{\sum(d_{ij} - \hat{d}_{ij})^2}{\sum d_{ij}^2}},$$

where d_{ij} is the distance between points i and j on the map and $\hat{d}_{ij}$ is the value closest to d_{ij}, subject to it being monotonically related to the dissimilarity between policies i and j. The stress is thus invariant under rigid motions of the configuration (rotation, translation and reflection) and also under change of scale (uniform stretching or shrinking).

At each iteration the scaling procedure is as follows:
1) move the mass centre of the set of points to the origin;
2) normalize the set;
3) determine the stress, gradient and step size;
4) move the points to decrease the stress.

Repeat the above procedure until a minimum stress has been found.

Torgerson metric MDS is used to find the initial configuration since it provides a solution in which the relative positions of the points are approximately correct and therefore better than a random configuration [2]. The gradient is used to determine the direction in which to move the points. The step size is determined by a formula which Kruskal derived empirically after numerous experiments. This formula is only valid when the set of policies has been normalized.

In our application of NMDS the dissimilarities can assume only a small number of discrete values resulting in large numbers of ties, i.e. dissimilarities of equal value. Kruskal [5] initially introduced two approaches to the treatment of ties, namely secondary (STT) and primary treatment of ties (PTT). If we refer to all the dissimilarities having a particular value as an equivalence class, STT requires that all the distances within each class must be of the same length. PTT allows the distances within a class to vary as long as the limits set by the adjacent classes are not exceeded. Kendall (in an appendix to Rivett [7]) extended these ideas by introducing tertiary treatment of ties (TTT) which allows some equivalence classes to be treated by STT while others are treated by PTT at the same time.

The best dimension in which to perform the scaling is determined experimentally by comparing the decrease in stress with increase in dimension and choosing a dimension where the stress is reasonably low, subject to the dimension being low enough to allow easy interpretation of the map. Two-dimensional scaling is often sufficient for our application, with one dimension indicating preference and the other accommodating the indifferences of the DM. If a problem results in a geometric structure of higher dimension, it is desirable to project it onto two dimensions to make the map more comprehensible; care however, should be taken that preference information is not lost in the projection.

3. DERIVATION OF QUANTITATIVE DATA

The DM is prompted to respond to each pair of policies, stating whether he is indifferent between them or which one he prefers. A "no response" option can be handled by NMDS, but is strongly discouraged since it severely impairs the derivation of quantitative data. His responses therefore consists mainly of signed binary data. The Ross ordering [9] is used to present the policies, since this ordering avoids suggestive repetitions, space and time errors and maintains the greatest possible spacing between pairs involving any given policy.

In general it has been shown that long distances give overall structure and short distances the fine structure of the map obtained by NMDS. Since we obtain only two equivalence classes directly from the responses (indifferences and preferences) we derive additional quantitative information using the successive neighbour method. This is an attempt to give more structure to the maps.

All pairs of policies that are indifferent make up the first equivalence class, called first neighbours. Pairs that are not indifferent, but which have both policies indifferent to a common third policy make up a second equivalence class, or second neighbours. Pairs not in one of the first two equivalence classes, but which have each policy indifferent to one of the policies in an indifferent pair, make up the third equivalence class, or third neighbours. This process may be continued to find more equivalence classes, i.e. fourth, fifth neighbours, etc. The successive neighbour method assumes that first neighbours are more similar than second neighbours, which are in turn more similar than third neighbours, and so on. The equivalence class in which a pair falls is used as the nonmetric distance between the policies. Note that if a graph is drawn of all the policies (represented by points) and indifferences (represented by links), the level of neighbourship between any two policies is the same as the shortest distance in the graph between those policies, where the unit of distance is equivalent to the length of an indifference.

The maximum level of neighbours to be used can be specified, the maximum possible level being one less than the number of policies. Kendall [4] observed that if we cannot determine accurately the distances between objects which are far apart and therefore merge all large distances into a single equivalence class which is treated by

STT, the furthest points are pulled closer together. He labelled this the "horseshoe effect". To counteract this effect, all the distances corresponding to neighbours of level greater than the maximum equivalence class specified by the user are grouped into two additional equivalence classes, namely intermediate neighbours and long range neighbours. The levels of neighbours to be used for long range neighbours can be specified and all the levels of neighbours not yet assigned to an equivalence class make up the intermediate neighbours class.

The nonmetric distances obtained from the successive neighbour method constitute the dissimilarity data which is the input to the NMDS algorithm. STT is applied to lower level neighbours, up to a specified level, and PTT to all the remaining levels of neighbours. Intermediate neighbours should receive PTT since this class comprises all the leftovers and therefore contains a wide range of distances. However, STT can be applied to long range neighbours as it tends to help in reducing the horseshoe effect by stretching the whole map.

4. UNCONNECTED DATA SETS

In the graph mentioned in the previous section, the indifferences provided by the DM frequently lead to an unconnected graph. This can happen if the DM, for example, groups the policies into two or more subsets, containing some indifferences within the subsets, but none between the subsets resulting in disjoint subsets. This is often the result of a natural separation of the policies into two or more groups of approximately equally desirable policies. Ideally this would result in a partially ordered graph, all policies in one subset either being preferred to or not preferred to all policies in the other subsets. In practice, however, the DM is often inconsistent and a complete partial ordering is not obtained.

When disjoint subsets are scaled, only minimum distances between subsets are defined because there are no links between the subsets and therefore no level of neighbourship. Note that two policies can only be neighbours on some level if they are connected via a series of links and that the preferences are not used in the derivation of dissimilarities. The minimum distance between disjoint subsets must therefore be larger than the distance between first neighbours. During scaling the stress will increase if disjoint subsets move too closely together. However, if they move further apart than the highest level of neighbours the distance between them does not influence the stress and therefore nothing prevents the subsets from moving apart indefinitely. Within a connected map there may be extreme policies with no upper limit defined on the distance between them, but since they are connected via a series of links, it would mean that one or more of the links will be stretched when the policies are pulled apart, increasing the stress, because STT is applied to links (i.e. equivalence class 1). The scaling of unconnected sets thus usually results in each subset contracting to a blob on the map, since the distances between subsets become very large in comparison to the distances within subsets.

This matter is not addressed much in the literature, since it happens very rarely in most applications. Other applications usually have a measure of proximity on a scale of say 1 to 9 for each pair of objects, which virtually eliminates the possibility of subsets drifting apart. The solution usually proposed, is to scale each subset separately. This is not desirable for our application, since we want the same distance scale to be used for all the subsets in order to get a more representative map. Rivett introduced artificial links to obtain a connected graph, but thereby decreased the accuracy of the map and thus of the resulting rank ordering.

To the best of the authors' knowledge none of the available MDS programs can scale unconnected data sets. The following procedure has therefore been developed to scale disjoint subsets concurrently. Each subset is scaled relative to its mass centre, which has been moved to the origin for convenience. This is equivalent to introducing an extra dimension and moving the subsets apart in this dimension. The policies can now occupy the same space in the original dimensions, but in the extra dimension each subset has a distinct coordinate which is fixed throughout the scaling. Since we do not want interaction between the disjoint subsets, we do not include this extra dimension in the calculation of the stress, gradient and step size. The scaling is now done analogous to the standard method, i.e.

1) determine the mass centre of each subset and move them so that the mass centre is at the origin;
2) normalize the entire set;
3) determine the stress, gradient and step size;
4) move the entire set to decrease the stress.

Repeat the above until the minimum stress has been found. The subsets are then positioned relative to one another using average preferences between subsets, or average inter-subset preferences, as described in the next section.

5. PREFERENCE INFORMATION

After the scaling has been done, the preference information is used to determine the decision axis of the map and thus to orientate the subsets and position them relative to each other along this axis. The decision axis, referred to as the average preference axis, is calculated for each subset separately by adding the unit vectors between policies which are not indifferent and then taking the average. These are relative vectors, i.e. between policies and directed towards the most preferred policy of the two.

Only preference within a subset is used to orientate the subset. Each subset is therefore orientated separately before they are placed relative to each other. The average preference between subsets is used to obtain a rank ordering for the subsets. For each pair of subsets the average preference is calculated by taking the number of preferences for the one subset, subtracting the number of preferences for the other subset and dividing by the total number of preferences. Using these average preferences the row sum procedure [3] is used to find a rank ordering for the subsets. The subsets are then placed according to this rank ordering along the decision axis with the ideal distance of a second neighbour pair between the closest alternatives of two consecutive subsets. Since the ordering of the subsets relative to each other is known from the preferences, only the ordering of policies within each subset has to be determined from the scaling. The disjoint subsets can therefore be referred to as partially ordered subsets.

If the scaling has been done in more than two dimensions, the subsets are orientated to retain the maximum preference distribution in the lower dimension before they are projected onto two dimensions, thereby eliminating the possibility of projecting the most preferred policy so that it is surrounded by less desirable policies.

6. EVALUATION

The method was evaluated by generating sets of numerically valued policies, e.g. numbers between 0 and 100, from the uniform, normal and U-shaped distributions [8]. Indifferences and preferences were assigned according to a normal distribution so as to introduce some inconsistencies. If the distance between two policies is less than a pseudo random number from the normal distribution an indifference is assigned and else a preferences is assigned. If a preference is assigned, it can be reversed with a small probability, imitating a real DM. Different DM's were imitated by adjusting the mean and variance of the normal distribution. Since the true rank ordering is known for these policies, the corresponding maps can be used to evaluate the performance of the method. The Spearman rank correlation coefficient is calculated for each map. Note that the stress indicates how well the map represents the responses of the DM whereas the Spearman rank correlation measures how close the rank ordering obtained from the map is to the correct rank ordering.

Different seeds were used to generate a number of data sets from each of the three distributions containing 5, 10, 15, 20, 25 and 50 policies in the interval (0,100). The mean of the normal distribution used to assign indifferences was varied from approximately half the average distance between two policies from the uniform distribution (for the particular number of policies in the data set) to twice the average distance, and, in the cases of 20 and 50 policies, to four or five times the average distance. The standard deviation was varied from approximately a third of the mean to the mean. Probabilities of 0, 0.01, 0.1 and 0.2 were used to reverse the order of preference. The resulting maps varied from being very sparsely linked (0.4 links per policy on average) to very densely linked (3.5 or more links per policy on average). The Spearman rank correlation coefficient was calculated for each map. Very high values were obtained in the instances where no preferences were reversed, generally ranging from 0.8 for data sets with 5 policies to above 0.96 for data sets with 50 policies. For only about 2% of the cases were rank correlations less than 0.7 obtained. These were for small data sets (5 and 10 policies) with very high inconsistency. In these cases the stress was also very high (more than 20%) and the supplied data so badly corrupted that the method could only roughly recapture the original structure. In general the stress was below 10%.

As expected, the reversal of preferences seems to have the greatest effect on sparsely linked data sets containing a number of partially ordered subsets. In these cases the subsets can easily be placed in the incorrect order relative to each other. The reversed preferences also influence the orientation of the decision axis, but the effect diminishes as the average number of links per policy increases. It was verified that the use of long range neighbours effectively counteracted the horseshoe effect.

Figure 1 shows the map produced from a data set of twenty policies from the uniform distribution. The policies were numbered in increasing order of value. A mean value of 10 and a standard deviation of 2.5 were used for the normal distribution according to which the indifferences and preferences were assigned. The method accurately

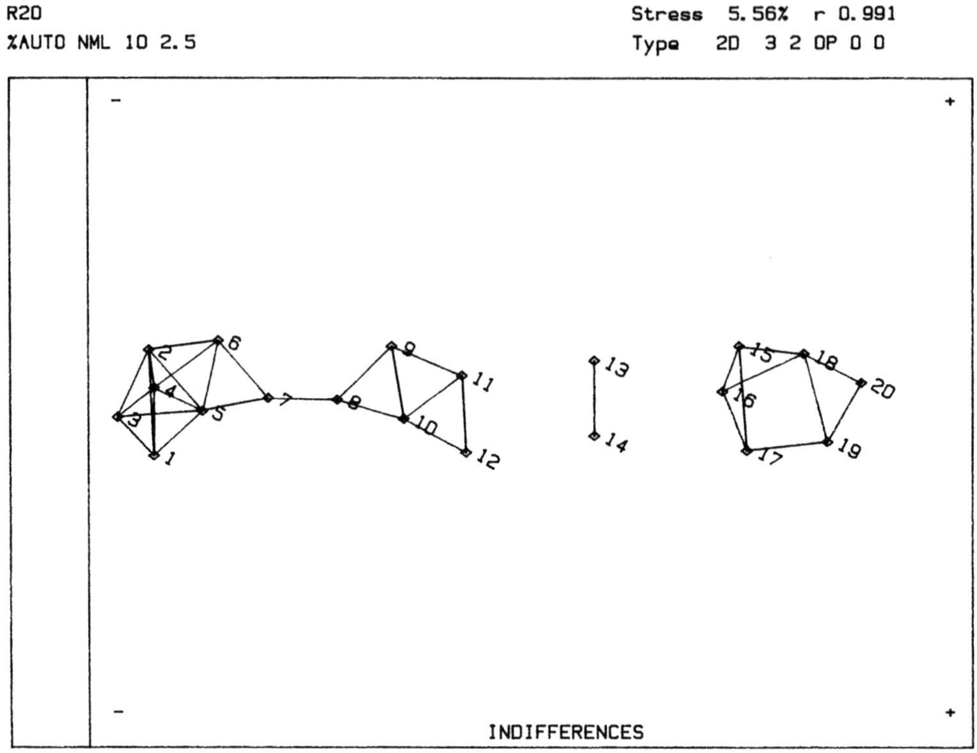

Figure 1. Numerically valued policy map, $\mu = 10$, $\sigma = 2.5$

reproduced the order of the policies as quantified by the Spearman rank correlation coefficient of 0.991. The stress of 5.56% is low enough that we may be confident that the method accurately reconstructed the map implied by the data.

The map contains some examples of common features. There are three disjoint subsets that reflects a partial ordering in the policies inherent in the DM's responses. Most inconsistencies can be accommodated in two dimensions without increasing the stress, for example, the truss-like structure 8-9-10-11-12. The subsets are orientated such that the average preference within subsets is aligned horizontally with more preferred being to the right. However, the subset containing only policies 13 and 14 was aligned vertically to indicate their indifference.

Note that the least preferred subset has the slightly crescent shape which is characteristic of long thin subsets. In severe cases this develops into a horseshoe and must be corrected by specifying long range neighbours, sometimes at the expense of slightly increasing the stress.

7. COMPARATIVE RESULTS

To further evaluate the performance of our method we chose the problem of selecting pumped storage sites as summarized by Stewart [10]. We used Stewart's criterion to obtain indifferences from the utilities, i.e. there exists a just noticeable difference (JND) between utilities, such that indifference would be stated between any pair of options which differ by less than JND in utility. If indifference could not be stated, we chose the option with the highest utility as the preferred one. The utilities are shown in table 1.

Option	1	2	3	4	5	6	7	8	9	10
Utility	0.931	0.885	0.846	0.820	0.809	0.779	0.732	0.697	0.694	0.196

Table 1. Problem of selecting pumped storage sites (taken from Stewart [10], table 8)

<table>
<tr><th></th><th colspan="2">Full data set</th><th colspan="4">Half data sets</th></tr>
<tr><th></th><th></th><th></th><th colspan="2">Even</th><th colspan="2">Odd</th></tr>
<tr><th>Rank</th><th>JND 0.1</th><th>JND 0.2</th><th>JND 0.1</th><th>JND 0.2</th><th>JND 0.1</th><th>JND 0.2</th></tr>
<tr><td>1</td><td>1</td><td>1</td><td>1</td><td>2</td><td>1</td><td>1</td></tr>
<tr><td>2</td><td rowspan="2">2,3</td><td>3</td><td>2</td><td>4</td><td>3</td><td>5</td></tr>
<tr><td>3</td><td rowspan="2">2,5</td><td rowspan="3">4,5,6</td><td>1</td><td>2</td><td>3</td></tr>
<tr><td>4</td><td rowspan="3">4,5,6</td><td>6</td><td>5</td><td>4</td></tr>
<tr><td>5</td><td rowspan="2">4,7</td><td>3</td><td rowspan="2">4,7</td><td rowspan="2">2,7</td></tr>
<tr><td>6</td><td>3</td><td>8</td></tr>
<tr><td>7</td><td>7</td><td>6</td><td>9</td><td>7</td><td>6</td><td>9</td></tr>
<tr><td>8</td><td rowspan="2">8,9</td><td rowspan="2">8,9</td><td>7</td><td>5</td><td rowspan="2">8,9</td><td>6</td></tr>
<tr><td>9</td><td>8</td><td>9</td><td>8</td></tr>
<tr><td>10</td><td>10</td><td>10</td><td>10</td><td>10</td><td>10</td><td>10</td></tr>
<tr><td>r</td><td>0.985</td><td>0.933</td><td>0.892</td><td>0.802</td><td>0.948</td><td>0.817</td></tr>
</table>

Table 2. Rank orderings and Spearman rank correlations for pumped storage sites.

Stewart's experimental design was repeated by using the same two values for JND (0.1 and 0.2) in combination with the full data set (all 45 pairs) and the two half data sets obtained by using only every second pair wise comparison (22 and 23 pairs respectively). In five of the six cases alternative 1 was ranked first and alternative 2 under the top 3. The Spearman rank correlation coefficient was calculated for each rank ordering compared with the utility ordering. The complete rank orderings and the correlations are shown in table 2. QualScal thus performs comparable to Stewart's method without needing a value for each alternative on each attribute. This is particularly remarkable considering that Stewart's "just noticeable" difference of 0.2 is greater than the difference in the utility of the first and seventh ranked alternative and therefore every response between the nine most preferred alternatives is an indifference except those between alternatives 1 and 8 and alternatives 1 and 9. Although QualScal requires all the pair wise comparisons, the information is much easier to provide than that required by Stewart's method.

French [1] discusses a particular example which, when NMDS is applied to it, results in the most preferred policy (number 4) being plotted in the centre of the map instead of at an extreme. We reproduce the decision problem in table 3. Each policy's performance is indicated against each of three objectives, with higher values implying better performance. Only differences of twenty or more are significant on each objective scale. Therefore policies i and j are considered to be indifferent if

$$\max\{|A_i - A_j| , |B_i - B_j| , |C_i - C_j|\} \leq 20.$$

Since the use of preference information should eliminate the possibility of mapping the most preferred policy in the centre of the map, we decided to derive some preferences from the data and apply our method to the problem. Two sets of preferences were derived. For both sets only differences larger than twenty were considered. For the first set two policies were regarded to be incomparable when they dominated each other on different objectives, for example policy 5 dominates policy 2 on objective B ($B_2 - B_5 = -25$), but policy 2 dominates policy 5 on objective C ($C_2 - C_5 = 30$). No responses were given for incomparable policies. For the second set the policy which dominated on the objective which produced the largest difference was chosen as the preferred policy, for example in the above instance policy 2 is preferred to policy 5 ($5 \prec 2$). Both data sets included the triad $7 \prec 5 \prec 1 \prec 7$.

	Objective			Indifferences				
Option	A	B	C					
1	35	40	70	2	3			
2	55	35	60	1	3	4	6	7
3	50	50	50	1	2	4	5	6
4	65	50	60	2	3	6		
5	55	60	30	3	6			
6	65	40	45	2	3	4	5	7
7	75	20	50	2	6			

Table 3. Decision problem taken from French [1]

Rank	Option	
	Set 1	Set 2
1	4	4
2	3	3
3	6	6
4	1,2	2
5		1
6	5	5
7	7	7

Table 4. Rank orderings

Applying our method to these data sets resulted in the rank orderings shown in table 4. Option 4 is clearly indicated to be the best. Therefore, although the problem has a three-dimensional geometric structure, our method resulted in a very good two-dimensional representation, especially if we keep in mind that only the rank ordering is of concern to us.

8. FUTURE WORK

An important aspect which still remains to be investigated at the time of writing is the sensitivity of the method to missing responses. This is of concern since the number of pairs to which the DM has to respond quickly become unmanageable as the number of policies increases.

We are convinced that areas exist in which this method can be applied advantageously. The simplicity of input makes the method well suited to problems where the

DM is unable to or unwilling to make more than a qualitative comparison between any two policies. The method is also useful to obtain a first ranking of policies that may be used to order the DM's thoughts before a more sophisticated, but more time consuming, method is used. The method requires more testing on real world applications, but initial results shows much promise and more detailed testing is planned.

REFERENCES

1. French S (1980) On the methodology of decision analysis III - on various decision aids. Working paper, Dept. of Decision Theory, Univ. of Manchester.
2. Greenacre MJ, Underhill LG (1982) Scaling a data matrix in a low- dimensional Euclidean space. In *Topics in applied multivariate analysis* (Hawkins DM (ed)) Cambridge Univ. Press, Cambridge, 183–268.
3. Huber PJ (1963) Pairwise comparison and ranking: optimum properties of the row sum procedure. *The annals of mathematical statistics*, **34**, 511–520.
4. Kendall DG (1971) Seriation from abundance matrices. In *Mathematics in the Archaeological and historical sciences* (Hodson FR, Kendall DG, Tautu P (eds)) Edinburgh Univ. Press, Edinburgh, 215–251.
5. Kruskal JB (1964a) Multidimensional scaling by optimising goodness of fit to a nonmetric hypothesis. *Psychometrika* **29(1)**, 1–27.
6. Kruskal JB (1964b) Nonmetric multidimensional scaling: a numerical method. *Psychometrika* **29(2)**, 115–129.
7. Rivett BHP (1977) Policy selection by structural mapping. *Proc. R. Soc. London A.* **354**, 407–423.
8. Rivett BHP (1980) Indifference mapping for multiple criteria decisions. *Omega* **8(1)**, 81–93.
9. Ross RT (1934) Optimum orders for the presentation of pairs in the method of paired comparisons. *J. Educ. Psychol.* **25**, 375–382.
10. Stewart TJ (1984) The inferring of preferences from indifferences in multicriteria decision analysis. *Omega* **12(2)**, 175–184.

MODELLING INACCURATE DETERMINATION, UNCERTAINTY, IMPRECISION USING MULTIPLE CRITERIA

Denis Bouyssou
ESSEC
B.P. 105
F-95021 Cergy, France

Abstract

The purpose of this paper is to study how the consideration of several criteria, as opposed to a more traditional mono-criterion approach, helps the modelling of imprecision, uncertainty and inaccurate determination (I.U.I.D.) in a decision-aid study. After a brief review of the main sources of inaccurate determination, uncertainty and imprecision that arises in a decision-aid situation, we show that the use of multiple criteria allows to build partial preference structures, to discuss in a powerful way the precision of the evaluation on each criteria and to create a clear language between the actors via the use of the evaluation tableau. We argue that this proves useful in dealing with I.U.I.D.

1- Introduction.

The use of multiple criteria in decision-aid models is often justified (see *e.g.* Zeleny (1982) or Schärlig (1985)) by the fact that the world is governed by multiple objectives and that any decision implies to balance "pros" and "cons". This widely-shared point of view can however be criticized (see Bouyssou (1987) and Roy (1988b)). Using a mono-criterion approach to decision-aid does not imply that one considers that "reality" is governed by a single criterion. It is well-known that, in this kind of models, multiple objectives are often taken into account *e.g.* via the use of constraints, sensivity analysis and "prices" allowing to convert heterogeneous consequences into a single unit. As emphasized by Roy (1988b), the use of multiple criteria does not simply appear as a generalization of traditionnal approaches but constitutes a new paradigm for analysing and helping decisions.

In this paper we wish to outline what we consider as an important justification for entering this new paradigm[1] : the management of imprecision, uncertainty and inaccurate determination

1 In this this paper we take this paradigm as a whole and do not discuss specific methods related to it.

(I.U.I.D) that is part of most decision situations. Our analysis follows that of Roy (1988a). He distinguished four main sources of I.U.I.D. that the analyst has to deal with. We briefly present them in section 2. In section 3, we recall the main originalities of models explicitely using several criteria. In section 4 we try to show how the consideration of several criteria helps the modelling of I.U.I.D. and leads to models that are significantly different from those deriving from the consideration of a unique criterion in this respect.

2- The four main sources of uncertainty, imprecision and inaccurate determination in decision models (Roy (1988a)).

a) The "map" is not the "territory".

Locating a plant, chosing an equipment, investing in new activities are crucial decisions for a firm. The purpose of decision-aid is to compare such complex alternatives. If one wants to use a formal model of decision-aid, the complexity of these alternatives and their consequences makes it often impossible to compare them directly. This comparison is made possible through the use of "maps" of these complex "territories". For an alternative, a map consists of a model of the consequences of its implementation (in order to describe an alternative it is possible to use several maps of different "scale" using *e.g.* a hierarchical model). These maps create a tractable language that allows an effective communication between the various actors of the decision process and provides an adequate basis for the comparison of the alternatives. However the establishment of the maps inevitably involves many simplifications, omissions and distorsions which introduce in the model an important source of arbitrariness. Indeed, there are often several and equally valid ways of building these maps. While forced to use maps in order to compare territories, the analyst has to make a tradeoff between the richness and the readability of the maps : the "richer" is one map the closer it is to the territory, but the more difficult it may be to compare it to other maps.

b) The "future" is not a "present" to come.

The alternatives that are to be compared will only be implemented in a more or less distant future. Thus, at the time of the study, the consequences of the implementation of an alternative are very often unpredictable for they depend on environmental factors and/or the strategy of other actors that are still unknown and may well be influenced by the implementation of that alternative. This is the most classical source of I.U.I.D. that is mentionned in every textbook on decision models.

Many efforts have been devoted to cope with this unpredictability using, *e.g.*, probability distributions, plausibility measures, scenarios, etc.

As Roy (1988a) mentionned, the unpredictability of the consequences of implementing an alternative also stems from the fact that the alternatives are not completely specified at the time of the study. When a firm tries to compare several sites for locating a new plant, the precise characteristics of each site may not have been completely investigated yet. Furthermore, the precise draft of the plant to be built may not be available and may well depend on the site chosen. Thus, even if one could predict with a very high precision the consequences of an alternative, an element of inaccurate determination would remain since the alternatives are still "projects".

c) The data are not the result of exact measurement.

The establishment of a map usually involves the consideration of two types of data. Data of type I are closely linked to the territory that the analyst wishes to describe. The modelling of uncertainty mentionned in the above paragraph will apply to this first type of data. For instance, suppose that an analyst has to evaluate the human consequences of building a polluting plant on a given site. He will have to cope with uncertainty since he will be forced to envisage various scenarios for the growth of the population in that area. He will also have to deal with imprecision since the present number of people living close to the projected plant is far from being perfectly known : counting houses on a map or on-site studies do not lead to precise evaluations. Thus, it is important to realize that many figures used in decision-aid models are only "order of magnitudes". This imprecision is often seen as stemming from the measurement techniques that are used. It also comes from the fact that, in many situations, the very definition of what "should" be measured is very imprecise. Using the same example as above, it is not clear how the analyst should take into account schools, hospitals, second homes, etc. This inaccurate determination of what is to be measured is certainly at least as important as the imprecision inherent to any kind of measurement.

Data of type II concern the way the first type of data is used in the construction of the map. Parameters like discounting rates or utility functions designed to capture an attitude toward risk are examples of this second type of data. They are more linked to a particular value system than to an alternative. In our siting example such data could consist of the weight assigned to each inhabitant that is function of the distance between his residence and the projected plant, the attitude toward risk of the firm concerning the amount of nuisance created for the riparians, etc. Though techniques have been created to assess these data, it is important to keep in mind that they are very often "created" as well as "measured" (see *e.g.* the work of McCord and de Neufville (1983) concerning utility functions).

d) The model is not the description of a real entity independent of the model.

Data of the second type are connected with certain aspects of the preference system(s) of the actor(s) involved in the decision process. It is well known that the questionning process used by the analyst in order to obtain these data may significantly influence the answers (see Bouyssou (1984)). This is all the more true since the preference system of an actor may not be completely structured at the time of the study : areas of firm conviction may well coexist with areas of hesitation and ambiguity in which the influence of the model on what is to be "captured" is overwhelming. Furthermore, the various actors may well disagree and, as a result of a discussion, some actors may change their mind on some point thus creating some "inconsistencies" with previously stated judgements. In such cases the management of these hesitations, contradictions and conflicts seems a prerequisite to any convincing decision-aid model. This is linked to what Roy and Bouyssou (1986) called a *constructive* attitude towards decision-aid, as opposed to a *descriptive* one, in which the role of the analyst is not to describe as accurately as possible supposedly pre-existing preferences but to provide information and tools that are useful for justifying, building and arguing preferences.

3- The multiple criteria approach to decision-aid.

From the point of view of the management of I.U.I.D., the main feature of an approach using multiple criteria is to break down the modelling process into two different phases : the construction of the criteria (which gives rise to the evaluation tableau) and the aggregation of these criteria (see Fig. 1). As advocated by Roy (1985), the analyst should use the smallest possible amount of type II data in the construction of the various criteria. We noted in section 2, that some data of type II such as utility functions have to be taken into account in order to build the criteria. However, sensitive information such as the tradeoffs between the various criteria are only introduced in the aggregation phase, contrary to what is usually done in a mono-criterion approach in which the construction of the unique criterion involves at the same time data of both types.

This approach is based on what could be called an "*act of faith*", *i.e.*, the belief that the explicit construction of several criteria will have a "positive role" in the modelling process. It rests on an underlying *assumption* stating that in most decision-aid studies it is possible to identify a small number of "points of view" (usually between three and no more than ten, at least at the upper level if a hierarchical model is used) around which it is possible to build a familly of criteria that is exhaustive and simple enough to be accepted as a basis of discussion by all the actors of the decision process.

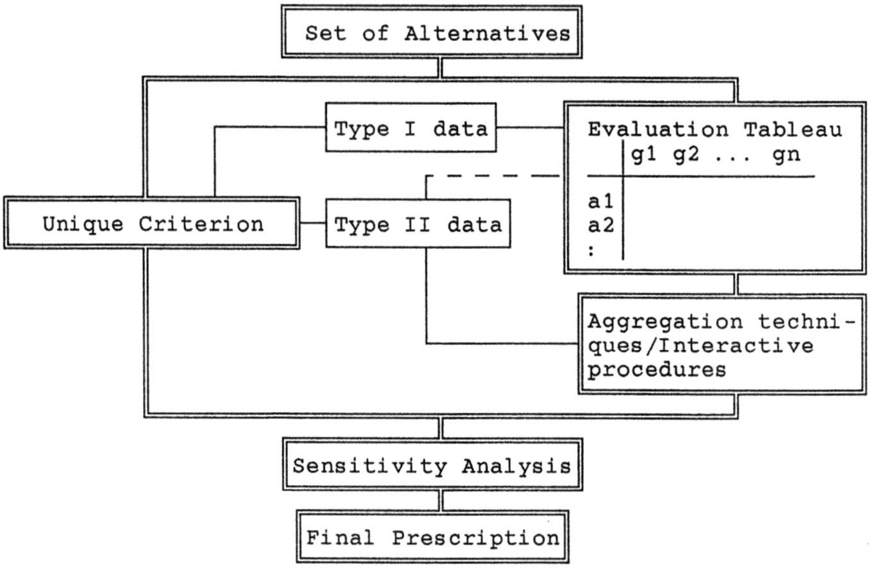

Figure 1 : Mono-criterion and multiple criteria approaches to decision-aid.

4- The management of imprecision, uncertainty and inaccurate determination.

We will center our discussion in this section around what we consider to be the three main originalities of the multiple criteria approach in the management of I.U.I.D.

a) The establishment of partial preference structures.

In a multiple criteria approach, a criterion is used to "sum up" evaluations on consequences related to a same point of view (see Roy (1985)), *e.g.* cost, safety, environment, etc. Formally a criterion can be defined as a function associating a real number $g_i(a)$ to each alternative "a", such that every actor in the decision process admits that if $g_i(a) \geq g_i(b)$ then alternative "a" is at least as good as alternative "b" on the point of view that underlies the definition of criterion[1] g_i.

The interest of such partial preference relations for the management of I.U.I.D. is tied to the fourth source we mentionned in section 2. These partial preference relations can be seen as the stable part of the preference structure of the actors and thus as a possible basis for discussion between the actors. The criteria are designed in order to aggregate consequences that are very "close" to one another for they relate to the same point of view and are as free as possible from potentially highly conflictual type II data such as tradeoffs or weights. In locating a plant it seems reasonable to admit that environmentalists could agree with technicians on the definition of an index of safety of an installation whereas, in most countries, they will definitely disagree on the possible tradeoffs between safety and costs.

As opposed to mono-criterion models which use "sensitive" information from the beginning, the multiple criteria approach aims at modelling what appears to be sufficienly stable in the perception of the actors in order to obtain a first basis for consensus. This may prove useful in conflictual decision or when the perception of the problem by the various actors is still very poor. Partial preference relations are often useful tools for the management of hesitation and conflicts. In this respect the "analytical" side of models using several criteria may be seen more as a volontarist attitude in order to obtain decomposable and, thus, stable models than as a philosophical commitment.

b) The discussion of the "precision" of the evaluations on the criteria.

In a mono-criterion approach it is not unusual to consider that if g(a), the evaluation of alternative "a" on the criterion, is greater than g(b) then "a" is considered as being strictly preferred to "b", even if the difference between the evaluations is very small. Given the first three sources of I.U.I.D. we mentionned, this mode of comparison may lead to "unconvincing" preference situations. Small differences in somewhat arbitrary maps do not imply that the territories really differ. Furthermore, the way the analyst has dealt with the imprecision and/or uncertainty affecting most data in order to obtain the evaluations on the unique criterion is certainly not the only sensible one. Other reasonable ways of doing could have lead to different evaluations and, possibly, to a reversal of the comparison of "a" and "b". In this type of models, the only way to test the significance of a conclusion is to perform a thorough sensitivity analysis. A sensitivity analysis should ideally combine all plausible values for the parameters. However, given the number of these parameters and the

1 The definition of such a partial preference relation implies a number of "independence" hypotheses.

complexity of the calculations leading to the map g(a), such a thorough sensitivity analysis can rarely be performed. It very often appears as a "one-dimensional" sensivity analysis testing the robustness of the conclusions by varying only one parameter at a time.

These difficulties are still present in a multiple criteria approach[1]. Yet, in this framework, the analyst may try to deal with these difficulties separately on each criterion. Since the model leading to the definition of each criterion is usually far less complex and uses much fewer parameters than the one that would lead to unique criterion this may be seen as an advantage. Let us note however that the analyst has often to introduce mostly "volontarist" hypotheses (such as probabilistic independence of some probability distribution for instance) in order to be able to cope with uncertainty and imprecision separately on each criterion.

Dealing with uncertainty and imprecision separately on each criterion, may prove useful from the point of view of sensitivity analyses. Furthermore, as argued by Roy (1985), the analyst may also try to define on each criterion a preference structure taking into account the fact that small differences may not be significant. This can be done using thresholds, a difference between two evaluations being significant only if it is "sufficiently" large. The evaluation of these thresholds is not an easy task. As long as the model leading to definition of the criterion is not too complex, simple reasonings may be used to give them a (not unreasonable) numerical value (see Bouyssou and Roy (1987)). Because of the complexity of models leading to a unique criterion such reasonings that rely on intuitive considerations and the evaluation of experts can rarely be used in this approach so that the "precision" of the evaluations on that criterion can only be questionned through a sensitivity analysis. However, in order for the criteria to be accepted by all actors, it is sometimes essential to discuss the "precision" of the evaluations of the alternatives before the last stage of the study : a criterion showing non-convincing preference situations is likely to be rejected by some actors (even if they are told that a sensitivity analysis will be performed).

c) The use of an evaluation tableau and the aggregation phase.

In many decision-aid contexts (see Roy (1985)), the evaluation tableau has an interest in itself[2], independently of the application of any formal technique leading to the establishment of a prescription. In the discrete case this evaluation tableau is the classical alternatives *vs* criteria matrix.

---

1 For instance, when constructing a criterion "cost", using the expected utility of a probability distribution defined on actualized cash flows, it is clear that a small difference in the value of that criterion should not be considered as representative of a strict preference : the probability distribution used is rarely the only reasonable one, the utility function may have been greatly influenced by the type of assessment method used, etc.

2 In some situations the analyst is only hired for the construction of the criteria (see, *e.g.*, Grassin (1986)).

In the continuous case, *i.e.*, when the set of alternatives is either infinite or very large, subsets of the evaluation tableau, like the so-called payoff matrix, are used by most procedures. The use of an evaluation tableau may prove useful in dealing with the sources of U.I.U.D. we mentionned.

This evaluation tableau first constitutes a language that is common to all the actors of the decision process. This map shared by all actors is undoubtedly richer than the one created by a unique criterion and, thus, closer to the territories that are to be compared. However, in spite of this richness, we are not aware of real-world situations in which this map has been rejected because of its complexity[1]. In fact it seems than in most real-world studies, it is possible to reach a consensus concerning the various points of view to be taken into account.

The map created by the evaluation tableau is operational in many respects. The establishment of a prescription implies the assessment of many data of type II. This assessment is usually based on "imaginary" alternatives (see Keeney and Raiffa (1976)). Using a mono-criterion approach one is bound in order to speak of such alternatives either to describe each of their consequences and characteristics, which is often very difficult, or to give their evaluations on the unique criterion in which case the alternatives have very little intuitive appeal. For instance in large linear programs, an alternative can either be described using a vector of hundreds of decision variables or by giving the value of the objective function[2]. The use of multiple criteria often allows to reach a convenient compromise for speaking of imaginary alternatives[3]. This gives the analyst a sound framework for assessing data of type II and discussing their "precision".

The evaluation tableau allows the actors to implement simple reasonnings such as dominance, the use of aspiration levels or a simple lexicographic method in order to justify and elaborate preferences. In face of an evaluation tableau, most actors will recognize that a decision will inevitably be the result of a compromise between several conflictual objectives. It is then difficult for the analyst working into this framework to convince people to accept his recommandations just because of the sophistication of the methods he uses. This may allow the analyst to avoid some difficulties that are frequently encountered when using a unique criterion (see Roy (1981) and also Ackoff (1979)) and, thus, gives him tools for the management of hesitation and conflicts. In this approach, the analyst often proposes an "optimal" solution. Confronted to that optimal solution, an actor is likely either to accept it without restriction because of its "scientificity" or to reject it

---

1 Tentatively, this might be explained observing that the number of criteria used is usually close to the "magical number seven plus or minus two" introduced by Miller (1956).

2 This type of difficulty has lead some authors to make a great use, when possible, of visualizations of alternatives in linear programs (see Jacquet-Lagrèze and Meziani (1988)).

3 Let us also mention that the use of multiple criteria may be useful in order to generate new alternatives as noted by Starr and Greenwood (1977).

because the many simplifications, ommissions and distorsions contained in the model are incompatible with his value system (on these aspects, see GRETU (1980)). The use of multiple criteria allows and, sometimes, forces the analyst to regard its model as support to reflexion, negotiation and creativity tolerating hesitations, ambiguities and iterations.

If the analysis goes further than the construction of an evaluation tableau then it is clear that highly sensitive data of type II will be needed either through the assessment of weights and/or tradeoffs or through a dialogue driven by an interactive procedure. By separating as much as possible this information from the rest of the data, an approach using multiple criteria allows to clearly locate conflicts between actors and perform thorough sensitivity analysis. This is true even if the aggregation method aims at building a unique criterion on the basis of the familly of criteria contained in the evaluation tableau (which is done for instance in Multiattribute Utility Theory, see Keeney and Raiffa (1976)). This particular form of aggregation should not be confounded with an approach directly aiming at building a unique criterion. In the latter case, data linked to the description of the alternatives are often inextricably mixed with data linked to a particular preference system.

Using multiple criteria can be seen as a "diplomacy of small steps", trying to model what can be modelled in spite of the presence of hesitations, conflicts and ambiguities. Many other intermediate steps could be envisaged apart from the evaluation tableau. It seems however that the use of multiple criteria gives to the analyst clear and sound tools to deal with I.U.I.D. Central to the the management of I.U.I.D. into this framework is what could be called a "volontarist analytical" approach which is only very partially covered by the classical "divide and conquer" approach. Division appears here less as a cartesian device than as a pragmatic way of building a convincing prescription for decision-aid.

REFERENCES

Ackoff R.L. (1979), The future of operational research is past, *JORS*, Vol. 30, pp. 93-104.

Bouyssou D. (1984), *Approches descriptives et constructives d'aide à la décision : fondements et comparaison*, Thèse de troisième cycle, Université de Paris - Dauphine, 270 pp.

Bouyssou D. (1987), *Méthodes d'aide multicritère à la décision : fondements et mise en oeuvre*, Document de travail, Université de Paris - Dauphine, 25 pp.

Bouyssou D. and B. Roy (1987), La notion de seuils de discrimination en analyse multicritère, *INFOR*, Vol. 25, pp. 302-313.

Grassin N. (1986), Constructing criteria "population" for the comparison of different options of high voltage line routes, *EJOR*, Vol. 26, pp. 42-47.

GRETU (1980), *Une étude économique a montré... mythes et réalités des études de transport*, Cujas, Paris.

Jacquet-Lagrèze E. and R. Meziani (1988), Linear programming and interactivity - A manpower scheduling DSS, in G.K. Rand (Ed.), *Operational Research '87*, pp. 176-189, North-Holland, Amsterdam.

Keeney R.L. et H. Raiffa (1976), *Decisions with multiple objectives - Preferences and value tradeoffs*, Wiley, New-york.

McCord M. and de Neufville R. (1983), Fundamental deficiency of expected utility analysis, in S. French et al (Eds.), *Multiobjective Decision Making*, pp. 279-305, Academic Press, London.

Miller G.A. (1956), The magical number seven plus or minus two - Some limits on our capacity for processing information, *The Psychological Review*, Vol. 63, pp. 81-97.

Roy B. (1981), The optimisation problem formulation : criticism and overstepping, *JORS*, Vol. 32, pp. 427-36.

Roy B. (1985), *Méthodologie multicritère d'aide à la décision*, Economica, Paris.

Roy B. (1988a), Main sources of inaccurate determination, uncertainty and imprecision in decision models, in B.R. Munier and M.K. Shakun (Eds.), *Compromise, negotiation and group decision*, pp. 43-62, Reidel, Dordrecht.

Roy B. (1988b), Des critères multiples en recherche opérationnelle : pourquoi ?, in G.K. Rand (Ed.), *Operational Research '87*, pp. 829-842, North-Holland, Amsterdam.

Roy B. et D. Bouyssou (1986), Comparison of two decision-aid models applied to a nuclear power plant siting example, *EJOR*, Vol. 25, pp. 200-215.

Schärlig A. (1985), *Décider sur plusieurs critères - Panorama de l'aide à la décision multicritère*, Presses Polytechniques Romandes, Lausanne.

Starr M.K. and L.H. Greenwood (1977), Normative generation of alternatives with multiple criteria evaluation, in M.K. Starr and M. Zeleny (Eds.), *Multiple criteria decision making*, pp. 111-128, North-Holland, Amsterdam.

Zeleny M. (1982), *Multiple criteria decision making*, McGraw-Hill, New-York.

THE MEETING SIMULATOR FOR STIMULATING ABDUCTION

Y. Sawaragi
The Japan Institute of Systems Research
Yoshida Ushinomiya-cho 4, Sakyo-ku, Kyoto 606, Japan
T. Nakagawa
System Sougou Kaihatsu Co., Ltd.,
Hongou 4-8-17, Bunkyo-ku, Tokyo 113, Japan
K. Hori
RCAST, University of Tokyo
Komaba 4-6-1, Meguro-ku, Tokyo 153, Japan
S. Inabayashi
System Sougou Kaihatsu Co., Ltd.,
Hongou 4-8-17, Bunkyo-ku, Tokyo 113, Japan

ABSTRACT

We usually collect opinions from several people in order to make decisions. The manager who makes decisions can often form new ideas through the observation of meetings where the views of the participants collide.

In this paper, we propose a meeting simulator. The purpose of this simulation system is to stimulate abductive reasoning through the presentation of the dynamic process involved in meetings. The user initially provides some clue concerning the discussion to the system, and then the system presents the user with an extended sample of discussion on this topic by several participants. This discussion is not deterministic, and the user can alter the discussion flow at any time by providing other clues on relevant topics. The system consists of a multi-agent-model which simulates the situation of several participants at a meeting. Each agent-model consists of discourse rules, association networks, memory organizing packets, deductive rules, and a working memory.

The system has been implemented on the AI shell ESPARON (developed by the S.S.K. Company) which incorporates distributed problem solving agents. We demonstrate the effectiveness of the system by presenting some actual examples of simulated meetings. The system is expected to work with other intelligent decision support systems.

INTRODUCTION

This paper describes a system which simulates the process of a meeting. The purpose of the system is to stimulate the user's abduction, as shown in Fig. 1. Abduction is one of the three types of human reasoning, i.e. deduction, induction and abduction. Deduction is the process whereby a conclusion is reached by reasoning from axioms and premises. Induction is the process whereby a theory is inferred from facts. Abduction is the process whereby a new idea is derived from observing existing ideas and facts.

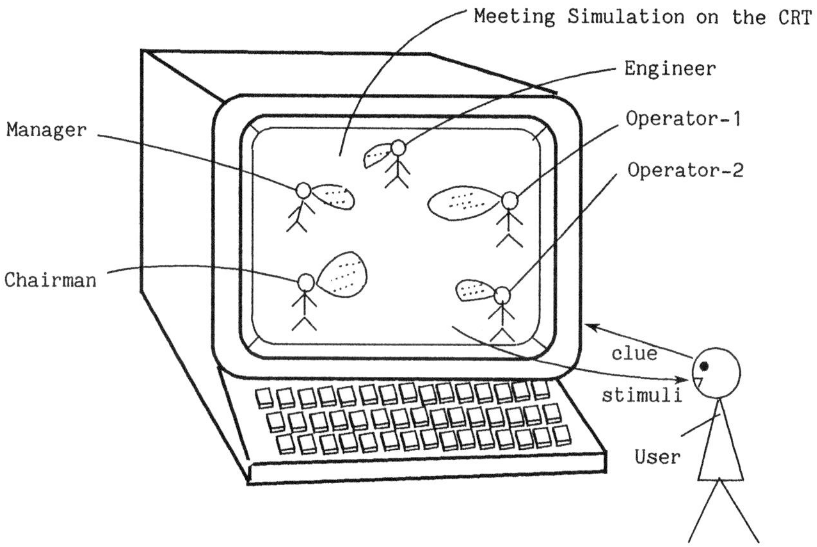

Fig. 1 Concept of Meeting Simulator

Existing decision support systems can aid users only in some areas of deductive and inductive inference. The problems treated have usually been "hard" ones, such as numerical analysis, and there have been no systems which support the user's abductive reasoning. We can see the importance of the abductive process by examining real decision making processes. The problems which should be treated by this process are "soft", fuzzy, and unstructured ones [1][2][3][4], which cannot be handled by existing OR/MS technologies. For example, the manager of a cement production company faced with a production cost problem naturally, analyzes the numerical data, but the most important phase of the manager's decision making is that involving the generation of a new idea, for example, the founding a new transportation company, through discussion with others. The meeting simulator proposed in this paper aids such subjective decision making in environments characterized by "fuzzy" knowledge, varying opinions, and uncertain facts.

The merit of using the meeting simulator rather than holding a real meeting is that the user can hypothetically change the flow of the discussion in the meeting, and can thereby extract varied and unusual opinions, which would not normally emerge in a real meeting.

The reason we use this form of "meeting", instead of a conventional knowledge base, is that we must incorporate differences of opinions among agents, and that we can expect the user of the system to be operating in a nearly real emotional environment which would promote the abductive processes.

CONFIGURATION OF THE MEETING SIMULATOR

The configuration of the meeting simulator is shown in Fig. 2. The system has been implemented on the AI shell ESPARON [5] [6], which incorporates distributed problem solving agents. The meeting simulator consists of several agents, and the number of agents is not limited.

Each agent consists of discourse rules, association networks, memory organizing packets (MOPs) [7], deductive rules, and a working memory.

Discourse rules are rules which prescribe what the agent should do during the discourse. For example, one rule tells an agent to search its knowledge if it receives a question from another agent. We can change the personality of each agent by changing the discourse rules. For example, we can make an agent aggressive or defensive. Some of the implemented discourse rules are shown in Fig. 3.

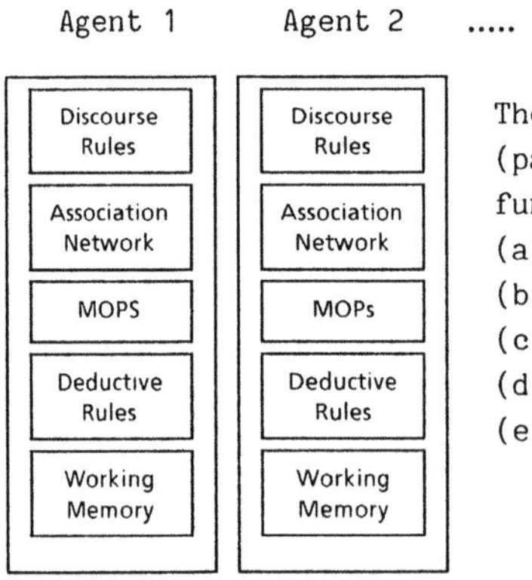

The meeting simulator consists of agents (participant, which have several of their own functions, as shown in Fig. 2).

(a) Discourse Rules
(b) Association Network
(c) MOPs
(d) Deductive Rules
(e) Working Memory

Fig. 2 Schematic Organization of Meeting Simulator

```
/* Operator1 Agent : Metarule Part */
metarule(op1,mr01,
   envelope(Message,From),
   [From ¥== op1, assertdata(message(Message,From))]).

metarule(op1,mr02,
   newdata(message(msg(Id,Sg,Fcs,Emt,S,V,C),Frm)),
   [asserted(MOP_list(Fcs)),
   cleardata(MOP(Fcs)),
   assertdata(subject(Fcs)),
   inference(01_MOP,_,fwa)]).
```

Fig. 3 Part of Implemented Discourse Rules

Association networks describe the associations among topics. The network consist of nodes and links, and each node is a pointer to a memory organizing packet (explained below). Each link has the value representing the strength of association between the linked topics. In the initial implementation, the structure of the network is represented by a logical form. An example of an association network is shown in Fig. 4. Each agent has its own association networks.

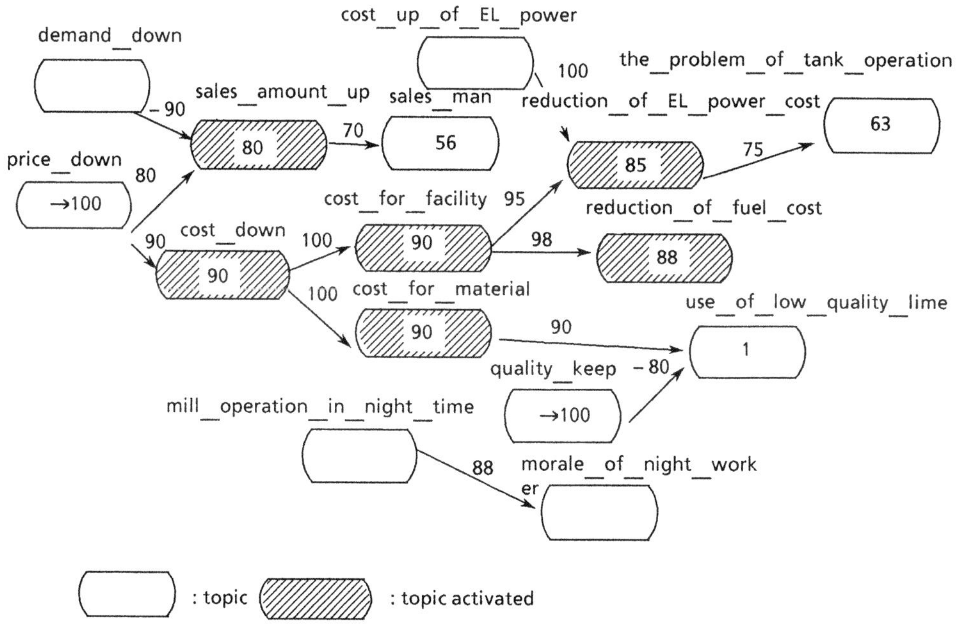

(The number beside arrow line indicates the weight of each link, and the number in the node shows the current activation level.)

Fig. 4 Example of Association Network

Memory organizing packets constitute each agent's store of expertise. Each memory organizing packet contains neutral knowledge and/or supporting opinion and/or opposing opinion with respect to a proposal. The context sensitivity of the opinions can be realized by putting the same topic in several different parts of the association network, and varying the memory organizing packets. An example of a memory organizing packet is shown in Fig. 5.

Deductive rules, if necessary, can be utilized in any agent subject to the discourse rules.

The function of the working memory is to store the list of potential topics and the temporal data.

Fig. 6 shows the structure of the internal representation of an opinion which is used in the internal communication among the agents. The internal representation consists of classification, focus, emotion, S, V, and C. The classification shows the types of utterances such as what-questions, why-questions, how-questions, expressions of support and objections. The focus indicates the topic under consideration, while the emotion shows the agent's emotional level. S is for subject, V is for verbs and C is for complements, corresponding to the content of an utterance. In the current implementation, the process of generating natural language text from the internal representation [8] is omitted, and the surface structures of the sentences are described directly, instead of the internal deep structure (S+V+C).

```
rule(m_MOP,r03,
   subject(reduction_of_fuel_cost),
   [Knowledge(101, "We mainly use coal in Kiln process."),
   knowledge(102, "The cost of fuel depends on the producing district."),
   knowledge(103, "The coal from China is rather low calory but cheep.")],100).
rule(m_MOP,r04,
   subject(reduction_of_fuel_cost),
   opinion(support, 10, the_use_of_low_calory_coal, 0,
         mangr, propose, the _use_of_low_calory_coat),100).
```

Fig. 5 Example of Memory Organizing Packet (MOP)

Classification	Focus	Emotion	S	V	C

Fig. 6 Internal Representation Used for Communication Among Agents

WORKING PROCESS OF THE MEETING SIMULATOR

In outline, the mechanism of the meeting simulator is as follows.

1. The user starts by presenting the system with the topics to be discussed. The system possesses the capability of displaying the list of topics for which the agents are prepared (Fig. 7).
2. The chairman agent transmits the topic to the agents.
3. Each agent receives the topic from the chairman and generates the purport of his utterance, if any.
4. The chairman distributes the received opinions to the agents and displays them to the user.
5. The user can change the flow of simulation at any time when he receives an opinion from an agent, and can introduce any available topic that he wishes to be discussed during the meeting.

```
ES>go
(User requests to the start of the meeting simulation.)

>>> You can select from these options following:
      1. request new topic (by order) = >
      2. continue about current topic = >
      3. stop the meeting = >
      == Your option ? 1.
(The system shows its menu, to get the first clue from the user.)

*** request new topic ***

If you want to see the moplist, please key in "m" m*.
(* User requests to see the list of MOP's.)
data(man, moplist(sales_amount_up),100).
data(man, moplist(reduction_of_FL_power_cost),100).
data(man, moplist(reduction_of_fuel_cost),100).
data(man, moplist(use_of_low_quality_lime),100).
date(man, moplist(morale_of_night_worker),100).
data(man, moplist(the_overcoming_of_unstable_condition),100).
                                   [abbrevation]

(The Mop's list from the agents.)

=== You can key in your topic : cost-down*
(* User's first clue to the simulator)

USER :   question/how//cost-down/0
   user/request/xxx
```

Fig. 7 Example of Simulated Meeting (at Beginning of Program)

In Fig. 8, we show an example of the dialogue during a simulated meeting. This includes the development of topics, emotional reactions, conflict of opinions, and mediation, all of which we believe, stimulate the user's abductive process.

```text
= = = You can key in your topic : reduction_of_EL_power_cost.
USER : question/how/reduction_of_EL_power_cost/0
       user/request/xxx
man : 101 Electric power utilization is under the contract with Tokyo_EL_Power Co.
man : 102 EL power charge consists of the contract fee and the charge depends on the amount.
man : 103 Cost reduction through optimizing of contract is available only once in a year.
man: 104 Cost reduction of EL power is available through improvement on adgustment of operation in
usual.
       Classification     Focus                                    Emotion
man → support/10/improvement_of_EL_power_operation/o
       mangr/limit/realize_cost_reduction_through_improvement_of_operation
         S      V                        C
>
eng : 101 Most part of EL power is consumed for mill process in this plant.
eng : 102   The other process operations depend on production rate.
            We can not control them easily.
eng : 103 We can use the cheaper power in night and on holiday.
eng → support/10/mill_operation_in_night_time/0
      engnr/propose/the_mill_operation_in_the_night_time
>
association by o1_net : increasing_night_workers
op1: 101    We are working on the system of 4 groups 3 shifts
op1:  102   Our goal is to reduce the night workers
op1→ object/10/objection_to_night_operation/3*
     oprt1/object/increasing_of_night_operation_of_mill_process
>
association by o2_net: support_to_night_operation
op2 : 101 We think the system of 4 groups 3 shifts is available and it is not unlawful.
op2 : 102 I support the increasing of night works for cost saving and stabilizing of our plant.
op2 → support/10/support_to_night_operation/4*
      oprt2/support/the_increasing_of_night_operation_of_mill
>
association by o1_net : distrust_to_old_workers
op1 : 101 The old worker can select easy job in night time
op1 : 102 They sometimes like to do night work for getting allowance
op1 → object/10/against_the_support_to_night_operation/8*
      oprt1/distrust/the_opinion_of_old_worker
>
association by o2_net : reproof_to_distrust_expression
op2 : 101 The purpose of discussion is to realize the goal of our company.
op2 : 102 The reproof about individual stand point is unsuitable
op2 → object/10/reproof_to_distrust_expression/9*
      oprt2/reprove/the_expression_of_distrust_without_proper_reason

chr : 101 We must discuss about our task without emotion.
chr : 102 Both of opinions are respectable when they are based on individual view point.
chr : 103 Now, we have to be moderate to avoid getting extremely.
chr → support/10/inhibit/0
      chrmn/inhibit/the_over_emotional_discussion
```

(The chairman inhibits the over emotional discussion.)

Fig. 8 A Sample Dialogue from a Meeting Simulation

IMPROVEMENTS FOR FURTHER IMPLEMENTATION

As an initial stage, we have established the association part of the meeting simulator as a special "asnet" function for the association networks based on forward reasoning in the ESPARON system. Asnet processes the association network model and generates several concepts with individual values of activation levels. Now, we are developing an association network model on a neural network[9] basis. This new model is expected to have several outstanding features, as followed.

1) Learning: Self-organization based on example data sets.
2) Parallel processing: Rapid execution of conceptual association.
3) Unexpectedness: Unexpected associations different from the given data sets.

We have experimented with back-propagation neuro-networks (BPN) with a view to realizing the features listed above, particularly seeking unexpectedness for more effective stimulation of human abduction.

1. Learning

At the beginning of the learning process, we define the numbers of PEs for each layer of the BPN model and some parameters. In a typical case, i.e. a manager's association network, we assumed the full-connection model of 121 concepts, and declared 121 input PEs, 50 hidden PEs and 121 output PEs. The number 121 corresponds to the concepts which the manager was to associate. The number of PEs in the hidden layer, 50, was defined provisionally, and the coefficients of learning and momentum for the BPN learning process were appropriately adjusted. The learning was interrupted and stopped, when the rate of accuracy reached 80%. This required about 2 hours on a personal computer and neuro-computing board, with 34 iterations. The total number of weights of the PEs in this model is more than 12,000. All weights are updated by the BPN method whenever a learning data set is input. The number of learning data sets is 121, corresponding to the number of concepts to be learned. Therefore, more than 12,000 parameters are recalculated 121 times in each iteration. An example of the CRT display in the learning process is shown in Fig. 9.

2. Evaluation

We can evaluate the effect of association created by the BPN model. Simulation reveals the behavior of association memory when some stimulus enters the input layer. We can also monitor the behavior of PEs within the hidden layer. The activation pattern of the hidden layer seems to indicate certain rather abstract concepts relating to some of the stimuli fed to the input layer. Fig. 10 shows the display image on the CRT during the simulation of association. After completing all of these BPN modeling procedures, the association memory model was successfully realized on a neural network.

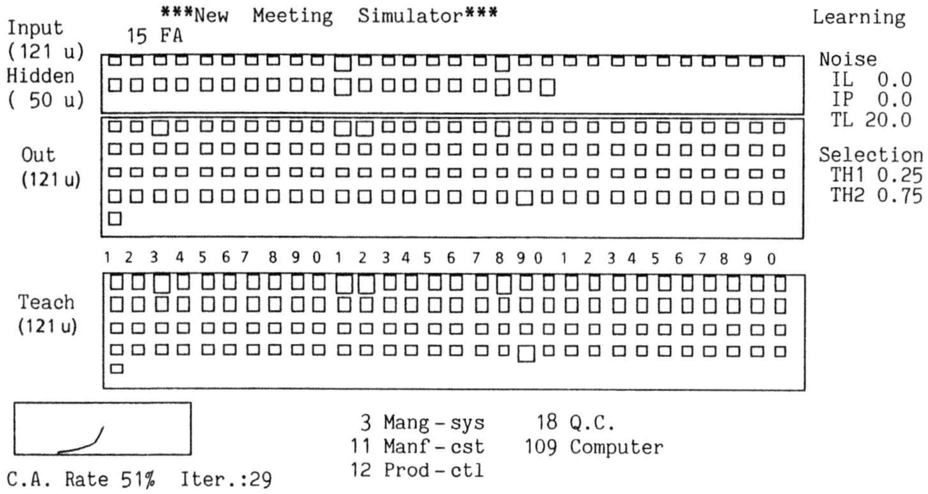

The activation status of each PE is represented by a white square, with the area of the square proportional to the activation value.

Fig. 9 An Example of CRT Display during Learning Process

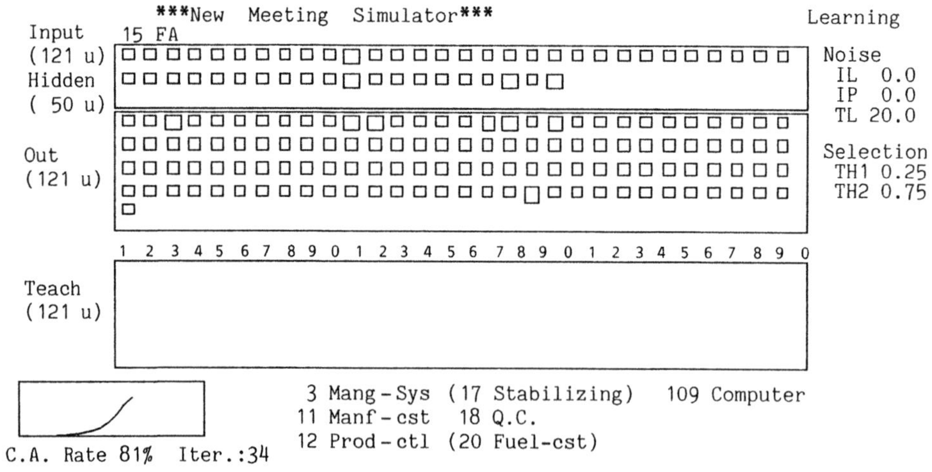

Fig 10. An Example of CRT Display during Association Process

3. A mechanism for conceptual learning in deep structure

In general, the BPN learning process can be adjusted by means of the learning factor and the momentum factor. Furthermore, some noise can be added to the training data sets to make the model more robust. However, the BPN model itself will not necessarily yield genuinely human-like association, which sometimes involves nonlogical association, unless some additional methodology is provided.

We have realized a mechanism for generation of unexpected concepts from the deep structure which is not learned directly in the BPN learning process. This mechanism depends upon the following three factors.

(1) Limited number of PEs in the hidden layer: The number of PEs in the hidden layer must be confined within some limit to prevent an excessive degree of freedom in the BPN learning process. An excessive number of PEs often makes the model meaningless. The appropriate number of PEs can be determined experimentally.

(2) Controlled learning rate: To leave the room of unexpected association, the BPN learning process is programmed to stop at some prescribed rate of accuracy. We have confirmed by experiments that unexpected free association such as human abduction is able to occur within the process of uncompleted learning with a noisy environment.

(3) Noise to the input: White noise is added to the input data to effect fluctuation of the output of the association model in the neighborhood of its original domain. An example of such apparently unexpected association from the deep structure under the given domain is as follows.

=Example=

Input		Learning	Association
Symbolic No. and	concept	Symbolic No. of output	Symbolic No. of output
25	Work at night	26, 27, 29	26, 27, 29, 55, 80
27	Labor problem	25, 28, 29, 31, 32	25, 28, 29, 31, 32, 55
33	Holiday system	29	29, 55
60	Generation gap (employees)	29, 55	29, 55
63	Labor market	29, 64, 65	29, 55, 64, 65
64	High educational level	29, 46, 51	29, 46, 51, 55, 59
73	Sense of values	29, 79,	29, 55, 73, 79

(note)29: Interests of young employees 55: Consciousness of employees

* Each number represents the corresponding concept.

This example shows that output 55 is frequently associated with inputs related to output 29, because when input 60 was entered both 29 and 55 were taught as concurrent outputs. This suggests the existence of some abstract concept in the hidden layer relating to both 29 and 55. In fact, we verified that these inputs generate output 55 with considerably greater frequency than other inputs. This is explained by the schematic illustration shown in Fig. 11.

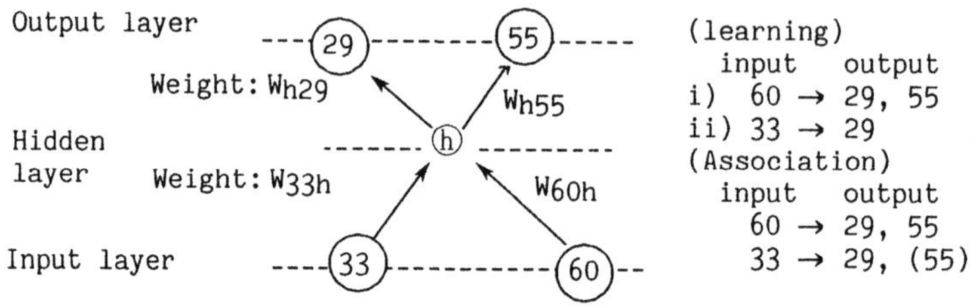

(Learning Process)
W_{60h}, W_{h29}, W_{h55} are increased by learning the data set (i).
W_{33h}, W_{h20} are increased, and W_{h55} is decreased by the data set (ii).
When the learning process is stopped, the weight W_{h55} keeps to remain with high probability, as a result of the BPN learning process.
In this case, the PE in the hidden layer labeled "h" is assumed to relate to both 29 and 55.

Fig. 11 Schematic Illustration of BPN Learning Process

4. Implementation of association memory on the meeting simulator

Our meeting simulation system has been developed on an expert shell called ESPARON. We have combined this new neural network association model with ESPARON. The association part in the structure of each participating agent is implemented by neural computing facilities. The ANZA board, which is a neural computing board developed by HNC Inc., can be utilized by linking this board with ESPARON. We have provided a software interface between ESPARON and ANZA. Hence, we can combine both of these functions on a single personal computer using MS/DOS. An overview of the implementation is shown in Fig. 12.

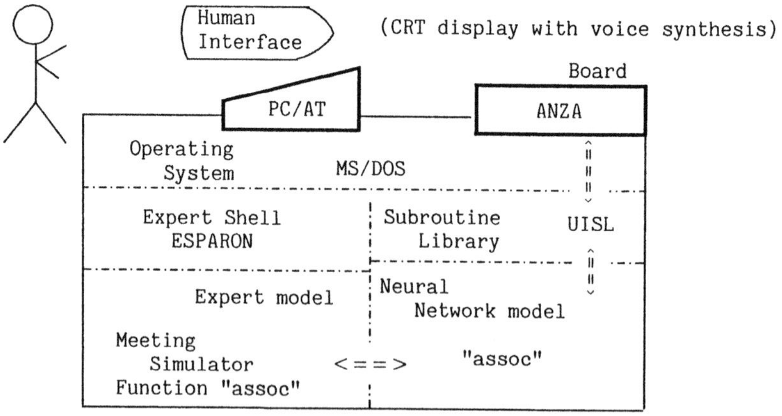

Fig. 12 Overview of Implementation of Meeting Simulator

CONCLUSION

We have described a meeting simulator for stimulating abduction which supports human activity that could not be handled by so-called expert systems.
This simulator handles the collaborative intellectual tasks performed by several agents, where each agent has its own role.
Of course, the ability of the computer system is determined by the given knowledge and data, but the user, whose hidden intellectual process is stimulated by the system, can develop new ideas.
Many problems require the collaboration of intelligences. The meeting simulator was devised as a stepping stone for access to human activity.
The meeting simulator described above is narrative and qualitative, but quantitative data and a mathematical models [10] [11] [12] etc. can also be connected to this system. Further extensions of the functions of the system and the connection of the meeting simulator with existing quantitative decision support systems are planned. Self-organization, self-learning, and evolution can be further developed and are under study in some practical domains.

REFERENCES

(1) E. Plaza et al: "Consensus and Knowledge Acquisition" Lecture Notes in Computer Science, Eds. B. B. Bouchon, R. R. Yager. P. 294 - P. 306 Spring-Verlag.

(2) J. Darzentas : "Knowledge Modeling in Fuzzy Expert Systems", ibid., P. 159 - P. 172

(3) Dik Gregory: "Delimiting Expert Systems" IEEE Transaction Vol. SMC-16 No. 6, 1986

(4) K. Niwa: "Knowledge-based Human-computer Cooperative Systems for Ill-structured Management Domains" IEEE Trans. Vol SMC-16 No. 3, 1986

(5) Technical Note on ESPARON, S.S.K. Co., Ltd., Tokyo, Japan

(6) H. Ogawa: "Hybrid Problem Solving Agents" Information Processing Society of Japan 1986

(7) R.C. Schank: Dynamic Memory, Cambridge Univ. Press 1982

(8) D.D. McDonald: "Natural Language Generation as a Computational Problem" in M. Brady and R.C. Berwick (eds): Computational Models of Discourse, MIT Press, 1983

(9) D.E. Rumelhart et al. (eds): "Parallel Distributed Processing" MIT Press, 1986

(10) Y. Sawaragi: "Towards Interactive and Intelligent Decision Support Systems" VIIth. International Conference on Multiple Criteria Decision Making Vol. 2 Aug. 1986 Kyoto Japan

(11) T. Nakagawa: "General Structure of Hierarchy Control and Management with both an Objective and Subjective Approach" Proceeding of the IIASA Conf. Sopron, Hungary, July 1986

(12) T. Nakagawa, ibid., Lecture Notes in Control and Identification Sciences, No. 105 eds. C.I. Bynes, A. Kurzhanski, Spring-Verlag

THE USE OF SYMBOLS IN MULTICRITERIA DECISION MAKING

David Cray
Decision Analysis Laboratory
School of Business
Carleton University
Ottawa, Canada K1S 5B6

Much of the literature dealing with multicriteria decision making uses formal or mathematical models to bring order out of the chaos of conflicting opinions, counterposed objectives and contradictory information that most important choices involve. Humans may, of course, use MCDM tools to achieve similar ends. Indeed it is the purpose of the varied approaches embodied in MCDM to help individuals and groups make such decisions more effectively. Alas, many groups do not have access to such decision aids when faced with a decision problem involving many criteria, yet they must still make choices. Simon (1947, 1972, 1978) and his colleagues (March and Simon, 1958; Newell and Simon, 1972) have demonstrated that individuals and groups faced with a complex decision will construct an area of bounded rationality which contains a limited amount of information, a few alternatives and a few criteria. Having simplified the problem in this way a decision can then be taken.

There are a number of different ways in which the complexity and ambiguity surrounding a decision can be reduced (MacCrimmon and Taylor, 1976). An individual or group may, for example, simply ignore ambiguity and treat the available choices and information as well-defined and exhaustive. As an alternative strategy they may seek more information as a means of clarifying ambiguous facts. This may also serve as a device for utilizing a third means of ambiguity reduction, delaying until the decision context is clarified. In strategic decisions, decisions which almost inevitably invoke multiple criteria, ambiguity may be
reduced by considering only one alternative at a time (Mintzberg, et al., 1976).

More recently there has been another approach used in understanding the processing of information in organizations, that is through the use of symbols. Much of this literature has dealt with introducing change (Das, 1987) gaining commitment (Kelly and Ibrahim, 1987) or managing meaning in organizations (Smircich, 1983). The literature also suggests that symbols may be utilized in reducing the ambiguity and complexity in certain decision making situations (Bate, 1984; Sapienza, 1985; Thompson and Wildavsky, 1986). A symbol is a representation of a fact or a state. The representation may be quite concrete and direct or it may be understood only through a complex process of association. In either case, the symbol can be seen as a way of rendering a great deal of information in a relatively comprehensible and coherent way. As such it would seem that symbols could be used in the Simonian reduction of the unbounded decision environment to one of bounded rationality.

There are two aspects of symbols that need to be taken into account if one wishes to use them to understand decision making. First, in any group setting, including organizations, the meaning of symbols is negotiated (Smircich, 1983, Smircich and Stubbart, 1985). Though some symbols may be relatively static, others are subject to frequent reinterpretation. This implies, as a second point, that symbols may not carry the same meaning for separate individuals or groups (Hirsch and Andrews, 1983). Examples of this can easily be seen where symbols are used to identify contending groups (e.g. sporting events, wars, elections). Less dramatic examples occur in almost any setting in which symbols are important.

If symbols are used, then, as a means of creating a decision arena with bounded rationality, two main questions arise. First, in what sort of decisions are symbols likely to be important? There are, of course, important costs associated with the use of symbols such as the loss of specificity, conflicting interpretations of symbols and potential redefinition over time. It would seem likely then, that symbols would not be used extensively in all decisions, but only in specific circumstances. The second issue addresses the use of symbols in MCDM. When symbols are employed in making strategic decisions, how do they effect the decision making process and its outcomes? One can frame a very large number of hypotheses around these two basic questions. The purpose of this paper is to generate a few such hypotheses by drawing on three cases in which symbols played an important role in the decision making process.

A METHODOLOGICAL ASIDE

The data for this paper were collected in two ways. The author was present during the meetings which are described in the first and third cases and had been concerned with the events which lead to the particular choices made. In both meetings he took a relatively minor part in the discussions. In neither, however, could he be described as a disinterested observer. The description of the events were checked for factual accuracy but not interpretation, with colleagues who were also present. The second case was collected by a colleague working on the Bradford Studies of strategic decision making (Hickson, et al., 1986) who conducted separate interviews with two of the participants. He read this version of the case to verify its descriptive accuracy. In all three cases, therefore, one faces the tradeoff so often found in behavioral research, intimate knowledge of process versus objectivity buttressed by disinterest in the outcome. A systematic examination of the hypotheses generated would, of course, require a much more rigorous methodological approach. However, the most effective strategies for studying the use of symbols in organizations is still a matter of some debate in the field.

THREE CASES

The Veto

The first case occurred at the annual meeting of an organization which oversees the collection and disbursement of strike funds for faculty unions at Canada's four year universities and colleges. This organization is familiarly called the Defense Fund. It was originally established by and remains connected to a larger organization, the Canadian Association of University Teachers (CAUT) which is composed of most faculty associations in Canada, nearly half of which are not unionized. Since the Defense Fund has only a rudimentary formal structure and no paid employees, most policy decisions are made at the annual general meeting. Three of the officers of the CAUT are, by virtue of their positions, also trustees of the Defense Fund. Indeed, they have a special status which gives them the power to veto certain changes in the Defense Fund's constitution no matter what the vote of the other thirty or so trustees may be.

A motion was brought before the annual meeting to replace one of the CAUT trustees with another officer from a newly formed bargaining cooperative, another offshoot of CAUT. This required a constitutional amendment. The CAUT trustees objected to the motion, not so much because of the substantive change which was quite logical, but because the amendment was seen to begin a separation process between the Defense Fund and CAUT. Separation was an issue which had arisen before, but it had never been put in the form of a motion and had not received general support from the trustees. The potential disassocation of the two organizations was a highly political issue, but it also had a substantive basis in legal and financial concerns. The motion, as it was framed, had little real import for the separation issue, but because it would reduce by a third (one of three) the number of trustees acting for CAUT, they resisted it.

As the debate became more heated and involved, a symbol emerged which eventually became the pivot around which the whole decision process turned. That symbol was the veto which could be exercised by the CAUT trustees if the motion were passed. For those who were committed to the separation of the two organizations it became a symbol for the unwarranted intrusion of a more conservative group over a more militant one. For the officers of CAUT and their supporters it became the symbol of an exercise of power that was being forced upon them by dissident members. For those who had entered the meeting with little commitment to either the issue or the underlying positions, the veto may have taken on other meanings, but in the debate it was these two basic interpretations which were the focus of the argument. After a long debate and some attempts at averting a showdown, the motion was carried and the CAUT officers were put into the position of having to decide whether or not to veto the motion.

This case also suggests two possible effects of the use of symbols in strategic making:

4. Decision making processes which employ powerful symbols will be of shorter duration than similar decisions which do not.
5. The use of powerful symbols in decision making processes will limit the alternatives considered.

The Scanner

The second decision is quite different from the first in that it occurred over a much longer period of time and was not confined to a formally defined group of people. This decision occurred in Great Britain in a large city of the industrial North which, for reasons of confidentiality, will be called Welton. The events leading to this decision were set in motion by the appearance of a new type of medical technology, the body scanner. As with many such expensive instruments the research hospitals, including in this case one located in a large city near Welton, were the first to acquire scanners. When the Welton Health District, spurred by the staff radiologists, submitted a request for its own scanner (it had been sending patients to a hospital in an adjoining health district for scans) the request was rejected on the grounds that the large sum of money required could not be spent until more solid evidence on the use of such machines could be collected.

The decision of the higher authority, the National Health Service (NHS), might have been accepted as final except for two symbolic aspects of the scanner. The rejection of the Welton request by the Health Service authorities (located in London) seemed to symbolize the general neglect of the North in favor of Southern concerns. Second, the scanner took on an almost mythic potency in its power to aid physicians in curing disease. In particular the radiologists and surgeons were excited about its potential for reducing the amount of exploratory surgery necessary. Doubts about the scanner's effectiveness and fears over potential dangers were seen not as real considerations, but merely as a device to deny adequate funding to the Welton District and others like it. A decision, urged by the radiologists and supported by the Health District administration, was therefore taken to attempt to raise the money for the scanner by voluntary contribution. This was unusual but not unprecedented. Other health districts had raised funds for other types of projects.

One important aspect of this particular decision is the variety of interpretations attached to the scanner as a symbol. For the radiologists in Welton the scanner was a symbol of their attempt to remain at the top of their profession. In an era of cutbacks and amalgamation, the professional status of a department could have a distinct impact on one's career. For the hospital and health district administrators the symbol was much the same but with less overt technical/professional content and more resource allocation implications. The scanner became a symbol of the right of less prestigous districts (Welton, in comparison to the neighboring

There are several points here which are suggestive for further study. First, there were several criteria which were important for making this decision. Some were fairly technical such as the handover of power from a position that was being phased out and the extent of legal responsibility by the Defense Fund for debts incurred by the parent organization. Others were overtly political, e.g. the relationship between the two organizations and the relative power of one over the other. Still other criteria were more emotional, such as the felt need by many at the meeting to maintain solidarity among all Canadian professors regardless of their organizational affiliation. These criteria for many of those at the meeting led to contradictory outcomes when considered singly.

The second point to consider is time pressure. The contentious motion could have been postponed and a motion to do so was introduced and defeated. The trustees felt that the matter needed to be decided at this meeting because the next meeting was a year away. Given the time pressure and the multiple criteria there was a need for some means of simplifying the choice, for putting a boundary around the decision arena so that rationality could operate.

The third point is that there was a highly skewed distribution of information among those at the meeting. It is probably safe to say that less than half the trustees were even aware that the issue was to be raised at the meeting. Less than a quarter knew the detailed political background of the broader dispute, though most knew the general outline. The ignorance of relevant information is hardly surprising since the Defense Fund has no permanent officers. The chairmanship rotates among trustees (all of whom are university professors or librarians) and the executive work is handled by an employee of CAUT who has a number of other, often more pressing, duties. If there are no crises, i.e. disbursements to support job action, then the members are unlikely to meet from one year to the next. The trustees are often not called upon to consider seriously the business of the Defense Fund for months on end. A few trustees take a keen interest in the operation and position of the Fund, but they are a small minority of those who may vote at the annual meeting. This combination of infrequent meetings, skewed information resources and important decisions is typical of a number of important bodies such as pension fund trustees, school boards, and boards of directors. The examination of this first case then, provides three possible hypotheses concerning the occurence of symbol usage in MCDM. A reliance on symbols as a means of reducing ambiguity and complexity is more likely to occur:

1. When a large number of non-comparable criteria are applicable.
2. When there is time pressure, especially when failure to make the decision will result in long delays.
3. When information among those deciding is unequally distributed.

district which had a scanner, is smaller, poorer and has a much larger immigrant population) to have equipment as good as that of richer regions. For the general population of the Welton Health District, the scanner also held the meaning of resource deprivation, but also, after the drive was launched, a symbol of what the district could do to help itself. As the area is rather depressed, this was an important quality. Finally, the symbolic content of the scanner to the higher levels of the Health Service in London was not very powerful or interesting. In rejecting the request for the scanner, the officials pointed out that they had purchased only a few scanners for evaluation and testing. In their view the technology was promising but unproven. Given the high cost of each machine, they could not justify further expenditures until the tests were complete. They may have recognized the symbolic content of the scanner for other groups, but it did not affect their decision.

This decision, as with many strategic decisions, is really a series of decisions, each one shaping the next. In each the symbolic content of the scanner was important. In the first decision, to make the application to the National Health Service, the radiologists and adminsitrators of Welton had symbolic elements strongly in mind. In rejecting the request the NHS recognized that granting Welton's request would lead to a flood of similar ones from other areas (Welton's was hardly the first as it was). The third decision, to attempt to raise money from the public for this equipment was innovative and not one which was taken lightly. It is probably not too much to say that, had the scanner not been such an evocative symbol, the decision to go public would not have been contemplated, much less taken.

The symbol for this decision was defined in a rather different manner than the veto. The definition of the scanner as a remarkable piece of equipment was put about by the manufacturer abetted by the popular press (the first workable models were made in the U.K.). When the Health Service authorized the purpose of a machine for evaluation it gave the symbol further power and clarity. In some ways the medical personnel at Welton and the citizens of the district were only reacting to a symbol defined by others. If such extravagant claims had not been made for the scanner, it is doubtful that the refusal of Welton's request would have been treated much differently than any other bureaucratic snub. This leads to two additional hypotheses about the role of symbols in shaping the decision process:

6. The more powerful the symbol, the more likely it is to alter the normal process by which groups make decisions.
7. The acceptance of strong externally-provided symbols by decision makers, may require the rejection of accompanying alternatives.

One can see an example of the principle underlying hypothesis seven in labor negotiations. If management claims a tenative contract as a victory, the union leadership is almost required to recommend that it be rejected.

The analysis of this case also provides an extension of hypothesis five above. The acceptance of the symbol of the scanner provided by the manufacturers made it nearly impossible for the Welton group to accept the negative outcome from London.

8. The use of powerful symbols in decision processes will limit the number of acceptable alternatives (as well as limiting the number of alternatives considered).

The Outside Member

The third decision to be considered involved the proposed reform of the graduate program in a School of Business. A few students had contacted The Graduate Committee for clarification of the regulations governing the composition of various examining boards. Since the program was in only its third year, several members of the Graduate Committee, as well as others in the School, thought a thorough review was in order. After a number of meetings two specific ideas emerged. One was to transfer some of the power of the Director of the School of Business (who will be referred to as the Director) in approving boards to the Director of Graduate Studies (who will be referred to as the Chair as this individual is also the Chair of the Graduate Committee) and to specify more clearly what the composition of the boards should be. The second idea was to give faculty members who participated in graduate supervision formal credit which could eventually lead to a reduction in their teaching loads.

The Director of the School opposed both moves on various grounds, one of which was that it would limit the Director's power and discretion. His concern was not so much for himself as he was stepping down in six months, but for the office of Director. A meeting between the Director and the Chair and a second during which the Director met with the whole of the Graduate Committee failed to achieve a compromise. Therefore, the issue was brought before the faculty as a whole. This in itself was unusual as the tradition in the School had been that program changes might be brought to faculty meetings for discussion but that decisions were made largely by consensus. Faculty meetings were mainly used for formal ratification.

The first motion made at the meeting was for an alteration of the requirements for the literature review in the thesis proposal. This proposal generated a lively debate, with the Graduate Committee taking one side, and a number of faculty members arguing for diverse alternatives. The Director took almost no part in the discussion as there was no political content in this symbol. The motion eventually passed.

The second motion made was to include in one examination board a member from the School who was outside the student's area of concentration. This motion was substantively innocuous as the small size of the School almost insured that this must occur. However, it carried the full symbolic weight of the reforms proposed by the Graduate Committee because it limited the Director's options in appointing this committee. The technical merits of the plan, along with the tedious descriptions of remotely possible calamities so popular at faculty meetings, soon gave way to a heated discussion of the whole thrust of the reforms. The discussion became a debate between the Director of the School and the Chair of the Graduate Committee. Eventually a vote was taken and the motion defeated. After more spirited debate, the Graduate Committee was instructed to reconsider its plans.

The political content of the symbol is quite clear in this example. Had the Graduate Committee won the vote over the use of the outside member, it might have won the battle over the other parts of its proposed changes. By prevailing, the Director of the School, who had been quite active and powerful during his long tenure, gave notice that he would preserve his influence until the end of his term. Any reforms which he opposed must await his departure. This was understood by virtually everyone who participated in the meeting. Since it was clear that this was what was at issue, what little technical interest there was in the reform had little influence on the outcome.

What very likely was not clear to many of those who voted was the political symbolism for those in the larger university community. The Director's term in office had been highly successful, enlarging and professionalizing the School and giving it political clout within the university. He had been nominated recently for Dean of the Faculty, but had declined to stand. He was widely thought to have long-term political ambitions within the unversity. If he had been seen to lose control over his own unit as the end of his term approached, his ambitions, if any, would have been put in jeopardy. It may well have been that his technical objections to the proposed reforms were sincerely felt, but there is no doubt that his own interpretation of the situation was in political terms.

It should be clear from this discussion, and indeed from the other two cases, that symbols used in decisions often have powerful political connotations. It would seem from these three examples as well as other work (Hickson, et al., 1986, Wilson, et al., 1986) that:

9. When a symbol contains technical as well as political significance, the political significance will have more impact on the final outcome.

Powerful political symbols also seem to have the function of reducing the number of alternatives in the sense that group members are often restricted to supporting one side or another (see Wilson, 1982 for an illustrative case study).

10. The more powerful the political content of a symbol, the fewer alternatives will be considered.

11. The more powerful the political content of a symbol, the more likely the alternatives considered will represent opposed choices.

IMPLICATIONS FOR MCDM

The discussion above, besides leading to some interesting hypotheses, has two direct implications for the study of MCDM. First, the type of decisions which may be considered as legitimate subjects for MCDM approaches should be broadened. Studies in MCDM have traditionally focused on technical/tactical decisions and avoided strategic decisions in which procedures, goals and information are less well defined. (By way of illustration one might examine the list of examples in Zeleny (1982:2-10).) Where strategic decisions have been subjected analysis using MCDM techniques, they have focused on the technical aspects while ignoring the ill-structured or political implications. While there have been some recent attempts to apply formal approaches to overtly political decisions (Kersten, Michalowski, Matwin and Szpakowicz, 1988) the tendency to ignore these political aspects is deeply rooted in the history of MCDM. The problem with this bias is that it defines a very large class of important and interesting decisions as being outside the realm of MCDM, and it leads MCDM anaylsts to ignore important considerations in the decisions they do examine.

Virtually every complex group decision relies on both technical and political criteria (Astley, et al., 1982). If one concentrates only on the technical (or 'rational') content of the decision, this can lead to a very different understanding of the decision process (Allison, 1971) and hence the specification of inaccurate models. As can be seen in the three cases discussed above, complex decisions, especially those for which there is no clear hierarchy among the criteria being used, tend to rely on political considerations which, in turn, may be reduced to a single binary symbol. These are not irrational or non-rational decisions. Rather they are decisions which have undergone a particular process of bounding rationality. There is a tendency among scholars and managers to equate rationality that they do not understood with politics. This does not mean the political considerations are therefore irrational and hence unanalyzable.

The second broad implication concerns the symbols embedded in MCDM models. If a problem is modelled and that model is incorporated into a tool for decision making then symbols are being used on two levels. At one level the model itself is a symbol of the problem and like all symbols it represents a selective focus on particular attributes. This symbolism is usually apparent to the designer of the system since he/she must choose the attributes to be included. The symbolism may not be apparent to the eventual user of such a system. As Haines says "Inforamtion is the symbolic representation of reality and such symbolic representation is often undertaken to enable decisions to be made" (Haines, 1978:27). When the information or information categories comprise the modellor's representation of reality, then the content of the symbol may have different meanings for designer and user which may limit the utility of the system.

At a deeper level casting the problem in an MCDM model invokes a separate set of symbols. For most models, especially those which are presented as a computer-based decision aid, one symbolic feature is that of a rational system. This has two implications for the user(s) of such a system. The information which is not included in the system (which the system does not call for or incorporate) is unimportant or non-rational. Second, the outcomes provided by the system are themselves rational. The selective focus mentioned above is seen a rational subset of all the criteria in the problem. The symbol of the MCDM model or a system built on such a model is rationality and rationality of a very explicit kind. If either the modellor or the user is unaware of how this symbolism is produced or its possible effects on the treatment of information, the channelling of the decision process and the bounding of rationality for the decision makers, then the full potential of MCDM is unlikely to be realized.

BIBLIOGRAPHY

Allison, Graham T., Essence of Decision, Boston: Little, Brown, 1971.

Astley, W. Graham, Runo Axelsson, Richard J. Butler, David J. Hickson and David C. Wilson, "Complexity and Cleavage: Dual Explanations of Decision Making", Journal of Management Studies, 19:357-75.

Bate, Paul, "The Impact of Organizational Culture on Approaches to Organizational Problem-solving", Organization Studies, 5:43-66, 1984.

Das, Hari, "Symbols of Organizational Change and Change in Organizational Symbols: A Look at Organizational Development through Organizational Symbols", paper presented at the Third International Conference on Organizational Symbolism and Corporate Culture, Milan, June, 1987.

Haines, George H. Jr., "Comments on Three Papers on 'The Effect of Information on Market Behavior'", pp. 27-31 in The Effect of Information on Consumer and Market Behavior, edited by Andrew A. Mitchell, Chicago: American Marketing Association, 1978.

Hickson, David J., Richard J. Butler, David Cray, Geoffrey R. Mallory and David C. Wilson, Top Decisions: Strategic Decision Making in Organizations, Oxford: Blackwell and San Francisco: Jossey-Bass, 1986.

Hirsch, Paul and John A.Y. Andrews, "Ambushes, Shootouts and Knights of the Roundtable: The Language of Corporate Takeovers", pp. 145-55 in Organizational Symbolism, edited by Louis R. Pondy, Peter J. Frost, Gareth Morgan and Thomas C. Dandridge, Greenwich, Connecticut: JAI Press, 1983.

Kelly, Joe and A.B. Ibrahim, "The Turnaround Manager Unties the Existential Knot of Executive Life", paper presented at the Third International Conference on Organizational Symbolism and Corporate Culture, Milan, June, 1987.

Kersten, Gregory E., Wojtek Michalowski, Stan Matwin and Stan Szpakowicz, "Rule-Based Formalism for Negotiation Support", Theory and Decisions, Vol. 22, 1988 (in print).

MacCrimmon, Kenneth R. and Ronald R. Taylor, "Decision Making and Problem Solving" pp. 1347-1453 in Handbook of Industrial and Organizational Psychology edited by Marvin D. Dunnette, Chicago: Rand McNally, 1976.

March, James G. and Herbert A. Simon, Organizations, New York: Wiley, 1958.

Mintzberg, Henry, D. Raisinghani and A Theoret, "The Structure of 'Unstructured' Decision Processes", Administrative Science Quarterly, 21:246-75, 1976.

Newell, Allen. and Herbert A. Simon, Human Problem Solving, Englewood Cliffs, N.J.: Prentice-Hall, 1972.

Sapienza, Alice M., "Believing is Seeing: How Culture Influences the Decisions Top Managers Make", pp. 66-83 in Gaining Control of the Corporate Culture, edited by Ralph H. Kilmann, Mary J. Sexton Roy Serpa and Associates, San Francisco: Jossey-Bass, 1985.

Simon, Herbert A., Administrative Behavior, New York: Macmillan, 1947.

Simon, Herbert A. "Theories of Bounded Rationality", pp. 161-176 in Decision and Organization, edited by C.B. Radner and R. Radner, Amsterdam: North-Holland, 1972.

Simon, Herbert A., "Rationality as Process and Product of Thought", American Economic Review, 68, no. 2:1-16, 1978.

Smircich, Linda, "Organizations as Shared Meanings", pp. 56-65 in Organizational Symbolism, edited by Louis R. Pondy, Peter J. Frost, Gareth Morgan and Thomas C. Dandridge, Greenwich, Connecticut: JAI Press, 1983.

Smircich, Linda and Charles Stubbart, "Strategic Management in and Enacted World", Academy of Management Review, 10:724-36, 1985.

Thompson, Michael and Aaron Wildavsky, "A Cultural Theory of Information Bias in Organizations", Journal of Management Studies, 23:273-86, 1986.

Wilson, David C., "Electricity and Resistance: A Case Study of Innovation and Politics", Organization Studies, 3:119-40, 1982.

Wilson, David C., Richard J. Butler, David Cray, David J. Hickson and Geoffrey R. Mallory, "Breaking the Bounds of Organization in Strategic Decision Making", Human Relations, 39:309-31, 1986.

Zeleny, Milan, Multiple Criteria Decision Making, New York: McGraw-Hill, 1982.

IDENTIFICATION AND DEVELOPMENT OF ALTERNATIVES: INTRODUCTION TO THE RECOGNITION OF PROCESS TYPOLOGIES

Maria Franca NORESE, Anna OSTANELLO
Politecnico di Torino, Dip.Automatica e Informatica
C.so Duca degli Abruzzi, 24 - 10129 TORINO

1. INTRODUCTION

Recent developments in Aiding Decision Making, in the organizational processes, tend to attach importance to some aspects of the modelling process normally neglected by the "classic" OR (cf Chekland, 1981; Grauer et alii, 1985; Landry et alii, 1983; Moscarola, 1984 ; Schneeweiss, 1987). New research approaches have been developed; in particular the analysis of actual Decision Aid Processes and of their results (formal representations, formulations of problems, operative tools, models) may help in recognizing the activities that allow the analyst - or model builder - to produce valid answers to a client's demand concerning a problem situation.

Using an empirical framework of the modelling process, this research seeks to refine an understanding of how the analyst carries out formalization activities within his/her interaction with the Client (Decision Aid (DA) Process or Analyst/ Client (A/C) Interaction). The approach adopted to derive the framework is based on DA Process reconstruction (cf Nutt, 1984; Witte, 1972). A set of actual cases (25 cases conducted by the Authors), where a Multicriteria Approach had been assumed, has been carefully selected: for each of them both the story and most of the activities had been registered during the process, memorized and verified.

The research shows that DA Process is made up of sets and sequences of activities whose structuring depends on a variety of factors related to the context situation and the analyst's approach. The recognition of these activities and of their structuring is conducted by assuming an integration of the rational and the political-organizational paradigms (cf Grauer et alii, 1985; Moscarola, 1984).

A Frame model is presented, which identifies and defines five "step-activities" called Routines. Different combinations of these routines at distinct levels of action - individual and collective - can produce different configurations of DA processes. Their structuring is not generally predictible in the actual interactions. Some characterized configurations, which have been recognized on the basis of both the internal structure and the sequencing of the routines, are discussed in the paper.

2. FRAME MODEL

The Frame model which is proposed assumes five different routines as basic steps, in the modelling process, leaving a phase recognition as a result of process reconstruction (cf Mintzberg et alii, 1976; Witte, 1972). The concept of routine is somehow equivalent to that of a task sub-process, where different "standard" activities are developed.

A routine can be essentially characterized by:

a) the main domain of application of concrete and abstract activities - "object of the action" of the actual routine; each routine can thus have its own "object";

b) the nature or the purpose of the prevailing activities.

The activities developed in different contextual situations, within the same routine, can be very different in terms of domain of inquiry, actor's involvement and information processing.

The routines can be considered both as "main" or "principal" (i.e. as contexts with identified objects of action) and "secondary" or "supporting" (i.e. as tools to support the action).

A Main Routine, MR, is intended to produce locally stable elements, that can either be assumed as an answer to some question/problem or can develop other routines.

A Secondary Routine, SR, is activated within Main Routines to promote and accelerate their results, in terms of stability, convergence, validity.

A MR can have a general structure varying in connection with the factors characterizing the context where it is applied. The combinations of the SRs into a MR define the structure of it. A "Concatenation" from a MR to one or more other routines can be activated as some locally stable results are achieved. A locally stabilized routine normally produces results to be used either as input to other routines or to end the process.

The five routines, assumed for the process representation, are named, respectively: Identification (R1), Structuration (R2), Control (R3), Development (R4) and Communication (R5). A characterization of them is given below by the domain, the main activities and some results.

Identification - R1

This routine has analogies with the "Identification phase" of Mintzberg et alii (1976), but here it is considered within the A/C Interaction. It is activated to recognize the problem situation in the main process, i.e. the organizational process, that generated the DA Process.

The principal domain of application of this routine is the main process, especially the decisional context.

The activities in R1 are mainly of inquiry, applied to catch C's motivations and possibly the function(s) of the intervention project. They can identify symptoms of dissatisfaction (if any) and/or nature of the (internal and external) stimula of

the proposed action (cf Roy, 1985).
The initial lack of formal representations or an oversimplified and highly structured demand can produce a high level complexity, not always perceivable by the analyst (A) at the beginning of the intervention.

One of the main results of R1 consists in a definition of a framework, allowing A to identify the operational context. It can help in the choice of a "particular angle of attack" (cf Landry et alii, 1983) so as to activate Relationship, Information and Value Systems with regard to the intervention project and to the recognized problem situation.

Structuration - R2

This routine is activated to make both the proposed action and, more generally, the actor's representations - that are in relation with the proposed action - operational. It can be seen as a synthesis of parts of the Intelligence and Design phases of the Simon's model (cf Simon, 1965).
This routine is present in almost all studies, since generally the main task of the DA Process is to test and develop the operational character of the proposed action. Sometimes this action, that has been made operational, is the only result of the intervention.

The main domains are the operational context and the decisional structure.
The principal activities in R2 are:
- exploration of lines of inquiry, sometimes identified by R1;
- search, activation and exploration of formal and informal information channels and relational structures;
- development and control of alternative ways of intervention in the operational context;
- direct and indirect search of alternatives to identify criteria, levels of aspiration, preferences, reference systems,...
By this routine, A tries to assess the logical validity (cf Landry et alii, 1983) of his/her answers; in other words he/she tries to give the model a recognized capacity of describing the problem situation, in relation to the identified actions, consistently with the actual operational context (cf Norese and Ostanello, 1984 a,b).
The use of supporting routines (especially of Communication and Development) is wide and determinant in defining the impact of R2 on the process. The way the supporting routines are activated, in terms of sets, sequences and action constraints, will characterize the purposes of this routine as more political or operational. R2 can indeed generate: modification of the demand; delays or temporary interruptions, with consequent changes in the operational context; breaks at the organizational level; interruption of the intervention.

Control - R3

This routine is the core of the validation process and aims to secure quality of information and stability of the representations. It is used both as a connection between routines with different kinds of logic and to identify and activate some following steps of the process or to induce local decisions.

A Control routine can be applied to two different domains:

- the cognitive-operational domain, for the purpose of understanding, verification and stabilization of the acquired or processed information ;
- the organizational-political domain, to "pilot" the interaction process (for instance, to enlarge the analyst's operational field and autonomy, to acquire legitimacy) and then the modelling process.

For cognitive and operational purposes R3 is activated to verify:

- pertinence of the collected and processed data (data validation);
- quality and efficiency of the data processing tools and of the solution methods (experimental validation);
- suitability of the language used to translate the conceptual model into a formal model (logical/experimental validation);
- pertinence and accuracy of the elicited and developed representations (logical/ conceptual validation) and of their usefulness with respect to the problem situation (operational validation).

Frequent activities of this validation process are:
evaluation and comparison of different and sometimes contrasting representations, to reduce ambiguity; evaluation and selection of operational context and information, of routines to activate, tools to use and results to communicate, to negotiate, to present to C.

For political-organizational purposes R3 is essentially activated as a supporting routine for instance to:

- inquire into the organizational context so to recognize the"true" nature of the problem and the real function of the intervention;
- find new resources, such as information sources, informal relationships,..;
- make evident the nature and state of the actual relationships;
- evaluate, choose and negotiate about the operational context and the developments of the project or possibly its interruptions;
- convince, coopt, defend, counter-attack, in cases of organizational conflicts that could generate risks of technical legitimacy loss.

Within this domain, Control routine is heavily supported by Communication.

Development - R4

This routine is similar to the Simon's Design phase (Simon, 1965), if activated as main routine. It is adopted to develop formal models of the alternatives and more often to produce formalized representations to help the stabilization of the main routines or simply to technically support Communication and Control, Identification and Structuration sub-processes.

The principal domain of action is the Information base which has been elaborated during the intervention.

The prevalent activities are strictly connected with the quality of the Information System -whose states can range between "preexisting and enforced" and "non-existing"- and the state of the alternatives to be developed. These alternatives indeed could have been recognized in R2 and can be in different states depending on the kind of interaction: only with C, with actors of the decision process, with other sources.

We distinguish the alternatives to be modelled as:

-"given or available", when they are partially developed and associated with a formal Information System, in the organization context, that can be partially, wholly or nothing usable;

-"existing, but not available"; they must be identified in the market "supply", on the basis of propositions and arguments of the actors,...; in this case a formal or informal Information System exists, at least partially, but it is not available in the organization context;

-"non-existing"; they must be elaborated both by design techniques and on the basis of the Information deduced from formal and informal sources.

Procedures of modelling (in the classical sense) are prevailing when the alternatives are "given". Some identified procedures to deal with other alternative states will be presented in the next Section.

Communication - R5

Communication can be assumed as a supporting routine (operational character). It may be considered as a principal routine when it is activated in strategic way, for instance to create consensus (political character).

The main domain of R5 is the operational context. The activities of R5 are essentially applied to: investigation, to find or confirm information; dissemination of information, stimula, provocations; explanation and clarification; influence, to reach consensus on formal representations or other organizational acts; justification; negotiation; persuasion.

The communication activities dominate every "phase" of the decision process (cf Witte, 1972); the role of communication is central to understanding and explaining "how groups make collective sense of their experience and how they come to take organized action" (Donnellon et alii, 1986). This role is indeed central to the A/C process, where otherwise no "shared-meanings" nor "shared-representations" (cf Norese and Ostanello, 1984a) would be reached and then neither routine nor process stabilizations would be achieved (cf De Sanctis and Gallupe, 1987).

3. DEVELOPMENT PROCEDURES

The Development routine (R4) has been identified as a step of the DA Process where a set of alternatives or, more generally, of actions is developed. By "action", we assume the B. Roy's definition: " Une action "a" est la représentation d'une éventuelle contribution à la décision globale susceptible, eu égard à l'état d'avancement du processus de décision, d'être envisagée de façon autonome et de servir de point d'application à l'aide à la décision (ce point d'application pouvant suffire à caractériser "a")"(Roy, 1985).

Adapting this definition to the DA Process, the concept of "action" can also be referred to the present state of the DA Process and thus to the present activated routine.

Then, within a routine, a representation of an actor - A, C or any other involved actor- can be an autonomous application point of A's activities and then can be an "action". It can assume different characters depending both on the evolution of the decision process and on the different "needs" and problems identified by R1 and/or made operative by R2.

Different action categorizations can be found in literature; an action can be, for instance: real or fictitius, realistic or unrealistic, ideal, of reference (cf Roy, 1985); concrete or abstract; given or available, existing, non-existing (cf Section 2, in this paper).

Development of actions helps in sensing, formulating and solving problems (cf Bourgine and Espinasse, 1987; Courbon, 1984) and it means formalizing abstract representations and making them stable and usable in particular contexts (of Identification, Structuration, Communication or Control).

Different structures of development processes can thus be produced where the Multicriteria Approach - more then other operational approaches - is very useful; their "shapes" can be profiled by combining routines of the Frame model.

3.1 Multicriteria Approach

The Multicriteria Approach (cf De Montgolfier and Bertier, 1978) is here intended above all as a supporting tool for the DA Process; it allows logics of connection between the development activities and the other routines to be outlined and their evolution to be controlled.

It constitutes then a useful representation tool suitable for reaching either satisfying levels of the developed actions or representations showing a certain kind of equilibrium in front of different conflicts (cf Roy, 1988), by different development processes. It allows, for instance:

- an arrangement of different representations (e.g.hypotheses, interpretations) into a formal scheme of direct linking between conceptual and formal models;
- abstract actions to be taken into account and developed, in a proper context, by relevant criteria of the context itself. In it the representations can be enriched and modified or stabilized and refined (Bourgine and Espinasse, 1987) and then these now more formalized actions can be used, for example, to explore either the action

space (action generation) or the criterion space in a different context (criteria generation).

The multidimensional representations generated by the Multicriteria Approach can be used by the analyst to support his/her activities in different contexts, both operational and decisional. The analyst's perception and representations can be modified, through the different interactions, however without substantial transformation of the model structure.

3.2 The identified procedures

A development procedure is a sub-process having an identified structure. Some procedures have been recognized and their "shapes" delined through the structuring of the routines; they have been called respectively: Evaluation, Identification and Elaboration. Their main features are identified in the "state" of the actions and of the operational context.

(Multicriteria) Evaluation

If the actions to be developed are "given or available", then the development activities apply to formalizing them in relation to the problem situation identified, generally, in R1 and to the operative constraints recognized in R2.
In this case the development process is not complicated; it is essentially a connection of two the basic routines R4 and R3.
If the Multicriteria Approach has been activated in the previous stages, for instance R1 and/or R2, then it can make a multidimensional development of the available alternatives easy and coherent (Multicriteria Evaluation, m(e)).

Identification

If the actions are "existing but not available", then they must be searched for and formalized by a more complex procedure. In this case the state of the actions, recognized and partially developed by other routines and procedures, is decisive in regard to the choice of both the approach and the development process.
a) Identification by search and screen, i(ss)
Mintzberg et alii (1976) define a "Search and Screen" routine to develop "ready-made solutions", by connecting two routines of the Development and Selection phases of their model.
When there are a large variety of existing alternatives (e.g. in the " market supply", in the organization environment) and the alternatives are not clearly "differentiated", then the development can become complex and require wide margins of time; it moves from one activity to another, to identify the operative context and explore it, to search for alternatives, to formalize and evaluate them, to identify and/or refine needs and potentialities and then to define a new set of actions, and so on, until the last "stable" set of developed actions is reached.

If this identification process is essentially devoted to recognizing criteria (of search, screen, representation and evaluation), then the Multicriteria Approach is winning.

b) Identification by multicriteria approach, i(ma)

It is adopted when the existing actions are "proposals" relating to the problem situation rather than "ready-made solutions".This development process is similar to the i(ss) for what concerns the structuring of the routines, but with different processed information and generally a different approach.

In this case indeed, the actions can be new and old proposals made by the actors involved in the decision process or, more generally, actor's representations concerning decision, problem formulations, possible solutions,...

i(ma) aims to recognize and formalize the dimensions of the problem, to distinguish between: individual and subjective points of view and collective and shared opinions, contextual solutions and more general ideas, forms of accordance, inclusion, conflictuality, incompatibility, incomparableness (cf Norese, 1988).

A standard shape of i(ss) and i(ma) is shown in Figure 1.

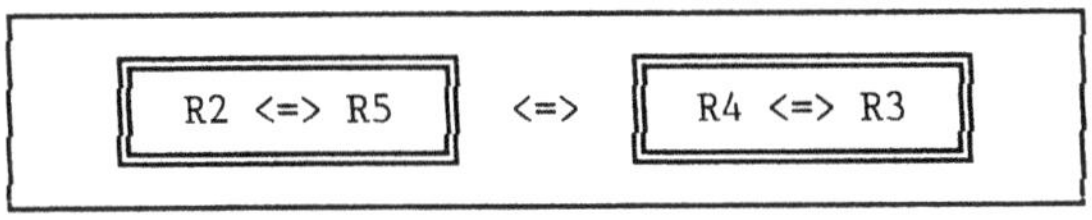

Figure 1. Standard shape of i(ss) and i(ma)

Elaboration

This development procedure is adopted if the actions are "non-existing" and at least one must be elaborated.

a) Elaboration by design, e(d)

An "Elaboration by design" is similar to the "Design Routine" of Mintzberg et alii. In these cases, the Multicriteria Approach can help in structuring and formalizing all the pieces of information that the analyst has searched for and recognized in previous routines.

b) Elaboration by multicriteria approach, e(ma)

It is especially activated in communication contexts. It helps in developing representations that must include and synthesize different elements - varying from structured data to partial analyses, verbal indications, informal interpretations - and in supporting activities of Communication and Control.

This process sometimes permits the actions to be elaborated on the basis of an iterative development of criteria that can be integrated, validated, destructurated or reformulated in a negotiation process so as to produce one or more shared actions.

A standard shape of e(d) and e(ma) is shown in Figure 2.

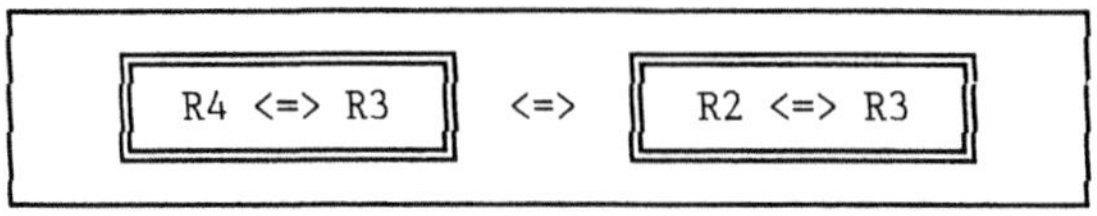

Figure 2. Standard shape of e(d) and e(ma), in Development or Communication contexts.

4. TOWARDS DECISION AID PROCESS TYPOLOGIES

A reconstruction of the research cases shows the near singularity of each DA Process. For instance, we did not even recognize a general common starting stage, nor a common ending stage (cf Nutt, 1984). Common general characters, essential for defining a typologization scheme, have however been recognized.

The evolution of the A/C Interaction and a greater or lesser complexity of the analyst's acting are strictly related to the complexity of both the technical/ operational and organizational/political situations the analyst is confronted with. These situations generate many different constraints on the analyst's action and induce him/her to adopt as much as possible pre-defined and structured operational instruments and behaviour, depending also on the analyst's experience.
Different "states" of the operational context and the decision structure ("Context situation") induce different critical levels of the analyst's work and can thus directly influence and condition the technical process. That determines hierarchization and recursivity of the routines.

An initial scheme to deline DA Process typologies has been defined; it correlates combinations of routines and of development procedures with identified context situations. A strict relationship between the nature of the Information System (IS) and the action development has been already pointed out (cf Section 3.). To proceed to a DA Process typology identification, we shall then characterize the context situations by: the state of the IS, the state of the decision context and of the main process.

For what concerns the IS, we take the following characterizing features: level of structuring of the available data (from "totally pre-defined" to "total absence of structured information"); information quality (from "clear, recognizable" to "confuse, ambiguous"); multiplicity of the sources.

For what concerns the main decision process and the decision context, the following factors are taken: multiplicity of levels of the action; conflictuality and organization stability conditions; margins of time; analyst's role and, consequently, margins of action in terms of autonomy, relationship with the decision structure.

Some DA Process basic typologies have thus been recognized and respectively named: "classical", "representation" and "action destructuration". Other typologies can be derived by combining routines, procedures and basic typologies.

Classical

This typology is characterized by: a high level of structuring of the problem formulation; a set of alternatives, given or easy to reach and to elaborate; the presence of a structured IS, with a unique information source.

Normally the starting routine is Development (R4). Identification and Structuration routines become inessential, for the presence of an action which has been formulated with a high detailing level and made operational,within other processes or at a different level of action, either by the Client or by others. This process is essentially coincident with the Evaluation procedure (cf Section 3.2).

Representation

This typology is the opposite of the previous one. It is indeed characterized by: non well-defined problem; absence of structured information with a multiplicity of information sources. A problem formulation and an operational character of the proposed action are the real objectives of the intervention.

In this case the starting conditions impose an inquiry oriented to exploit all the existing information channels and to activate new ones.

Thus a starting stage consists in a cycling of Identification and Structuration (IS) steps with iteratively activated Communication and Control (CC) routines; provisory results, formal first representations, are submitted for verification to the Client (negotiation of the problem terms) or to the actorial context.

At this stage, Development routine frequently appears to support other routines, by offering formal representations so as to stimulate, explore, control the acquired results, communicate, verify the pertinence and the completeness of the collected information, refine and negotiate the terms of the problem.

As a stabilization of the problem formulation is achieved, this process can end or induce a new stage (normally of action Development) or sometimes it can activate several new sub-processes (with new actors, new perceived problem situations,..).

A shape of this type of process is shown in Figure 3.

IS		CC
R1/R3 <=> R2/R4	<==>	R5/R4 <=> R3/R4

Figure 3. Representation process

Action destructuration

This typology is somehow in the midst of the two previous ones. It is recognized whenever initial assumptions taken as in the "classical" process, result ill-founded.

The starting routine of Development -widely supported by Development procedures and

Control routines - cannot be stabilized so that no operational character of the proposed action can be developed.

A Communication routine is thus activated to restructure the action. It is supported by formal representations previously developed in order to: point out the reduced coherence between the actual problem situation and the enforced conceptual model; induce some evolution within the problem representation; formulate hypotheses for the elaboration of a new IS.

This stage of the DA Process - that we call "Communication for Structuration" (CS) - aims to redefine the operational context so as to permit either the stabilization of the Development routine or, if impossible, the activation of a new stage of "Communication for Identification" (CI). This possibly activated stage marks an evolution of the process and acts as a connection between different levels of problem identification (for instance, between different action levels from technical/operational to organizational/political).

This stage can produce new Structuration and Development routines or it can activate a new process - based on a newly proposed action - or even to end the DA Process.

A shape of this kind of process is shown in Figure 4.

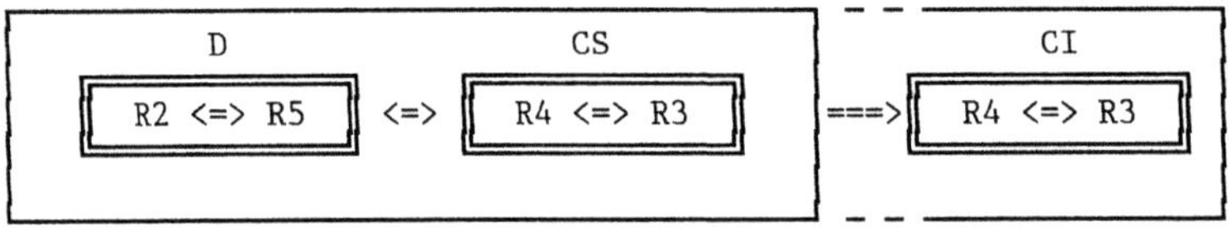

Figure 4. Destructuration process

5. CONCLUDING REMARKS

A Frame model has been proposed to represent the analyst's activities within decision aid processes. By it some procedures of action development have been reconstructed, pointing out the role played by different possible states of the proposed action and the potentialities offered by the Multicriteria Approach. These procedures show that action development is a recurrent task within the DA Process and seldom coincide with the process itself. Typologies of DA Process have also be deduced.

The necessity of integrating organizational and political dimensions in the process of model building, within actual organization contexts, is one of the main assumptions of our research. Decision aid process must not be confused with sequences of modelling activities (in a "rational" sense), although these are both relevant and essential for communication and expected by the client himself.

The analyst's need of adequate operational instruments has been recognized; more work has to be done to support him/her, confronting with decision aid situations made complex by the nesting of technical, organizational and political problems. Our research is leading in that direction.

A more elaborate scheme to profile DA Processes is being developed; it will permit the explicit correlation of routines and development procedures with context situations, distinguishing between individual (of the analyst) and collective (analyst/client) levels of activities.
It is mainly at the collective level that the Multicriteria Approach, conceived as an operational instrument, shows its usefulness.

REFERENCES

Bourgine, P., and Espinasse, B. (1987), "Aide à la décision : une approche constructiviste", in: AFCET, Aide à la décision dans l'organisation, Paris, 47-55.
Checkland, P. (1981), Systems Thinking, Systems Practice, Wiley, New York.
Courbon, J.C. (1984), "Transparency of data, information and models in Decision Support Systems", in: J.P.Brans (ed.), Operational Research '84, North-Holland, Amsterdam, 90-102.
De Montgolfier, J., and Bertier, P.(1978), Approche multicritère des problèmes de décision, Editions Hommes et Techniques, Paris.
De Sanctis, G., and Gallupe, R.B. (1987), "A foundation for the study of group Decision Support Systems", Management Science 33, 5, 589-609.
Donnellon, A., Gray, B., and Bourgon, M.G. (1986), "Communication, Meaning and Organized Action", Administrative Science Quarterly 31, 43-55.
Grauer, M., Thompson, M., and Wierzbicki, A.P. (eds.) (1985), Plural Rationality and Interactice Decision Processes, Lecture Notes in Economics and Mathematical Systems, Springer-Verlag, Berlin, 248.
Landry, M., Malouin, J.L., and Oral, M. (1983), "Model validation in Operations Research", European Journal of Operational Research 17, 207-220.
Mintzberg, H., Raisinghani, D., and Théorêt, A. (1976), "The structure of the unstructured decision process", Administrative Science Quarterly 21, 246-276.
Moscarola, J. (1984), "Organizational decision process in ORASA intervention", in: R.Tomlinson and I.Kiss (eds.), Rethinking the Process of Operational Research and Systems Analysis, Pergamon Press, Oxford, 169-186.
Norese, M.F., and Ostanello, A. (1984a), "Un cas de modélisation partagée pour un problème de décision collective", in EIASM: Actes 19èmes Journées du Groupe EURO sur " Aide à la Décision Multicritère", Liège, mars 1984.
Norese, M.F., and Ostanello, A. (1984b), "Planning processes and technical interventions: an integrated approach", Sistemi Urbani 2, 247-259.
Norese, M.F. (1988), "A multidimensional model by a multiactor system", in: B.R.Munier and M.F.Shakun (eds.), Compromise, Negotiation and Group Decision, Reidel, Dordrecht, 263-276.
Nutt, P. (1984), "Types of organizational decision process", Administrative Science Quarterly 29, 414-450.
Roy, B. (1985), Méthodologie Multicritère d'Aide à la Décision, Economica, Paris.
Roy, B. (1988), "Why use multiple criteria in Operations Research ?", in: Operational Research '87, G.K.Rand (ed.), North Holland, Amsterdam.
Simon, H.A. (1965), The Shape of Automation, Harper and Row, N.Y.
Schneeweiss, Ch. (1987), "On a formalization of the process of quantitative model building", European Journal of Operational Research 29, 24-41.
Witte, E. (1972), "Field research on complex decision making - The phase theorem", International Studies of Management and Organization, 156-182.

ON-LINE GROUP DECISION SUPPORT

Tony Gear and Martin Read
Decision Dynamics Limited
Parsons Walk
PO Box 142
Wigan, Lancs, WN1 1RW

"..........., *social innovation has clearly become management's new dimension*".

(Peter Drucker, 1986)

ABSTRACT

The paper describes an on-line system which enables each member of a decision-making group to communicate individual decisions and judgements during the course of discussion and be provided with immediate processed feedback information. Some early and very promising trials of the system concept are summarized, and directions for further development are outlined.

INTRODUCTION

The application, in modern organisations, of management by objectives (MBO) and information technology (IT), has enabled the maintenance of reasonable levels of overall productivity. However, this process leads to individual specialisation, in skill and knowledge terms, and exacerbates the problems of effective communication between people in small groups who meet for purposes of decision-making, problem-solving or idea generation.

To counter these communication difficulties, leading to lowered morale, many organisations have introduced techniques such as team briefing and team-building. It is now timely to see if information technology, in some suitable form, can also aid the inter-personal processes operating in small groups and teams meeting for various tasks.

Recent research findings support the contention that the acceptance of group goals enhances group performance significantly, as does the provision of task feedback information related to performance, eg. Matsui et al (1987). The findings from this study suggest that effectiveness of task feedback in group goal setting will be maximised if the feedback involves not only individual, but also group, performance information. The experiments for this study were for the simplest form of group (ie. two-member teams). However, there is every likelyhood that the benefits will be at least as significant for larger groups. There have been a number of demonstrations in the laboratory and in the field to show that feedback enhances performance on the simplest tasks (ie. single criterion activities). Again, there seems every likelyhood of improvements on more complex multi-dimensional tasks. Locke et al. (1981) has reviewed results in this field. Ilgen (1987) found that individuals receiving information about performance on tasks demanding quality output under time pressures outperformed groups who only received feedback on one measure, speed or quality.

This paper is concerned with the consequences of structuring feedback to a group working as a team on some complex task. The contention is that by providing the team with an on-line means of displaying feelings and/or performance, that this will considerably augment unstructured feedback systems, with the overall result of increasing the performance of the group.

The paper builds upon the work of Gear, Marsh and Sergent (1985). They provided each team member with hand-held hardware units to record individual feelings, linked to a central processor which would feedback and display group feelings during a meeting. Analysis of a trial of this system strongly supported a hypothesis that the feedback of group feelings was influencing subsequent group behaviour to a highly significant degree. Further, there appeared to be increased levels of such factors as: commitment to decisions, satisfaction with the group, cooperation, trust and perceived influence. It was felt that this trial warranted further development of hardware and software to form a group decision support system (GDSS) that could be tested on groups with real decision problems.

Experimental applications of various methodologies to problems of decision making in groups have been reported, eg. Gear, Lockett and Muhlemann (1982). They tend to suffer from a mis-match between the speed at which groups would like to request feedback at interim stages of the process, as compared with the capability of the hardware and software technology actually available.

In the following sections, a novel approach is described which allows groups to function at their own rate, with information flows to, and from, the support system. Applications of the system to a number of real-life problems in group situations are summarized and conclusions drawn.

FEEDBACK MODEL

The system comprises a set of hand-held devices, one for each group member.

Signals from each unit are transmitted to a remote receiver and fed into a micro-processor. The signals are analysed in accordance with the decision model which is being used. Processed information is displayed on a screen to each group member. The screen is used to :

- request information from each group member via a hand-held unit during the group process
- display interim and final processed information
- provide expert advice and menu options during the group process

APPLICATIONS

Experiments have been conducted in two key areas:
process support to help a group discuss an issue, solve a problem, or generate ideas or options
decision making to help a group rank or select from a list of options in order of preference

Process Support

Periodically, each group member is asked (by the display) to record on their hand-held unit a response to the question "how well do you feel the group is progressing on the task?" Responses are entered by the selection of one of a set of numeric keys from 1-9 on each hand unit.

The scale used is:

1	:	negligable progress
3	:	
5	:	reasonable progress
7	:	
9	:	excellent progress

The scores selected by each group member are used to form group average and range information, which are displayed to the group. Previous results are also displayed so that trends are also clearly apparent. The group can then utilise this information in its subsequent work.

Decision making

In this application, details of the number of group members, number of options, and number of attributes involved, are first keyed into the micro-processor. Tests to date have involved the use of scoring and pairwise comparative methods (for a detailed description of the latter see, for example, Saaty (1980)). As we are centrally concerned with the enhancement of group processes through the use of the support system with feedback, issues relating to possible shortcomings of these models are largely irrelevant to this paper.

At each stage of the process, group members respond to questions appearing on the display. During the process the group is kept informed, via screen feedback, of the results of mathematical analysis of individual consistency of response and of differences of opinion.

Also, at each stage a menu of options is displayed, allowing the group leader or chairman to opt to repeat a stage, continue the process, or seek more detailed display information. These options are called via this person's hand unit.

The display also advises the group regarding a recommended menu option to select, using expert opinion pre-programmed into the software. In this sense, the system is operating as an expert group decision support system.

The general structure of the software model is shown in figure 1.

FIELD TRIALS

A number of applications of the system have been undertaken working with management groups, and involving live issues rather than artificial ones. The objective was to identify problems and record perceptions of the usefulness of the online guidance and feedback.

These experiments are now summarized in anonymous form:

TRIAL 'A'

This involved a group of 5 of the directorate of a Polytechnic. The problem involved the method of introduction of a management information system. The options are shown below:

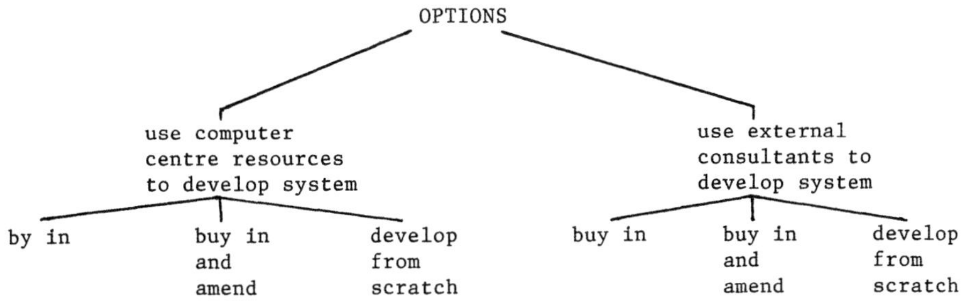

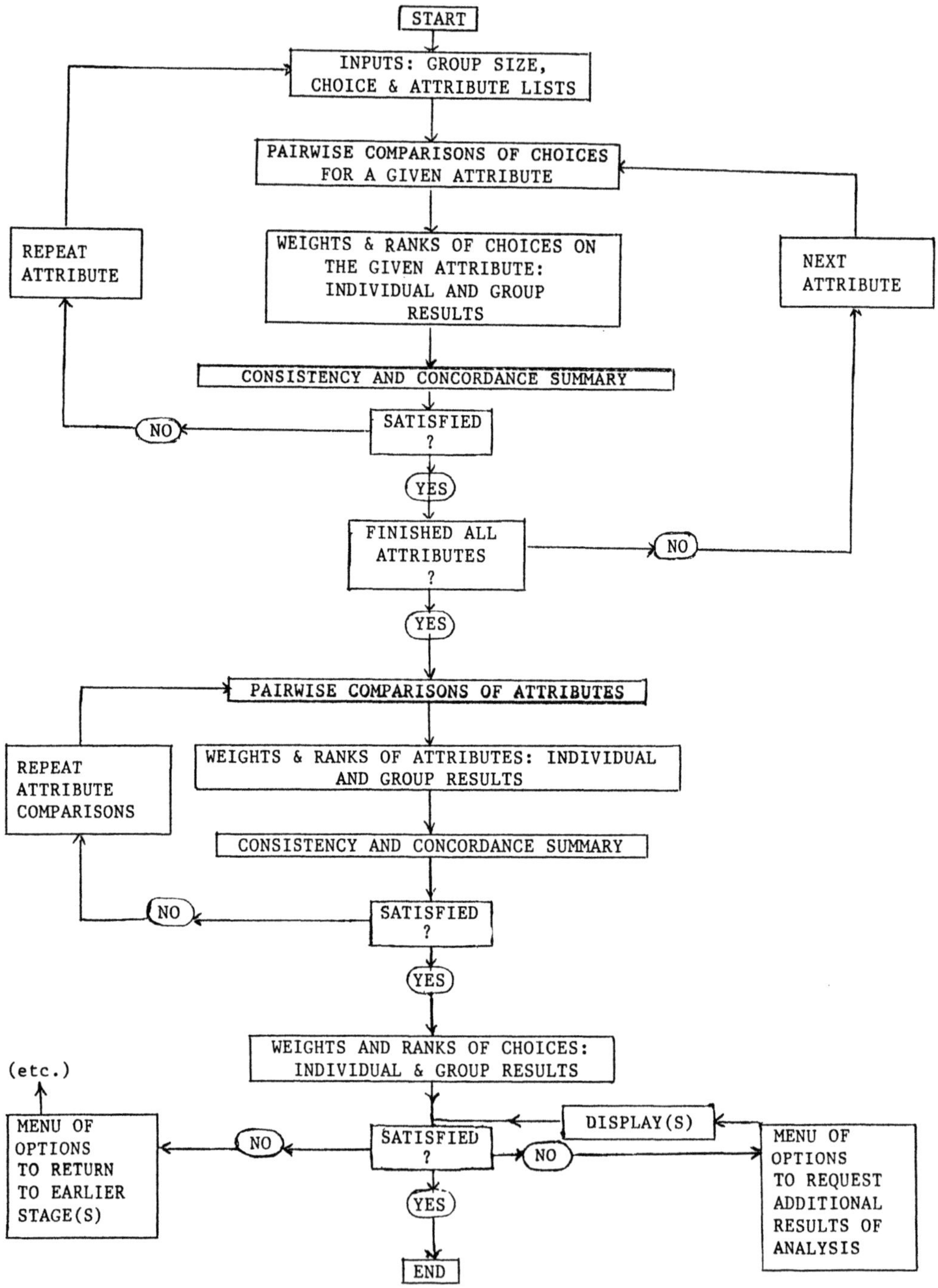

Figure 1: Structure of Software (Summary)

The criteria under consideration were:

- cost
- control of system
- quality of system
- speed of system
- maintenance charges
- staffing requirements (number)
- staffing requirements (conditions of service)

After a first run of the system some of the criteria were seen to be unimportant. A second run then resulted in a considerable degree of consensus for one preferred option. Total run time was 2½ hours, including the definition of options and criteria.

TRIAL 'B'

This involved a group of 6 managers of regional business innovation centres (BIC's) in the North of England. The decision involved the siting of a new satellite business innovation centre in the North. Four options based on town names were short-listed. The attribute list was :

- level of usage
- environment
- agency link
- facilities

A single run of the system involving a run time of 1½ hours was sufficient to achieve considerable consensus for one of the locations.

TRIAL 'C'

This involved a group of 6 store managers of retail outlets of a Hi-fi and TV chain in the North of England, concerned with a series of market analyses.

The first issue selected by the group for analysis was to compare the brand names of a number of major manufacturers of Hi-fi equipment, not all stocked by the chain. A holistic set of pairwise comparisons gave high consistency and good agreement in ½ hour.

A more detailed analysis was then undertaken of the same brand names in Hi-fi, on each of a number of attributes, listed as follows:

- visual appeal
- brand loyalty
- price
- performance

At this stage differences of opinion emerged and became the subject of intensive and sometimes heated discussion, with some re-runs of sub-sets of comparisons.

The group then decided to compare their judgements of customers' perceptions of their chain with the chains of major competitors. After some discussions, 5 chains, including their own, were selected. These were first analysed holistically, and then at the attribute level, defind as follows:

- price (personal perception)
- price (customers' perception)
- level of service
- customer loyalty
- store atmosphere

Agreement was high for the holistic analysis. At the attribute level, agreement was not always high, and resulted in some re-runs of stages, including intense debate. The comparisons of the relative importance of the attributes as factors contributing to a chain's image resulted in considerable differences of opinion. These were highlighted on the screen display, showing some clusterings of opinion, leading to a lively debate. The differences of opinion came as a surprise to the group.

Overall, the meeting lasted for 4 hours, and the group were keen enough to run through the lunch period. The market reasearch consultancy firm, on whose behalf the application was undertaken, stated that both they and their clients were very pleased with the insights gaind and results of the analysis.

IMPLICATIONS

The experiments in cognitive feedback with working groups have resulted in a great deal of support for the project. Some quotations from professionals in a variety of managerial and educational fields are listed below:

John Harter, Management Consultant, MANTRA.
"I think this is a very exciting possibility, and I look forward to working with you ....".

Brian Booth, Deputy Director, LANCASHIRE POLYTECHNIC.
"Your system .... should have considerable commercial potential".

Barry Davies, Senior Lecturer, School of Retail Marketing, MANCHESTER POLYTECHNIC.
"The system .... is able to make a major contribution to management decision-making".

Eric Davies, Managing Partner, SALES RESEARCH AND DEVELOPMENT, Market Research Consultants and Management Training.
"We believe this is an exciting project and can envisage a nationwide network of marketing research companies offering this service to clients".

Brian Yeomans, Principal Lecturer in Education Studies, TRENT POLYTECHNIC.
"I can see tremendous potential for the system in educational contexts, if it were developed to meet the sorts of needs outlined".

John Barnard, Director, Human Resources Division, P-E INBUCON, Management Consultants.
"..........tremendous potential as a teaching and research tool in a broad range of fields, notably management".

Andrew Monger, Manager, Long Term R & D Studies, BRITISH GAS.
"I consider that this is a valuable tool for quantifying subjective preferences and would find a range of further applications".

John Gillespie, PROJECT SOFTWARE INC, NY. USA. Management Consultants.
"all of us here believe that the device you are developing has a broad range of applications and is such a good idea that, in hindsight, it seems amazing that no-one has thought of it before."

Nick Marsh, Director, ALKEMI, Auckland, New Zealand. Management Training and Consultancy.
"I agree with you that there are large possibilities for the concept internationally."

It is clear that sufficient results have been generated to warrant the development of theory to explain the effects of feedback and to proceed to an improved research design to test these effects. The ad hoc trials undertaken to date represent anecdotal case reports, nevertheless they do represent an important first stage in any new development. The type of studies which are now required and aspects to consider have been defined in Gear et al (1985).

CONCLUSIONS

There is no doubt in the minds of the authors, as a result of the applications outlined in this paper, and a number of other trials, that the on-line means of providing guidance, supplying judgements, and receiving processed feedback, greatly enhances the group's progress. Furthermore, groups appear to enjoy using the system, and appreciate the speed and ease with which problems can be set up, run, modified and re-run, in timely response to need. This may also be related to the individual autonomy and equalization of judgemental inputs that the system allows.

The system concept is very flexible. Other applications, involving group discussion and debate for their resolution, can easily be handled with appropriately tailored or customized software. Alternative decision or behavioural methodologies can be utilised, as desired, by the use of appropriate software. The system is adaptable to most micro-processors, by simply plugging into the serial port.

Applications under active development by the authors include :

- decision making in groups
- management training in groups
- classroom support in education
- business and leisure games
- therapy groups
- market research

The relevance of the on-line system concept to improve group performance has been amply demonstrated. Interested parties with one of the above or other applications in mind, or concerned to undertake further research and development of the concept, are invited to communicate directly with the authors.

REFERENCES

DRUCKER, PETER 1986 "The Frontiers of Management"
P354 last sentence of book.
Heinemann, London

GEAR, TONY E., LOCKETT, ALAN GEOFFREY, MUHLEMANN, ALAN DAVE (1982)
"A Unified Approach to the Acquisition of Subjective Data in R&D."
IEEE Series on Engineering Management.
Vol. EM-29 No.1. pp. 11-19

GEAR, TONY E., MARSH, NICHOLAS R., SERGENT, PETER (1985)
"Semi-Automated Feedback and Team Behaviour."
Human Relations, Vol 38. No.8. pp. 707-721

ILGEN, DANIEL R., MOORE, CAROL F (1987)
"Types and Choices of Performance Feedback"
J of Applied Psychology Vol. 72 No.3
pp 401-406

LOCKE, EA., SHAW KN., SAARI LM., LATHAM GP. (1981)
"Goal Setting and Task Peformance: 1969-1980
Psychological Bulletin 90, pp 125-152

MATSUI, TAMAO,. KAKUYAMA, TAKASHI,. ONGLATCO, MARY LOU UY (1987)
"Effects of Goals and Feedback on Performance in Groups." J of Applied Psychology[2]
Vol 72 No.3 pp 407-415

SAATY, TL (1980)
"The Analytic Hierarchy Process"
McGraw-Hill, New York

The Effect of Decision Maker Characteristics on Group Decision Making

Lori S. Franz
College of Business Administration
University of Missouri
Columbia, MO 65211 USA

Gary R. Reeves
College of Business Administration
University of South Carolina
Columbia, SC 29208 USA

Juan J. Gonzales
College of Business Administration
University of Texas at San Antonio
San Antonio, TX 78285 USA

Interactive group decision making models assume that rational and predictable decision makers will provide inputs to the decision process. Unfortunately, we know that all decision makers will not act in accordance with such assumptions. It is intuitively obvious that groups with one or more inflexible decision makers will have more difficulty achieving consensus than groups consisting of flexible decision makers. Likewise, indecisive or inconsistent decision makers may have difficulty attaining consensus. How decision maker characteristics effect interactive group decision making systems is unclear. Whether individual differences in decision type will interfere with the convergence of the interactive approach is unknown, however, one would expect some differences in performance among the various interactive group approaches.

Many interactive computer guided approaches to the group multiple objective decision problem have been suggested in the MCDM literature. Numerous choices exist for decision facilitators who wish to select an interactive group decision making approach. Among the promising approaches are those offered by Franz, Reeves, and Gonzales (1986); Korhonen, Wallenius, and Zionts (1982); Tell (1978); and Thiriez and Houri (1976).

The choice of preferred decision aiding approach should be determined by the problem type and the characteristics of the users. User characteristics which should be considered include both the willingness of the decision makers to meet the time, technology and input demands of the interactive approach and the users' personality characteristics. Assuming an appropriate problem for the procedure and a group of users who are willing to interact with the system in the way prescribed, the choice of approach may be dependent on knowledge of the approach's performance under certain behavioral assumptions. Very little is known, however, about the characteristics of the available group decision models under use conditions with other than a set of rational and equal decision makers. Clearly, the operating characteristics of an approach may be entirely different with a set of rational decision makers than with a group consisting of one or more decision makers exhibiting behaviors which would be classified as inconsistent or inflexible. Although it may be unrealistic to assess groups for decision maker types before the solution approach is chosen, all approaches should be tested to see that they continue to operate in a manner consistent with rationality assumptions when behavioral variables are introduced.

This paper proposes a methodology to assess the effect of different decision making characteristics on the interactive group decision process. The evaluation procedure will be illustrated via a test of the multiple decision Maker SIMOLP method (Franz, Reeves and Gonzales, 1986). The group SIMOLP procedure will be identified henceforth as GSIMOLP. Rational DMs will be simulated with various utility functions with respect to test problems. Intensity of preference measures simulated from the DMs utility functions will be used as a basis for studying the group process. Decision maker characteristics of inflexibility and inconsistency will be examined by simulating operations consistent with the decision maker characteristics. It will be demonstrated that data can be collected on the convergence characteristics of the methodology under varying decision maker assumptions for a set of established test problems.

DECISION MAKER CHARACTERISTICS

Two types of decision makers which have characteristics impacting on the convergence of any group decision process will be considered initially. They are the inconsistent decision maker, and the inflexible decision maker. Since the achievement of consensus is dependent upon the participation and agreement of all group members, these characteristics, which may impede the progress of a given approach toward consensus, will be considered first.

<u>The Inconsistent Decision Maker.</u> It is commonly accepted that decision makers may not maintain consistent objectives or preferences for outcomes during the decision process. The reasons cited are manyfold. Johnson (1976) states that objectives are not constant; rather they change to reflect new situations and new awareness of problem ramifications. This would be somewhat expected in any

iterative group decision process, where learning would be assumed to occur.

The Inflexible Decision Maker. The nature of multiobjective problems requires that several viewpoints toward the solution exist. The introduction of multiple decision makers suggests that some decision makers may hold biased views toward certain aspects of the problems. For instance, multiple decision makers might consist of vice presidents of marketing, production, finance and so forth. With such a group of individuals, compromise solutions could be hampered by an individual with a single dimensional view of the problem. Inflexibility can also be introduced in cases where a particular group member is "self oriented". Here the member may respond to a problem with unwarranted rigidity and may persist in a stereotyped response referred to as fixation (Patton and Griffin, 1978). Either of these responses where an unwillingness to compromise exists, would define an inflexible decision maker.

METHODOLOGY

The interactive multiple objective multiple decision maker model (GSIMOLP) of Franz, Reeves and Gonzales (1986) will be used to demonstrate the analysis of effect of DM characteristics on the operation of the methodology. This model is based on the Simplified Interactive Multiple Objective Linear Programming (SIMOLP) method of Reeves and Franz (1985) which has been modified to allow for consideration of compromise solutions generated by multiple DMs. SIMOLP has the advantage of generating a set of increasingly similar solutions to facilitate the decision making process. Additionally, SIMOLP requires no special software, only an existing LP code. In order to take into account the relative importance of various objectives as seen by the DMs, the ordinal ranking procedure of Cook and Kress (1985) is used to drive the solution toward a consensus.

Consider the following MOLP problem:

$$\begin{aligned} &\max(c^1x = z_1)\\ &\max(c^2x = z_2)\\ &\quad\vdots\\ &\max(c^kx = z_k)\\ &\text{s.t.}\quad x \in S \end{aligned}$$

where $S = (x \in R^n \quad x \geq 0,\ Ax = b,\ b \in R^m)$. There are L decision makers involved in the decision process.

STEP 1: Generate initial solutions. Solve the k single objective linear programming problems

$$\max z_i$$
$$\text{s.t. } x \in S$$

to obtain k efficient points, x^i, i = 1,2,...k, to the original problem and their associated images, the k nondominated criterion vectors, z^i, i = 1,2,...k, respectively. Let E* = (x^i, i = 1,2,...k) and N* = (z^i, i = 1, 2, ...k).

STEP 2: Generate the compromise solution. Set i = i + 1. Form the hyperplane, z_i, which passes through the k elements of N*. Solve the LP problem

$$\max z_i$$
$$\text{s.t. } x \in S$$

to obtain the efficient point, x^i, and its associated image, the nondominated vector z^i. This point serves as a compromise solution for the DMs to evaluate. Form the set N** where N** is the union of N* and z^i.

STEP 3: Generate the group consensus ranking. Have each DM examine the members of N**, ranking them ordinally, using the Cook and Kress procedure to denote intensity of preference. Form a consensus ranking from the ordinal ranking of the individual decision makers using the assignment methodology.

STEP 4: Drop the group's least preferred solution. Store the top k members of N** in N*.

STEP 5: Go to STEP 2.

The process repeats until a consensus is reached or until the process generates no further compromise solutions for consideration. A consensus is considered to have been reached when all DMs agree on the first solution in the ranking. If the point is reached where the hyperplane no longer identifies new compromise solutions, then a consensus ranking of the remaining points is used to determine the group preferred solution. Specifics on determining the hyperplane weights, clarification of the hyperplane generation procedure, and consensus ranking procedure can be found in Reeves and Franz (1985) and Franz, Reeves and Gonzales (1986).

To form a consensus ranking from the ordinal rankings of the decision makers, an adaptation of the ordinal ranking with intensity of preference procedure of Cook and Kress (1985) is employed. The ordinal ranking procedure establishes the existence of a unique distance function which measures the disagreement of the compromise or consensus ranking from the individual rankings of the group members.

An approach to developing a consensus ranking for use in guiding multiple decision makers toward solutions to multiple objective problems can be derived through an extension of the assignment method. Cook and Seiford (1978) showed that a consensus ranking (without intensity of preference) could be established for a problem in which m committee members have given ordinal rankings of n objects. The consensus can be derived by solving

$$\min Z = \sum\sum d_{ij}x_{ij}$$

subject to:

$$\sum x_{ik} = 1 \quad \text{for all } k,$$

$$\sum x_{ik} = 1 \quad \text{for all } i,$$

$$x_{ik} \geq 0 \quad \text{for all } i,k.$$

where d_{ik} is the sum of deviations between a selected ranking k and the m given member rankings of object i. Cook and Seiford establish that the distance function is unique and has the property of maintaining transitivity. They note that the assignment method provides a simple method which can be used to solve problems with complete rankings, ie. no ties. Franz, Reeves and Gonzales (1986) showed that the methodology can easily be extended to include both of the above possibilities by expanding the assignment matrix of Cook and Seiford (1978) to include more columns. Dummy rows can be added to balance the problem and enable its solution.

<u>The Assumption of a Rational DM.</u> Initial testing of the GSIMOLP model assumed a set of rational DMs. This testing was accomplished by generating an unchanging utility function for each DM. The more similar the group utility functions, the faster the methodology to converge upon an group compromise solution. Previous testing showed that with a set of decision makers with dissimilar utility functions, a group compromise solution was generated within four and eight iterations for 90% of the test problems. In only one out of ten test cases, did the GSIMOLP methodology fail to identify an acceptable group compromise, forcing the decision makers to rely on a group consensus ranking at that point (Franz, Reeves and Gonzales, 1986).

To consider the impact of an inflexible DM, one of the DMs will be randomly chosen as inflexible. For this case, that DMs highest weight will be retained, with all other weights for that DM set to zero. The test problems will be rerun and the number of iterations until group compromise is reached will be used as a measure of comparison.

The inconsistent DM will be modeled as an individual who oscillates between two utility functions. The inconsistent DM will be randomly selected and his or her original utility function retained. Additionally, a second utility function will be generated for this individual, based on a transformation of the original function. The utility functions will alternately be activated for each iteration of the solution procedure. The number of iterations to achieve group consensus will be used as the comparison measure.

To test the sensitivity of the GSIMOLP model to behavioral variations in DMs, two variations of the original model will be evaluated with the same DM assumptions as described above. From this, the sensitivity of changes in methodology to DM characteristics will be assessed. The procedural variations for a group of DMs are as follows:

<u>GSIMOLP2: GSIMOLP Procedure Variation 1</u>.

Step 1. Generate initial solutions.
Step 2. Generate compromise solution.
Step 3. Generate individual DM rankings.
Step 4. Have each DM drop his or her least preferred solution.
Step 5. Generate compromise solution for each DM.
Step 6. Have each DM rank all active solutions, his/hers + those of the other DMs.
Step 7. Generate a consensus ranking of all solutions, having the group retain the k most preferred solutions.
Step 8. Go to Step 2.

<u>GSIMOLP3: GSIMOLP Procedure Variation 2</u>.

Step 1. Generate initial solutions.
Step 2. Generate initial compromise solution.
Step 3. Generate individual DM rankings.
Step 4. Have each DM drop his or her least preferred solution.
Step 5. Generate compromise solution for each DM.
Step 6. Have each DM rank all active solutions, his/hers + those of the other DMs.
Step 7. Have each DM retain his/her k most preferred solutions.
Step 8. Go to Step 5.

<u>The Impact of DM characteristics on Preferred Methodology</u>. The original GSIMOLP methodology and the two variations above exhibit different orientations toward the individual decision maker versus the group as a whole. It can be seen that the original procedure allows only one group decision path to be explored and offers individual DMs little flexibility in exploring options. It is hypothesized that this

methodology, which limits the DM the most, will be affected by the DM characteristics the least.

Variation 1, GSIMOLP2, provides some flexibility to the DM to explore different decision paths. Some movement toward group consensus is required at each iteration. It is hypothesized that the presence of a indecisive DM will hasten the convergence of the procedure, while the presence of an inflexible DM will have little impact on the solution procedure.

GSIMOLP3, the second variation, offers the maximum opportunity for individual DMs to explore different decision paths, with no forced movement toward group consensus. It is hypothesized that this methodology will be greatly impeded by the presence of DMs which do not fit the rationality assumptions.

RESULTS

Computational results with each of the procedures were derived by analyzing ten randomly generated multiple objective linear programming test problems (Steuer, 1983) run with each of the three procedures outlined above. Each problem and procedure was tested against ten variations in DM characteristics as described above, including a set of simulated rational decision makers, a set of rational decision makers with more similar utility values, and variations which chose each of four different decision makers as alternately inflexible and inconsistent. This generated a 300 problem solution set from which an initial analysis was performed. From this problem set, preliminary conclusions regarding the sensitivity of the GSIMOLP procedure and its variations to DM characteristics can be drawn. The analysis performed assumed that there was at most one decision maker in the group who was either inflexible or inconsistent. Interactions between decision maker characteristics were not tested, nor were the impact of variations of utility function values on the DM characteristics investigated.

All three procedures of GSIMOLP terminate when the methodology fails to generate new consensus decisions for consideration, yielding a subset of solutions which are acceptable to the group. The assumption is that the decision makers will select the solution in accordance with their consensus ranking or negotiated solution from the subset of solutions identified as acceptable in that final iteration. For this reason the measure of number of iterations to develop the final subset of solutions is used as an indicator of each procedure's efficiency. The mean iterations for each set of ten test problems for each of the conditions is presented in the table below.

Mean Iterations to Process Termination

	GSIMOLP	GSIMOLP2	GSIMOLP3
Rational DMs	6.1	5.6	5.9
Rational DMs with less divergent pref	6.3	3.6	6.8
Inflexible DM1	5.3	3.6	6.8
Inflexible DM2	5.6	4.5	6.4
Inflexible DM3	5.7	3.8	6.7
Inflexible DM4	5.1	5.5	6.0
Inconsistent DM1	6.4	6.0	6.9
Inconsistent DM2	5.2	4.1	7.0
Inconsistent DM3	6.4	4.4	7.9
Inconsistent DM4	6.6	4.5	7.6

From the above data it can be seen that both variations of GSIMOLP converge on the final set of solutions in less iterations on the average than did the original GSIMOLP methodology. It does not appear that the procedures will necessarily converge in less iterations when there is less variance in the way that decision makers specify their preferences for the possible outcomes, although this appeared to be the situation in the test of GSIMOLP2. In cases where an inflexible DM is part of the group process, the first variation, GSIMOLP2, and the original GSIMOLP appear to have advantages over the second variation in terms of iterations required. This shorter time to convergence and lessened interaction of DMs with the procedure would have to be evaluated against the loss of individual orientation which is offered by GSIMOLP3. The ability to maintain more individual control over the set of considered alternative may outweigh the time required for some decision makers.

In cases where an inconsistent decision maker participates in the group process, the first variation, GSIMOLP2, appears considerably more attractive in its ability to reduce the set of solutions to an acceptable one in relatively few iterations.

This initial evaluation of the three GSIMOLP procedures has shown that all of the procedures are relatively insensitive to the impact of the tested decision maker characteristics on the time to convergence. Results suggest that while a viable set of compromise solutions can be identified by all of the procedures, an analyst may wish to consider the use of a particular methodology based on his or her beliefs about the types of individuals in the group and the time they are willing to devote to the process.

REFERENCES

Armstrong, R. D., Cook, W. D., and L. M. Seiford, "Priority Ranking and Consensus Formation: The Case of Ties," Management Science, Vol. 28, No. 6, 1982, pp. 638-645.

Cohon, J. L., Multiobjective Programming and Planning, Academic Press, New York, 1978.

Cook, W. D. and M. Kress, "Ordinal Ranking with Intensity of Preference," Management Science, Vol. 31, No. 1, 1985, pp. 26-32.

Cook, W. D. and L. M. Seiford, "Priority Ranking and Consensus Formation," Management Science, Vol. 24, No. 16, 1978, pp. 1721-1732.

Franz, L. S., Elicitation of Decision Maker Preference Structure through Interactive Goal Programming, unpublished Ph.D. thesis, University of Nebraska, 1980.

Franz, L. S., Reeves, G. R., and J. J. Gonzales, "An Interactive Approach to Multiple Objective Multiple Decision Maker Problems," in Toward Interactive and Intelligent Decision Support Systems, (Y. Sawaragi, K. Inoue and H. Nakayama, eds.), Springer-Verlag, Berlin, 1986, pp. 172-181.

Geoffrion, A. M., Dyer, J. S., and A. Feinberg, "An Interactive Approach for Multicriteria Optimization with an Application to the Operation of an Academic Department," Management Science, Vol. 19, 1972, pp. 357-368.

Johnson, E., "Experiences in Multiobjective Management Processes," in M. Zeleny (Ed.) Multiple Criteria Decision Making, Springer-Verlag, New York, 1976, pp. 135-152.

Korhonen, P. J., Wallenius, J. and S. Zionts, "Two Interactive Procedures for Multicriterion Optimization with Multiple Decision Makers," University of Jyvaskyala, Working Paper 14, 1982.

Patton, B. R. and K. Giffin, Decision-Making Group Interaction, Harper and Row, New York, 1978.

Reeves, G. R. and L. S. Franz, "A Simplified Interactive Multiple Objective Linear Programming Procedure," Computers and Operations Research, Vol. 12, No. 6, 1985, pp. 589-601.

Steuer, R. E., Set of MOLP Test Problems, Private Communication, January, 1983.

Tell, B., "An Approach to Solving Multi-person Multiple Criteria Decision Making Problems," in Multiple Criteria Decision Making, (S. Zionts, Ed.), Springer-Verlag, Berlin, 1978.

Thiriez, H. and D. Houri, "Multi-person Multi-Criteria Decision-Making: A Sample Approach," in Multiple Criteria Decision-Making (H. Thiriez and S. Zionts, Eds.), Springer-Verlag, Berlin, 1976.

Zionts, S. and J. Wallenius, "An Interactive Programming Method for Solving the Multiple Criteria Problem," Management Science, Vol. 22, 1976, pp. 652-66.

GESTALT SYSTEM OF HOLISTIC GRAPHICS: NEW MANAGEMENT SUPPORT VIEW OF MCDM

Eero Kasanen
Department of Economics, University of Oulu,
90101 Oulu, Finland

Ralf Östermark
Department of Business Administration, Åbo Academy, Henriksgatan 7,
20500 Åbo 50, Finland

Milan Zeleny
Graduate School of Business Administration,
Fordham University at Lincoln
Center, New York, New York 10023, U.S.A.

ABSTRACT

We present a holistic system of unaggregated multidimensional "profiles", avoiding the information-destroying reduction of multicriterion dimensionality to a single "number". Visualization of system information is proposed, transforming numerical data into graphical images. This respects and enhances human ability (and preference) to reason directly on the basis of a graphical Gestalt.

The system is driven by attainable and non-attainable ideals, aspiration levels or referential structures and their purposeful redefinition under changing circumstances.

Management support systems, of which any MCDM technology must be an integral part, both impose and reflect system-organizational requirements and conditions which should not be ignored by methodological developments or applications. Referential or aspirational structures, their statement, displacement and propagation, are powerful strategic tools which enhance the flexibility, responsiveness and competitiveness of a modern enterprise.

1. PROFILES AND AGGREGATES

In designing interactive Gestalt decision support, some faulty, ineffective or outdated artifacts of MCDM conventional wisdom must be avoided:

- Presenting the decision maker with only one solution at a time;
- Asking decision maker large number of hypothetical questions unrelated to given context;
- Interacting with mathematical structures that are irrelevant to decision-maker's needs and purposes;
- Specifying criteria weights a priori, before establishing and exploring their value potentials in a given context;
- Assuming implicit (or even explicit) the device of utility "function", just because it simplifies calculations;
- Ignoring prescriptive ideals, references and aspirations so that decision-maker's mediocrity can be accommodated and enhanced.

We use the concept of profile (shape, gestalt, image) to emphasize a fundamental departure from the conventional handling of information. By profile we refer to an unaggregated set of data, measurements or estimates. For example, a set of estimated future returns over discrete (or continuous) time is a profile. Its net present value (NPV) or return on investment (ROI) are aggregates which reduce profiles to points in time. Any formula (from simple averages to quasipyramid or semicube utility functions) reduces a profile of extensive variety to a single number of no variety at all. Such collapse of informational variety implies loss of profile information all the way to zero.

In Figure 1, comparison of differential profiles of future returns is presented. Observe that not only can we reduce fundamentally different profiles to single numbers, but also, due to specific interplay of measurements and discounting factors, all such numbers can turn out to be (almost) identical. Undifferentiated aggregates (NPV=500) record the loss of information originally present in the profiles. The decision-making process comes to halt or worse: if the inherent information loss is unappreciated or ignored, the decision becomes implicitly (and perhaps invisibly) arbitrary and therefore **very** damaging.

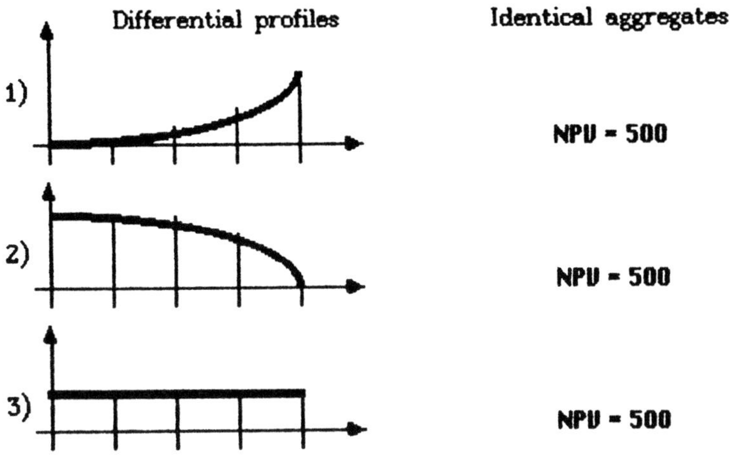

Figure 1.
Comparison of profiles and their aggregates.

Additional aggravation stems from the fact that in order to calculate the information-poor aggregates (NPVs, utility functions) we have to construct the information-rich profiles first. Extensive and expensive, multidimensional profiles are created for the sole purpose of collapsing them into single-dimensional aggregates which carry little or no information.

A manager contemplating Figure 1 would be struck by information-rich managerial and organizational implications of the left-side profiles. He would not fail to see that <u>none</u> of this crucial knowledge has made its way into the right-hand side aggregates.

2. VISUALIZATION OF SYSTEM INFORMATION

Relevant information, constraints, trade-offs and opportunities must be made transparent through a simple, holistic and safely comprehensible pattern responsive to changes, simulations or "what-if" scenarios. Visualization is the way of communicating effectively with the end-users of management information: professional decision makers.

If we take the traditional bar-diagram visualization of ideal and anti-ideal points (like IDEA, Ref. 1) and turn its horizontal axis upon itself

(to form a circle), a compact and information-rich visual star structure (resembling a spider web) is obtained.
The spider-web or "star structure" (see Ref. 3) comes close to system requirements of visualization. It provides a powerful progress-display tool for monitoring "shrinkage towards the target" in better-run Japanese companies, as in Figure 2 (Ref. 2).

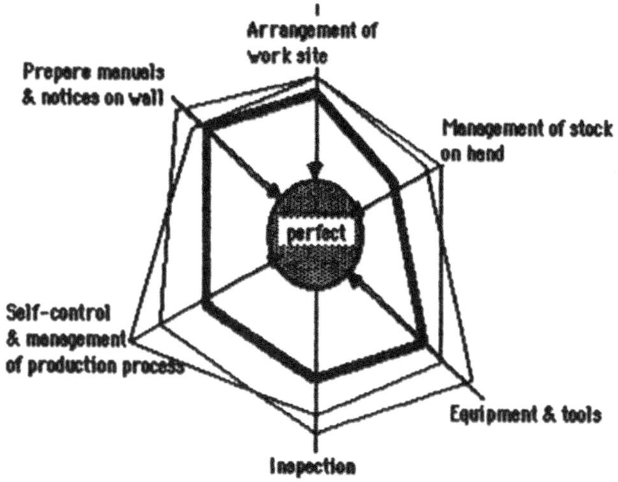

Figure 2.
Spider-Web Chart for Rating Work Center Performance
(Adopted from Ref. 2. p. 53)

The diagram in Figure 2 is appealing not only to managers, but also (more importantly) to workers at operational levels. Each of the six factors displayed may contain sub-structures of additional sub-factors, to be visualized only if and when needed. With the use of such "nested" system displays, every operator can become an informed data collector diagnostician, engineering technician, quality technician, and trainer. <u>That</u> is the purpose of a management support system: to allow people to become more flexible and perform more functions; not to sink even deeper into their already narrow specialties (see Refs. 8).

All underlying aspects of the MCDM model-variables, constraints, criteria, aspirations, uncertainty, fuzziness, trade-offs, sensitivity, resource usage and assumptions - can become effectively functional through this transparent visualization device. (An outline of the nondominated set in two dimensions would be an example of the desired visualization level.)

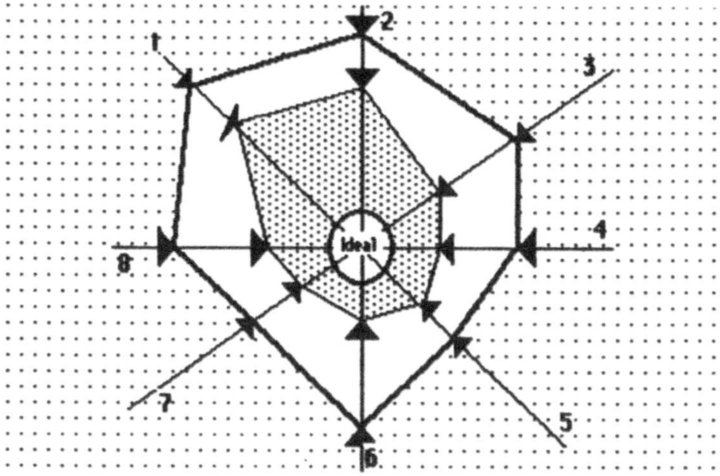

Figure 3.
Star diagram of eight-criteria display

In Figure 3, graphical representation of a complex decision-making situation is displayed: eight separate criteria are monitored in terms of their recorded progression (arrows) towards the ideal structure. Each solution is represented by one full polyhedral cycle around the web.

Any changes in variables or constraints are immediately reflected in the shape and location of the web cycles. For example, a shift from dominated (heavy lines) to a dominant solution is displayed in Figure 3. The shaded space indicates the remaining dissimilarity (magnitude of deviation) from the ideal; the white space "between" solutions records the magnitude of change.

Only decision-relevant information is displayed: status quo, ideal or aspirational goal structures, the magnitude and shape of progression, and a base for holistic comparisons. Feasibility constraints and other limitations on the number of options and their achievable levels remain implicit. Any number of variables, any number of constraints and the practical number of criteria (32 dimensions still preserve diagram's clarity) can be displayed in a simple operational web-diagram without loss of information.

The underlying set of constraints defines limiting structures (polyhedral cycles) which form a constraining envelope (e.g., intersection of nondominated polyhedral cycles) of currently feasible ideal-similar profiles. Further progress towards the ideal structure is possible only through de novo redesign of constraints (Refs, 6,7), which represents the next stage of our software development.

3. EXAMPLE OF A FINANCIAL MODEL

The Model architecture of a commercial bank was designed and programmed to support managerial interaction using both simulation and optimization methodology. The software was programmed mainly with IFPS (Interactive Financial Planning System) and FORTRAN. The model base consists of elements needed for in depth analysis and forecasting at corporate as well as concern levels. The database query, simulation and optimization procedures are completely menu-driven for ease of use. For a more detailed description of the system, (see Ref. 13).

In the practical example we have reversed the direction of improvement (outwards) for the eight separate criteria specified for the banking environment. The resource utilization profile is displayed to establish the feasibility of a given compromise (most ideal-similar profile).

In Figure 4, two alternative options (two web cycles) are displayed. The heavily drawn option tends towards dealing in securities while the fuzzily drawn one is oriented more towards exploiting available markets. Each axis is normalized between 0 and 1. The circle represents maximum resource utilization (the ideal) and the origin stands for inferior utilization (anti-ideal). The remaining space within the circle is available managerial slack or the opportunity space.

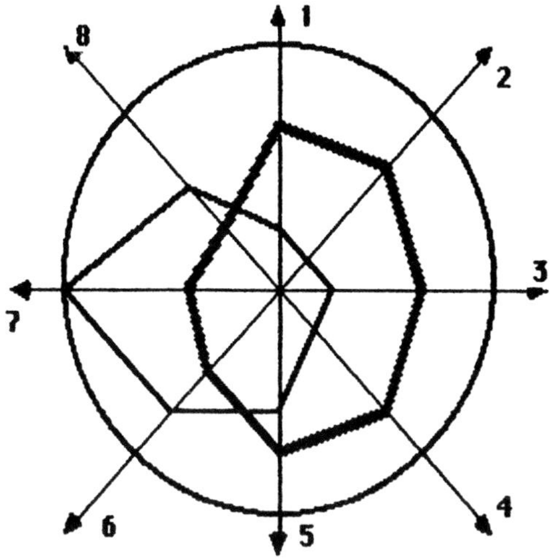

1. Ability to pay (current ratio > 0.2)
2. Market share in public deposits (max 30%)
3. Exploitation of stock markets (max issue per year 5% of capital stock)
4. Reservations (max)
5. Depreciation (max)
6. Commissions (max growth 4%)
7. Dealing in securities (max growth 15%)
8. Personnel costs (max growth 5%)

Figure 4.
Spider-web of resource utilization in banking environment.

4. ALGORITHMIC INTERFACE

There are several stages in the proposed Gestalt interface:

1. Establish logical minima for all criteria and let the system calculate the attainable maxima of all criteria. The logical minimum is composed of the minimal values of the objectives, below which the manager will never want to go in any trade-off situation. The maximum levels indicate what can be achieved if we concentrate only on one objective at a time. Mathematically we solve the problem

$$f_i^{MAX} = \text{Max } f_i(x)$$
$$g(x) < 0$$
$$f(x) > f^{MIN}$$

The crucial idea of the conflict zone is that all the meaningful managerial aspirations and all the efficient solutions can be represented as profiles within the initial zone. It provides a holistic picture, Gestalt, of the decision making situation that will be displayed throughout the session.

The situation can be illustrated with the Star structure as follows

Figure 5. Initial conflict zone

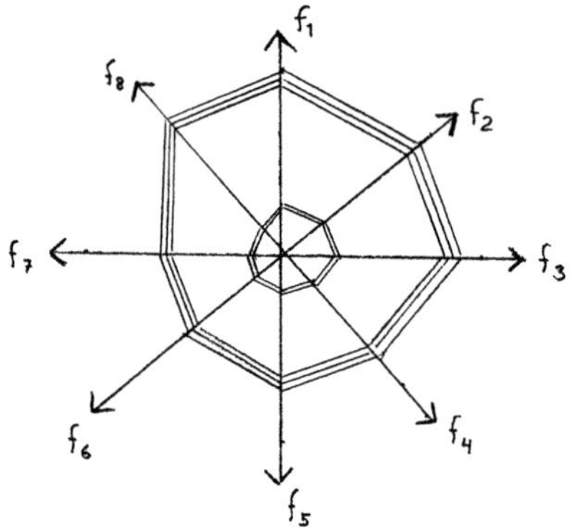

==== logical minima, f_i^{MIN}

==== attainable maxima, f_i^{MAN}

:::: initial conflict zone

2. State an aspiration profile ($f_1^A, \ldots, f_n^A$) within the initial conflict zone and the model will calculate the most similar efficient profile ($f_1^S, \ldots, f_n^S$).

The stated managerial aspiration profile may turn out to be efficient, inefficient, feasible or infeasible. There is no need to constrain managerial creativity. It is the task of the mathematical algorithm to transform any meaningful aspiration profile into an efficient solution that, in some sense, is close to the stated preferences. Decision makers do not want to answer hypothetical questions on their potential preference structures but they are probably willing to see where their own aspirations would lead.

Figure 6. Aspiration profile and the most similar efficient solution

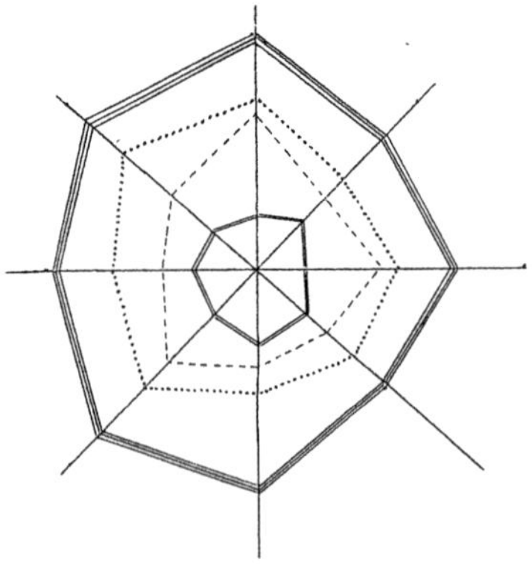

..... aspiration profile f^A

----- the most similar efficient solution f^S

The question of " similarity " or " closeness " has produced a wide variety of methods and concepts in the MCDM literature. Two well-known approaches are:

(1) the L_p distance from an ideal point (Ref. 10) and

(2) the achievement scalarizing function (see Ref. 11).

Faithful to the Gestalt approach Kasanen & Östermark have defined similarity by minimization of the angle between the aspiration vector and an efficient point (see Ref. 9).

Maybe there is no correct formal method that would correspond to changing managerial concepts of similarity. Most probably the exact algorithm is of secondary importance, anyway. Model ergonomics, responsiveness to managerial questions, holistic be presentation of the decision making latitude and validity of the underlying model have greater impact on the success of management support. What is needed, however, is a robust, quick and reliable algorithm that can produce a similar efficient solution to any given meaningful aspiration.

Let us introduce our new algorithm that is based on the idea of finding the efficient point that lies closest to the extended aspiration vector. As the problem is too hard to solve as such we have to operationalize the concept of closeness suitably.

2.1. Find the point y where the extended aspiration vector intersects the cone that starts at the ideal point.

$$y \in (b(f^A - f^{MIN}) + f^{MIN} \; ; \; b \in R^+) \text{ and}$$
$$\text{Boundary } (Z \; ; \; Z \geq I_i),$$

where I_i = i:th coordinate of the ideal point. The point y is easy to calculate with any line search algorithm. If the aspiration vector has zeros in some columns then the existence of the intersection point can be secured by small perturbation of the zero columns. Point y is an over-ideal (or ideal) point, by definition.

2.2. Find the efficient point that lies closest to the point y by some L_p norm. For this problem the solution always exists and it can be calculated with standard optimization packages (see Ref. 12, p. 90).

3. Explore the local trade-off opportunities via <u>a p-level conflict zone</u>, through establishing lower and upper boundaries around the current non-

dominated profile f^S. The p-level conflict zone is generated by lowering the normalized aspirations by p per cent. The bounds show what can be achieved if all other objectives are lowered by p per cent, thus revealing the local shape of the efficient frontier. The p-level conflict zone shows in a holistic way the trade-off opportunities. Mathematically we solve the problem

$$f^L = (1-p)(f^S - f^{MIN}) + f^{MIN}$$

$$f_i^U = \text{Max } f_i(x)$$

$$g(x) < 0$$

$$f(x) > f^L$$

Figure 7. p-level conflict zone

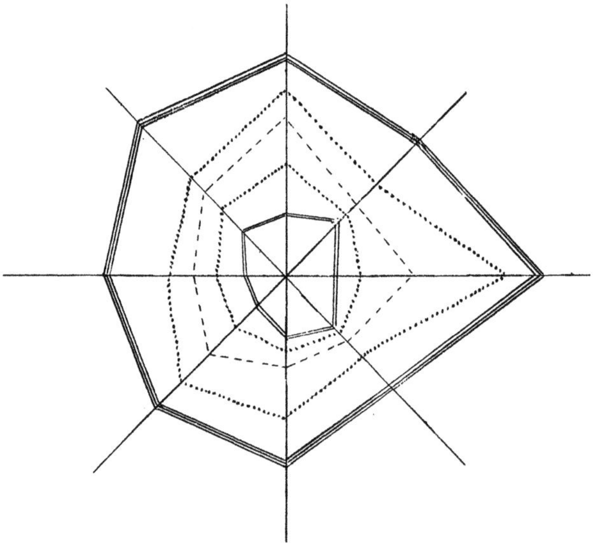

---- current solution f^S

.... boundaries of the p-level conflict zone

The conflict zone provides information for the definition of the next managerial aspiration profile f^A.

Steps 2 and 3 are repeated as many times as required. Note that the manager does not have to answer any hypothetical or tedious questions; the manager states aspirations and looks at the trade-off opportunities around efficient solutions.

4. Optional: perform sensitivity analysis (via Monte Carlo simulation) on the currently similar nondominated profile by exploring the r-level statistical zone. All external parameters are redefined as normally distributed stochastic variables, 1000 realization runs simulated and r

estimated, for example by:

$$r = (f_i(x) - f_i^*)f_i^*$$

where $f_i(x)$ is ith criterion as a function of the vector of stochastic variables x, and f_i^* is the ith criterion value attained before the simulation. There is (1-r) probability that each criterion (separately) stays within upper and lower r-bounds if the corresponding profile is selected (decision taken).

5. All of the above steps are displayed and interfaced through the spider-web visualization device.

5. STRATEGIC ISSUES

The proposed Gestalt approach does not define, characterize or "compute" the solution (see all of the failed OR/MS methodology). Nor does it establish a feedback-like interactive learning procedure for the decision maker (see Ref. 4). It can be best viewed as a canalizing or path-finding system, a support which, by constraint setting and resetting, continually redefines the space through which the decision maker can move towards the desired solution.

Metaphorically, one can a priori calculate the optimal trajectory between one's unsure hand and the object of desire, and then "execute" it. Or one could define a movement-correcting feedback which would guide (and teach) the hand by minimizing its successive deviations from the object. Finally, one can move through a constraint-defined path (or canal) and correct only if "bumping" into one of the constraints. The first two approaches are inflexible, agonizingly slow, mechanistic (engineering or cybernetic analogy) and mostly "besides the point". It is the last paradigm we choose to exploit.

Decision support systems should not be separated from their organizational context and embedding. In fact, decision support must be an integral part of a broader management support which reflects and enhances the strategically requisite organization. There is little wisdom in creating decision support systems that further entrench (or institutionalize) organizational rigidity, politicize hierarchies, ossify power structures, deepen already extreme task, labor and knowledge specialization, separate decisions from actions and decision makers from decision takers. Such organizations are neither competitive nor self-sustaining and their continued "propping up" simply spreads their inefficiency over broader spheres of humanity.

Management support does not make decisions. It provides the knowledge necessary for making appropriate decisions at appropriate levels. The appropriate levels are those where the corresponding action is taken: strategic decisions at the levels of strategy formulation, operational decisions at the levels of "doing". In this sense, decision information (not necessarily control information) should flow downwards, not upwards, through organizational hierarchy. Knowledge-based management support systems should explicitly reflect this requirement.

There exist "information" systems which actually collect data at the lowest levels, ship them up into the center of decision "makers" (who have little or no knowledge of the lower-level local conditions), wait for their "decisions" to be made, formulated and written into lower-level directives, ship it all back down to the lowest levels and let the "non-decision makers" implement it under new and already changed local conditions. It is virtually impossible to contemplate designing organizations processing information inefficiently more. Developing support systems that would support "it" is at least a dubious activity.

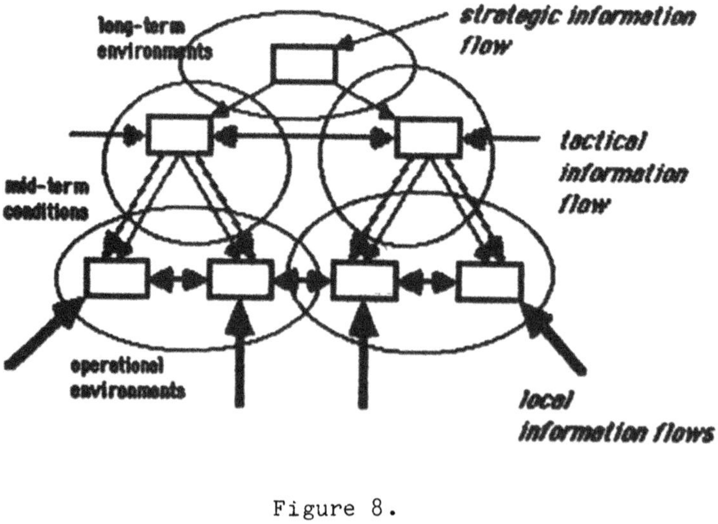

Figure 8.
Information flows required by a flexible, responsive and competitive organization.

In Figure 8, note that all decision information travels downwards and horizontally. Strategic decisions create a framework (rules of the game milieu) for making necessary tactical and operational decisions. Local decisions are properly made by those who can best possess necessary local knowledge, in proper place, at appropriate time.

Proper use of locally operational knowledge increases organizational flexibility and responsiveness to external and internal fluctuations. This enhanced flexibility is necessary for coping with the ever increasing uncertainty, turbulence and changeability of environmental conditions.

Increasing organizational flexibility provides the only effective answer to planning for future uncertainties. So called "planning", concerned more with predicting future consequences of current and future actions, is not planning proper, but forecasting. Forecasting and predicting what is definitionally unpredictable cannot be planning.

Planning is a continuous broadening of requisite organizational ability to cope with ever wider ranges of relevant circumstances.

Metaphorically: a totally flexible corporation would not have to forecast at all because it would be able to respond to any and all fluctuations just-in-time. Creating such a corporation should be the aspiration profile of any planning process striving for excellence and optimality.

A truly effective framework for dealing with uncertainty in dynamic multiobjective planning is conceptually self-evident: enhanced flexibility. Why it should continue to remain unrecognized or ignored goes beyond the narrow scope of this paper.

In the MCDM framework, the information which travels downward is the ideals, referential and aspirational structures. These structures provide "models" or norms for the lower decision-making levels to pursue "as closely as possible": it is the closeness of their "matching" that determines the quality and is explicitly rewarded. Further explicit interreference with local decision making would be counterproductive. The actual pursuit of referential structures, actual alternatives chosen and implemented, and actual constraints imposed or relaxed belong to decision makers at operational levels. Providing decision-making "models" and creating the conditions for their effective self-motivated pursuit leads to Gestalt management support.

Strategic and tactical shifts and reorientations are expressed through the referential or aspirational adjustment and redefinition. Knowledgable and purposeful displacements of ideals and aspirations effectively convey strategic changes in emphasis, thrust, weights of importance and evaluations of trade-offs. Organization-wide change in policy can be broadcast to possibly thousands of distributed decision-autonomous work centers through resetting operational ideals.

REFERENCES

1. M. Zeleny, "Interactive Decision Evolution Aid (IDEA)", in: Multiple Criteria Decsion Making, McGraw-Hill, New York, 1982, pp. 369-373.
2. R.J. Schnoeberger, World Class Manufacturing Casebook, Free Press, New York, 1987, pp. 52-54.
3. R. Östermark, Linear Goal and Multiobjective Programming in Financial Planning, Företagsekonomiska institutionen, Ser. A:234, Åbo Akademi, Åbo, 1986.
4. A. Lewandowski, A. Wierzbicki, "Aspiration Based Decision Analysis and Support", Part 1, WP-88-03,IIASA, January 1988.
5. E. Konarzewska-Gubala, "On the Use of Bipolar Reference Objectives in Multiple Criteria Decision Making", in: Toward Interactive and Intelligent Decision Support Systems, Vol 1, edited by Y. Sawaragi,
K. Inoue and H. Nakayama, Springer-Verlag, New York, 1987, pp. 132-141.
6. M. Zeleny, "Optimal System Design with Multiple Criteria: De Novo Programming Approach", Engineering Costs and Production Economics, 10 (1986), pp. 89-94.
7. M. Zeleny, "Systems Approach to Multiple Criteria Decison Making: Metaoptimum", in: Toward Interactive and Intelligent Decision Support Systems, Vol. 1., edited by Y. Sawaragi, K. Inoue and H. Nakayama, Springer-Verlag, New York, 1987, pp. 28-37.
8. M. Zeleny, "Management Support Systems: Towards Integrated Knowledge Management",Human Systems Management, 7 (1987) 1, pp. 59-70.
9. M. Kasanen, R. Östermark, The Managerial Viewpoint in Interactive Programming with Multiple Objectives. Kybernetes 16/1987.
10. M. Zeleny, Multiple Criteria Decision making, McGraw-Hill, New York, 1982.
11. A. Wierzbicki: " On the completeness and constructiveness of parametric characterizations to vector optimization problems " OR Spectrum 2, pp. 73 - 88 (1986).
12. Gearhart, W.,B.: Analysis of Compromise Programming, in: MCDM: Past Decade and Future Trends, edited by M. Zeleny, JAI Press, 1984, pp. 85-100.
13. Designing Financial Model Architectures. A Study of Managerial Decision Support. Publications of the Turku School of Economics, 2:19.

THE USE OF MULTICRITERIA TECHNIQUES TO RANK HIGHWAY INVESTMENT PROPOSALS

A D Pearman
P J Mackie
A D May
D Simon

Institute for Transport Studies
University of Leeds
England

ABSTRACT

Local authority transport planners are annually faced with the problem of selecting, from a wide range of proposals, a sub-set of projects which is in some sense "best" relative to their investment budget constraint. The choice is a complex one, for at least three reasons. The range of proposals is often wide in terms of project type, cost and impacts. Many of the impacts are difficult to quantify, both in theory and practice. Finally, there is not a single decision maker, but a series of groups, each with legitimate interests in what proposals should go ahead.

Since the mid 1970s, many British local authorities have devised priority assessment techniques (PATs) as decision support tools to help in the project prioritisation process. The PATs differ widely in structure, complexity, type of information required, cost band of schemes to which they are applied, and role in the planning process. However, they share a common objective, namely to reduce multivariate information on diverse project types to a common basis so as to facilitate comparison, to establish priorities and hence to optimise the allocation of scarce capital resources.

This paper reports the findings of a two-year, ESRC-financed study of PATs. It first outlines briefly the characteristics of those PATs which have been devised and then sets out the main conclusions of a comparative analysis of a representative sample of them. The bulk of the paper, however, is concerned to describe a multicriteria decision support tool which, using the insights gained from the analysis of existing PATs, has been generated as the final stage of the project. The development of the hierarchical value tree is explained, together with a discussion of the problems encountered - the size of the tree, the wide range of project costs, the intangible nature of many impacts, the difficulty in eliciting weights, etc. Finally, progress with software development for computer implementation is described.

1. INTRODUCTION

One of the problems to which decision analysis and other multiple criteria evaluation techniques are commonly applied is resource allocation. This is particularly the case in the public sector, where the need to take account of a wide range of diverse impacts and to do so in a way which is explicit and open to discussion is often present. The research reported in this paper is concerned with such a resource allocation problem, specifically, the question of how local transport planning authorities should allocate funds between competing capital projects.

There are a number of factors specific to local authority transport priority assessment techniques (PATs) which, from the point of view of decision analysis, make it a non-standard and therefore interesting area of application. Among the factors which need to be borne in mind are:

(a) PATs are "production line" systems, which must be capable of application to a wide range of problem types. Unlike most multicriteria decision making exercises, they cannot be tailored to the specific choice in hand.

(b) The use and interpretation of the PAT will be in the hands of non-specialists in decision analysis.

(c) A wide range of scheme costs may be involved for which different levels of project evaluation may be cost-effective.

(d) There are potentially a large number of relevant dimensions of impact, some of which are environmental or social and so not straightforwardly quantifiable.

(e) A PAT may be used in one of several roles in the planning process. For example, there is often a wish to assess both the perceived severity of the transport problem being addressed as well as the quality of potential solutions to the problem.

2. BACKGROUND

Reorganisation of British local government in 1974 brought with it a number of conflicting pressures for highway authorities and the professionals who advised them. Many authorities inherited large highway programmes at a time when financial problems, reaction against highway construction and concern over planning blight meant that many of the schemes would not be implemented. They therefore required procedures for selecting preferred projects from large pools of disparate schemes spread over wide geographical areas.

Several local authorities responded, largely independently, by developing a range of priority assessment techniques. The purposes for which they were designed included problem assessment, comparison of alternative solutions to problems, coarse sieves to eliminate less urgent or attractive schemes, and establishing detailed priorities between schemes. While the methods themselves differed, their documentation suggests a degree of uniformity in the criteria on which they were developed: inclusion of the full range of transport policy objectives; assessment of problem severity as well as solution efficacy in most cases; adaptability to different forms of data source; reliance on readily available data; simplicity of execution; adapatability to different value judgements and priorities among objectives; identification of output as a decision making guide rather than an apparently 'right' answer.

The priority assessment techniques developed by local authorities are, with few exceptions, of the points-scoring variety. A series of variables is defined relating to traffic, safety, environment and other relevant issues. These may be measured objectively or subjectively. Each variable is given a score for the respective problems and/or solutions by the evaluating officer. These are then summed to produce an aggregate numerical score for each problem or project. Additionally, each variable and/or group of variables can be given a weight to reflect policy or other considerations.

Analysis of 24 PATs reported to us by local authorities in Great Britain revealed a wide variety of practice within this general framework, notably with regard to the number of variables used, the scoring systems employed and the weighting systems. Of the 24, six were chosen for more detailed analysis, by applying them to six projected highway schemes, ranging in cost from £95K to £5.8m. This analysis revealed substantial differences in performance, both in terms of project ranking and the percentage of points allocated under different major impact headings. A fuller account is available in Simon et al. (1987).

While it is inevitable that local authorities have different requirements, it does seem that some standardisation of overall approach would be advantageous. This does not imply that all authorities should use identical PATs, variables and weights. Political judgement in this sphere is rightly a local matter. However, our analysis and discussions with representatives of co-operating authorities led us to conclude that there is a good case at least for ensuring that techniques used possess desirable and broadly compatible, logical properties.

3. A MULTI-ATTRIBUTE VALUE FUNCTION APPROACH TO PRIORITY ASSESSMENT

There are two main aspects to the development of a PAT. The first, which is principally a transport planning/engineering issue and will not be discussed here in detail, is the choice of appropriate scheme objectives and attribute scales to measure those objectives, such that the range of likely scheme alternatives may be adequately described and differentiated without imposing data collection requirements which are unrealistic relative to the scale of project involved. The second is the choice of an appropriate method to permit the sifting or ranking of alternatives in a way which recognises the many different dimensions of project impact which are potentially relevant. The two aspects are, of course, to a degree, interdependent.

As the previous sections of the paper have indicated, existing practice has typically been to use some, more or less crude, form of additive linear multi-attribute value function, based on points-scoring, to rank projects. Depending on the stage at which a PAT is used in the overall project evaluation and choice process, ranking is not the only objective which might be considered. In certain applications, it may be more appropriate to screen projects in the first instance. The objective is to delete from consideration unacceptable alternatives, rather than to identify the best, which is the subject of a subsequent, separate analysis.

Although it is clear that none of the local authorities which we contacted used formal quantitative analysis to decide precisely which transport projects to implement, it also seemed to be the case that they were prepared for the most part to screen more informally, on the basis of social, political and professional judgement. Once a project reached the stage of deserving any substantial formal evaluation, the perceived need was for an analysis more akin to ranking than screening. For this reason, because the independent judgement of many authorities had gone towards points-scoring, because too radical a departure from the status quo might not be readily accepted, and because our conception of the most defensible form of analysis was to implement a quasi cost-benefit analysis, we opted to design an evaluation procedure which used a multi-attribute value function to aggregate the net benefits of each project, while calculating separately the cost implications.

As has already been outlined, PATs are production line systems, applied more or less mechanistically to all candidate projects. Such an approach is contrary to the normal one adopted in decision analysis, which emphasises the specific tailoring of the analysis to the choice problem in hand, and re-iteration through the various stages of the decision analysis to fine-tune the process. Our need was to define a single problem structure capable of handling a wide range of project types and cost scales, albeit perhaps with different (but consistent) modes of operation, dependent on project size or type. The framework within which a PAT is formulated must be capable, if necessary, of accommodating all the significant impacts likely to accrue in the wake of a transport scheme costing perhaps several million pounds.

The requirement for comprehensiveness could in principle demand the development of an unmanageably complex set of objectives and attributes. In practice, this turned out not to be the case. It was, however, inevitable that a number of the objectives could not be quantified other than subjectively, a factor which is potentially problematic given the wide range of project sizes which a PAT may have to handle. For example, subjective assessments of the visual disamenity arising from a minor junction improvement and that from a small by-pass may be very difficult to make in a consistent fashion. Faced with a choice between implementing separate PATs for different scales of project (which would have undermined project compatibility, a major objective) or developing operating procedures for a single PAT which would endeavour to guarantee reasonable consistency, we opted for the latter.

One other important aspect of the structuring phase of the problem is to determine the overall goal of the multi-attribute evaluation. In this respect also PATs are unusual in that there are the two separate but related evaluation exercises which may have to be undertaken - one to assess the severity of a problem at a given site, the other to assess the quality of proposed solutions to the problem. It is by no means clear that the attributes (or attribute weights) appropriate to measure problem severity will necessarily be exactly the same as for assessment of solution quality. However, both for economies in data collection and because it seemed desirable that most of the dimensions which come together to suggest the existence of a traffic problem ought also to be addressed in comparing solutions to that problem, it was decided that for each solution objective there ought to be the possibility of an equivalent problem objective. Nonetheless, the attributes used to measure these objectives could well be different. For example, solutions could be assessed in absolute terms, the difference between "with project" and "without project" scores, whereas problem severity could be measured through rates - present performance relative to design standard for the road, or some other "ideal point". Weights attached to the attributes could easily vary independently of the differing measurement methods. For example, accidents might well figure more prominently in a social-political assessment of problem severity than they do in a resource-allocation oriented evaluation of solutions.

The procedure adopted for assessing the decision maker's preference structure for the PAT design process departs in a number of respects from the pure theory of multi-attribute evaluation. There is, of course, not a single decision maker but a series of interested parties; nor is the PAT designed to make decisions, merely to provide some decision support. Our PAT responds to such features through its facility to use and compare the consequences of employing different weighting systems, as well as undertaking sensitivity analysis. The system also has no formal recognition of uncertainty and so uses multi-attribute value functions, not utility functions. Purely practical considerations imply that formalising the uncertainty associated with typical schemes processed in this way is not feasible. Sensitivity analysis on impact scores provides an informal alternative which is adequate.

The final major departure from the pure theory of multi-attribute evaluation is in specifying the preference structure which will be the basis of evaluation. The type of structure used is a value tree with up to three hierarchical levels, derived by a heuristic mixture of bottom-up and top-down structuring, strongly influenced by our knowledge of existing PAT procedures. The latter typically exhibit concern with some if not all of the four areas of impact - safety; traffic; environment; development and planning. Thereafter, within those sub-headings, there are a host of (subjectively and objectively assessed) lowest-level objectives, the choice between which is essentially one based on professional judgement, seeking to balance ease/cost of assessment when delegated to individual officers, and comprehensiveness within the major head. The PAT we have designed is principally intended to emphasise that impacts under the four major heads should always be taken into account and to suggest an operationalisable series of lowest-level objectives/attributes for doing so. Altogether, 32 lowest-level attributes selected from the set of possible candidates are used to describe and differentiate schemes in the size range which concerns us (see Figure 1). This is just about within the range that can sensibly be handled by a multi-attribute value model.

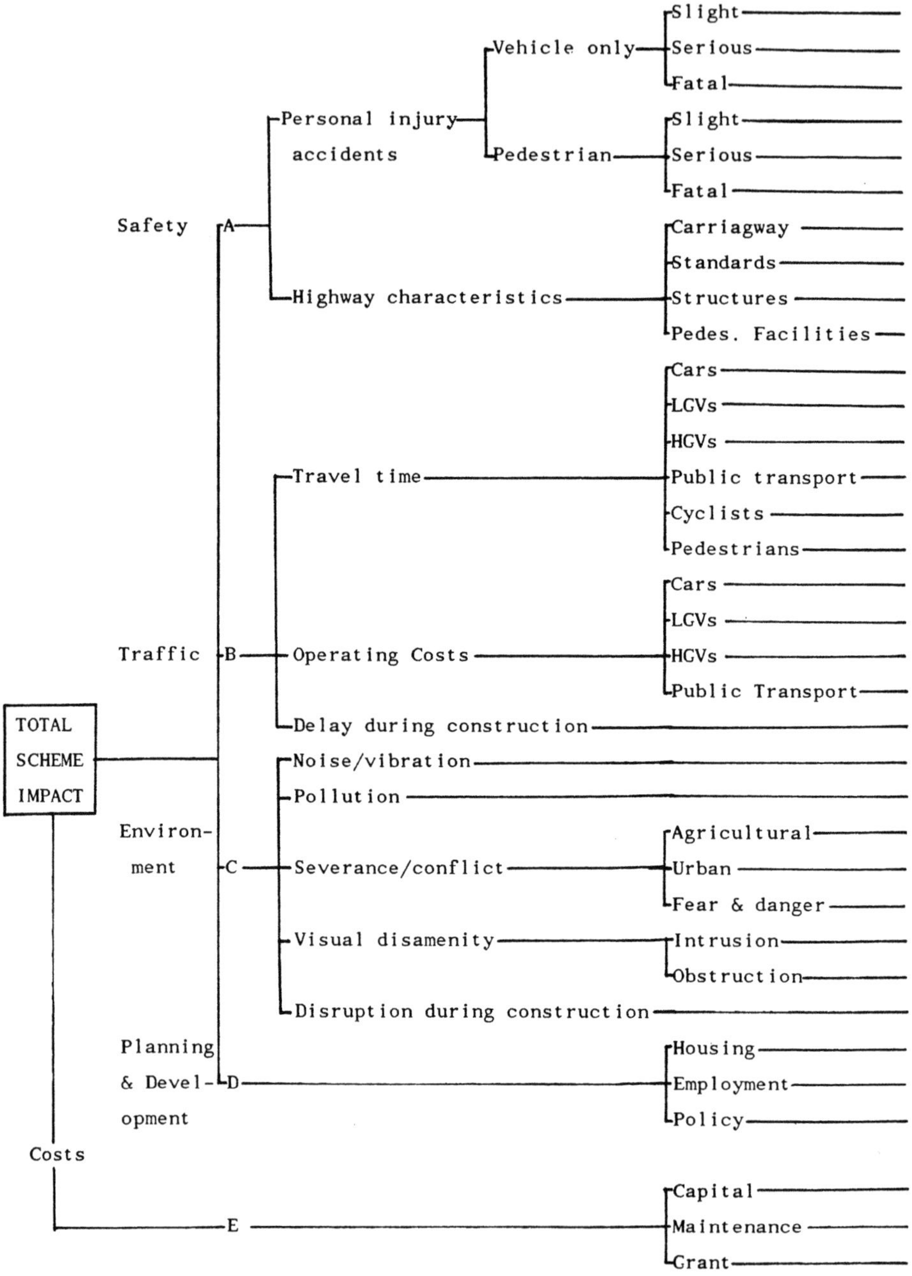

Figure 1 The COMPASS Value Tree

In some ways the strongest and least easily adjustable imposition of the multi-attribute structure selected for our PAT is the linear additive form of the multi-attribute value function:

$$V_i = \sum_{j=1}^{n} w_j v_j(X_{ij})$$

hence implying mutual preferential independence. This implies that preferences between alternative levels of attribute k are independent of the levels achieved with respect to all other attributes (j=1,....n; j≠k). Although, as is so often the case, no formal testing of this assumption has taken place, in our view it is an adequate approximation, viewed in the context of this particular application. There is a good deal of evidence to support the use of linear models in circumstances similar to ours (Dyer and Larsen, 1985).

The individual attributes involve both objective and subjective assessments, which itself poses a further problem in the light of the production line mode of use for PATs. Standard decision analysis usually scales the preference function in the light of the known magnitudes of the impacts of each alternative on each attribute. Single-value attribute functions, $v_j(X_{ij})$ would normally be assessed on a zero to one scale prior to evaluating the scaling constants which "stick" the single-value functions together to give the multi-attribute function. The attribute functions and scaling constants would then be checked for consistency. In defining a PAT, however, the consistency checks cannot in practice be made; nor is it possible to guarantee that all schemes to be evaluated will record attribute scores within the range for which the value functions were initially calibrated. While this is not a significant problem for variables which are naturally quantifiable and assessed with a linear value function (say time savings) it does present problems with non-linear value functions which might be only locally accurate and with subjectively assessed variables with "closed" scales. At present, we have no better method of handling this question of extreme scoring than to "star" such instances to draw their attention to the user and to score them at the relevant extreme of the scale concerned, 0 or 1.

The final issue in the preference structuring phase of the evaluation process is the determination of the scaling constants, w_j. This has not proved at all straightforward in practice - because there are a large number of attributes, measured in disparate units and with objective and subjective evaluations alongside each other. Although it proved effective to develop the structure of the value tree hierarchically, attempts to assess the weights by sequential assessments at the different levels in the tree hierarchy have so far foundered on the difficulty of trading off highly aggregated attributes such as safety and environment. Assessing the required trade-offs at the lowest level in the hierarchy has proved more practicable, either using the ratio technique (von Winterfeldt and Edwards, 1986) or more informally with reference to "standard" Department of Transport money valuations of items such as accidents, time savings, etc. At present, our PAT presents the decision maker with a default set of weights which he/she is free to adjust. The intention is to provide more formal weight elicitation techniques so that the decision maker can derive an acceptable set of weights independently. However, some further experimentation and research will be necessary before this facility is implemented in practice.

A prototype version of the Leeds PAT is now available and is in the process of being tested by potential users. It is based on the Lotus 1-2-3 spreadsheet package, a choice which was made partly to speed up development as compared to writing a complete package from scratch and partly on the grounds that the software would be available to the vast majority of local authorities. It endeavours to place particular emphasis on straightforward data input requirements, ready availability of sensitivity testing and graphical output to enhance the user's comprehension of the relative strengths and weaknesses of the highway schemes being compared. It is also designed to bear in mind that for smaller projects it may well not always prove practicable to measure all impacts at the full level of detail required by the lowest level of the value tree hierarchy. If this is so, the tree structure of the valuation permits the user to "retreat" up the tree and perform the evaluation in terms, say, of the four main headings only. This may be operationalised in two ways. First, a lowest-level variable may be chosen as a proxy for all in its sub-group. For example, noise may be chosen under the environment heading. Then its achievement score attracts the weight of its whole heading (environment), rather than just its own. Alternatively, a subjective assessment on a zero to one scale may be made at the first level in the hierarchy. Although clearly there is the danger of some inconsistency with either approach, it should be remembered that these will typically be low cost, low benefit schemes where the costs of accurate assessment could substantially influence the true value of undertaking the project. It is also the case that less accurate measurement of chosen (proxy) attributes may be acceptable for less expensive projects.

4. CONCLUSIONS

This paper has sought to report the first two phases and the initial elements of the third phase of a project designed to promote greater consistency between the selection of small- and medium-scale transport improvement projects and the underlying preferences of the relevant decision making body. The first phase of the study revealed a considerable heterogeneity of method, at least in terms of detailed implementation, albeit within a general points-scoring approach to project assessment. The second phase provided evidence to suggest considerable inconsistencies between the results of the project selection techniques although it was impossible, formally, to disentangle the effects of differences in preferences and differences in the techniques' intended mode of operation from genuine differences in the values of the schemes tested. The third and final phase of the study, is intended to create a logically defensible PAT based on the ability to provide up to three decision indicators in relation to any proposal. These are respectively multi-attribute assessments of problem severity and solution quality, and a measure of financial implications to the authority in terms of capital cost and maintenance cost changes.

The PAT acts as decision support, rather than as a tool for decision making. The system is designed to store and, if required, present data on options in the matrix framework fashion familiar to transport planners. It is be sufficiently flexible to permit individual authorities to adjust both the set of objectives which is used to assess projects, the attributes used to measure the attributes and the weights associated with those attributes. Finally it has graphical output with the necessary facilities for sensitivity testing to ensure that it can form a functional input to the wider political-social-technical decision making process which underlies transport project selection.

ACKNOWLEDGEMENTS

We are grateful to the Economic and Social Research Council for financing this project, and to numerous British local authorities for their support and co-operation - particularly the six, the comparison of whose techniques forms the basis for this work.

REFERENCES

Dyer, J.S. and J. B. Larsen (1985) Using Multiple Objectives to Approximate Normative Models, *Annals of Operations Research*, Vol. 2, pp. 39-58.

Simon, D., P.J. Mackie, A.D. May and A.D. Pearman (1987) Assessing Local Authority Priorities, in *Highway Appraisal and Design,* Proceedings of the PTRC Summer Annual Meeting, Seminar E, pp.255-66, PTRC, London.

von Winterfeldt, D. and W. Edwards (1986) *Decision Analysis and Behavioral Research,* Cambridge U.P., Cambridge.

Linear Goal Programming in Estimation of Classification Probability

E.U. Choo, W.C. Wedley and K.F. Lam
Faculty of Business Administration
Simon Fraser University
Burnaby, B.C.
Canada V5A 1S6.

INTRODUCTION

Classification of objects into groups is a common practice in everyday life. Normally, the scores of some pertinent criteria about the objects are available and form the most important factor in the classification decision. Discriminant analysis [4], logistic regression [1,3], and linear goal programming [2,5] are alternative techniques used to determine the criterion weights. Classification with these techniques is based upon the weighted sum of the criterion scores of each object. The classification itself ignores more discriminating information about objects within groups. For example, dividing companies into bankrupt and non-bankrupt groups ignores the fact that the non-bankrupt group contains companies of varying qualities. Thus, it is more appropriate to determine the group membership probabilities for each object instead of just the pure classification into one of the groups. These probabilities can readily be used to discriminate objects within the same group.

When the group membership probabilities of past classification data are available, they should be used in the determination of the criterion weights. Working with only categorical dependent variables, we round off the group membership probabilities and lose the information which may be useful for discriminating objects within groups. This is the disadvantage of statistical discriminant analysis and the (0,1)-goal programming model [2]. In this work, we introduce a continuous linear goal programming model which works directly with the conditional probabilities of group membership rather than the simple group category to which an object belongs. With the use of conditional probabilities, the strength of group membership is measured and further discrimination within groups is possible. We then compare the classification power of this model with four other methods of classification: discrimanant analysis, logistic regression, (0,1)-goal programming and multiple regression. Empirical data from the actual experience of an M.B.A. admission committee is used to test the effectiveness in minimizing the error of misclassifications.

MODEL FORMULATION

Consider a classification problem with k criteria $C_1, C_2, \ldots, C_k$ and two parent populations. The criteria $C_1, C_2, \ldots, C_k$ are known as the discriminating variables in statistical discriminant analysis and as independent variables in the logistic and multiple regression. Classification problems with more than two parent populations can be similarly formulated. Let A be a mxk matrix representing the criterion scores of a known sample of m objects, where a_{ij} is the C_j value of the i-th object in the sample and let p_i be the probability that the i-th object belongs to one of the parent populations. The rows of matrix A are arranged in the descending order of the p_i values. Let $w_1, w_2, \ldots, w_k$ be the criterion weights to be determined. The weighted sum function $w_1a_{i1}+w_2a_{i2}+\ldots+w_ka_{ik}$ is used to distinguish the likelihood of belonging to the population and is known as the classification function. For perfect situation, the weighted sums $w_1a_{i1}+w_2a_{i2}+\ldots+w_ka_{ik}$ have the same ordering as the p_i values and thus we would expect the following :

$$\sum_{j=1}^{k} w_j a_{ij} \geq \sum_{j=1}^{k} w_j a_{tj}, \quad \forall 1 \leq i < t \leq m . \tag{1}$$

By the transitivity of inequalities, the m(m-1)/2 inequalities in (1) can be replaced by the following m-1 inequalities :

$$\sum_{j=1}^{k} w_j a_{ij} \geq \sum_{j=1}^{k} w_j a_{(i+1)j}, \quad \forall 1 \leq i \leq m-1 . \tag{2}$$

Since perfect ordering is not always possible, we introduce deviational variables $d_1, d_2, \ldots, d_{m-1}$ into (2) to allow more flexibility :

$$\sum_{j=1}^{k} w_j a_{ij} + d_i \geq \sum_{j=1}^{k} w_j a_{(i+1)j}, \quad \forall 1 \leq i \leq m-1 . \tag{3}$$

The objective is to minimize the sum of d_i. When all the d_i are zero, all the constraints in (1) are satisfied and hence the weighted sums $w_1a_{i1}+w_2a_{i2}+...+w_ka_{ik}$ have the same order as the membership probabilities p_i. Unfortunately, when some d_i is positive, the number of positive d_i may not represent the number of violations of constraints in (1). For example, suppose we have all $d_i=0$ except $d_3>0$. It is possible that d_3 is large enough for $w_1a_{41}+w_2a_{42}+...+w_ka_{4k}'$ to be larger than $w_1a_{11}+w_2a_{12}+...+w_ka_{1k}$. Then there would be three constraints violations in (1). An improvement is to modify the constraints in (3) as follows :

$$\sum_{j=1}^{k} w_j a_{ij} + d_i - d_{i-1} - d_{i+1} \geq \sum_{j=1}^{k} w_j a_{(i+1)j}, \quad \forall 1 \leq i \leq m-1 . \tag{4}$$

We note that each d_i appears in three consecutive constraints in (4). When $d_i>0$ and $d_{i-2}=d_{i-1}=d_{i+1}=d_{i+2}=0$, then $\sum_{j=1}^{k} w_j a_{(i-1)j} \geq \sum_{j=1}^{k} w_j a_{ij} + d_i \geq \sum_{j=1}^{k} w_j a_{(i+1)j} \geq \sum_{j=1}^{k} w_j a_{(i+2)j} + d_i$. This has the effect of minimizing the number of constraint violations in (1) corresponding to each $d_i>0$. Furthermore, the deviational variables d_i should be weighted by the difference p_i-p_{i+1} which reflects the desirability of having $\sum_{j=1}^{k} w_j a_{ij} \geq \sum_{j=1}^{k} w_j a_{(i+1)j}$. Also, the difference $\sum_{j=1}^{k} w_j a_{ij} - \sum_{j=1}^{k} w_j a_{(i+1)j}$ should be at least p_i-p_{i+1}. The model formulation for preserving the order of membership probabilities is given below :

$$\text{(GP)} \quad \text{MINIMIZE} \quad \sum_{i=1}^{m-1} (p_i-p_{i+1}) d_i$$

$$\text{S.T.} \quad \sum_{j=1}^{k} w_j a_{ij} + d_i - d_{i-1} - d_{i+1} - (p_i-p_{i+1})$$

$$\geq \sum_{j=1}^{k} w_j a_{(i+1)j} \quad , \quad \forall 1 \leq i \leq m-1 .$$

The continuous goal programming problem (GP) has m-1 constraints and m+k-1 variables.

The criterion weights w_1, w_2, ..., w_k obtained from the solution of (GP) can be used to rank any new object j by comparing the magnitude of its weighted sum $w_1a_{j1}+w_2a_{j2}+\dots+w_ka_{jk}$ with the weighted sums of the m sampled objects. The estimation of its membership probability is obtained by linear interpolation between the probability values p_1, p_2, ..., p_m.

Multiple Populations

For the classification problems with more than two parent populations, we need one classification function for each parent population. Suppose there are q parent populations. Then we need q criterion weights for each criterion, one for each parent population. Let $w_{t1}a_{i1}+w_{t2}a_{i2}+\dots+w_{tk}a_{ik}$ denote the classification function for the t-th parent population. Let A be the matrix of criterion scores of a sample of m objects. Let p_{tu} be the membership probabilities of the u-th object in the t-th population for u=1,2,...,m and t=1,2,...,q. For the situation of perfect ordering, (1) is replaced by :

$$\sum_{j=1}^{k} w_{tj}a_{uj} \geq \sum_{j=1}^{k} w_{sj}a_{vj} \quad , \text{ whenever } p_{tu} \geq p_{sv} \;, \quad \forall\ 1 \leq s,t \leq q \ \&\ 1 \leq u,v \leq m.$$

To alllow for imperfect ordering, (3) is replaced by :

$$\sum_{j=1}^{k} w_{tj}a_{uj} + d_r \geq \sum_{j=1}^{k} w_{sj}a_{vj} \quad , \text{ whenever } p_{tu} \geq p_{sv} \;, \quad \forall\ 1 \leq s,t \leq q \ \&\ 1 \leq u,v \leq m.$$

There are qm membership probabilities p_{tu} which are linearly ordered with qm-1 deviational variables d_r. The model formulation for preserving the order of membership probabilities with q parent populations takes the following form :

$$\text{(GPM)} \quad \text{MINIMIZE} \quad \sum\, [p_{tu}-p_{sv}]d_r$$

$$\text{S.T.} \quad \sum_{j=1}^{k} w_{tj}a_{uj} + d_r - d_{r-1} - d_{r+1} - (p_{tu}-p_{sv})$$

$$\geq \sum_{j=1}^{k} w_{sj}a_{vj} \quad , \text{ whenever } p_{tu} \geq p_{sv} \;, \quad \forall\ 1 \leq s,t \leq q \ \&\ 1 \leq u,v \leq m.$$

The problem (GPM) has qm-1 constraints and q(m+k)-1 variables. The consecutive ordering of p_{tu} are enforced by the minimization of d_r in

(GPM). An alternative simpler model is to solve (GP) q times independently, one for each parent population.

Additional Features

In practice, additional constraints can be added to obtain better and more meaningful criterion weights. This includes the positive or negative sign constraints on the weights of certain criteria. Higher weights can be imposed on more important criteria. Also, extreme values of criterion weights can be avoided by bounding the ratios of the criterion weights.

EMPIRICAL EVIDENCE

The decision making situation chosen for testing the model (GP) against other different methods is the admission of students to the 1987 class of the Executive M. B. A. Program in the Faculty of Business Administration, Simon Fraser University, B. C. Canada. A committee of four professors was given the task of reviewing all applicant files and making decisions of admission or rejection.

Prior to reviewing the applicant files, the committee convened to discuss evaluation procedures. Although there was much information in the applicant files, the committee agreed that four main criteria were of importance. These were:

(1) mangerial experience
(2) undergraduate preparation
(3) letters of reference
(4) scores on the Graduate Management Admission Test (GMAT).

No prior weightings were established for these criteria. Nevertheless, an evaluation form was prepared which required each committee member to rate each applicant on each criterion. The ratings were semantic indicators as to whether the applicant was poor, fair, good or excellent on the dimension being considered. As individuals, each committee member read the files and made the ratings. Later, when the members met as a group, these ratings were useful for establishing a consensus. Of 86 applications for the 1987 class, 37 were admitted.

The relevant independent variables used as criteria are presented in Table 1. The empirical study is base on 68 cases with complete information. There are 41 cases with probability of admission greater than or equal to 0.50. A list of the 68 cases is given in Appendix I.

Table 1. List of Criteria and Labels

Name	Description	Type	Label
JOBL	Job level	C	1=Non-business, 2=Consultant, 3=Low, 4=Middle, 5=Top
YMGT	Years in management	R	
JOBS	Job mobility	I	Number of job title switches in last 5 years
DEG	Highest education	C	1=Non-university, 2=Some university, 3=technical college, 4=CPA/CGA, 5=CA/non-science graduate/non-business graduate, 6=B.Sc., 7=B.Bus., 8=Masters degree, 9=Ph.D.
YEDD	Years out of school	I	Number of years since formal education
LETT	Employer's letter	C	0=No. 1 to 6 indicating strength of reference.
QGMAT	Quantitative score	R	Quantitative GMAT score
TGMAT	Total GMAT score	R	
LETA	Average reference	C	Average strength of references.
PROB	Probability	R	Acceptance probability.

C = Categorical variable, R = Ratio variable, I = Integer variable.

The dependent variable (probability of admission) was not actually used in the decision process of the committee. Instead, it was determined afterwards. Each committee member used the Analytic Hierarchy Process (AHP) [6] to determine weights for the four main criteria and the four semantic indicators used to rate each criterion. For example, each member had to think of prototype candidates with poor, fair, good and excellent records on managerial experience. Then with the managerial experience criterion in mind, they undertook AHP paired comparisons between prototype candidates to establish importance weights for the semantic indicators. In a like manner, priority weight z_{ij} was established for each indicator i with respect to each criterion j. Finally, AHP comparisons were carried out to determine weight C_j for each criterion j.

Using the indicator weights as absolute measures, each candidate k has a score $s_k = \sum C_j(\delta_{1jk}z_{1j}+\delta_{2jk}z_{2j}+\delta_{3jk}z_{3j}+\delta_{4jk}z_{4j})$, where $\delta_{ijk}=1$ means the candidate k was rated with indicator i with respect to criterion j, and $\delta_{ijk}=0$ otherwise. This score s_k can be used to measure the desirability of a candidate k. Saaty [7] calls this use of AHP as measurement with absolute values. These scores s_k are converted into probabilities of acceptance by rescaling them between 1, the probablilty for an imaginary candidate who scores excellent on all indicators, and 0, the probability for another imaginary candidate who scores poor on all dimensions.

Although it is possible to calculate the membership probabilities perceived by each committee member, we have used the average group ratings, importance weights, and indicator values to emulate the group discussion and consensus process which actually occurred in the committee. Thus, we are using group scores to generate the committee's probability of acceptance for each candidate. The average criterion weights and indicator values which were used in the study are given in Table 2. The resulting acceptance probabilities are in the last column of Appendix I.

Table 2. Group Criteria and Indicator Weights

Criterion	Criterion Weight	Group Indicator Weights			
		Poor	Fair	Good	Excellent
Managerial Experience	.305	.076	.148	.283	.493
Academic Preparation	.280	.056	.138	.323	.483
Letters of Reference	.095	.081	.158	.313	.448
G. M. A. T.	.320	.054	.137	.282	.527

Highest Potential Score = .305(.493)+.280(.483)+.095(.448)+.320(.527)
= .497, which is given a probability of 1.
Lowest Potential Score = .305(.076)+.280(.056)+.095(.081)+.320(.054)
= .064, which is given a probability of 0.

EMPIRICAL RESULTS

The data were randomly divided into development and validation samples. For the goal programming and discriminant analysis methods, criterion weights were derived from a development sample of size 15. Then, the remaining holdout sample of size 53 was used to measure the success of each method. Success is defined herein as the ability of each method to correctly identify the candidates with high (greater than or equal to 0.50) probability of admission. In contrast, multiple regression and logistic regression are applied to all 68 cases to obtain the criterion weights that would provide better classification results. The development sample is listed in Appendix II and a listing of the SPSSX [8] programme used is given in Appendix III.

The continuous goal programming problem (GP), with positive sign constraints on criteria 1, 2, 8 and 9, was solved with the development sample to yield an objective function value of zero. The criterion weights were applied to the validation sample with a result of 9 misclassifications. Discriminant analysis was next applied to the development sample with perfect classification but the criterion weights generated 13 misclassifications in the validation sample. The (0,1)-goal programming model, with positive sign constraints on criteria 1, 2, 5, 6, 7, 8 and 9 and based on the development sample, has a perfect hit ratio but 15 misclassifications in the validation sample. Both multiple regression and logistic regression on the 68 cases resulted in 12 misclassifications. Finally, discriminant analysis was applied to the 68 cases with the result of only 7 misclassifications. All the criterion weights obtained by the different methods are given in Table 3 while all the results of classification are given in Table 4. The methods used are listed as follows:

(1) Continuous goal programming (GP) on 15 cases.
(2) Discriminant analysis on 15 cases.
(3) (0,1)-goal programming [2] on 15 cases.
(4) Multiple regression on 68 cases.
(5) Logistic regression on 68 cases.
(6) Discriminant analysis on 68 cases.

The discrimanant analysis on the 68 cases has the best

classification results of 89.71%. This shows that the criterion weights obtained from the 15 sampled cases by (GP) have pretty good prediction power at 83.02%. We note that the development sample of size 15 is rather small for discriminant analysis and the (0,1)-goal programming which use the categorical indicator instead of the actual membership probabilities. The goal programme (GP), as formulated, worked well with 15 cases where $p_i - p_{i+1}$ is around 1/15.

Table 3. Criterion Weights of the Six Methods

CRITERIA	CRITERION WEIGHTS					
	CONTINUOUS GOAL PROGRAM	DISCRIMINANT ANALYSIS 15 CASES	(0,1) GOAL PROGRAM	MULTIPLE REGRESSION 68 CASES	LOGISTIC REGRESSION 68 CASES	DISCRIMINANT ANALYSIS 68 CASES
JOBL	31.3236	-5.934	0.00	-0.00993	-0.11621	0.08566
YMGT	0.00	-0.104	0.00	0.01141	0.07286	0.14039
JOBS	73.8502	3.564	1.301	0.02303	0.13761	-0.00953
DEG	-50.477	-3.0	-0.865	0.02473	0.13057	-0.16678
YEDD	6.4464	0.71	0.00	0.003	0.01346	0.04925
LETT	21.8124	0.584	0.00	0.00393	-0.01541	0.03364
QGMAT	33.8257	0.869	0.031	0.00192	0.01402	0.11623
TGMAT	0.6147	0.026	0.015	0.00164	0.00826	0.02466
LETA	20.3587	4.4	1.492	0.01419	0.06715	0.37585
CONSTANT	0.00	-42.015	-14.28	-0.79443	-6.40524	-20.56293

DISCUSSION

The continuous goal programming problem (GP) based on the 68 cases has no feasible solution. This is due to the presence of d_{i-1} and d_{i+1} in (4). When all the 68 cases are used, most of the $p_i - p_{i+1}$ values are too small to require the consecutive ordering of p_i values to be preserved. The infeasibility of (GP) implies that some of the sample cases have incorrect probabilities and should be deleted. For the development sample of 15 cases, the constraints are more meaningful since the $p_i - p_{i+1}$ values are larger. In view of this, it may be better to use $p_i - p_{i+n}$ instead of $p_i - p_{i+1}$ to ensure that the probability ordering is the correct one to enforce. To avoid the difficulty of choosing n, one may require that $p_i - p_{i+n} \geq \varepsilon$ for some predetermined constant $\varepsilon > 0$. An alternative is to use constraints (3) whenever $p_i - p_{i+1}$ is insignificant. The solution would enable the identification and deletion of the cases which are likely to have incorrect probabilities. The problem (GP) can then be reformulated and solved again to obtain improved criterion weights.

Table 4. Classification Results of the Six Methods

METHOD USED	NUMBER OF CASES	"YES" AND PREDICTED "YES"	"NO" AND PREDICTED "NO"	"YES" AND PREDICTED "NO"	"NO" AND PREDICTED "YES"	CORRECT PREDICTION PERCENTAGE
CONTINUOUS GOAL PROGRAM	15 53	5 10	10 34	0 8	0 1	100.00 83.02
DISCRIMINANT ANALYSIS	15 53	5 11	10 29	0 7	0 6	100.00 75.47
(0,1) GOAL PROGRAM	15 53	5 7	10 31	0 11	0 4	100.00 71.7
MULTIPLE REGRESSION	68	19	37	4	8	82.35
LOGISTIC REGRESSION	68	20	36	3	9	82.35
DISCRIMINANT ANALYSIS	68	20	41	3	4	89.71

SUMMARY AND FURTHER RESEARCH

The formulation of (GP) is based on the preservation of the membership probabilities of sample cases. It has worked well on the M.B.A. admission data. Moreover, the signs of the criterion weights have a better intuitive meaning. The different possible variations of using (GP) needs to be explored in terms of the quality of solutions and the difficulty in execution. Linear programming sensitivity analysis may be used to improve the development sample by deleting cases with incorrect membership probabilities. It will be interesting to see how well (GP) can be used as a general tool of probability estimation in other applications.

Appendix I : Listing of 68 Cases

Criterion values are listed in the following order:

JOBL	YMGT	JOBS	DEG	YEDD	LETT	QGMAT	TGMAT	LETA	PROB
4	5	4	7	17	3	37	600	4	.86
4	9	0	6	13	2	33	600	5	.58
5	6	2	4	4	5	38	640	5	.6
3	8	0	2	2	3	41	650	4	.58
1	8	0	3	17	5	25	460	1	.19
4	10	1	6	10	4	36	560	5	.82
3	0	3	7	4	3	40	590	4	.49
3	0	1	5	8	0	38	630	3	.63
4	8	2	6	9	4	40	610	5	.91
1	16	2	6	16	0	39	614	3	.99
3	3	1	6	6	3	41	530	4	.31
4	7	2	2	6	4	35	580	5	.64
4	6	1	7	10	3	27	450	4	.52
4	8	1	5	12	5	35	530	3	.49
1	0	0	6	10	4	30	600	3	.49
3	0	2	3	8	3	32	520	3	.27
1	7	3	6	8	4	30	560	1	.49
2	7	1	6	6	0	23	510	4	.49
3	0	0	3	22	4	23	460	3	.17
3	0	2	5	10	0	26	460	3	.28
3	12	1	4	2	4	27	480	6	.19
3	6	0	3	3	4	15	350	5	.11
3	2	1	6	12	3	34	470	3	.28
3	0	0	7	2	1	29	590	3	.47
3	0	2	2	8	3	25	450	3	.2
2	2	3	6	14	0	43	690	4	.67
1	7	1	5	7	4	26	510	2	.36
6	15	0	5	1	0	32	640	3	.53
3	3	2	2	6	3	44	700	5	.61
1	9	1	9	20	4	25	450	1	.49
1	0	1	5	10	3	28	510	2	.25
3	4	5	4	4	4	36	560	4	.49
3	0	5	6	4	4	27	470	3	.27
2	3	4	5	6	5	40	610	5	.54
3	2	1	3	5	2	34	520	5	.34
4	6	1	8	6	5	25	490	4	.47
1	2	1	3	9	4	21	530	1	.23
3	3	1	6	14	4	30	480	4	.15
3	0	2	4	7	2	45	650	3	.57
4	10	3	7	17	4	38	590	5	.63
4	10	0	4	4	0	27	490	4	.29
4	14	1	6	17	2	32	580	5	.35
3	0	3	6	6	2	35	500	4	.37
4	2	1	5	16	2	46	680	4	.55
4	7	1	6	11	6	45	640	4	.81
3	0	0	6	13	2	31	430	5	.27
3	1	1	7	5	4	39	560	5	.28
1	4	3	5	18	3	21	450	1	.4
3	0	1	8	1	4	35	570	4	.56
4	3	3	7	7	5	45	710	6	.91

Listing of 68 Cases (continued)

JOBL	YMGT	JOBS	DEG	YEDD	LETT	QGMAT	TGMAT	LETA	PROB
1	3	2	6	8	4	28	430	1	.12
4	2	1	8	6	5	52	780	4	.85
1	0	0	5	16	3	29	470	1	.29
2	5	1	6	8	0	28	390	4	.1
1	5	0	7	1	4	29	450	3	.26
3	6	1	1	25	3	21	400	5	.11
1	14	1	3	34	3	14	370	1	.2
4	3	2	6	7	4	32	560	3	.38
4	9	2	8	9	4	45	600	5	.76
1	18	1	5	18	3	26	470	1	.37
2	14	2	6	24	0	34	520	5	.49
1	5	1	5	13	3	31	540	1	.43
1	0	2	3	8	4	19	410	3	.19
4	3	2	7	7	2	37	610	4	.32
4	11	0	6	13	4	38	500	4	.42
4	5	2	5	7	5	39	620	3	.41
1	8	0	6	11	5	36	570	1	.56
4	5	1	6	9	5	30	600	3	.41

Appendix II : Listing of the Development sample of 15 Cases

Criterion values are listed in the following order:

JOBL	YMGT	JOBS	DEG	YEDD	LETT	QGMAT	TGMAT	LETA	PROB
2	5	1	6	8	0	28	390	4	.1
4	3	2	6	7	4	32	560	3	.38
4	5	2	5	7	5	39	620	3	.41
4	5	4	7	17	3	37	600	4	.86
3	0	2	5	10	0	26	460	3	.28
1	2	1	3	9	4	21	530	1	.23
1	3	2	6	8	4	28	430	1	.12
4	11	0	6	13	4	38	500	4	.42
4	10	3	7	17	4	38	590	5	.63
3	2	1	6	12	3	34	470	3	.28
4	7	2	2	6	4	35	580	5	.64
3	8	0	2	2	3	41	650	4	.58
3	0	3	6	6	2	35	500	4	.37
4	14	1	6	17	2	32	580	5	.35
4	8	2	6	9	4	40	610	5	.91

Appendix III : Listing of SPSSX Programme for 68 cases

```
$RUN *SPSSX SPRINT=-OUTX
FILE HANDLE OSS / NAME="-Q"
DATA LIST FREE FILE=OSS /
                JOBL,YMGT,JOBS,DEG,YEDD,LETT,
                QGMAT,TGMAT,LETA,PROB
```

Listing of SPSSX Programme (continued)

```
VARIABLE LABELS JOBL   "JOB LEVEL" /
                YMGT   "YEARS IN MANAGEMENT" /
                JOBS   "JOB SWITCHES" /
                DEG    "DEGREE" /
                YEDD   "YEARS SINCE EDUCATION" /
                LETT   "LETTER OF SUPPORT FROM COMPANY" /
                QGMAT  "QUAN GMAT SCORE" /
                TGMAT  "TOTAL GMAT SCORE" /
                LETA   "AVERAGE OF LET1,LET2,LET3" /
                PROB   "ACCEPTANCE PROBABILITY 0,1 "/
COMPUTE         CLASS=(PROB GE 0.50)/
DISCRIMINANT    GROUPS=CLASS(0,1)/
                VARIABLES=JOBL TO LETA/
                ANALYSIS=JOBL TO LETA/
                PRIORS=SIZE
STATISTICS      11 12 13 14
REGRESSION      VARIABLES=JOBL TO PROB/
                DEP=PROB/ENTER
COMPUTE         LODD=LN(PROB/(1-PROB))
REGRESSION      VARS=JOBL TO LETA LODD/
                DEP=LODD/ENTER
FINISH
```

REFERENCES

[1] Anderson J.A. Separate Sample Logistic Discrimination, *Biometrika*, Vol.59, 1972, 19-35.

[2] Choo E.U. and Wedley W.C. Optimal Criterion Weights in Repetitive Multicriteria Decision Making, *Journal of Operational Research Society*, Vol.36, 1985, 983-992.

[3] Cox D.R. *The Analysis of Binary Data*, London:Methuen, 1970.

[4] Fisher R.A. The Use of Multiple Measurements in Taxonomic Problems, *Annals of Eugenics*, Vol.7, 1936, 179-188.

[5] Glover F., Keene S. and Duea B. A New Class of Models for the Discriminant Problem, *Decision Sciences*, Vol.19, 1988, 269-280.

[6] Saaty T.L. *The Analytic Hierarchy Process*, New York, McGraw-Hill, 1980.

[7] Saaty T.L. Rank generation, preservation, and reversal in the analytic hierarchy decision process, *Decision Sciences*, Vol.18, 1987, 157-177.

[8] *SPSSX User's Guide*, Second Edition, McGraw-Hill, 1986.

FORMAL TENDER EVALUATION

K R Backwell
PA Consulting Group
Rochester House
33 Greycoat Street
London SW1P 2QF

INTRODUCTION

We all buy things (professional services, house extensions, long-term financial investments) which demand more than cursory consideration if we are to get what we want. If we can find more than one supplier, then obtaining several quotations and taking the cheapest is accepted as the best way to get a keen price.

In commercial life, this process is called competitive tendering. It is a key part of the purchasing undertaken by all organisations, in both the private and public sectors. Perhaps £10 billion changes hands this way each year with many thousands of contracts let on a competitive basis, some of them for multi-million pound projects. Yet the value of the contract is not always reflected in the rigour of the tender assessment process.

This paper is about evaluating tenders. It covers the process of tendering in outline, but its main focus is the importance of multi-criteria decision making in the evaluation of significant tenders. It concentrates on large tender evaluations as these usually involve groups of people and hence are multi-stakeholder, multi-criteria decisions. Finally, it relates practical tips and warns of potential pitfalls.

APPLICABILITY

Most organisations now use a "formal" multi-criteria approach in assessing large-scale tenders. By formal I mean that the purchase is sufficiently large and important to warrant:

- the involvement of all stakeholders affected by the purchase
- the setting up of a discrete project to manage the tender process.

This approach differs from both autocratic decision making and the evaluation of low value tenders, where the effort of a "formal" process is not justifiable. Figure 1 illustrates the class of tenders covered.

My experience lies in the evaluation of tenders for information technology products and services including tenders for software, hardware and related consultancy services. The scope is therefore limited to this special class of tender, although many of the lessons are undoubtedly applicable elsewhere.

Figure 1 : Applicability of "Formal" Evaluation

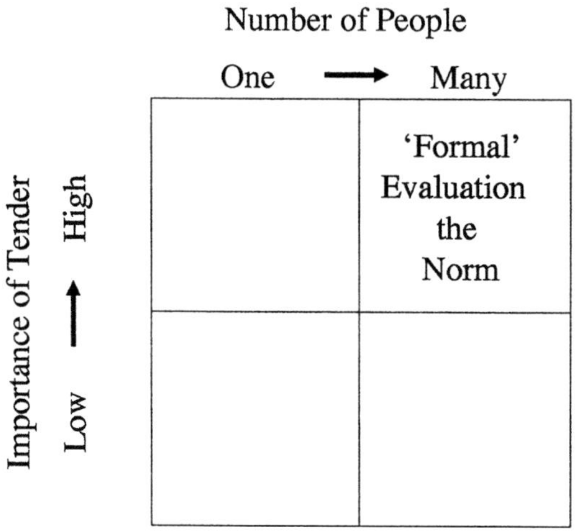

THE TENDERING PROCESS

The steps that a buying organisation will typically follow are shown in Figure 2. In the public sector, the process can be more prolonged if the tender must be widely publicised under the rules set down by the European Community (EC) and in the General Agreement on Tarriffs and Trade (GATT) (known as the EC/GATT rules). In these cases there may be a hundred or more interested suppliers initially, so it can take longer to get to a "short" list. The detail of the process also differs somewhat from the private sector but the fundamentals are the same.

In the following sections, the main steps in a tendering process are described in outline so that tender evaluation can be set in context. Later sections deal with tender evaluation methods and lessons from using the multi-criteria approach.

PLANNING

As with any project, the tendering process starts with detailed planning. Terms of reference and the scope of the project should be set at this point in some detail, yet it is surprising how often this is done only cursorily. These should be related to the organisation's wider objectives as these will be used later in defining the evaluation criteria.

There should already be a project manager and an organisation to oversee direction and progress, in which case it only remains to identify the people and experience needed to do the work. The structure and content of the invitation to tender, the method of evaluation to be adopted, the timetable for each step in the process and an estimate of the number of candidate suppliers are all determined as part of the plan.

Figure 2 : Steps in Evaluation Process

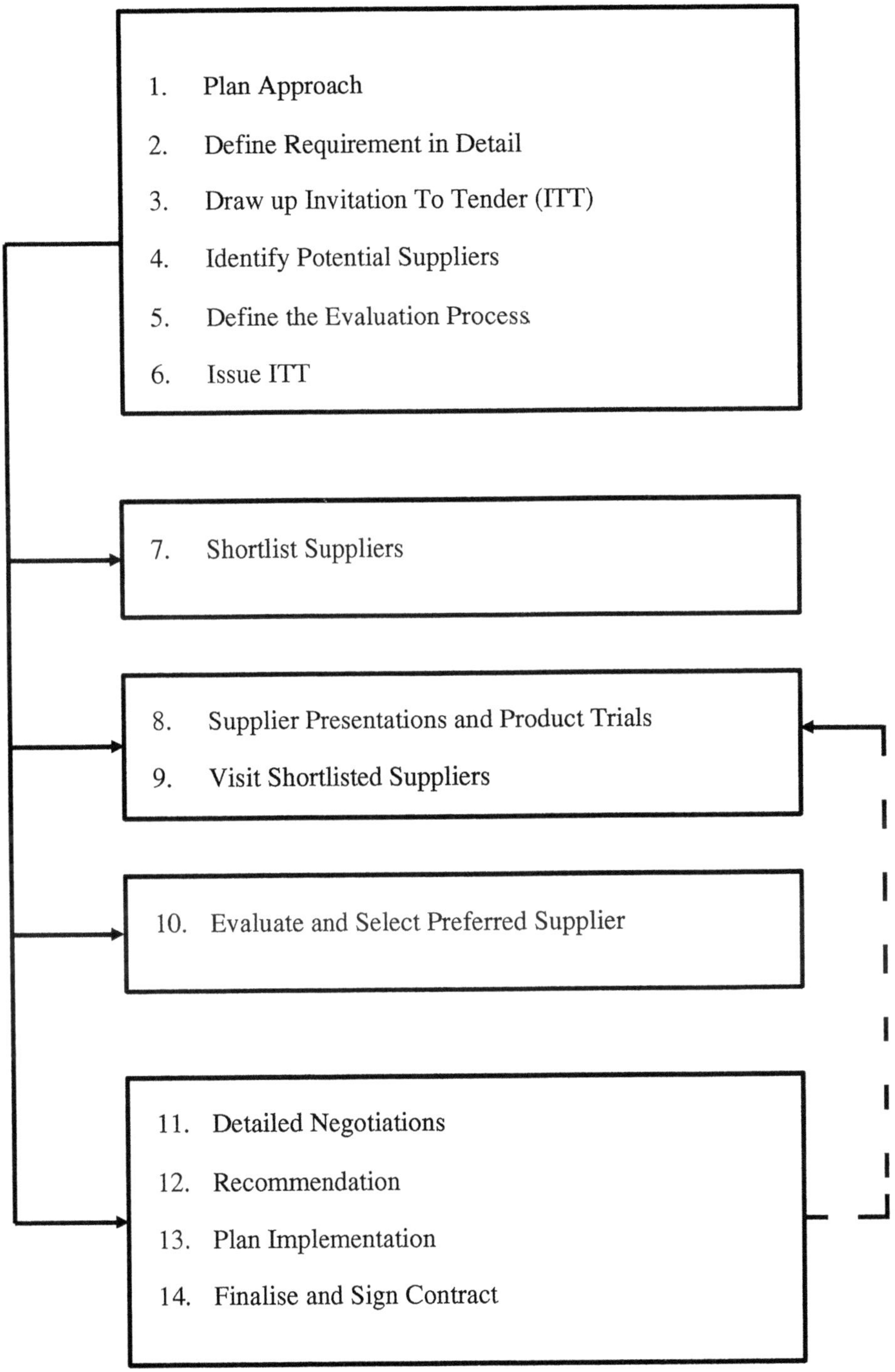

DEFINE NEED

The basic requirement, plus any desirable addition, now needs to be settled. The potential beneficiaries of the project need here to establish what they expect and to obtain the details of the problems to be resolved.

Depending on the application this can be a complex task, but as a minimum the requirements are usually subdivided into mandatory and desirable categories. A useful tip is to take sufficient time to define the requirements in detail. This way the suppliers do not have to make their own assumptions and their proposals can be evaluated on a common basis.

DRAW UP INVITATION TO TENDER

The defined requirements are then combined with a formal invitation to tender which sets out the form of the tender expected, terms and conditions to be complied with and target dates for tendering.

IDENTIFY POTENTIAL SUPPLIERS

An initial screening of potential suppliers may be needed to produce a worthwhile shortlist. The aim is to limit the effort involved in the evaluation but retain acceptable suppliers.

Depending on the size of the contract, starting with a limited short list will encourage reputable suppliers to tender because the selection looks less like a lottery.

DEFINE THE EVALUATION PROCESS

The method to be used in the evaluation should be considered prior to the issuing of the ITT. If the evaluation criteria are defined at this point it acts as a check that the full objectives are catered for in the specification and it helps to structure the form of the response requested. The treatment of cost and contractual issues should also be resolved.

If there is to be a shortlist, a separate list of "stopper" criteria (which include the mandatory requirements) may be developed which the tenders must meet if they are not to be rejected.

ISSUE ITT

The ITT is then issued. Meetings with the suppliers to clear up any misunderstandings and to provide any further information follow quickly.

SHORTLIST SUPPLIERS

Following receipt of the tenders, the suppliers may be further short-listed by applying the "stopper" criteria. The tenders which do not fully meet these criteria are rejected.

SUPPLIER PRESENTATIONS AND PRODUCT TRIALS

An important part of the process, supplier presentations and product trials give the evaluation team the opportunity to see the best the supplier can offer. It also gives the suppliers the opportunity to emphasise why their service/product is better than their competitors'.

REFERENCE SITE VISITS

If the product or service has been sold to other clients, then the evaluation team should visit reference sites to discover any customer problems and the supplier's performance in responding to them. The aim is to probe the claims made in the suppliers' proposals.

EVALUATION AND SELECTION

Evaluation is the core of the whole process. It is used to identify which is the most attractive tender and why. As the assessment of the tenders will involve subjective judgment, the most commonly used approach is MCDM. One or more 'workshops' are usually planned for the assessment and the criteria weighting to arrive at an overall ranking. Other methods based on graphical or mapping techniques are sometimes used to highlight differences and commonalities between tenders.

Whichever method is used:

- it must help to reach a common understanding of why a particular tender is "best"
- it should be seen to lead to an unbiased conclusion
- it should make clear the reasons for the selection to a wider audience than just the evaluation team.

BEYOND EVALUATION

After the evaluation there is still the negotiation, recommendation, implementation planning and signing of the contract. Sometimes a final hitch means there is also a need to re-open discussions with the second choice supplier.

Essentially the steps are as follows.

Negotiation

There are inevitably aspects of the suppliers' tenders which are open to negotiation. These generally include the price and often, for example, the names of specific people to support the implementation, changes in equipment performance specifications and contractual details.

Recommendation

A single "best" tender may have emerged but often the evaluation needs to be refined because of changes arising from the negotiation.

Implementation Planning

A more detailed implementation plan is needed so that the project can go ahead. This may even be drawn up in conjunction with the supplier.

Contract Finalisation and Signing

The signed contract formalises all that has been discussed and promised between the seller and the buyer.

TENDER EVALUATION : CHOICE OF METHOD

The tendering process centres on the evaluation because the basic objective is to establish which tender is the "best". Thus, whichever form of evaluation is used, the aim is to assess the tenders and determine an overall ranking.

There are many different approaches to evaluation, some differing only in degree, but all sharing a common feature: the need for the buyer to make subjective judgements when assessing the tenders and the suppliers. The best methods minimise the extent to which opinion or "gut feel" affect the outcome but none can eradicate them completely without missing vital elements of the decision.

The evaluation can be conducted by one manager if:

- he/she is prepared to read and judge all the tender documents, interview all suppliers, assess all products etc
- he/she has sufficient knowledge of all aspects of the purchase
- this is an acceptable style of decision making for the organisation.

In terms of time and effort this method is certainly efficient, but it is usually impractical for a computer-related tender. The second condition is rarely met and the third unusual as the managers of each affected function will expect to have a say in the selection.

GROUP METHODS

Most organisations prefer a group process in evaluating tenders because:

- a consultative approach means the final choice will be more widely acceptable which contributes to greater commitment to the implementation
- wider experience can be brought to bear on the decision so specific technical requirements can be assessed by those most competent to do so.

The opponents of group methods usually cite the additional effort and time required to reach a conclusion. This is a real risk as the evaluation can become over elaborate if the work is not carefully planned and coordinated.

There are several methods which can be used by a group to assess tenders. The most common, which lead to a final ranking of the tenders, are described below. Other methods rely on presentation and mapping techniques to feedback individual assessments of tender performance and criteria importance. These are not often used for tender evaluation except to highlight differences of opinion or understanding so they are not covered here.

Lowest Cost Compliant Bid

A relatively common method is to use mandatory conditions as a first 'filter' to reduce the number of tenderers to a compliant list and then select the lowest cost bid.

For straightforward purchases this method is useful and popular. In more complex cases, trying to define all the important supplier attributes as mandatory requirements can be counter productive. For example, "shoe-horning" the requirements into a mandatory form can lead to all the suppliers being nominally excluded because none meet the buyers original expectations on all fronts.

This reflects reality as buyers are usually faced with trades-off between a product's features and other relevant attributes such as the ability to provide after-sales support, a programme of product updating and financial stability.

Conversion to a Monetary Scale

To help get a better decision, the procedure for lowest cost compliancy can be modified so that a value is placed on supplier and product attributes. This means converting these attributes to a "money-equivalent" and then accepting the lowest price bid. Similar problems to the lowest cost compliant bid process arise because valuing intangible benefits and incorporating the assessment of risk can be both lengthy and sometimes unsatisfactory exercises.

This will be a familiar problem to anyone who has tried to justify expenditure on new computer equipment for managers where the benefits lie in better informed decisions, better quality reports, improved communications and so on. The problem can also be compounded by the need to estimate the worth of, for example, the advantage of buying a less complex product (which is therefore easier to use but also offers fewer functions).

Multi-Criteria Judgemental Scale

In a multi-criteria evaluation, a panel of stakeholders is asked to assess how well each of the suppliers and their tenders match up against a set of performance criteria. The relative importance of each criterion is also agreed using a weighting method and an overall score derived using simple linear multiplication and addition. The supplier with the highest score is then favourite to win the contract.

This method is usually described as seeking to establish the "utility" of the attributes in question. It has grown in popularity in the commercial world primarily because:

- it is easy to understand
- the underlying sums can be quickly verified
- it provides a clear framework for the evaluation team
- it helps the evaluators to concentrate on single attributes rather than tackling the whole problem at once
- the alternative methods have proved to be less workable in practice.

Consultants have also played a part in the growth in popularity of this method by being quick to recognise its value and encourage its use.

PRACTICAL EXPERIENCE OF MULTI-CRITERIA EVALUATION

Multi-criteria evaluation is particularly important for selecting tenders for high value, complex projects where the evaluation effort is clearly warranted and widely different skills need to be brought to bear.

Its real value lies in helping to explain clearly:

- the steps to be followed by the evaluation team and why they are necessary

- the supporting reasoning which lies behind the decision to others in the buying company (and probably also to the losing tenderers).

COMMON PITFALLS

There are many pitfalls but perhaps the most important one to avoid is placing undue emphasis on applying the method in "recipe book" form. The selection of who is to be involved and how they should contribute is equally, if not more, important. Others to beware of are:

- not recognising the importance of developing a common understanding of the issues involved
- ignoring the subjective element and concentrating exclusively on assessing the features of the tender that can be "measured"
- weak structuring and definition of the assessment criteria
- focussing only on the tender and not considering the suppliers ability to deliver and to provide future support
- wasting time debating issues that do not change the choice of supplier
- not recording the reasoning behind the assessments at the time they are made
- misunderstanding the purpose of the criteria weights and how best to derive them
- the use of spurious accuracy to convince those not directly involved in the evaluation that a thorough job has been done.

GOOD PRACTICE

There is no substitute for experience, but guidelines which everyone can understand are a good starting point. Other good principles include:

- agree on the approach to be used as early as possible as this can affect the way the invitation to tender is worded
- always involve the person with responsibility for the successful conclusion of the project (ideally the problem "owner" should be on the stakeholder panel)
- carefully plan who should be involved and when (seek out, and use to the maximum, experience of both the technical requirement and the evaluation method)
- describe the steps to be followed and agree the plan with all involved
- structure the assessment criteria
- use sensitivity analysis to build confidence in the result and to assess the "robustness" of the decision.

A contingency to provide scope for iteration is useful in case negotiations with the preferred supplier fall through and the shortlist needs to be reactivated.

Also, the multi-criteria evaluation process is speeded up if purpose-built or commercially available software tools are used. Suitable software should help make sensitivity analysis of the results easier.

FREQUENT QUESTIONS

There are three questions which seem to crop up frequently:

Q1. Is there any way of checking that the criteria cover everything relevant?

A. To check for consistency and completeness, a good method is to assess the overall objectives of the purchase and then interpret these as a list of criteria. This list should then be organised into a hierarchical form so that related criteria are grouped together.

By inspection, duplications and gaps are then more readily seen.

The hierarchical form is useful even if the structure is subsequently dispensed with. It also encourages a clear definition of the criteria so that all the evaluation team have a common understanding of the scope of each one.

As an example of a hierarchical structure, Figure 3. is an an early draft of the criteria structure proposed for the evaluation of a telecommunications network tender.

Q2. Should the criteria weights be set in advance of the assessment to avoid bias?

A. Both methods can be used although there is less effort involved in assessing weights after the tenders have been rated against the individual criteria.

In most instances however there is a preference for the weights to be set in advance so that the importance of the criteria are fixed before the tenders have been received. This means that the scoring scale must also be well defined in advance otherwise the comparative value of each criterion may not be properly represented.

Q3. Should cost be included with the other criteria or treated separately?

A. Some organisations, particularly if adopting the 'monetary scale' method, are keen to bring everything down to pounds and pence. My own preference is to keep the financial evaluation separate and only combine it with the subjective assessment at the end.

Figure 3 : Example Evaluation Criteria

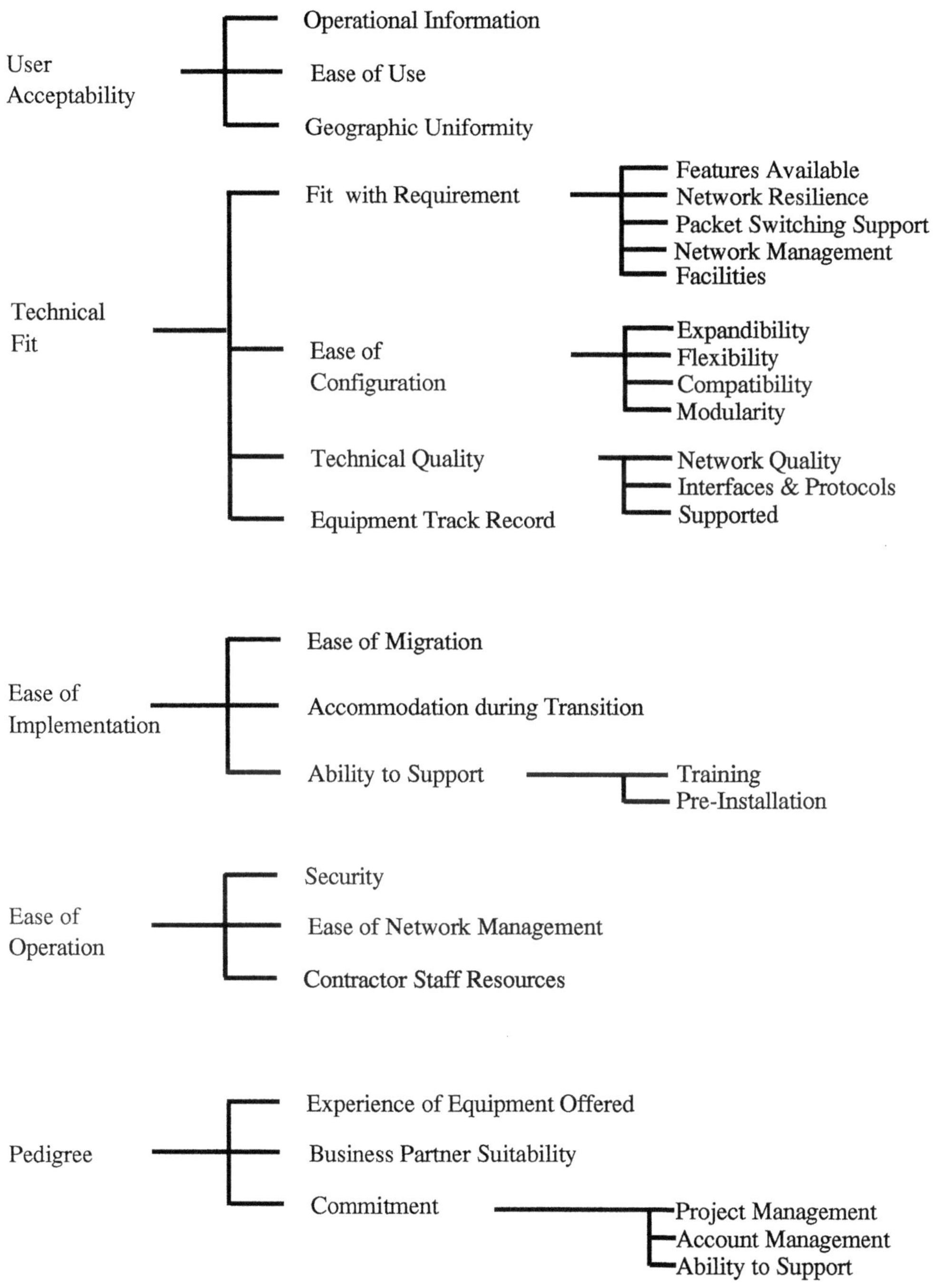

Figure 4. illustrates a graphical method for doing this. The example is fictitious but serves to illustrate how the net present value of the tender (financial benefits net of costs discounted over the lifetime of the purchase) can be compared with the rating assessed for each of the suppliers and their tenders.

All bids which lie significantly above and to the left of any other bid on the graph can be ignored. Sometimes there is a clear cut winner, but more often the bids reflect the old maxim that 'you get what you pay for'. To select the best bid, the remaining contenders are compared with each other. The bids which give less on the rating scale for each increment in net present value are usually rejected.

Figure 4 : Selecting on Cost Vs Rating

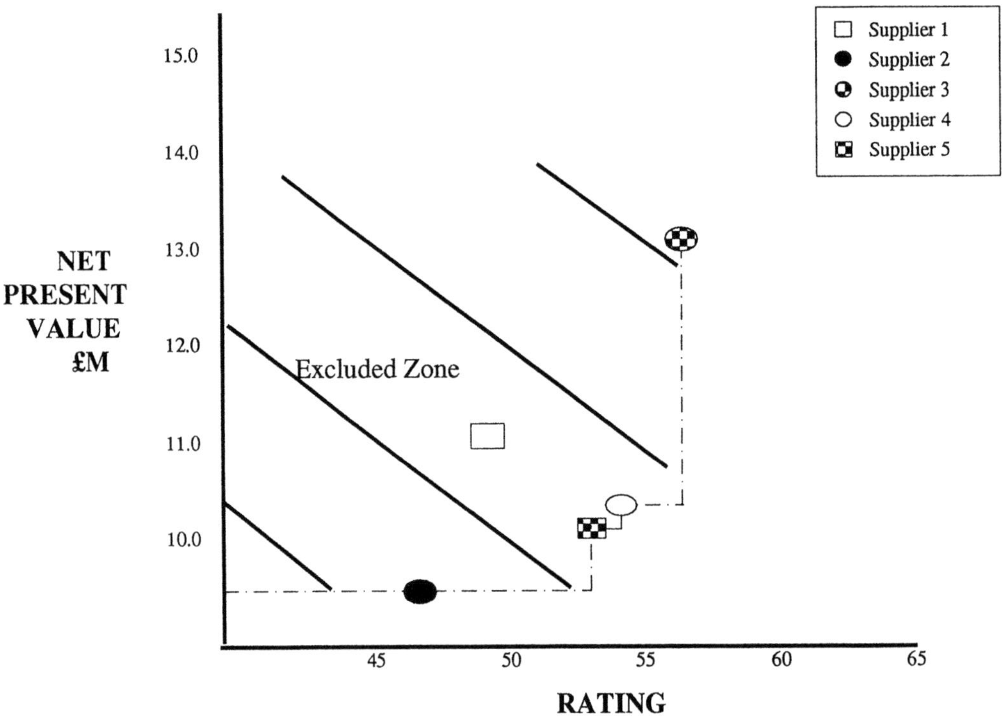

CONCLUSION

To sum up, there are several methods in common use for tender evaluation. The multi-criteria approach tends to be used most often for high value contracts but to achieve its full potential it requires experience of applying the method, and the technical requirement, plus careful management throughout.

Ranking, Monitoring and Control of Research Projects in the Pharmaceutical Industry

Gerd Islei
Manchester Business School
University of Manchester

Barry Cox
ICI Pharmaceuticals
Macclesfield

Mike Stratford
ICI Corporate Management Services
Northwich

Abstract

Following a group exercise to rank projects on a multi-attribute basis using the AHP in a particular area of pharmaceutical research, the work has been extended using MAUA to cover several different aspects

a) the setting of target completion dates for the four stages in research which are used as 'milestones' for monitoring progress
b) the regular review of weighted scores of attributes which are used to assess the need for management action e.g. reallocation of effort, project termination
c) with experience the generation of standard times for stages which will be used to make better forecasts of when compounds will be entering drug development
d) reduction of ambiguity by using structured attributes and detailed word models of the value-scales

This study describes the various stages of the model development and how the system is used to monitor and control research projects in the R & D environment of a large pharmaceutical company.

Introduction

Although R & D portfolio selection has been regarded as an important area of application for multicriteria techniques (Hansen (1982), Gear et al (1982), Haimes and Chankong (1984), Saaty and Kearns (1985)), published evidence describing successful cases does not match these expectations.

The early models tended to approach the problem as one of optimisation, usually with a single objective (Gear et al (1971), Wolfe (1973)). Cost benefit analysis was followed by some modest attempts with more sophisticated methodologies (invariably involving facets of multi-objective programming and aiming at a single decision maker) (Hwang and Masud (1979), Spronk and Zionts (1984)). These models usually require clearly defined systems with unambiguous objective functions. If a problem can be structured this way, then the methods provide extremely powerful results. Hardly any interaction between the decision maker and the decision analyst is needed (in other words the involvement of the manager in the modelling process is only peripheral). Obviously these conditions can only be met in a narrow setting with (technical) criteria that can be easily quantified.

More recently attention has shifted to the use of judgmental models based on subjective data and a variety of techniques have been developed (for an overview see, for example, Keeney and Raiffa (1976), Roy and Vincke (1981), Von Winterfeldt and Edwards (1986) Sawaragi et al (1987)). The most promising of these are usually easily comprehended and evaluated. But they also take into account human attitudes and limitations and support interactive procedures (Saaty (1980), Roy and Vincke (1981), Islei and Lockett (1988)). They emphasise a conceptual framework which is flexible so as to respond to the way practitioners think about the problem while, at the same time, enabling them to transcend habitual patterns (Yu 1986)). "Good" procedures permit frequent feedback between the decision maker's expectations/preferences and the model's prediction in order to maintain practical significance. They allow preferences to develop during the judgmental process.

The case we describe here makes use of various techniques and we show how a simple judgmental model developed over several years to suit the decision maker's requirements. We outline how the model progressed by refining and substructuring important aspects, and sketch the role played by the decision analyst.

Case Background

This case involved the Manchester Business School in a decision task concerning the selection of a research portfolio at a large pharmaceutical company (see Lockett et al (1986) for a full description). The collaboration was initiated against the following background:

A senior manager had been put in charge of a large research section and a vital part of his job concerns the selection of suitable research portfolio. His predecessor had used some form of cost/benefit analysis to assess project worth. This, however, was considered unsatisfactory.

The aim was:

- To develop a model which enabled the senior manager to assess and monitor the progress of research projects

- To include his team of project managers in the formulation and evaluation of the model
- To use this exercise as a means for pulling the team together

Therefore, the senior manager and his team of project managers faced the task of clarifying and assessing a complicated choice problem. But additionally they wanted to develop a judgmental model which could be used subsequently as a tool for project appraisal on a regular basis.
The researchers from the Manchester Business School acted as facilitators who helped the group of managers through unfamiliar routines of structuring the decision task.

Initially it was important to establish a shared understanding of the problem at hand. To develop a judgmental model together proved to be the basis for the group's commitment and the ultimate success of the process.

The final structural hierarchy used in this modelling task had 8 attributes and altogether 10 projects were evaluated. Individual assessments were obtained by using a computer package based on the Analytic Hierarchy Process (AHP) and we refer to Lockett et al (1986) for the discussion. The following Table gives the attribute weights for the participants.

	Project Manager							
Attributes	1	2	3	4	5	6	AVERAGE	Sen.Manager
Biological feasibility ┐	16.3	26.1	10.5	28.3	23.6	18.6	21.6	28.7
Clinical feasibility ┤	6.2	6.2	1.5	3.1	9.0	2.3	4.7	4.3
Chemical feasibility ┘	2.4	2.4	5.0	7.3	5.2	7.4	5.0	4.5
Therapeutic need	32.7	26.0	34.3	25.7	15.2	28.4	27.0	3.7
Competitive position	9.1	7.8	2.8	5.4	3.4	17.5	7.7	14.0
General Support	4.6	5.8	6.8	8.4	12.3	2.7	6.8	5.5
Product champion	19.1	3.1	14.6	10.1	13.9	11.6	12.1	20.7
Ancillary uses	2.9	5.3	4.3	4.8	6.7	6.0	5.0	2.0
Competence of staff	4.6	9.7	4.7	2.4	5.7	3.9	4.7	13.9
Research investment	2.1	7.6	6.5	4.4	4.9	1.5	6.5	6.5

TABLE Attribute Weights

Note : The column AVERAGE shows the mean of the individual results of the 6 project managers. The results of the senior manager are not included

For our purpose it suffices to highlight some features of the results and to note a few of the comments made:

- The attribute Technical Feasibility, which was considered as most important by the senior manager, had been subdivided in this exercise into Biological, Chemical and Clinical Feasibility

- There were remarkable differences in the evaluation between the senior manager and his project managers, which can be seen, for example, when comparing the weights for attributes Therapeutic Need and Competence of Staff
- The group found the AHP procedure, requiring all pairwise comparisons, tedious and also regarded the practical relevance of Saaty's consistency measure to be insignificant. Some considered the spread of their individual results to be too large (see also Islei (1986)). However, they liked the adaptive format of Saaty's approach which structures a decision task as value trees ("analytic hierarchies" in Saaty's terminology).
- All participants were pleased with the insights gained in the decision process. It enabled them to develop their understanding of the factors which are important in research planning.

Progression of the Model

Following this group exercise to establish a simple model as a framework for monitoring and control of the research projects several changes to the procedure were examined. A key aspect for all subsequent development was to keep the time of management involvement to a minimum and to make the judgmental model as transparent as possible. We also recognised that our system which was to be used on a regular basis by the decision maker should not depend significantly on particular skills/expertise of the decision analysts and should eventually become self-sufficient.

One feature which in our experience requires careful explanation by the decision analyst is the ratio scale used in AHP. We found that decision makers invariably relate their comparisons to additive preference scales and therefore any attempt to match new pairwise comparisons with assessments already made would have a larger (multiplicative) than intended (additive) effect on the result (Islei (1986), Lockett and Stratford (1987)). Additionally the eigenvector method adds weight to these misjudgments (Islei (1986)).
Parallel to this it was essential to reduce the data requirements of Saaty's technique which were considered excessive†.

The decision making team at ICI were encouraged by the results of their experiments (see Lockett and Stratford (1987)) with a version of Edward's swing weight technique (Von Winterfeldt and Edwards (1986)) for assessing project scores. In addition the senior manager who initiated the project felt more at ease making judgments within clearly defined reference frames rather than judging pairwise comparisons. Thus the procedure was suitably modified to take account of these requirements.

---

† As a consequence we developed at the Manchester Business School an approach to judgmental modelling based on Geometric Least Square (Islei and Lockett (1988)) which avoids some of these shortcomings. The computer package (called JAS) based on our technique requests comparisons within reference frames and allows the user to examine his preferences and inconsistences largely unassisted (Liddell (1988)).

To facilitate the scoring task for these standard reference frames anchor points were established via detailed word models. This was also seen as having the advantage of making the assessments more transparent and thus easier to communicate. In other words, the changes were considered an important aid to justify the judgments.

Some modifications were also made to the attribute structure. The original subdivisions of Technical Feasibility into Biological, Chemical and Clinical factors was dropped. Instead the senior manager favoured a partition of his most important attribute into the following two components:

(1) Median times as estimated for achieving completion
(2) Technical difficulties as expected during development.

All attribute weights, however, were assumed to be stable over the planning horizon and therefore did not change during this study†.

After making appropriate adjustments to the model the system in its various versions has been used regularly every three months to appraise the performance of the research projects in the portfolio. Figure 1 gives a flow chart of the decision making procedure.

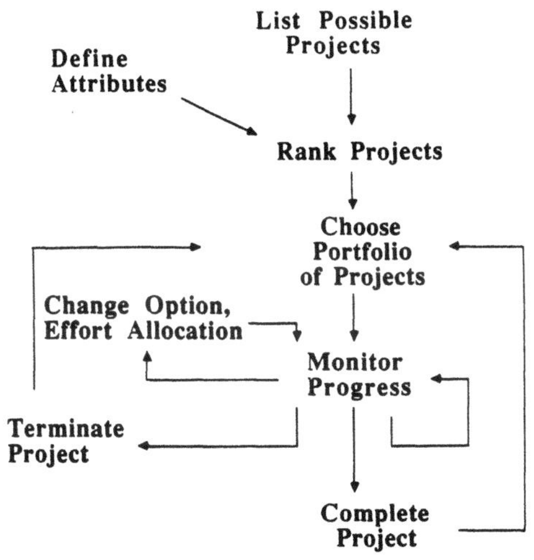

Figure 1 Flowchart of management decisions

† We tested this assumption several times using JAS and found that the attribute weights remained fairly constant. Similarly we noted that the remarkable difference of assessing the attribute weights between the senior manager and his project managers also did not change significantly (forthcoming paper).

To summarise the model development: The initial process involved the decision analysts and the users in a dialogue which enabled all participants to develop their understanding of the modelling task. It helped the senior manager to identify the relevant criteria for his decision making and to move the project selection from a narrow cost efficiency concept to a more balanced approach to improving the overall effectiveness of the R & D appraisal. The new system has been gradually transformed from a very simple and robust judgmental model to an effective personalised management tool which has now been used over a period of several years. This system, however, was not a finished product but an approach which evolved and adapted as the decision maker's understanding and need developed.

Monitoring Performance

- The Acid Test -

We will now describe some of the results which illustrate prominent features of the new system.

An important step in the model development was to replace the early subdivision of the attribute Technical Feasibility (into Biological, Chemical and Clinical) which was felt to lack relevance, with a supplementary structure containing time estimates as one component.

These are derived as follows:

Technical Feasibility in this pharmaceutical research progresses through 4 distinct stages

Stage 1: Exploratory research, development of test
(Mean time ~ 1.5 years)
Stage 2: Establishment of screening cascades
(Mean time ~ 1.5 year)
Stage 3: Identification of lead component
(Mean time ~ 3.5 years)
Stage 4: Identification of development candidate
(Mean time ~ 1 year)

in other words the overall lead time for a project is on average 7-8 years.

To obtain the time estimate of Technical Feasibility for the various projects the manager assesses the probability of completion within the time frame for each of the stages. The probability-time-curve represents the cumulative probability of these four stages (see Figure 2)

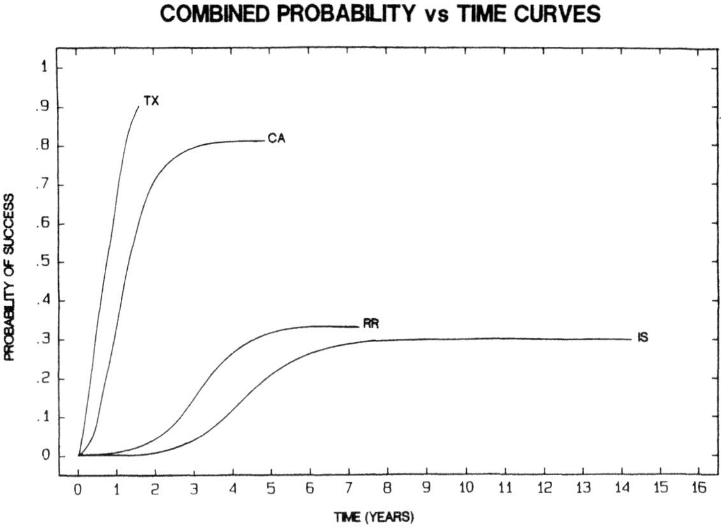

Figure 2 Probability time-curve
- the status quo ante -

These probability-time-curves are then used by the decision maker to make value judgments about the time element of Technical Feasibility. Thus monitoring these assessments provides a basis on which time estimates of the various projects can be compared with the actual times achieved. This has been applied to great effect by the senior manager to adjust an imbalance in his R & D portfolio.

The above Figure 2 represents the position when the decision maker started to make use of the systems. This contrasts with Figure 3 which shows the situation about 2 years later.

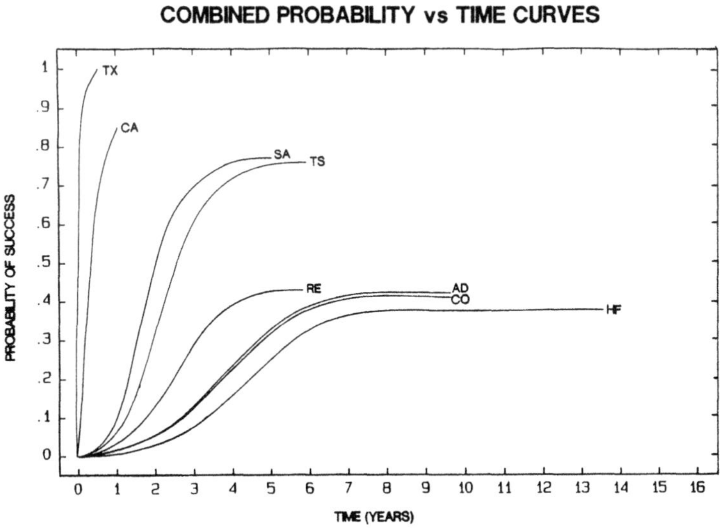

Figure 3 Probability time-curve
- the status quo-

Quite clearly the portfolio in Figure 3 is better balanced and this ensured that subsequently the research section was able to complete projects much more evenly. Although the decision maker in this situation had only partial control over the development of the particular options, he was able to improve the overall performance of his portfolio by focusing on a critical aspect of the project selection.

The next example is illustrated in Figure 4. It shows the scores of project CA for the four most important attributes over a time horizon of about 3 years.

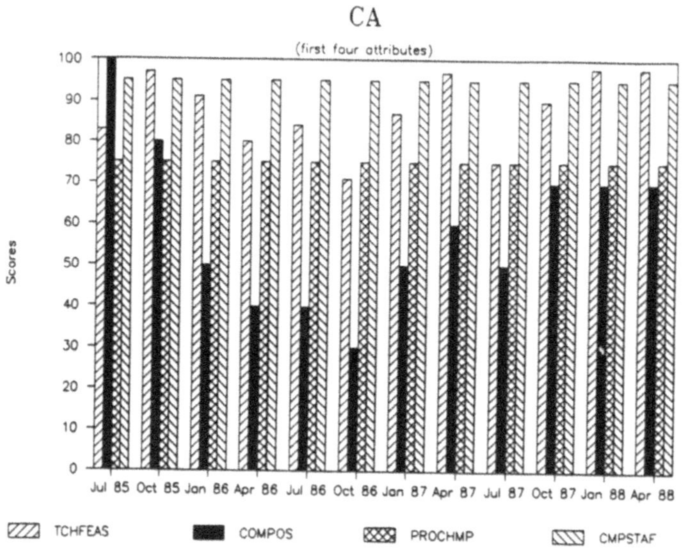

Figure 4 Attribute scores for project CA

TCHFEAS : Technical Feasibility
COMPOS : Competitive Position
PROCHMP : Product Champion
CMPSTAF : Competence of Staff

CA was a new project with apparently no competing product in development by any other pharmaceutical company. This situation, however, changed dramatically in 1986 when some compound was announced by a competitor which had very similar properties. It resulted in a significant drop in Competitive Position of their own project. But fortunately the competing compound, being chemically different, ran into some technical difficulties which enabled their own product to gain ground without any management interference.

The development of project HF over the same period (see Figure 5) required a very different approach by the senior manager.

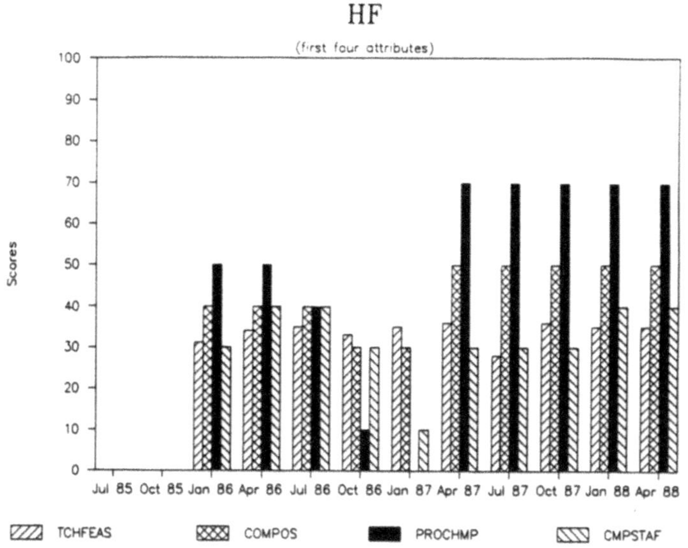

Figure 5 Attribute scores for project HF

It is not too difficult to realise that in this case the manager in charge was lacking the personality, skills and attitude which are necessary to champion a research project to successful completion. This can be seen by looking at the progression of PROCHMP over the period Jan.86-Jan.87. Apparently his performance also had a detrimental effect on the research staff working under him (see CMPSTAF). Eventually he was replaced (Apr. 87) by a different project manager whose performance not only improved the achievements of the team but also affected the Competitive Position positively.

Conclusion

In this paper we presented an application of R & D portfolio modelling in the pharmaceutical industry. We described how a very simple and robust judgmental model was developed and transformed into an effective management tool which permits a detailed performance appraisal. It provides the basis for monitoring the progression of the projects and thus enables the senior manager to focus on critical developments which may require specific management attention/intervention.

In the early stage of the systems development the advice and guidance of the analyst was essential. However, gradually the models and associated concepts were being subsumed into the overall planning process. They no longer need decision analysts for day-to-day support; advice is now sought only for key developments.

We also showed how the use of AHP in the initial model was subsequently replaced by an approach based on a version of Edward's swing weight technique for obtaining scores of the projects' performances.This was complemented by the application of the Judgmental Analysis System (JAS) developed at the Manchester Business School which has proved successful for assessing long term strategic components of the model such as attribute weights.

Overall the system has been very effective, so much so that the approach has now been adopted by other departments in the organisation.

References

Gear, A.E., Lockett, A.G., and Pearson, A.W. (1971)
"An Analysis of some portfolio selection models for R & D", IEEE Trans. Eng. Manag. Vol. EM-18, 66-77.

Gear, A.E., Lockett, A.G., and Muhlemann, A.P. (1982)
"A unified approach to the acquisition of subjective data in R & D", IEEE Trans. Eng. Mang. Vol EM-29, 11-19.

Haimes, Y.Y. and Chankong, V. (eds.) (1985)
Decision Making with Multiple Objectives
Lecture Notes in Economics and Mathematical Systems, Vol. 242, Springer, Berlin.

Hwang, C.L. and Masud, A.S.M. (eds.) (1979)
Multiple Objective Decision Making - Methods and Applications
Lecture Notes in Economics and Mathematical Systems, Vol. 164, Springer, Berlin.

Hansen, P. (ed.) (1983)
Essays and Surveys on Multiple Criteria Decision Making
Lecture Notes in Economics and Mathematical Systems, Vol. 209, Springer, Berlin.

Islei, G. (1986)
"An approach to measuring consistency of preference vector derivations using least square distance" in : J. Jahn and W. Krabs (eds.), **Recent Advances and Historical Development of Vector Optimization**, Lecture Notes in Economics and Mathematical Systems, Vol. 294, Springer, Berlin, 265-287.

Islei, G. and Lockett, A.G. (1988),
"Judgemental modelling based on geometric least square", European Journal of Operations Research, 36, 27-35.

Keeney, R.L. and Raiffa, H. (1976)
Decisions with Multiple Objectives
John Wiley and Sons, New York.

Liddell, J. (1988)
"Decisions, decisions..."
Management Topics, IBM United Kingdom Ltd., Issue 31, 20-23.

Lockett, A.G., Hetherington, B., Yallup. P., Stratford, M. and Cox, B. (1986)
"Modelling a research portfolio using AHP : a group decision process"
R & D Mgmt. 16, 151-160.

Lockett, A.G. and Stratford, M. (1987)
"Ranking of research projects : experiments with two methods", OMEGA, 15, 395-400.

Roy, B. and Vincke, B. (1981)
"Multicriteria analysis : survey and new directions", European Journal of Operations Research, 8, 207-218.

Saaty, T.L. (1980)
The Analytic Hierarchy Process
McGraw-Hill, New York.

Saaty, T.L. and Kearns, K.P. (1985)
Analytic Planning
Pergamon Press, Oxford.

Sawaragi, Y., Inoue, K. and Nakayama, H. (eds.) (1987)
Towards Interactive and Intelligent Decision Support Systems
Lecture Notes in Economics and Mathematical Systems, Vol. 286, Springer, Berlin.

Spronk, J. and Zionts, S. (eds.) (1984)
Special Issue on Multiple Criteria Decision Making,
Management Science, 30, 1265-1387,

Von Winterfeldt, D. and Edwards, W. (1986)
Decision Analysis and Behavioural Research
Cambridge University Press, Cambridge.

Wolfe, J.N. (ed.) (1973)
Cost Benefit and Cost Effectiveness
Allen & Unwin, London.

Yu, P.L. (1986)
"Foundation of effective goal setting"
in : J.Jahn and W. Krabs (eds.) **Recent Advances and Historical Development of Vector Optimization**, Lecture Notes in Economics and Mathematical Systems, Vol. 294, Springer, Berlin, 317-342.

THE ANALYSIS OF LIQUID MILK SUPPLY TO URBAN AGGLOMERATION BY BICRITERION LINEAR TRANSPORTATION MODEL WITH THE MPSX/370 PACKAGE

W. Szymanowski[1] and W. Ogryczak[2]

[1]Agricultural University of Warsaw
Warszawa, Poland

[2]Institute of Informatics
Warsaw University
Warsaw, Poland

Abstract

In the paper we formulate a bicriterion linear transportation model to analyse the raw milk supply for an urban agglomeration. The model has been verified for Warsaw. To solve the bicriterion model the lexicographic as well as the parametric methods were used. The model can be considered as an element of a Multicriterial Decision Support System of raw milk supply for Warsaw.

1. Introduction

The rapid growth of the urban population as well as the nutrition value of milk have caused a fast increase in the milk production and consumption. In 1984 the consumption of liquid milk in Polish big cities was on the level of 1.2 milliard litres, that is 46 percent of total amount of liquid milk produced in Poland. The amount of the collected raw milk around large agglomerations lags behind the increase in the consumption of liquid milk in big cities. Warsaw was the best example of this problem. In 1984, the modelled system of milk supply to Warsaw consisted of four stages. The system covered 38 regional milk cooperatives in 9 districts of Poland, which supplied about 210 million litres of raw milk to four city milk plants in Warsaw. The raw milk was transported each day from more than 63 thousand suppliers through almost 18000 permanent milk reception stations. 50 percent of that amount was delivered from the Warsaw District The expansion of the milk transfer area to supply Warsaw resulted in an increase of the transportation radius as well as in higher transportartion costs.

The forecast of the Central Union of Dairy Cooperatives determines that in 1990 raw milk collected around Warsaw will cover 20.7 percent of total amount of milk consumption. Till the 2 000, the gap between supply and demand for the raw milk will be much deeper. It also have an influence on the energy consumption, because the dairy industry is one of the most energy intensive industries between food industries in Poland.

The aim of the paper is to show the possibilities of the liquid milk supply to Warsaw by the optimization methods. This analysis have been achieved by a bicriterion linear transportation model. To predict the future possible transfer area of supplying Warsaw for 1990 we have made ex-post analysis for the period of 1984-1986. The model was solved by lexicographic as well as by parametric methods. The bicriterion analysis was performed with MPSX/370 package, which is regarded as a standard commercial tool to solve linear programming problems.

2. The statement of the model

The model of the milk marketing system for urban agglomeration is presented on Fig. 1. One of the main functions of this system is transportation of goods from the producers to the final customers. This transportation process consists of the following homogenous three parts /see: [2] :

1. the collection of milk as a raw material or semiproducts to the milk processing plants,
2. the transportation of milk as a raw material or semiproducts between milk processing plants and city milk plants,
3. the distribution of liquid milk or liquid and solid milk products from the city milk plants to the retailing customers.

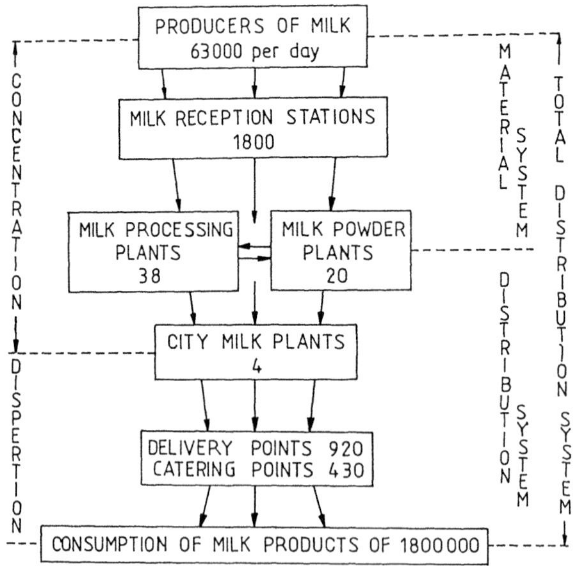

Elements of total physical milk marketing system for Warsaw

The analysis of the liquid milk supply to Warsaw concerns inter-plant transportation process. To formulate a bicriterion linear transportation model of this process we have to accept the following assumptions:

- regions supplying milk reception stations as well as supplying milk processing plants are stable, therefore collection process of raw milk is out of our optimization process /see: [1] /,
- the distribution of final dairy products is analysed separately by Vehicle Routing and Scheduling Models, so it is out of our optimization process /see: [6]/,
- three kinds of means of transportation separately are taken into considaration: 5 000, 10 000 and 18 000 litres road tankers. Raw milk within Warsaw District is collected exclusively by 5 000 litres road tankers,
- there are two models: a "Summer" and a "Winter" model, which describes seasonal fluctuations in the purchase of raw milk and the variations of its volume necessary to produce liquid milk in Warsaw. The "Summer" model uses data for July. The "Winter" model uses data for January. In the latter a possibility of milk regeneration from milk powder has been assumed,
- milk transported to Warsaw is standardized and is a surplus of milk purchase over and above the local market needs, which provides liquid dairy products or liquid and solid dairy products.

First we formulate the "Summer" model, which will be also a basis to develope the statement of the "Winter" model. The main purpose of the "Summer" version of the bicriterion linear transportation model is to determine the destinations and the amount of raw milk /in litres/ transported to Warsaw by different types of road tankers:

$$x^{r}_{l_1 m} \geqslant 0 \quad \text{for} \quad l_1 = 1,2,\ldots,38; \quad m = 1,2,3,4; \quad r = 1,2,3; \qquad /1/$$

where: l_1 - index of milk processing plant;

m - index of city milk plant in Warsaw;

r - index of type of road tankers.

Total amount of raw milk transported from each milk processing plant to all city milk plants in Warsaw cannot be greatrer than the raw milk surplus a_{l_1} over the l_1 local market needs.

$$\sum_{r=1}^{3} \sum_{m=1}^{4} x^{r}_{l_1 m} \leqslant a_{l_1} \qquad \text{for } l_1 = 1,2,\ldots,38; \qquad /2/$$

where: a_{l_1} - difference between the raw milk supply and the demand of the local market.

Similarly, the total amount of raw milk transported from several milk processing plants to each city milk plant in Warsaw is equal to its demand b_m of milk needed to produce the liquid milk.

$$\sum_{r=1}^{3} \sum_{l_1=1}^{38} x^{r}_{l_1 m} = b \qquad \text{for} \quad M = 1,2,3,4; \qquad /3/$$

There are two criteria for the optimization. The first one is the minimization of the total transportation costs and the costs of purchase of the raw milk transported to Warsaw.

The second one is the minimization of the total energy consumption including energy used in the transportation process and in the several stages of the liquid milk production process. These two criteria are independent each of other, because the first one is created by unit price of raw milk paid to each milk processing plant as well as by the transportation tarrifs paid to carrier for the raw milk transported to Warsaw. The second criterion is created by the energy consumption used in several stages of the liquid milk production process.

$$\underset{\left\{x^{r}_{l_1 m}\right\}}{\text{minimize}} \quad f_1 = \sum_{r=1}^{3} \sum_{l_1=1}^{38} \sum_{m=1}^{4} \left(KF_{l_1} + KT^{r}_{l_1 m}\right) x^{r}_{l_1 m} \qquad /4/$$

$$\underset{\left\{x^{r}_{l_1 m}\right\}}{\text{minimize}} \quad f_2 = \sum_{r=1}^{3} \sum_{l_1=1}^{38} \sum_{m=1}^{4} \left(EF_{l_1} + ET^{r}_{l_1 m} + EF_m\right) x^{r}_{l_1 m} \qquad /5/$$

where: KF_{l_1} - unit price of raw milk paid to each milk processing plant,

$KT^{r}_{l_1 m}$ - unit transportation cost of raw milk transported between milk processing plant l_1 and city milk plant m in Warsaw by r kind of road tankers,

EF_{l_1}, EF_m - energy consumption on the several stages of unit liquid milk production,

$ET^{r}_{l_1 m}$ - energy consumption of raw milk unit transported between milk processing plant l_1 and city milk plant m in Warsaw by r kind of road tankers.

The equations /1/-/5/ describe the "Summer" model of liquid milk supply to Warsaw. The "Winter" model is constructed on the basis of the "Summer" model /1/-/5/. The main purpose of the "Winter" version of the bicriterion linear transportation model is to determine the destinations and the amount of raw milk as well as a raw milk to produce the milk powder for the liquid milk regeneration in Warsaw:

$$x^r_{l_1 m} \geq 0, \quad y_{l_2 m} \geq 0 \quad \text{for } l_1 = 1,2,\ldots,38; \; m = 1,2,3,4; \; r = 1,2,3; \; l_2 = 1,2,\ldots,20; \qquad /1a/$$

where: y_{l_2} - amount of raw milk to produce milk powder for liquid milk regeneration in m city milk plant in Warsaw transported by 5 ton vehicles.

Total amount of raw milk to produce the milk-powder for the liquid milk regeneration in all city milk plants in Warsaw is equaled to the raw milk surplus over the local market needs /see: /2a//.

$$\sum_{m=1}^{4} y_{l_2 m} \leq a_{l_2} \qquad \text{for } l_2 = 1,2,\ldots,20; \qquad /2a/$$

where: a_{l_2} - storing capacity of milk-powder plant of raw milk.

Similarly, the total amount of raw milk as well as the amount of raw milk to produce the milk-powder for the liquid milk regeneration in each city milk plant m in Warsaw is equal to its demand. The demand of liquid milk b_m is given in litres:

$$\sum_{l_1=1}^{38} \sum_{r=1}^{3} x^r_{l_1 m} + \sum_{l_2=1}^{20} y_{l_2 m} = b_m \qquad \text{for } m = 1,2,3,4; \qquad /3a/$$

The "Winter" model fulfills the following two criteria of optimization:

$$\underset{\left\{x^r_{l_1 m}, y_{l_2 m}\right\}}{\text{minimize}} \; f_1 = \sum_{r=1}^{3} \sum_{l_1=1}^{38} \sum_{m=1}^{4} \left(KF_{l_1} + KT^r_{l_1 m}\right) x^r_{l_1 m} + \sum_{l_2=1}^{20} \sum_{m=1}^{4} \left(KF_{l_2} + KT_{l_2 m} + \overline{KF}_m\right) y_{l_2 m} \qquad /4a/$$

$$\underset{\left\{x^r_{l_1m},\ y_{l_2m}\right\}}{\text{minimize}} \quad \sum_{r=1}^{3} \sum_{l_1=1}^{38} \sum_{m=1}^{4} \left(EF_{l_1} + ET_{l_1m} + EF_m\right) x^r_{l_1m} \ + \sum_{l_2=1}^{20} \sum_{m=1}^{4} \left(CF_{l_2} + ET_{l_2m} + \overline{EF}_m\right) y_{l_2m} \qquad /5a/$$

where: KF_{l_2} - unit price of milk-powder paid to each milk-powder plant,

KT_{l_2m} - unit transportation cost of milk-powder transported between milk-powder plant l_2 and city milk plant m,

$\overline{KF}_m$ - unit storing cost of milk-powder paid by each city milk plant m,

EF_{l_2} - energy consumption of storing raw milk to produce a unit of milk-powder in milk-powder plant l_2,

ET_{l_2m} - energy consumption of raw milk unit transported between milk-powder plant l_2 and city milk plant m by 5 ton vehicles,

$\overline{EF}_m$ - energy consumption to regenerate a unit of liquid milk from milk-powder plant l_2 in city milk plant m in Warsaw.

The first criterion effects the minimization of the total cost of purchase of the raw milk as well as the milk-powder to regenerate the liquid milk in Warsaw and their transportation costs. The second criterion leads to the minimization of the total energy consumption in the transportation process as well as in the several stages of the liquid milk production in Warsaw.

3. Simulations for 1984 and 1990 models

The "Winter" and the "Summer" models are of a different size. The "Summer" model consists of 245 decision variables and 42 constraints, whereas the "Winter" model contains 259 variables and 60 constraints.

For ex-post optimization the period of three years 1984-1986 has been selected. Exactly, July, as the period of the minimum level of the milk transfer area, has been used in the "Summer" model and January, as the period of the maximum level, has been used in the "Winter" model. Over 110 experiments were made for 1984-1986: 36 experiments for each of three years. Half of them with the "Winter" model and half with the "Summer" one. The experiments includes:

- three types of the road tankers: 5 000, 10 000, and 18 000 litres,
- two kinds of local markets offering the liquid milk products or liquid and solid ones,
- three kinds of the optimization techniques: the lexicographic method for both hierarchies of the objective functions and the parametric analysis to identify objective functions and the parametric analysis to identify all the efficient vertices. The computations were performed with the MPSX/370 package see [3].

The results of the experiments for the lexicographic method were compared with the real routes of the raw milk transported to Warsaw in 1984 for the "Winter" and "Summer" models. To get more comparable results we assumed: raw milk was transported only by 18 000 litres road tankers and the local markets were offering both: the liquid and the solid milk products. The results are presented in Table 1.

Table 1

The results of ex-post analysis for "Winter" and "Summer" models for 1984 obtained by the lexicographic method

Criteria	"Winter" model			"Summer" model		
	Computer solution	Real solution	% of sav.	Computer solution	Real solution	% of sav.
"Energy" /MJ/1000l/	718,1	809,5	12.8	732,5	830,6	13.4
"Cost" /zł/1000l/	27897	35185	26.1	18998	27936	47.0

The parametric analysis allows us to create all the Pareto - optimal efficient solutions to the model /1/ - /5/ or /1a/ - /5a/. The efficient frontier were generated by the standard parametric procedure of the MPSX/370 package by solving the following problem:

$$\text{minimize} \left\{ f_{k_1} + \alpha\left(f_{k_2} - f_{k_1}\right),\ \alpha\in[0,\ 1] \ \text{subject to: /1/-/3/ or /1a/ - /3a/} \right\} \qquad /6/$$

The bicriterion transportation models /1/-/5/ and /1a/-/5a/ were used to determine the future destinations and amount of the raw milk transported to Warsaw. Optimization was made for 1990 including period of the maximum level of the milk transfer area to Warsaw in the summer /in July/ and in the minimum one in the winter /in January/. For the 1990 - model the additional assumptions - were taken into considaration:

- two level of the milk consumption in Warsaw: at the level of 96 and 120 litres per capita annually respectively;

- two alternatives of the forecasted raw milk purchase prices and transportation tariffs;
- two kinds of the road tankers /5 000 and 18 000 litres/ respectively;
- two kinds of the local markets providing the liquid milk products or the liquid and solid ones respectively;
- a various Warsaw supplying area covering nine districts or only the Warsaw Macroregion /6 districts/ respectively;
- a stable use of energy in the forecasted period including transportation energy and processing energy in several stages of liquid milk production.

For the 1990 model, 48 experiments have been made including 36 experiments for the "Winter" model and 12 for the "Summer" one. The experiments used three kinds of optimization techniques: lexicographic method with two hierarchies of the objective functions as well as the parametric analysis which treated both the objective functions simultaneously.

4. Conclutions

The main application of the model is the analysis of the circumstances that will have influence in the choice of the best raw milk supply routes to Warsaw for both criteria functions: "energy" and "cost" in the future. The most important circumstances concerns:

- changes in the level and in the structure of the milk consumption in Warsaw,
- seasonal fluctuation of the milk purchase level,
- changes in the forecasted prices of raw milk purchase and in the forecasted transportation tariffs,
- changes in the depth of the local markets, i.e. in the level of raw milk surplus over and above the local market needs, predestined for Warsaw,
- changes in the policy of acquiring new vehicles for the raw milk transportation to Warsaw.

The bicriterion transportation model and the procedures of its solution discussed in this paper can be implemented on the microcomputer type IBM. PC/AT as the first step to construct the Multicriterial Decision Support System /MCDSS/ for the raw milk supply for Warsaw.

[1] P. Cegiełła, W. Szymanowski, J. Prokuratorski: "COLOS: A collection optimization system as an application of the vehicle scheduling problem". EJOR 27 /1986/, pp. 82-90.

[2] "Costs of transport in relation to dairy factory structure". Bulletin of International Dairy Federation. Brussels. Document No 162, 1983.

[3] "IBM Mathematical programming system extended/370 - MPSX/370: program reference manual", IBM SH-19-1095, 1979.

[4] I. Nykowski: "Bicriterion linear transportation problem" /in Polish/. Przegląd Statystyczny 39 /1986/, pp. 9-25.

[5] W. Ogryczak: "Mathematical programming with the MPSX package /in Polish/, Warsaw University Press, Warsaw 1989.

[6] W. Szymanowski, P. Cegiełła, J. Prokuratorski: "The application of discrete programming to the design of delivery systems in large urban agglomerations", EJOR 10 /1982/, pp. 100-105.

[7] W. Szymanowski, W. Ogryczak: "Models of the foodstuff distribution organization channels for large urban agglomerations", Preprints of the 5-th IFAC/IFIP/IFORS Internationale Conference "Control on Transportation Systems", Vienna 1986, pp. 75-82.

DESIRABLE PROPERTIES OF INTERACTIVE MULTI-OBJECTIVE PROGRAMMING METHODS

HANS G. DAELLENBACH
UNIVERSITY OF CANTERBURY, CHRISTCHURCH, N.Z.
and
JOHN T. BUCHANAN
UNIVERSITY OF WAIKATO, HAMILTON, N.Z.

ABSTRACT

Based on the results of a laboratory experiment and the literature on MCDM and decision making behaviour, this paper draws a number of conclusions as to the general strategy of interactive MCDM methods and some of the desirable properties they should exhibit. It postulates that MCDM is decision making under ambiguity. As a result the traditional problem solving paradigm underlying management science has to be augmented to cater for the special characteristics of decision making under ambiguity.

1. INTRODUCTION

In a recent paper [1987] we described a laboratory experiment which compares the performance, from a user's point of view, of four interactive methods for solving multiple objective programming (MOP) problems. The methods used were the method of Zionts and Wallenius (ZW) [1976, 1983], the Surrogate Worth tradeoff methods of Haimes and Hall (SWT) [1974], the method of Steuer and Choo (SC) [1983], and a naive method (Buchanan [1985]) using a MAXMIN formulation. The naive method maximizes the minimum percentage achievement level (see Posner and Wu [1981]), where at each iteration the decision maker (DM) specifies a desired solution and the method finds the closest efficient solution whose individual objective values will be in about the same proportions as the desired solution. These four methods were chosen because they provide a good cross section of the various approaches to MOP problems. The SWT method requires pairwise cardinal tradeoff information, and the ZW method pairwise ordinal tradeoff information - both terminating "internally". The naive and the SC method ask for holistic assessments and both are terminated by the DM. For the ZW and naive methods the DM only gets to see a local picture, while the SC method gives a more global picture, with the SWT method somewhere in between.

Twenty-four staff members and graduate students in business administration and operations research each used these four methods to solve a three-objective production scheduling problem under four different scenarios as to future sales and cost environment - the problem used by Wallenius [1975], adapted for N.Z. conditions. 75 percent of the participants had at least two years of practical commercial experience. The reader is referred to the original paper for a complete description of the experimental design and statistical analysis made. Section 2 briefly summarizes the major results of this experiment.

Two a priori unexpected findings came out of this experiment. First, users preferred methods that allowed them sufficient time to explore the solution space before homing in on the preferred solution. Second, users wanted to be in control of the solution search, rather than relinquish control to the "computer". These desires can be explained in part by an intolerance to ambiguity (Gimpl [1984,1985]). Section 3 postulates that MCDM should be classified as decision making under ambiguity and explores some of the implications of this on interactive MOP methods (IMOPM). Section 4 draws some conclusions from this discussion as to desirable properties for IMOPMs.

2. SUMMARY OF RESULTS OF LABORATORY EXPERIMENT

Each method was assessed by each participant using six criteria: (1) the relative preference for using each method; (2) the participant's confidence that the solution method actually found her/his preferred solution; (3) the ease of using the method; (4) the ease of understanding the overall logic the method uses to find the preferred solution; (5) CPU time; and (6) solution search elapse time. All criteria were rated on a scale from 0 to 10. Figure 1 shows the histograms of the ratings obtained for the first four criteria, and Figure 2 depicts the rankings derived from these ratings. The average rating or average ranking is shown with each corresponding histogram. An asterisk next to an average rating or ranking denotes a statistically significant difference from all other methods at the 5 percent level of significance.

Note that these results are not intended as an indictment or an indication of superiority of one or another IMOPM, but rather as a measure of how the participants view the approach and principles used by the methods. Furthermore, the results are strictly speaking only valid for this particular experiment, which involved only three objectives. Different rankings might have been found for problems with more objectives.

The ZW method fared worst under the first two criteria, preference and confidence in solution, but showed the lowest elapsed times - only 50% of the next higher one (SC method). In general, the participants felt that the method did not involve them sufficiently and that they had little control over the solution method. As a result, they had doubts that the solution found was "their" solution, rather than the "computer's" solution. The method did not seem to give the participants sufficient time to become confident about their own preference structure before it produced the "preferred" solution. Thus, it seems that two of the method's strengths, namely needing to ask comparatively few paired tradeoff questions and making maximum use of the information provided to infer a preference structure, resulted in lower ratings on preference and confidence. Some participants found it difficult to choose between very close alternatives, or alternatives that were very far apart, or between two alternatives that were both considered bad.

As expected, the SWT method was found the most difficult and tedious to use, since it required cardinal information. It also was ranked lowest in terms of ease of understanding its logic. For many it took on the character of a "black box". However, participants appreciated the graphical displays and the opportunity to edit and change their assessments.

The SC method ranked highest on all four criteria reported here. This suggests that most participants had a global approach to decision making. They liked getting an initial broad overview of the solution space and then being able to narrow down the search to smaller and smaller areas. Some participants found it difficult to assimilate all the information contained in the six solutions presented at each iteration. This difficulty is likely to increase for more than three objectives, as confirmed by anecdotal information reported by one of the referees of our 1987 paper.

The above point suggests that there is a conflict between the desire for global information and the user's ability to process it. This difficulty may though be a consequence of the perceived "pressure" under which the user of an interactive program may feel he/she is while sitting in front of the console. If encouraged to take sufficient time - hours or days rather than seconds or minutes - to digest the information, this difficulty may largely disappear. This is at least the findings of Martinson [1977], who developed an experimental "interactive" Minmax MOLP package for use in U.S. Bureau of Land Management district offices for the development of multiple-use land district schemes. During the solution phase, managers were supplied hard copy of land allocations and percentage goal achievement levels for several sets of goal importance weights. Typically the number of goals was in the range of 5 to 10. This output was carefully studied and discussed among the district staff, as well as with the analyst, which lead to updating runs with one or several new weight structures. This process continued until a preferred solution was identified. Given the ample time between successive updating runs, the users did not experience difficulties to digest the wealth of information supplied by the program after each updating run. It was also during these extended periods between updating runs that the participants slowly and deliberately firmed up their goal structure for each district scheme studied. Users particularly appreciated to have percentage achievement levels reported at each updating runs, rather than just the absolute values. They felt that the former made comparisons between different solutions more meaningful.

A number of other points were repeatedly made by participants. Many wanted more flexibility, in terms of being able to change previous assessments after seeing their effect, and being able to backtrack to change responses given earlier, but later perceived to be inconsistent at a subsequent iteration. Several wanted the facility to constrain one of the objectives to a given level. Some "guiding influence" provided by the solution method was preferred to none, while still leaving the user in overall control.

3. MCDM - DECISION MAKING UNDER AMBIGUITY

Traditional decision making is classified according to decision making under certainty and decision making under risk or uncertainty (where the latter often refers to the absence of information about the underlying probability distributions). There is a third category - decision making under ambiguity (Gimpl [1984, 1985]). Budner [1962] defines ambiguity as involving situations that exhibit one or more of the following features:

(1) They are characterized by a lack of familiarity, viz. new situations with few or no familiar cues.

(2) They are highly complex, i.e., involve processing many

interactions and cues.

(3) They involve contradictory elements and cues that each suggest different responses.

MCDM problems usually exhibit the latter two and often all three features. Gimpl adds that ambiguous situations are also characterized by lack of goal formation and hence, by analogy, formation of a preference structure. A DM may not be able to form goals until he/she has a sufficient "feel" for what is available or possible. Means and ends are not separable in the DM's mind. Any MCDM method which is based on the premise that the DM has formed explicit goals and that all the method has to do is to externalize her/his goal preference structure or the underlying utility function may thus not serve the DM well. This point is also made by a number of other researchers (Fischoff, Slovic, and Lichtenstein [1980], Baum and Carlson [1974], and French [1984]). French puts it eloquently: "Briefly, I believe that a good decision aid should help the decision maker explore not just the problem, but also himself."

Budner [1962] suggests that persons intolerant of ambiguity react to novel, complex, and/or contradictory situations by repression or denial, anxiety and discomfort, destructive or reconstructive behaviour, or avoidance of problem solving. Persons intolerant of ambiguity have been shown to cling longer to inappropriate or wrong beliefs, to find it harder to shift to goals consistent with rapid change in their environment, and hence stick to old patterns of decision making longer. Destructive and reconstructive behaviour leads to premature goal setting and a demand for uniformity (Frenkel-Brunswick [1949]).

MCDM aids for decision making under ambiguity thus have to expand the traditional O.R. paradigm of problem formulation - identification of the DM(s), her/his/their objectives, the available courses of action, and the environment - followed by mathematical modelling, derivation of a solution, etc.. (See also Keen [1977].) In view of the above findings, it may be important that the DM is restrained from premature goal setting (Gimpl [1984]). Instead, as Schon [1983] suggests under ambiguity the DM should perform experiments: (1) exploratory experiments to become familiar with the situation - moves made to learn more about the "topography" of the situation, without any pre-conceptions, and (2) directed experiments - deliberate testing moves undertaken with an end in mind, such as finding how far it is to the top of the "church hill". Gimpl argues that it may in fact be optimal to use random moves for the exploratory experiments. Within the IMOPM framework, exploratory experiments will allow the DM to become more familiar with decision space (the feasible region) and the attribute space, possibly prior to goal formation. Directed experiments will lead to goal and preference structure formation, and ultimately to finding the preferred solution. So MCDM methods should not aim to be only normative in a narrow sense, but to be also at least partially descriptive for thorough exploration of the problem.

Note in passing that the SC method uses this kind of strategy. If the efficient solutions presented to the DM are expressed in the form of attribute achievement levels, the method starts out with a randomly generated set of solutions which allows the DM to acquire some familiarity with the problem, i.e., remove some of the ambiguity, possibly prior to any firm goal and preference structure formation. This is then followed by directed moves to explore smaller and smaller neighbourhoods. This could well explain its high rating in the laboratory experiments.

It is also well recognized in the literature on organizational behaviour, that as the decision maker learns more about the problem her/his preferences may change. The satisfaction-search-aspiration model developed by March and Simon [1958, pp. 48-52] corroborates that under limited bounded rationality goal formation is not independent of how the DM learns about the solution space. They state that the level of satisfaction is an increasing function of the expected value of rewards and a decreasing function of the level of aspiration. The lower the level of satisfaction experienced by a subject, the higher will be her/his level of search, provided the search environment is perceived as benign. Increase in search in turn tends to increase the expected level of rewards and an increase in the latter will result in an increase in the level of aspiration, which closes the cycle of interactions. Some years later March [1978] stated that this type of decision making behaviour is a form of intelligence to be refined by technology rather than corrected by it (as flawed).

Again the implication is that IMOPMs should not only support the DM in finding the preferred solution, but help her/him in the formation of goals. Furthermore, the manner in which this process is approached may in fact bias goal formation and, consequently, what is produced by the method as "the preferred solution".

This conclusion is re-inforced by French's concern [1984] about the anchoring effect of the initial solution (see Kahneman, Slovic, and Tversky [1982]). In revising an initial judgment, a subject tends to pull back to that judgment. This problem is well-recognized in subjective probability assessment. French asks whether an arbitrary choice for the initial solution may not have a similar anchoring effect and if the user of an IMOPM would end up with a different preferred alternative had he/she started from a different initial solution.

In the same paper, French also raises the potential effects of intransitivities and other departures of the DM's stated preferences from the assigned value-function model and notes that many of the IMOPMs will 'converge' before such departures become apparent. This is echoed by the concern raised by a number of participants in the MCDM experiment about low level of involvement in and lack of control over the solution process, particularly with the ZW method.

4. INFERENCES ON SOME DESIRABLE PROPERTIES OF INTERACTIVE MOP METHODS

Oppenheimer [1978], Soland [1979], and other researchers have produced lists of properties that IMOPMs should satisfy to be effective tools (such as being relatively easy for the DM to use, not demanding too much or unnecessarily complicated information, not being overly "programmed", not imposing a rigid preconceived mathematical structure on the DM, being robust in terms of the mathematical properties assumed by the model, using graphical displays effectively, etc.). The findings of the IMOPM laboratory experiment and the issues raised above lead us to add several other desirable properties to those lists:

(1) IMOPMs should allow the DM to operate in a dual mode: either (a) to remain in control of the "solution" procedure or (b) to let the method take over control of the process, with the option of switching back and forth between the two modes at each iteration.

(2) As starting point in the solution procedure, the IMOPM should list the "best" (maximum or minimum) feasible achievement levels for all attributes specified by the DM. This will give the DM an indication of the "ideal solution" (Zeleny [1976]). It has a dual purpose: it is likely to provide a "desirable" anchor and it will lead to set a high initial aspiration level.

(3) As an initial phase, the IMOPM should allow the DM to explore the solution space in terms of attributes as well as a subset of decision variables identified as "important" by the DM. The purpose of this phase is two-fold: to let the DM get a feel for what is possible and how the various attributes interact with each other, and to facilitate both goal and preference structure formation or consolidation. By this process the DM will be able to decide which of the attributes studied will assume the character of goals or objectives. This requires two facilities:

(a) The method should start out by presenting a "representative" subset of efficient solutions in terms of attributes (note goals may not have been firmed up yet).

(b) It should then ask the DM to submit a series of trial solutions and then find one or several adjacent efficient solutions for each.

Facility (b) captures the idea of random exploratory moves, as well as directed testing moves. It should also be available to the DM under the control mode suggested in (1) at any time during the solution process. This would allow backtracking to restart the solution process at any desired solution or changing responses provided earlier in the process.

(4) The IMOPM should allow the DM to constrain one or more attributes (or goals) to stated (feasible) levels for one or several iterations (or moves).

(5) The IMOPM should have the option of reporting goal achievement levels for any solution in either absolute or percentage terms or both. For a method that uses explicitly or implicitly a weighted utility function, it would also be useful to report the weights associated with the goal achievement levels at each solution.

It goes without saying that the presence of these facilities will tend increase the elapsed time of the solution process. For repetitive problems, this could become a disadvantage. It can though be easily overcome by making these facilities optionally available for the DM to invoke as need be. For one-shot problems or important problems, slowing down the convergence rate to the preferred solution may in fact be essential in terms of producing a more researched and better solution. By giving the DM more time to come to grips with the complexities of the problem, he/she will be able to better understand her/his own preference structure. As a result the DM will have more confidence that the solution found is her/his preferred solution.

This point may be so obvious to most researchers in MCDM that it does not need any elaboration, although speed of convergence and the small number of responses required are often cited as strong features of an IMOPM. However, our major doubt about IMOPMs has always been the assumption that a DM is capable of giving virtually instant, valid responses while sitting at a computer terminal - responses that represent her/his best judgment. We find it incongruous to imagine a DM, faced with an important MCDM problem, answering tradeoff questions or identifying a new trial solution within a time frame measured in seconds or minutes after having seen an updated solution. If the answers are that easy to give, he/she will hardly need a sophisticated interactive program to find the preferred solution. If the problem is important, involving crucial and far-reaching tactical or strategic decisions, he/she may want to evaluate, sleep over, re-evaluate, and discuss it with other people involved or affected before supplying any answers to the program. Hence, the usefulness of IMOPMs that can lead to "finding the preferred solution" by sitting at a terminal for an hour or so may be rather limited.

For problems where the DM has to make decisions on behalf of a constituency (political or public sector body, a major corporation) or problems which involve group decision making, the deliberate and unhurried evaluation is even more crucial. The process of formation of goals and preference structure, representative for the constituent body or the individual member of the group, and the group interactions needed to produce a group consensus has to be a highly deliberate and unhurried process. It may take weeks or months. At the conclusion of this process the DM(s) will want to say: "We have thoroughly looked at this problem from every angle and carefully weighed all the consequences, and after considerable soul-searching we have decided to ...."

The meaning of "interactive" in MCDM should thus not be taken as synonymous with "interactive" in a computer science context of real-time response. For important problems worth solving it will be of little importance if the IMOPM program has to be restarted where the DM(s) left off hours (days, weeks) earlier.

REFERENCES

1 Baum, S. and R.C. Carlson (1974) " Multigoal optimization in management science," Omega, 2: 607-623.

2 Buchanan, J.T. (1985) "Solution methods for multiple objective decision models", Ph.D. Thesis, University of Canterbury, Christchurch, New Zealand.

3 Buchanan, J.T. and H.G. Daellenbach (1987) "Comparative evaluation of interactive solution methods for multiple objective decision models", European Journal of O.R., 29: 353-9.

4 Budner, Stanley, (1962) "Intolerance of ambiguity as a personality variable", Journal of Personality, 30: 29-50.

5 Fischoff, B., P. Slovic and S. Lichenstein (1980) "Knowing what you want: measuring labile values." In T.S. Wallsten (ed.), Cognitive processes in choice and decision behaviour, Hillsdale, New Jersey: Lawrence Erlbaum Associates. 117-141.

6 French, S. (1984) " Interactive multi-objective programming: its aims, applications and demands", Journal of the Operational Research Society, 35: 827-834.

7 Frenkel-Brunswick, (1949) "Intolerance of ambiguity as an emotional and perceptual personality variable", Journal of Personality, vol.8, 108-143.

8 Gimpl, Martin L, Stephen R. Dakin (1984) "Management and Magic", California Management Review, vol.27: 125-136.

9 Gimpl, M.L. (1985) "Intolerance of ambiguity and the need for forecasts," Paper presented at the 5th International Symposium on Forecasting, Montreal, Canada, 1985. Department of Business Administration, University of Canterbury, Christchurch, New Zealand.

10 Haimes, Y.Y., and W.A. Hall (1974) "Multiobjectives in water resources systems analysis: the surrogate worth tradeoff method," Water Resources Research, 10:615-624.

11 Kahneman, D., P. Slovic and A. Tversky (1982) Judgement under Uncertainty: Heuristics and Biases. Cambridge University Press.

12 Keen, P.G.W. (1977) "The evolving concept of optimality", In M.K. Starr and M. Zeleny (eds.), Multiple criteria decision making. TIMS studies in the management sciences, 6. Amsterdam: North Holland, 31-58.

13 March, J.G. (1978) "Bounded rationality, ambiguity and the engineering of choice", Bell Journal of Economics, 9: 587-608.

14 March, J.G. and H.A. Simon (1958) Organisations. New York: John Wiley.

15 Martinson, F.K. (1977) An application of multiple objective linear programming to the formulation of management plans for multiple-use public lands, Ph.D. Thesis, College of Business and Administration, University of Colorado, Colorado, USA.

16 Oppenheimer, K.R. (1978) "A proxy approach to multi-attribute decision making," Management Science, 24:675-689.

17 Posner, M.E. and C-T. Wu (1981) "Linear max-min programming", Mathematical Programming. 20:166-172.

18 Schon, Donald A. (1983) The Reflective Practitioner, Basic Books.

19 Soland, R.M. (1979) "Multicriteria optimization: a general characterization of efficient solutions", Decision Sciences, 10: 26-38.

20 Steuer, R.E. and E.U. Choo (1983) "An interactive weighted Tchebycheff procedure for multiple objective programming", Mathematical Programming, 265: 326-344.

21 Wallenius, J. (1975) "Comparative evaluation of some interactive approaches to multicriterion optimization", Management Science, 21: 1387-1396.

22 Zeleny, M. (1976) "The theory of the displaced ideal." In M. Zeleny (ed.), Multiple criteria decision making. Kyoto, 1975 Lecture Notes in Economics and Mathematical Systems. New York: Springer-Verlag, 123, 153-206.

23 Zionts, S. and J. Wallenius (1976) "An interactive programming method for solving the multiple criteria problem", Management Science, 22: 652-663.

24 Zionts, S. and J. Wallenius (1983) "An interactive multiple objective linear programming method for a class of underlying nonlinear utility functions", Management Science, 29: 519-529.

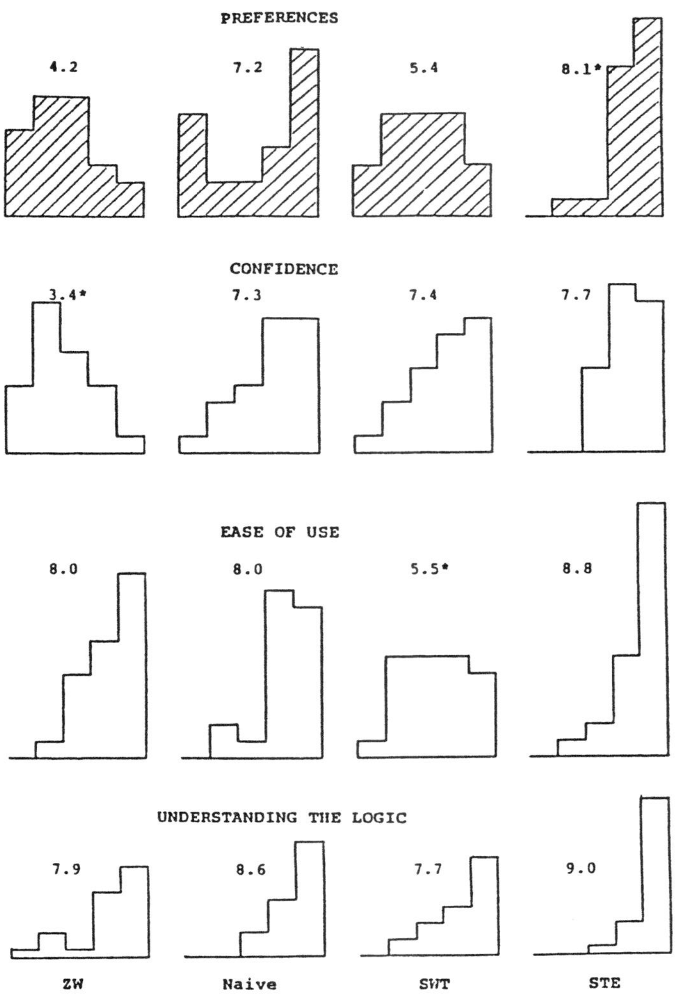

Figure 1: Ratings

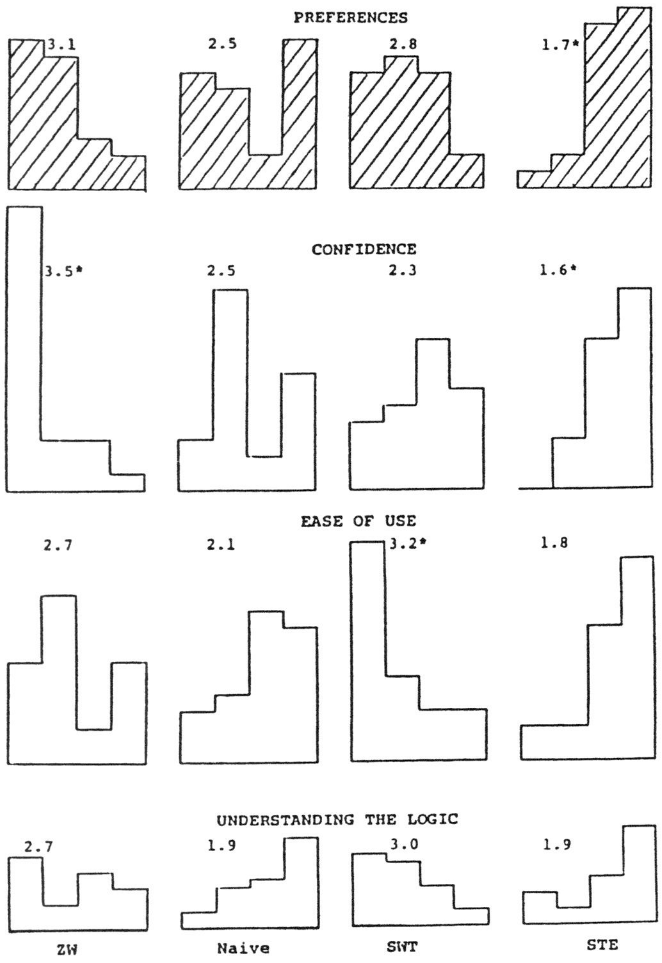

Figure 2: Rankings

MULTICRITERIA ISSUES IN MARKETING: A SALES RESOURCE ALLOCATION EXAMPLE AND POTENTIAL AREAS OF FUTURE RESEARCH

Lorraine R. Gardiner*, Antonie Stam**, and Erich A. Joachimsthaler***

* Department of Management, 322 Thach Hall, Auburn University, Auburn, AL 36849, U.S.A.

** Department of Management Sciences and Information Technology, Brooks Hall, University of Georgia, Athens, GA 30602, U.S.A.

*** IESE, Avenida Pearson 21, 08034 Barcelona, Spain.

ABSTRACT

Numerous problems in the marketing area have a multicriteria nature, yet have been modeled predominately with single criterion models. This paper briefly discusses four types of multicriteria marketing problems, and some of the research that has been conducted regarding them. Interactive multiple objective programming techniques offer potential decision support capibilities for decision makers faced with these problems. A sales resource allocation example is presented to demonstrate the type of support these techniques can provide.

1. INTRODUCTION

Though many decisions in marketing demand consideration of multiple criteria (Eliashberg and Michie 1984; Wind 1981), marketing researchers typically offer single criterion marketing models to provide normative recommendations to managers. Three notable exceptions are the goal programming-based marketing models by Zoltners and Dodson (1983), McClure and Wells (1987), and an application of the Analytic Hierarchy Process of Saaty (1988) by Wind and Saaty (1980). However the latter technique does not explicitly take boundary or feasibility constraints into account, limiting their applicability to constrained problems.

This paper describes four marketing problems which have a multicriteria nature: (1) target market portfolio selection, (2) order selection, (3) distribution channel design, and (4) sales resource allocation. We will discuss modeling these problems as multiple objective programming (MOP) problems and using interactive MOP (IMOP) techniques for decision support.

A sales force allocation example is presented to demonstrate the contribution that an MOP formulation and an IMOP solution procedure can make to the resolution of these types of marketing problems. More specifically, a prototype interactive decision support aid is discussed which was developed to address the sales force allocation problem of a California hospital office systems distributor. The problem is solved using the Augmented Weighted Tchebycheff Procedure (AWTP) of Steuer and Choo (1983). The small scale model is a predecessor to the full scale model which is currently being developed.

The remainder of the paper is organized as follows. Section 2 contains a discussion of four types of multicriteria marketing decision problems. The sales force allocation example is described in Section 3. Section 4 provides concluding remarks.

2. MULTICRITERIA MARKETING DECISION PROBLEMS

Multiple criteria typically need to be considered to solve the marketing problems described in this section. IMOP procedures offer unprecedented decision support for this purpose, in particular for marketing problems where the set of feasible alternatives can be defined by a set of mathematical functions.

2.1 Target Market Portfolio Selection.

Firms which offer multiple end-use products or services frequently do not have adequate resources to serve all their potential markets. When this is the case, the firm's management must decide which markets it will attempt to serve. The selection of target markets involves the consideration of criteria such as sales revenue, growth potential, profit contribution, cyclical nature of demand, production capacity, and budgetary constraints.

While many firms make market selection decisions one market at a time, Zoltners and Dodson (1983) suggest that a portfolio approach might be more appropriate. In the market selection system that they propose, relevant criteria are aggregated for prospective market portfolios. The system offers two basic capabilities to the decision maker. First, the system can generate a market portfolio which is either (a) optimal with respect to an Archimedian goal programming formulation, (b) feasible with respect to prespecified goal levels, or (c) optimal with respect to one objective and feasible in terms of all other goal levels. Second, the decision maker may input a market portfolio of interest and receive the relevant aggregated criterion levels.

To the authors' knowledge, the use of IMOP techniques has not been investigated for market selection problems. It would appear that these procedures offer attractive features to marketers for solving these problems. For example, IMOP procedures would offer the decision maker the ability to sample the set of feasible portfolios in a manner which is guided by his/her preferences. In this way, more information would be provided regarding potential market combinations than the decision maker receives from techniques which generate only a single suggested portfolio.

2.2 Order Selection.

Order selection problems are similar to market selection problems in that some firms attract more orders than they have resources to accept. Management in these firms must then decide which orders to accept. A problem of this type is presented in the Harvard Business School case of Titanium Fabrication Division (Cline and Shapiro 1978). A portfolio approach similar to Zoltners and Dodson (1983) for market selection may also be appropriate for order selection.

Both corporate and functional criteria need to be considered in order selection decisions. Corporate criteria include profitability, growth rate, and geographic concentration of customers. Shapiro (1987) indicates that order selection has its greatest functional level impact on the marketing and manufacturing areas. Marketing criteria for order selection may include profitability, market share, growth rate, and nonpayment risk while manufacturing criteria may be related to productivity, capital investment, quality, cost, and service. Thus order selection problems require interfunctional cooperation in the decision process. IMOP methodology, combined with group decision making techniques, offers a means for manufacturing, marketing, and corporate representatives to evaluate order portfolios with respect to both corporate and functional criteria.

2.3 Distribution Channel Design.

A distribution system is a network of organizations and people, called channel intermediaries (e.g., wholesalers, manufacturer representatives), that links a manufacturer to the various customer segments. Designing such a system requires making two important decisions: what type of channel intermediaries to include in the system and how the network of intermediaries should be linked together.

Existing literature in marketing addressing these two decisions divides into descriptive and prescriptive approaches. Descriptive research generally studies the nature of the decision problems: the 'why,' 'how,' and 'who' in channel design decisions. This research is predicated upon the assumption that channel design involves consideration of the multiple, often conflicting, criteria of the manufacturer as well as the channel intermediaries (e.g., Eliashberg and Michie 1984). Prescriptive research establishes normative guidelines as to how managers should design the distribution channel system. A case in point is the decision model by Rangan, Zoltners, and Becker (1986). This decision model, however, as do all other normative modeling efforts in this area, yields recommendations to managers assuming that channel designers have only one objective in mind - to maximize profit. As descriptive research shows, however, channel designers build a distribution system considering multiple short-run and long-run objectives and goals including maximizing sales volume, achieving reliable and prompt delivery, minimizing costs, and maximizing coverage in key trade areas. The overriding goal is to evaluate the tradeoffs among the multiple criteria so as to manage the unavoidable channel conflict. IMOP procedures offer a number of advantages over the noninteractive single objective optimization approaches utilized to solve channel design problems today.

2.4 Sales Resource Allocation.

Sales resource allocation decisions include the determination of sales territories, assignment of accounts or territories to salespeople, and establishment of time allocations for sales, sales support, and prospecting activities. Criteria which may be relevant to making these decisions are expected sales from existing accounts, expected future sales due to new account development, sales personnel preferences, and maintenance of desired service levels. Extensive research has been conducted on models for sales resource allocation problems. Twenty-five sales resource allocation models are discussed in reviews by Zoltners and Sinha (1980) and Lilien and Kotler (1983). Lodish (1976) and Zoltners and Sinha (1980) set out models which are related to the sales resource allocation example given in Section 3. With the two exceptions outlined below, these models have only allowed explicit consideration of a single objective, usually related to profit or sales.

McClure and Wells (1987) set out a goal programming formulation for a sales resource allocation problem which includes goals related to sales, sales effort, and salesperson preference for accounts. Stam, Joachimsthaler, and Gardiner (1987) investigate an MOP formulation of a sales resource allocation problem and the use of an IMOP solution procedure. This investigation is discussed in Section 3.

3. SALES RESOURCE ALLOCATION EXAMPLE

This example is based on the sales force allocation problem of a California distributor of hospital office systems. A detailed description of this application can be found in Stam, Joachimsthaler, and Gardiner (1987). The sales manager for this firm is faced with the decision of how best to assign the firm's three most productive salespeople to 41 priority accounts in a large metropolitan area and establish time allocation priorities for the salespeople's activities. In this particular situation, travel time considerations are not relevant to the assignment decision. The time allocation priorities are determined for three sales-related activities: sales to existing accounts, service to existing accounts, and new account development. In determining the assignments and time allocation recommendations, the sales manager considers the relative effectiveness of salespeople, relative importance of the accounts, and constraints on the salespeople's time.

The proposed interactive decision aid consists of an MOP model and an IMOP solution procedure. The MOP model is designed to allow explicit consideration of five criteria: expected sales to existing accounts based on the relative effectiveness of salespeople, expected sales adjusted to reflect account importance, achievement of service level goals, current sales opportunity loss due to service and prospecting activity, and expected future sales due to new account development. The AWTP by Steuer and Choo (1983) was selected as the IMOP procedure in this marketing decision aid due to its sampling properties, applicability to mixed integer programming problems, and overall appeal to users. For a discussion of its appeal to users, see Buchanan and Daellenbach (1987).

The model and data described in this example are taken from the prototype model used in the development of a full scale model. This smaller scale model covers the assignment of the three salespeople to ten accounts. The prototype was used to familiarize the sales manager with the solution procedure, help the manager gain insights about the problem, and fine-tune the model. The full scale model and decision aid are currently being developed and will be discussed in a later paper.

The multiple objective programming formulation is

Objectives:

$$\text{maximize } z_1 = \sum_i \sum_j e_{ij} r_j(t_{ij}) \quad [1]$$

$$\text{maximize } z_2 = \sum_i \sum_j e_{ij} a_j r_j(t_{ij}) \quad [2]$$

$$\text{minimize } z_3 = \sum_j b_j d_j^- \quad [3]$$

$$\text{minimize } z_4 = \sum_i c_i (T_i - \sum_j t_{ij}) \quad [4]$$

$$\text{maximize } z_5 = \sum_i p_i n_i \quad [5]$$

Subject to:

$$\sum_j t_{ij} \leq T_i \qquad \text{for } i=1,\ 2,\ 3 \quad [6]$$

$$\sum_j s_{ij} \leq S_i \qquad \text{for } i=1,\ 2,\ 3 \quad [7]$$

$$\sum_j (t_{ij} + s_{ij}) + n_i \leq R_i \qquad \text{for } i=1,\ 2,\ 3 \quad [8]$$

$$\sum_i t_{ij} \geq L_j \sum_i x_{ij} \qquad \text{for } j=1,\ldots,10 \quad [9]$$

$$t_{ij} \leq U_j x_{ij} \qquad \text{for } i=1,\ 2,\ 3;\ j=1,\ldots,10 \quad [10]$$

$$s_{ij} - S_i x_{ij} \leq 0 \qquad \text{for } i=1,\ 2,\ 3;\ j=1,\ldots,10 \quad [11]$$

$$\sum_i s_{ij} \geq H_j \sum_i x_{ij} \qquad \text{for } j=1,\ldots,10 \quad [12]$$

$$\sum_i s_{ij} + d_j^- \geq SL_j \qquad \text{for } j=1,\ldots,10 \qquad [13]$$

$$n_i \leq N_i \qquad \text{for } i=1,\ 2,\ 3 \qquad [14]$$

$$\sum_i x_{ij} \leq 1 \qquad \text{for } j=1,\ldots,10 \qquad [15]$$

$$x_{ij} = \begin{cases} 1, & \text{if salesperson } i \text{ is assigned to account } j \\ & (i=1,\ 2,\ 3;\ j=1,\ldots,10) \\ 0, & \text{otherwise} \end{cases} \qquad [16]$$

Objective [1] maximizes expected sales in dollars due to the account assignments and sales time allocations. The e_{ij} are coefficients indicating the relative effectiveness of salesperson i selling to account j. The $r_j(t_{ij})$ are sales response functions where the t_{ij} are variables indicating the time allocated to salesperson i for direct selling activity to account j during the planning period. The planning period used in the prototype model is one week. The relative effectiveness coefficients and sales response functions are set out in Lodish (1976). Objective [2] maximizes expected sales in dollars, adjusted by account-specific adjustment factors a_j (Lodish 1976), which reflect the relative importance of the accounts. In objective [3], the weighted sum of deviations from service level goals in hours is minimized. The b_j weights reflect account service priorities. The deviational variables d_j^- appear in [13] and represent the service underachievement (in hours per week) for each account. Objective [4] minimizes the current sales opportunity loss in dollars due to service and new account prospecting activities. The c_i are the estimated hourly costs in dollars of the sales personnel not engaging in direct selling activities to existing accounts. The T_i are upper bounds on direct selling time for each salesperson. Objective [5] maximizes the expected sales in dollars in the next planning period due to current prospecting for new accounts. The n_i are variables representing the time spent on new account development by each salesperson and the p_i are coefficients which estimate the hourly contribution to sales in the next period due to current prospecting.

Constraint sets [6] and [7] limit the amount of time available to each salesperson for direct selling and service. The s_{ij} indicate the time allocated to salesperson i for service to account j, and the S_i are the upper bounds for time devoted to service by each salesperson. Prospecting time for each sales person is constrained by upper bounds N_i in [14]. The total amount of time available to each salesperson for sales, service, and prospecting activities is constrained in [8].

Constraint sets [9] and [10] define the domains of the sales response functions. Constraint set [11] ensures that service to an account is provided by the person selling to the account. Constraint set [12] provides a lower limit of H_j hours of service to each account. Service level goal constraints are set out in [13]. The SL_j are service level goals (in hours) for each account. The account assignment constraints [15] ensure that no more than one salesperson is assigned to an account. Finally, [16] defines the 0-1 integer account assignment variables x_{ij}. A detailed description of the numerical parameter values used in the prototype study can be found in Stam, Joachimsthaler, and Gardiner (1987).

An example decision making session with three iterations is presented below to illustrate the type of decision support provided by the prototype decision aid. The AWTP provides the decision maker with a widely dispersed sample of nondominated criterion vectors in the first iteration. In subsequent iterations, samples of nondominated criterion vectors are generated from increasingly smaller neighborhoods about the preferred solution from the previous iteration. In addition to the criterion value information presented at each iteration, the decision maker may also access assignment and time allocation information.

Table 1 contains the criterion values generated in the three iterations of the AWTP. In the first iteration, the decision maker selected Solution 1-2 as the preferred solution and, for future solutions, placed a lower bound of 1,067 on expected future sales due to current prospecting activities and an upper bound of 8.67 on the weighted sum of deviations from service level goals (in hours).

The solutions in Iteration 2 were generated based on the selection of Solution 1-2 in the first iteration and the bounds determined by the decision maker. Solution 2-4 was selected by the decision maker. Lower bounds of 3 hours were added for the amount of time spent prospecting by the each salesperson. Additionally, lower bounds of $64,000 and $16,000 were placed on the values of objectives [1] and [2], respectively.

Finally, the solutions in Iteration 3 were generated and Solution 3-5 was preferred as the final solution. The final recommended assignments and time allocations of salespeople to accounts associated with Solution 3-5 are given in Table 2.

As can be seen from the example decision process, the MOP model for the sales resource allocation problem allows explicit consideration of multiple criteria relevant to the decision. Using the AWTP, the decision maker has an opportunity to interact with the model, learn about the types of solutions which are feasible, and evaluate tradeoffs among the objectives in the process of reaching a final decision.

Table 1: Three Iterations using the AWTP.

	Solution					
Iteration 1	1-1	1-2*	1-3	1-4	1-5	1-6
z_1	68,368	66,720	50,736	64,272	76,512	36,336
z_2	15,760	16,864	11,144	13,708	17,324	7,712
z_3	4.75	10.90	0.60	9.57	7.65	2.40
z_4	7,600	5,248	26,880	17,376	0	38,480
z_5	0	1,728	0	2,803	0	2,640
Iteration 2	2-1	2-2	2-3	2-4*	2-5	2-6
z_1	58,608	63,232	46,848	73,184	58,496	44,480
z_2	14,976	13,188	11,068	16,328	15,512	11,928
z_3	6.35	8.67	1.81	8.67	8.67	8.67
z_4	10,560	18,768	29,328	5,344	3,968	18,592
z_5	1,067	2,669	1,067	1,067	1,067	2,891
Iteration 3	3-1	3-2	3-3	3-4	3-5*	3-6
z_1	64,992	65,264	70,576	66,320	72,928	70,304
z_2	16,000	16,000	16,000	16,704	16,236	16,128
z_3	7.58	8.67	8.00	8.67	8.67	8.67
z_4	8,064	9,184	7,424	5,936	1,067	4,528
z_5	1,067	1,475	1,067	1,067	5,552	1,067

*: Most Perferred Solution in the Iteration.

z_1: (Max) Expected Sales (\$/week)
z_2: (Max) Expected Adjusted Sales (\$/week)
z_3: (Min) Service Underachievement (weighted hours/week)
z_4: (Min) Sales Opportunity Loss (\$/week)
z_5: (Max) Expected Sales Next Week due to Prospecting (\$)

Table 2: Final Assignments and Strategic Plan for Solution 3-5

Sales-person	Account 1	2	3	4	5	6	7	8	9	10	$\sum_j s_{ij}$	$\sum_j t_{ij}$	n_i
1			X_5^1				X_1^{10}	X_1^8		X_1^1	8	20	3
2					X_1^9	X_1^{11}			$X_{1.1}^{4.9}$		3.1	24.9	3
3	$X_{1.4}^6$	X_1^{10}		X_1^8							3.4	24	3.6

X_s^t = allocation of t sales hours and s service hours.

4. CONCLUDING REMARKS

Interactive multiple objective programming can be utilized to solve numerous problems in marketing. As shown in the sales resource allocation example, IMOP procedures provide the decision maker information with respect to the impact of candidate decisions on relevant objectives which is not typically available. At present, IMOP procedures are virtually untapped resources for supporting marketing decision making. Further research is necessary to investigate the application of these techniques to marketing problems.

REFERENCES

Buchanan, J. T. and H. G. Daellenbach (1987). "A Comparative Evaluation of Interactive Solution Methods for Multiple Objective Decision Models," *European Journal of Operational Research*, Vol. 29 (June), pp. 353-359.

Cline, C. E. and B. P. Shapiro (1978). "Titanium Fabrication Division (A)," Harvard Business School Case (578-208), revised version 1/87.

Eliashberg, J. and D. A. Michie (1984). "Multiple Business Goal Sets as Determinants of Marketing Channel Conflict: An Empirical Study," *Journal of Marketing Research*, Vol. 21 (February), pp. 75-88.

Lilien, G. L. and P. Kotler (1983). *Marketing Decision Making: A Model Building Approach*, New York: Harper and Row, pp. 558-599.

Lodish, L. M. (1976). "Assigning Salesmen to Accounts to Maximize Profits," *Journal of Marketing Research*, Vol. 13 (November), pp. 440-444.

McClure, R. M. and C. E. Wells (1987). "Incorporating Sales Force Preferences in a Goal Programming Model for the Sales Resource Allocation Problem," *Decision Sciences*, Vol. 18 (Fall), pp. 677-681.

Rangan, V. K., A. A. Zoltners, and R. J. Becker (1986). "The Channel Intermediary Selection Decision: A Model and an Application," *Management Science*, Vol. 32 (September), pp. 1114-1122.

Shapiro, B. P. (1987). "Functional Integration: Getting All of the Troops to Work Together," Note (9-587-122), Harvard Business School.

Saaty, T. L., The Analytic Hierarchy Process (1988). *Multicriteria Decision Making: The Analytic Hierarchy Process*, University of Pittsburgh, Pittsburgh, PA.

Stam, A., E. A. Joachimsthaler, and L. R. Gardiner (1987). "Interactive Decision Support for Sales Resource Allocation with Multiple Objectives: An Exploratory Study," Working Paper, College of Business Administration, University of Georgia.

Steuer, R. E. and E.-U. Choo (1983). "An Interactive Weighted Tchebycheff Procedure for Multiple Objective Programming," *Mathematical Programming*, Vol. 26, pp. 326-344.

Wind, Y. (1981). "Marketing Oriented Strategic Planning Models," in Schultz, R. L. and A. A. Zoltners (eds.), *Marketing Decision Models*, New York: North Holland, pp. 207-250.

Wind, Y. and T. L. Saaty (1980). "Marketing Applications of the Analytic Hierarchy Process," Management Science, Vol 26 (July), pp. 641-658.

Zoltners, A. A. and J. A. Dodson (1983). "A Market Selection Model for Multiple End-Use Products," *Journal of Marketing*, Vol. 47 (Spring), pp. 76-88.

Zoltners, A. A. and P. Sinha (1980). "Integer Programming Models for Sales Resource Allocation," *Management Science*, Vol. 26 (March), pp. 242-260.

AN INTERACTIVE MULTI-CRITERIA MODEL FOR AUDIT WORKLOAD PLANNING

J. C. Gardner, R. J. Huefner, and V. Lotfi
Jacobs Management Center
State University of New York at Buffalo

Buffalo, NY 14260

INTRODUCTION

Planning the staffing of expected audit workload for a CPA firm can be a complex problem subject to a number of objectives and constraints. It is likely no solution fully satisfies all objectives; rather, a decision maker makes trade-offs, achieving more of one objective at the expense of another. Planning models aid in this process by providing alternative configurations and assessing the degree to which each objective is achieved. The audit staffing decision has sufficient complexity to benefit from the development of formal planning models. In this study, a multiple objective linear programming (MOLP) model appropriate to the audit planning decision is developed and tested. The model is an interactive one, leading the decision maker to the eventual schedule via a series of steps that depict various tradeoffs among a variety of objectives, subject to the constraints faced by the firm. Through the interim steps, the model assesses the scheduler's relative preferences among the various objectives, and refines the weights attributed to each.

Audit staff planning models using multiple objectives have been considered by several accounting researchers. Early studies by Killough and Souders [1973], Bailey, et al. [1974], and Welling [1977] used goal programming to generate staff planning solutions. However, goal programming potentially has severe implementation problems due to the model's requirement of prior specification of the objective weights. More recently Balachandran and Steuer [1982] (B & S) presented a single-period MOLP interactive model for the audit staffing problem. They justify the model by arguing that decision makers may be unwilling or unable to develop the ex-ante weights needed for the goal programming approach.

Our study extends the B & S study in two general ways. First, a multi- period (four quarters in a year) versus a single period MOLP model is proposed. The multi-period model, which includes a "busy season", represents a more realistic scheduling problem faced by an audit firm and thus is more likely to be useful in practice than a single period model. Second, the purpose of the B & S study was to present and explain the essential aspects of a simplified MOLP audit scheduling model. The purpose of this study is to not only present and explain a multi-period audit scheduling model but to empirically apply the model. That is, instead of the authors selecting a preferred solution at each iteration, a set of audit schedulers independently provide decisions at each iteration. In this way we can gain insights into the feasibility of the model proposed along with its derived solutions. Hence we were able to develop the implicit weights on the various objectives from the responses of actual decision makers, and to compare the objective values chosen along various dimensions (national vs. local firm partners, and managing/non-managing partner status). The outcomes of this phase demonstrate some promise for the eventual application of such models in practice.

The remainder of the paper is presented in five sections. In the next section the multi-period model of the audit staff planning problem is presented. A discussion of the particular staff planning case used in this study is next, followed by a presentation of the application procedures. The analysis of the results is then presented and the paper concludes with a discussion of the conclusions and implications of the study.

PROBLEM FORMULATION

The audit firm, consisting of N levels of personnel (ranging, for example, from supervising partner to junior accountant) faces a P-period projected workload.

Notation

x_{ijst} Number of hours of work scheduled for staff of level i, actually performed by staff of level j, scheduled to be completed in period t, actually completed in period s.

a_{ij} 1 if staff at level j can be assigned level i work, and 0 otherwise.

sw_{ti} Work of job type i scheduled but not completed in period t.

w_{ti} Number of hours of workload of job type i anticipated for period t.

pe_i Allowed percentage of interim work (work performed prior to scheduled period)

E_{tj} Number of staff of type j available during period t

$A^{+,-}$ Number of staff of type j to be hired/dismissed at the start of period t.

D_{tj} Actual number of professional development hours of staff level j in period t.

W_j Number of regular hours available each period from each employee at level j.

O_{tj}, I_{tj} Actual number of overtime (idle time) hours of staff at level j in period t.

pr_j Maximum allowable overtime for staff level j, as a percentage of regular time.

g_j Required ratio of personnel at one level to the total personnel at lower levels.

Q_j Target professional development hours for personnel at staff level j.

sp_j Shortfall in the achievement of the target professional development.

b_i Net billing rate per hour of job type i.

c_j Quarterly base salary at staff level j.

r_j Overtime rate per hour at staff level j.

h_j, f_j Hiring (firing) costs per position of type j.

FO Fixed Overhead

The Model

a- Objectives: The proposed MOLP includes seven objectives as follows.

1- Profit: $$\text{Max } Z_1 = \sum_s \sum_t \sum_i b_i \sum_j a_{ij} \; x_{ij} - \sum_j c_j \sum_t E_{tj} - \sum_j \sum_t O_{tj} - \sum_j \sum_t (h_j A^+_{tj} + f_j A^-_{tj}) - FO$$

2- Lateness of Work: $$\text{Min } Z_2 = \sum_i \sum_j \sum_{s=2} \sum_{t<s}^{P} a_{ij} x_{ijst}$$

3- Work declined: $$\text{Min } Z_3 = \sum_t \sum_i sw_{ti}$$

4- Staff augmentation: $$\text{Min } Z_4 = \sum_t \sum_j A^+_{tj}$$

5- Staff reduction: $$\text{Min } Z_5 = \sum_t \sum_j A^-_{tj}$$

6- Underutilization and idle time: $$\text{Min } Z_6 = \sum_{i=2}^{N} \sum_{j<i} \sum_s \sum_t a_{ij} x_{ijst} + \sum_t \sum_j I_{tj}$$

7- Professional development: $$\text{Min } Z_7 = \sum_j sp_j$$

b- Constraints: The above objectives are subject to the following seven sets of constraints.

1- Workload $$\sum_s \sum_j a_{ij} \; x_{ijst} + sw_{ti} = w_{ti} \qquad i=1,2,\ldots,N \quad t=1,2,\ldots,P$$

2- Interim Work $$\sum_j a_{ij} \; X_{ijt,t+1} \leq pe_i \; w_{t+1,i} \qquad i=1,2,\ldots,N \quad t=1,2,3$$

3- Staff $E_{t+1,j} = E_{tj} + A^{+}_{t+1,j} - A^{-}_{t+1,j}$ $\quad j=3,4,\ldots,N \quad t=0,1,\ldots,P-1$

4- Utilization $\sum_{s}\sum_{i} a_{ij}\, x_{ijts} + D_{tj} = H_j\, E_{tj} + O_{tj} - I_{tj}$ $\quad j=1,2,\ldots,N \quad t=1,2,\ldots,P$

5- Overtime $O_{tj} \leq pr_j\, H_j\, E_{tj}$ $\quad j=1,2,\ldots,N \quad t=1,2,\ldots,P$

6- Supervision $E_{tj} \geq (1/g_j) \sum_{i} E_{ti}$ $\quad j=1,2,\ldots,N-1 \quad t=1,2,\ldots,P$

7- Professional development $\sum_{t} D_{tj} + sp_j = \sum_{t} Q_j\, E_{tj}$ $\quad j=1,2,\ldots,N$

SOLUTION METHOD

We utilized a revised version of Steuer's Filter [1977] method to solve the proposed model. The filter method has been shown (see for example Gibson [1987], and Gibson and Reinstein [1986]) to be easy to implement, accurate, fast convergent, and require small computational effort.

The Filter method is an interactive MOLP solution procedure based on a weighting-vector space reduction approach. At each iteration a large set of weight-vectors is generated at random (with a pre-specified range). The weight-vectors are then filtered to get a reduced, well-dispersed set. The reduced weight-set is used to form several weighted-sum objective linear programs (LPs). The LPs are solved, one at a time, and the solutions (objective vectors) are stored. Again, a filtering process is applied to the solutions from the LPs to obtain a smaller, well-dispensed solution-set. The reduced-solution set is then presented to the decision maker. The decision maker is asked to select a most- preferred solution (from the reduced solution set). The weights associated with the most preferred solution are used to reduce the weight ranges.

The above process is repeated several times until the weight-ranges are reduced to a pre-specified width (determined by the administrator in consultation with the decision maker). The most-preferred solution from the last iteration is then recommended for implementation.

CASE PROBLEM

In order to evaluate the feasibility and performance of the proposed model, a case problem was developed. The values for various parameters were obtained by interviewing two professionals who had had scheduling/staffing responsibility with national accounting firms. Based on these estimates the following case problem was developed.

Case Data

The audit firm has six levels of personnel, as follows:

LEVEL	DESCRIPTION	INITIAL NUMBER	AVAILABLE HOURS	PD REQUIRED	BILLING RATE	ANNUAL PAY
1	Supervising Partner	1	800	40	$225	$125,000
2	Engagement Partner	3	900	40	150	85,000
3	Manager	9	1,400	75	110	40,000
4	Senior	18	1,800	100	80	27,000
5	Semisenior	18	1,800	80	60	22,000
6	Junior	27	1,800	80	50	18,000

Other costs include:

LEVEL	DESCRIPTION	HIRING COST	FIRING COST	ALLOWABLE OVERTIME	OVERTIME PAY	DESIGN RATIO
1	Supervising Partner	-	-	50%	-	1
2	Engagement Partner	-	-	50%	-	1
3	Manager	$20,000	$60,000	20%	150%	4
4	Senior	$10,000	$30,000	20%	150%	3
5	Semisenior	$ 5,000	$15,000	20%	150%	1
6	Junior	$ 2,000	$ 6,000	10%	150%	2

The projected workload, in hours, for the four quarters of the coming year is as follows:

	LEVEL1	LEVEL2	LEVEL3	LEVEL4	LEVEL5	LEVEL6	TOTAL
Quarter 1	132	446	2,079	5,346	5,346	8,019	21,368
Quarter 2	352	1,188	5,544	14,256	14,256	32,384	56,980
Quarter 3	220	743	3,465	8,910	8,910	13,365	35,613
Quarter 4	176	594	2,772	7,128	7,128	10,692	28,490
Total	880	2,971	13,860	35,640	35,640	53,460	142,451

Normal work hours are assumed to be divided equally over the year, making the available hours at each level as 200, 675, 3,150, 8,100, and 12,150 hours per quarter, for levels one through six, respectively.

Overall workload is 142,451 hours, while normal work hours available from present staff is 129,500. The workload estimates were designed with a busy season in Quarter 2 (where workload exceeds normal available hours of the present workforce by about 75 percent) and a small excess workload (10 percent) in Quarter 3. On the other hand, workload in Quarters 1 and 4 is below the normal available hours, permitting workload re-arrangement to these periods as well as the assignment of variable-timing tasks (professional development) to these periods. Overall workload for the year is 10 percent above the normal available hours of present staff.

Our case problem consists of 7 objectives, 154 constraints, and 308 variables. We utilized a revised version of the filter method to solve the stated MOLP problem. Before applying the interactive procedure, as required by the method, all of the objectives were normalized. When presenting the alternatives to the DM, the objective values were unnormalized to present the actual values. The interactive procedure was programming in FORTRAN on an IBM 3081 GX under CMS. Portions of the code consisting of the weight generation and filtering subroutines were obtained from Steuer. The composite LPs were solved by the Lindo software package (Schrage [1984]).

SOLVING THE MODEL

Eighteen partners, each from a different CPA firm, were given the case problem and descriptions of the objectives and the interactive process. They were advised that the model would refine the weights placed on the objectives based on their choices at each iteration, and that there was no "right" solution, as each person would assign different degrees of importance to the various objectives.

Each solution set consisted of seven iterations to reach the participant's optimal solution to the problem. Each partner performed three solution sets to allow for familiarization with the problem and the decision process. The seed number used to generate the solutions to the first iteration of each set was identical for each participant, so that each had the same starting point.

The partners were selected from both national and local/regional firms in equal numbers. All participants were engaged primarily in the firm's audit practice, and/or were managing partners of the office. Levels of experience were diverse, ranging from a newly elected partner to one with 22 years service as a partner.

ANALYSIS OF RESULTS

The output of the MOLP process produces three measures: the most preferred alternative solution, the associated weights, and the weight ranges. Our analysis will focus on the objective values in the most preferred solution, as these are able to be more accurately interpreted.

To analyze these results, a measurement scale is needed to standardize the values. Keeney and Raiffa [1976, chapter 6] suggest an efficient way to analyze solutions to multi-attribute utility function problems is to examine the relative location of each objective solution within the range of solutions actually encountered by the participant group. This range is shown in Table 1, along with the relative location of the mean solution within the range.

We observe that, for the one objective to be maximized (profit), the mean solution chosen ($3,122,473) was fairly high in the range of solutions encountered, at the 72% point. the other six objectives were to be minimized. The mean solutions fell much lower in the range of solutions at the 2.7% point to work declined at the 35.6% point.

TABLE 1

Values of Objective Solutions of Audit Partner Participants, Range Encountered, and Theoretical Maximum (n = 18)

OBJECTIVE	MEAN VALUE OF SOLUTIONS	THEORETICAL MAXIMUM VALUE	RANGE ENCOUNTERED HIGH	RANGE ENCOUNTERED LOW	MEAN LOCATION IN RANGE
Profit	$3,122,473	$3,287,971.50	$3,269,874	$2,697,874	72.0%
Work completed late	2,663.36 hrs	113,960 hrs	12,078.77 hrs	0 hrs	22.0
Work declined	1.65 percent	100 percent	4.639 percent	0 pct.	35.6
Staff added	2.41 people	84.25 people	12.717 people	0 peo.	19.0
Staff reduced	.65 people	72 people	21.32 people	0 peo.	3.0
Underutilization of staff	1,950 hours	352,467 hours	19,483 hours	0 hrs.	10.0
Professional development shortage	138.23 hours	15,484.53 hrs	5,069 hrs	0 hrs.	2.7

Using this approach to standardize our results, we proceed to analyze the partners' choices according to various attributes of the individuals or their firms. In particular, we compare results of partners:

- from national firms (n = 9) vs. local/regional firms (n = 9)
- who are managing partners (n = 9; 5 national, 4 local/regional) vs. those who are not (n = 9; 4 national, 5 local/regional)

National versus Regional Firms

The relative importance placed on each objective by an audit partner could be influenced by the environment within which he/she operates. In a national/regional firm dichotomization, factors such as differences in firm strategy and/or type and scope of client responsibilities may result in different objective preferences. The results of this comparison are shown below.

OBJECTIVE	NATIONAL	LOCAL/REGIONAL	DIFFERENCE
Profit	.7122	.7269	(.0147)
Work completed late	.1843	.2567	(.0724)
Work declined	.5127	.1993	.3134
Staff added	.0798	.2990	(.2192)
Staff reduced	.0000	.0606	(.0606)
Underutilization	.1313	.0690	.0623
P.D. shortage	.0545	.0000	.0545

While profit level is similar between national and local/regional firm partners (71.22% for nationals vs. 72.69% for regionals), the local/regional firm (versus national firm) partners are: 1) more tolerant of work performed late, 2) much less willing to decline work, 3) more apt to add and/or reduce staff, and 4) less tolerant to staff underutilization and professional development shortages (although the average PD level for national firms is caused by the preference of one partner). The largest differences were in work declined and staff additions; these may be reflective of a difference in the audit work and personnel hiring markets faced by the two groups. That is, national firms may be able to afford to be more flexible (choosey) in the audit work they are willing to perform and find it relatively more costly to hire and train new personnel. (Statistical tests on group differences were not performed due to the small size of the comparison groups.)

Managing vs. Non-Managing Partner

Since a managing partner's responsibilities include more strategic decision making than a typical audit partner, a dichotomization along this dimension may be insightful. In the table below, the objective results of managing partners and non-managing partners are compared.

OBJECTIVE	MANAGING	NON-MANAGING	DIFFERENCE
Profit	.6914	.7476	(.0562)
Work completed late	.1834	.2576	(.0742)
Work declined	.4632	.2488	.2144
Staff added	.1505	.2284	(.0779)
Staff reduced	.0357	.0249	.0108
Underutilization	.0905	.1098	(.0193)
P.D. shortage	.0000	.0545	(.0545)

Work declined shows the largest difference (.2144); all others have a difference less than .08. Managing partners are more willing to decline work and realize a lower profit, less work performed late, less staff augmentation and less professional development shortage. These results are consistent with the hypothesis that managing partners may be more conservative (more risk adverse) than non-managing partners in choosing between objectives. That is, they accept a lower profit in order to choose and keep "good" clients (greater work declined) whose work they do "on time" (less lateness) with the planned staff (lower augmentation).

CONCLUSIONS AND IMPLICATIONS

A multi-period audit staff planning model was developed, extending the single-period models of previous studies. Formulation of a multi-period model enables consideration of a busy season and strategies to cope with the peak workload.

Audit partners then used the model to make decisions in the context of a detailed case problem. A high degree of consistency of decisions occurred. In particular, it was shown that almost all partners prefer not to reduce audit staff nor fall short of professional development targets. Some variations in other objective value solutions did occur. The largest differences were observed when comparing partners of national and regional firms; national firm partners were less willing to perform work late and augment staff and were more willing to turn down audit work. Dividing the sample of partners by managing versus non-managing status showed that and, to a greater extent, managing partners were more willing to decline audit work.

On a broader scale, modeling a decision problem necessarily simplifies the problem situation. It is important to move from showing that a model can be formulated to developing a model that is potentially useful in practice. This requires:

1. the ability to develop a sufficiently complex model so as to be a realistic representation of the practical situation. We moved in this direction, extending from single-period to multi-period model, and incorporating (based in part on input from practitioners) such factors as a busy season and interim work.

2. the ability to tailor the model to individual firm situations. It would be fairly easy to change the coefficients and/or constraints in the underlying model to reflect the characteristics and policies of a particular firm. The model would also be easy to revise and re-run as the situation develops (e.g., changes in the workload estimate, unexpected staff changes, etc.).

3. the ability of appropriate decision makers to understand and use the model. Our experiences with the eighteen partners were very positive. They quickly understood the nature of the model, readily adapted to its use, and were enthusiastic about the experience when they had finished the exercise. The general consistency of results across a variety of individuals also suggests that the model was understandable and fairly easy to use. For a more extensive development of this point, see Gardner, Huefner, and Lotfi [1990].

In modeling a decision problem of this sort, the objective typically is not automating the decision or turning it over to a subordinate, but rather to serve as an aid to the decision maker, enabling an exploration of various combinations of solutions and a honing-in on a planning solution that reflects an acceptable trade-off. Several other support uses could also be envisioned. First, in a multiple office firm, the interactive model could be adapted to individual offices and local office decisions analyzed by the firm's strategic decision makers to insure that overall firm goals are being followed. Second, the model could be used as a standard for individual audit schedulers, with variances analyzed between the model outcomes and actual experience. Third, the interactive model could be used for training purposes. In particular, the results of persons being trained as audit schedulers could be compared to experienced personnel and/or the actual experience of the firm. The differences in objective solutions explained by the length of scheduling/staffing experience observed in the correlation results suggest this type of application would be especially useful.

REFERENCES

Bailey, A. D., W. J. Boe and T. Schnack, "The Audit Staff Assignment Problem: A Comment," *The Accounting Review*, (July 1974), pp. 572-74.

Balachandran, K. R. and R. E. Steuer, "An Interactive Model for the CPA Firm Audit Staff Planning Problem with Multiple Objectives," *The Accounting Review*, (January 1982), pp. 125-40.

Gardner, J. C., R. J. Huefner, and V. Lotfi, "A Multiperiod Audit Staff Planning Model Using Multiple Objectives: Development and Evaluation, *Decision Sciences*, forthcoming Winter 1990.

Gibson, M. H. L., Bernardo, J. J., Chung, C. and R. Badinelli, "A Comparison of Interactive Multiple Objective Decision Making Procedures," *Computers and Operations Research*, (Vol. 14, 1987), pp. 97-105.

Gibson, M. H. L. and A. Reinstein, "The CPA Firm Audit Staff Planning Problem with Multiple Objectives", *Proceedings of the Decision Sciences Institute*, 1986.

Keeney, R. L. and H. Raiffa, *Decisions with Multiple Objectives: Preferences and Value Tradeoffs* (John Wiley and Sons, 1976).

Killough, L. N. and T. L. Souders, "A Goal Programming Model for Public Accounting Firms," *The Accounting Review*, (April 1973), pp. 268-79.

Schrage, L., *Linear Integer, and Quadratic Programming with Lindo*, (The Scientific Press, Palo Alto, 1984).

Steuer, R. E., "An Interactive Multiple Objective Linear Programming Procedure," *TIMS Studies in the Management Science*, (Vol. 6, 1977). pp. 225-239.

Steuer, R. E., *Multiple Criteria Optimization: Theory, Computation, and Application*, (John Wiley & Sons, New York, p. 361, 1986).

Welling, P., "A Goal Programming Model for Human Resource Accounting in a CPA Firm," *Accounting, Organizations and Society*, (No. 4, 1977). pp. 307-16.

NMIS: A NATIONAL MODEL FOR INSPECTION SELECTION WITH LIMITED RESOURCES

BY

Ronald K. Klimberg
School of Management
Boston University
Boston, MA

Charles ReVelle and Jared Cohon
Department of Geography and Environmental Engineering
The Johns Hopkins University
Baltimore, MD

ABSTRACT

A two-stage decision support system, called the national model for inspection selection (NMIS), was developed to solve an inspection allocation problem faced by the United States Food and Drug Administration (FDA). The problem is to determine the 'best' level of resources to devote to inspection at the national level and to determine how to divide these resources effectively among the districts, satisfying at the same time, the mandated inspection frequencies at the district level. NMIS solves this inspection selection problem at both the district and national level, for the entire range of possible resources. Decision-makers have utilized NMIS's results to evaluate the cost-effectiveness of their decisions and to justify more inspection resources.

1. Introduction

The FDA is the U.S. federal agency responsible for regulating the quality, safety, and efficacy of all drugs to be used for human consumption. One of FDA's primary regulatory activities relative to the pharmaceutical industry is the routine surveillance of all marketed human drugs. These surveillance activities include the monitoring of the manufacturing process, the storage, and the distribution of all such drugs. During any of these stages, from the production until it ultimately reaches the consumer, there are potential variations/errors which may occur and which may result in a significant public health hazard. FDA attempts to minimize the possibility of a public health hazard by intermittently inspecting the procedures and processes being used by all drug firms and by collecting and analyzing drug samples.

Each year headquarters plans how to allocate financial and inspectional resources among the districts for these drug inspections. Their concerns focus on allocating the inspection resources among the districts in a manner that provides the drug inspection surveillance program with an overall consistent strategy. Given the allocations from headquarters, each district, of which there are 24 offices across the country, must select the particular drug firms and processes within these firms to inspect. An added restriction is FDA's legal obligation to inspect each drug firm in some way at least once every two years. Without central coordination of planning, the type of decision-making employed by FDA is not guaranteed to lead to the most efficient choices, although experience is likely to provide decision-makers with the capacity for choices that are not bad.

2. The First Stage: The District Level Subproblem

An initial formulation of FDA's inspection selection problem for drug manufacturers was a single-objective, zero-one, programming model for a two year planning period. The objective was the maximization of the expected number of violative inspections to be generated (VIGs). The size of this formulation was very large for an integer program, on the order of about 3,000 constraints and 12,000 variables. We felt the computational effort and cost to solve this problem formulation would have been prohibitive, Crowder et al (1983).

We took two steps to improve the likelihood of obtaining an integer solution. First, a smaller problem would greatly reduce the amount of computational effort to produce an all zero-one optimal solution. Since our problem formulation has the necessary block angular structure required by decomposition methods, we decomposed the problem into 24 district level subproblems and one overarching national problem. This problem, when solved as a relaxed linear program, still would not likely provide all zero-one variables in the optimal solution due to the non-unit nature of the a_js. In the second step, we increase the likelihood of all-integer termination by bringing the total resources variables into the objective function. The normal method of dualizing constraints would have brought the entire constraint into the objective function, Fisher (1981). Unlike the typical formulation, in our case, the total yearly resources are unknown, and therefore, we only brought the total yearly resources variables into the objective function.

The district level subproblem was now a much smaller two objective, zero-one, programming model. It was formulated as follows:

$$\text{MAX} \quad Z_1 = \sum_{j\varepsilon D} (p_j x_j + p_j y_j) \qquad (1)$$

$$Z_2 = -(b_1 + b_2)$$

S.T.

$$\sum_{j\varepsilon D} a_j x_j - b_1 \leq 0 \qquad (2)$$

$$\sum_{j\varepsilon D} a_j y_j - b_2 \leq 0 \qquad (3)$$

$$\sum_{j\varepsilon Q_i} x_j + \sum_{j\varepsilon Q_i} y_j \geq 1 \qquad i\varepsilon I \qquad (4)$$

$$\sum_{j\varepsilon Q_i} x_j \geq 1 \qquad i\varepsilon F \qquad (5)$$

$$b_2 - (1 - \alpha)\, b_1 \geq 0 \quad (6)$$

$$b_2 - (1 + \beta)\, b_1 \leq 0 \quad (7)$$

$$x_j,\ y_j = 0,\ 1 \qquad j \varepsilon D$$

$$\alpha,\ \beta \geq 0$$

where:

$$x_j = \begin{cases} 1 & \text{if process j is to be inspected in the first year;} \\ 0 & \text{otherwise;} \end{cases}$$

$$y_j = \begin{cases} 1 & \text{if process j is to be inspected in the second year;} \\ 0 & \text{otherwise;} \end{cases}$$

b_1 = the amount of inspection time allocated in the first year for the district;

b_2 = the amount of inspection time allocated in the second year for the district;

a_j = the amount of time required to inspect process j;

p_j = the estimated probability of detecting a violation by inspection for process j;

Q_i = the set of processes j in establishment i;

I = the set of establishments in the district;

D = the set of all processes j in the district; and

F = the set of establishments not inspected in the previous year, (i.e., prior to the application of the two-year planning model).

The objectives are (1): maximize the expected number of violations generated (VIGs), and (2): minimize the total inspection time utilized. Equations (2) and (3) are the yearly resource constraints on time available for inspection. The biennial requirement that each firm must have at least one inspection in the two year period is expressed by (4). A consequence of the biennial inspection obligation is (5) which requires all drug firms not inspected in the year preceding the planning exercise to have at least one inspection of at least one process in the first planning year. The last two equations, (6) and (7), require the district's inspection resources devoted to the first year to be relatively close to the district's second year inspection resources.

Using a small hypothetical data set, we relaxed the integer requirements and applied the weighting method of multiobjective linear programming to test the difficulty of producing all zero-one solutions, Cohon (1978). Several patterns were noted. Most solutions appeared to consist of those inspections with the greatest p_j/a_j ratio. As resources increased, the next inspection added to the solution was the inspection of the process with the largest p_j/a_j ratio (of the variables not yet in the solution). Whenever the result of applying relaxed LP had a non-all-integer solution there were at most one or two fractional variables. These non-integer variables always came from the variables with the smallest p_j/a_j ratio of the inspections in the solution. These non-integer values would change to one as resources were increased. Each of these observations is characteristic of the heuristic commonly used to solve the knapsack problem, Toyoda (1975), Nauss (1977), Loulou (1979). With one exception, the entire order, in which new inspections enter into the solution can be determined by the p_j/a_j ratios. This exception occurs when the weight on the objective of minimum resources is relatively large in comparison to the objective to maximize VIGs. In such a situation, inefficient inspections, inspections with relatively poor p_j/a_j ratios, can be found in the solution. For example, if the weight for the maximization objective was equal to zero, the problem would then be a single objective problem of minimizing resources. The minimum time solution would only have one inspection for each firm. This inspection would be an inspection of the process taking the least amount of time to perform. Within a particular firm, it is highly unlikely that this process is also the process within the firm with the maximum p_j/a_j ratio. As resources are increased from the minimum time solution, or as more weight is assigned to the objective of maximizing VIGs, these inefficient inspections will be eventually forced out of the solution and replaced by more efficient inspections within each firm. (Remember each firm must have at least one inspection of one process in the two year period). When this efficiency transfer of inspections occurs is determined by the relationship:

$$\frac{(p_k - p_j)}{(a_k - a_j)} \qquad (8)$$

where: j = the index of the process within firm i with the inefficient ratio and in the solution, and;

k = the index of the process within firm i with the maximum p_k/a_k ratio.

Every supported noninferior point (supported noninferior points are points on the convex hull of the noninferior set), can be generated by using the knapsack heuristic, i.e. the p_j/a_j ratios and the above criterion for efficient shifts, (8). The procedure would start with the minimum time solution. New inspections are added one at a time according to their p_j/a_j ratio and (8) until every drug process is inspected once a year. A district level tradeoff curve corresponding to each incremental solution can be produced which demonstrates the relationship between the number of VIGs and the amount of inspection resources to be used for the district.

We have yet to consider how to assign inspections to each year. Given a solution by the above heuristic, we would initially assign the most efficient process of the set of processes designated to be inspected within all those firms requiring a first year inspection, (5), to be inspected in the first year. Afterwards, the remaining designated inspections are assigned to the first year, in order of their p_j/a_j ratio until (6) and (7) are satisfied.

This procedure will maximize the number of VIGs in the first planning year.

3. <u>Second Stage: Decomposed National Problem</u>

The decomposed national inspection selection problem is to determine how to maximize the national total number of VIGs with a given level of national inspection resources. A formulation of the decomposed national problem.

$$\text{MAX} \quad \sum_{k=1}^{k} \sum_{t=1}^{n_k} v_{kt} x_{kt}$$

S.T.

$$\sum_{k=1}^{k} \sum_{t=1}^{n_k} b_{kt} x_{kt} = B \tag{9}$$

$$\sum_{t=1}^{n_k} x_{kt} = 1 \quad k=1, \ldots, K \tag{10}$$

where:

K,k = the set and index of all districts (k=1, . . . , K);

t = the index of resources levels (t=1, . . . , n_k);

n_k = the total number of resource levels in district k;

b_{kt} = the amount of resources used in district k with resource level t for both years;

v_{kt} = the number of expected violations to be detected in district k with resource level t for both years;

$$x_{kt} = \begin{cases} 1 & \text{if resource level t is allocated to district k;} \\ 0 & \text{otherwise} \end{cases}$$

The problem is a relatively small zero-one program. Equation (9) requires the total resources to be used among the districts to be equal to a national level of inspectional resources, B. The second constraint, (10), requires each district to be allocated some level of inspection resources.

The heuristic procedure to construct the district tradeoff curves produces a concave function, i.e. a curve increasing at a decreasing rate. We exploit this fact to develop an approach we call the 'greedy' algorithm to solve the decomposed national inspection selection problem. Although, this algorithm does not appear in texts, it is well known among operations researchers, Everett (1963), Nemhauser (1966), Toyoda (1975). Using the district tradeoff curves, the 'greedy' algorithm allows us to generate a corresponding national tradeoff curve for the entire range of possible alternatives. The first step to develop the national tradeoff curve is to find the national minimum resource level. This initial national level simply corresponds to each district being assigned its minimum resource level. Subsequent national resource levels are produced by examining the next incremental resource level on each district tradeoff curve. The district containing the largest incremental resource level is then chosen as the next resource level to be added to the national tradeoff curve. This procedure continues until every process in each drug firm is planned to be inspected twice.

Interactive computer programs have been developed for each stage of NMIS. The first step to running the decision support system is to execute NMIS's first stage for each district. Each district tradeoff curve may be graphed at a CRT terminal. A file containing the set of supported noninferior points describing the district tradeoff curve must then be saved. After generating and saving all the district tradeoff curves, NMIS's second stage may be executed. Utilizing all the district tradeoff curves as input and the 'greedy' algorithm, the second stage of NMIS produces a national tradeoff curve. Decision-makers would examine the national tradeoff curve and the various alternatives would be discussed. Once a consensus is reached on a particular national inspection resource level, each district's allocation is determined by re-executing NMIS's second stage. These district resource levels are the allocation levels necessary to obtain the expected national number of VIGs specified on the national tradeoff curve. Executing the first stage of NMIS again, an inspection plan identifying the particular firms and processes within each district to inspect over the next two years while still satisfying the inspection frequency requirements is produced. District management would apply these inspection plans as a tool in their selection of which firms and processes to inspect.

4. Application and Conclusions

During 1987, the NMIS decision support system was first introduced to FDA management. Headquarters management was very intrigued with its application and tradeoff curves. In particular, they were most interested in the national tradeoff curve and recent past national trends. The Commissioner of Food and Drugs responded by utilizing these results to justify the need for more drug inspection resources. Other top Headquarters management have employed the national tradeoff curve to estimate the amount of inspection resources to plan for in future years, up to 1990. On the other hand, district management were most interested in determining their level of performance on their tradeoff curve, as well as other districts level of performances.

NMIS provides FDA management with a powerful tool to assist them in evaluating and controlling their decisions with the limited resources available. The consequences of different alternatives may also be explored and used to justify the need for more drug inspection resources. With further application, the district inspection plans and listings of probability estimates will be used more by district and headquarters management as a tool to assist them in comparing the drug industry performance. NMIS is presently being updated with last year's regulatory data. It is expected to become an integral part of FDA's strategic planning of the surveillance of the pharmaceutical industry. Furthermore, these algorithms may be applicable to solving other real world problems.

References

Cohon, J.L. Multiobjective Programming and Planning, Academic Press, 1978.

Crowder, H., Johnson, E.L. and Padberg, M., "Solving Large-Scale Zero-One Linear Programming Problems", Operations Research, Vol. 31, No. 5, 1983.

Everett, H. "Generalized Lagrange Multiplier Method for Solving Problems of Optimal Allocation of Resources", Operations Research, Vol. 11, 1963.

Fisher, M.L., "LaGrangian Relaxation Method for Solving Integer Programming Problems", Management Science, Vol. 27, No. 1, January 1981.

Klimberg, R.K., Inspectional Scheduling: A Two-Stage Decision Support System Using Multiple Objectives, Doctoral Thesis, The Johns Hopkins University, 1987.

Loulou, R. and Michaelides, E., "New Greedy-Like Heuristics for the Multidimensional 0-1 Knapsack Problem, Operations Research, Vol. 27, No. 6, 1979.

Nauss, R., "An Efficient Algorithm for the 0-1 Knapsack Problem", Management Science, Vol. 23, No. 1, 1977.

Nemhauser, G.L., Introduction to Dynamic Programming, Wiley, 1966.

Toyoda, Y., "A Simplified Algorithm for Obtaining Approximate Solutions to Zero-One Programming Problems", Management Science, Vol. 21, 1975.

MULTIPLE CRITERIA DECISION SUPPORT SYSTEMS ON MICROCOMPUTERS AN APPLICATION IN PRODUCTION PLANNING

Birsen Karpak
Department of Management
Youngstown State University
Youngstown, OH 44555

ABSTRACT

In this paper we compare four interactive multiple criteria decision support systems on a microcomputer. VIG (Visual Interactive Goal programming) recently developed by Korhonen(1987) is an interactive decision support system providing nondominated solutions to the decision maker. The other three interactive methods considered are contracting cone method (CCM) and interactive Tchebycheff method (ITM) both developed by Steuer, and modified Belenson - Kapur method, which was originally developed by Belenson - Kapur (1973) and modified by the author (Karpak, 1978) to fit the cases where there are alternative solutions. All four of them are applied to a production planning problem in a manufacturing company.

1. INTRODUCTION

We have been solving different production planning problems in different manufacturing organizations with different multiple criteria methods. Our original experience was on a mainframe computer. Recently we have been solving problems on a microcomputer. We believe that if these methods are available on a microcomputer, their usage will proliferate. We had favorable experience with contracting cone method (CCM) and modified Belenson - Kapur method on mainframe (Karpak, Kuruuzum 1987). We compared the CCM, The ITM and modified Belenson - Kapur method in our previous study (Karpak, 1988b). We briefly give the results obtained for reference.

As software we used VINO (Cunningham, Schrage 1987) except

for Pareto Race. Data entry is quite facilitated in LOTUS based VINO.

The paper is organized as follows: section 2 describes the production system. In section 3, a production planning with the CCM is given. Section 4 illustrates the results obtained with the ITM. Implementation of the modified two person zero-sum approach is given in section 5. Section 6 gives results obtained by Pareto Race. Section 7 contains a summary and discussion.

2. PRODUCTION SYSTEM

The production system considered is quite similar to the system in our 1987 study (Karpak, Kuruuzum 1987) and exactly the same as in our recent study (Karpak, 1988a). We have explained the differences in Karpak 1988b. For the sake of traceability we shall briefly redescribe the system.

The company manufactures 96 different products. They may be grouped as D type fuses, HRC fuse cartridges, HRC load switches, cylindrical fuse cartridges, porcelain lampholders, fixation materials, ceramic insulators and English type fuses.

The portfolio of resources, usage of the resources by each product and sales forecasts for each product are given in Karpak, 1988a.

In our earlier approach, we had selected 95 percent service level for our target sales and obtained maximum marketable product and minimum demand levels to be satisfied from the marketing manager of the firm. In our previous model, when we solved the problem with maximum marketable amounts, we found a considerable excess of many resources. We were able to produce a maximum of almost all products. Our problem seemed to be a marketing problem rather than a production planning problem. When we presented the results, management wanted us to find the most profitable products. They were ready to increase marketing efforts for those products produced in excess of the maximum demands.

The company is primarily interested in profit maximization. However, they found too much idle capacity in certain resources when maximizing profit. In order to be able to use these resources, the glazing and assembly line departments had to work overtime. Management wanted to minimize overtime. They were eager to keep manufacturing costs as low as possible. Conflicting objectives of the system are:

1. Maximization of profit
2. Minimization of manufacturing cost
3. Minimization of overtime in the glazing department
4. Minimization of overtime in the assembly departments

3. PRODUCTION PLANNING WITH THE CONTRACTING CONE METHOD

The contracting cone method(CCM) basically reduces the criterion cone at each iteration (Steuer 1977 and 1986). Assuming we have k objective functions, initially the method generates 2k + 1 nondominated extreme point solutions corresponding to the following weighting vectors:

$\lambda_1 = (1,0,...,0)$
$\lambda_2 = (0,1,...,0)$
.
.
$\lambda_k = (0,0,...,1)$
$\lambda_{k+1}=(1/k^2,r,r,...,r)$
$\lambda_{k+2}=(r,1/k^2,r,...,r)$
.
.
$\lambda_{k+k}=(r,r,...,r,1/k^2)$
$\lambda_{2k+1}=(1/k,1/k,...,1/k)$

where $r = (k+1)/k^2$

For each weighting vector λ, the $\lambda' Cx$ linear programming problem is maximized and the decision maker is presented with 2k+1 solutions. In the next iteration the volume of the cone is reduced by 1/k around the solution selected by the decision maker.

Interaction with the decision maker and contraction of the

criterion cone continues until the decision maker is satisfied with a proposed solution.

First we solved the problem with the original 2k+1 λ's. Instead of 0, we have used $\epsilon > 0$ in order to eliminate dominated solutions. ϵ is taken as 0.0001. We contracted the criterion cone around the solution selected by the decision maker. Objective function values obtained with this contracted cone are given in table I

TABLE I
Objective Function Values With Contracted Gradient Cone

PROFIT (TL millions)	Manufacturing Cost (TL millions)	Overtime in Glazing (hours)	Overtime in Assembly Lines (hours)
3,894	1,115	552	0
3,688	1,186	0	0
2,209	696	0	0
6,116	1,857	1,248	78,624
4,752	1,111	1,173	30,552
6,111	1,783	672	70,523
5,789	1,843	1,248	1,250
5,481	1,750	1,248	78,624
4,980	1,854	552	0

Of the solutions our decision maker selected the solution yielding a profit of 4.7 billion Turkish Lira, 1,173 hours overtime in glazing, and 30,552 hours overtime in assembly departments. The decision maker was satisfied with the solution. He did not want any further solutions.

4. RESULTS OBTAINED WITH INTERACTIVE TCHEBYCHEFF

We used the same λ values for interactive Tchebycheff method(ITM) as we used for CCM though Steuer recommends randomly generating certain number of λ and then selecting the most dispersed ones using the FILTER algorithm.(Steuer, 1986 p.448)

The ITM could briefly be explained as follows: Let J^{max} denote the indices of the maximization objectives and J^{min} the indices of the minimization objectives. Then the Tchebycheff first finds z^* ideal criterion vector where

$$z_i^* = \begin{cases} \max\{c^i x + \epsilon_i \mid x \in S\} & i \in J^{max} \\ \min\{c^i x - \epsilon_i \mid x \in S\} & i \in J^{min} \end{cases}$$

The ϵ_i are numerically significant small positive numbers and S is the feasible region. Then we form the weighted Tchebycheff program(WTP)

$$\begin{aligned} &\min \{ \alpha + \rho \sum_{i=1}^{k} w_i \} \\ &\text{s.t.} \quad \alpha \geq \lambda_i w_i \qquad 1 \leq i \leq k \\ &\quad w_i = z_i^* - z_i \qquad i \in J^{max} \\ &\quad w_i = z_i - z_i^* \qquad i \in J^{min} \\ &\quad z_i = c^i x \qquad 1 \leq i \leq k \\ &\quad x \in S \end{aligned}$$

in which ρ is sufficiently small positive number. (See Silverman, Steuer, Whisman, 1987 p.120 and Steuer & Choo 1983). We generate initial 2k+1 nondominated solutions using the same λ's as we used for CCM method and then reduce the Λ space around the solution which our decision maker selected. We continue generating nondominated solutions until our decision maker is satisfied with one of them.

We have generated the initial 2k+1 solutions using the same initial 2k+1 λ values as we used in the CCM. We reduced the weight space around the solution preferred by the decision maker and found the following results:

TABLE II
Objective Function Values in Reduced Weight Space Using Interactive Tchebycheff

PROFIT (TL millions)	Manufacturing Cost (TL millions)	Overtime in Glazing (hours)	Overtime in Assembly Lines (hours)*
5,958	1,508	1,248	78,624
2,485	699	29	-5,865
2,471	699	4	-6,296
2,280	697	31	-1,703
4,115	939	1,248	24,305
5,444	1,103	1,248	78,624
5,202	1,249	1,248	54,843
5,352	1,606	1,248	45,876
5,150	1,229	1,248	53,316

*(-) sign in assembly overtime signifies slack capacity.

Of the solutions the decision maker selected the one corresponding to λ_5.; i.e. the solution yielding a profit of 4.1 billion Turkish Lira, 1,248 hours overtime in glazing, and 24,305 hours overtime in assembly.

5. RESULTS WITH MODIFIED TWO PERSON ZERO-SUM METHOD

In modified two person zero-sum method, initially we optimize each objective function successively and find the values of others at this point after eliminating alternative solutions in case they exist (Karpak,Kuruuzum 1987). Entries of the pay-off matrix are normalized and an optimal weighting vector λ'is obtained by applying linear programming to this normalized pay-off matrix (Belenson,Kapur 1973). $\lambda^T Cx$ is maximized in feasible space to obtain nondominated solutions, where $n_i = \lambda'/M_i$, M_i is the maximum value of objective i and $\lambda_i = n_i/(\sum_{i=1}^{k} n_i)$.

The solution vector is x^*. If Cx^* is the optimum from decision maker's point of view the solution has been obtained; if not, unsatisfied objectives may be increased if the decision maker accepts some decrease in other objectives. In this decision the analyst may help the decision maker by giving the explicit trade-offs between objectives with the help of sensitivity analysis.

We optimized the objective functions successively and obtained the pay-off table after eliminating dominated solutions. We have normalized the coefficients of each objective function and then applied the modified two person zero-sum method to obtain the following λ values:

$\lambda_1^* = .04,$ $\lambda_2^* = .24,$ $\lambda^*_3 = .40,$ $\lambda_4^* = .32$

The compromised function $.04c^1x - .24c^2x - .40c^3x - .32c^4x$ has been maximized over the same feasible region. The solution is as follows:

Profit	=	4,115 millions Turkish Lira
Manufacturing cost	=	1,110 millions Turkish Lira
Overtime in glazing	=	552 hours
Overtime in Assembly lines	=	33,543 hours

When we presented our compromised solution to the manager, he was satisfied with that and did not want any further solutions. We believe the decision maker would have asked for some other solutions if he had not had experience with the other two methods.

When we compare the optimal solutions (from the manager's point of view)obtained with these three methods, we see that all of them are fairly close to each other.

5. PRODUCTION PLANNING with PARETO RACE

VIG is an interactive software developed by Korhonen based on Pareto Race which is developed by the same author. It is not necessary to know Pareto Race which is a cornerstone of the software to be able to use it. We therefore confine our discussion to our experience with VIG. VIG generates efficient extreme as well as nonextreme solutions. It provides an initial solution on the screen and gives an opportunity to change this solution dynamically on the screen. Decision maker is able to see values of his/her goals as well as bar charts corresponding to these values.

Our initial solution with VIG is given in figure 1.

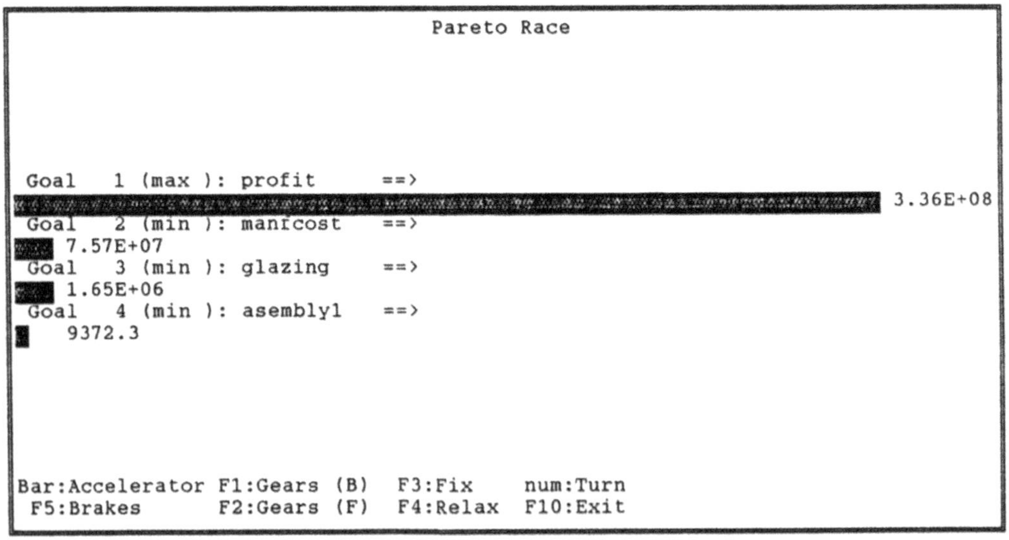

Figure 1 Initial Solution with VIG

Initial solution provided us only 3.36 billion Turkish Lira profit with 757 million TL manufacturing cost, and used only half of the capacities of both glazing as well as assembly departments. It was unacceptable. By using the space bar we started traveling on the efficient hyper-surface. We reached the optimal solution from decision maker's point of view within five minutes. In the course of reaching this optimal solution, VIG generated over 100 efficient solutions. Number of efficient solutions provided by VIG depends on the size of the efficient hyper-surface of our problem and speed we select. We monitor the speed by using function keys. "F1" increases speed in the backward direction, "F2" in the forward direction. Optimum solution is given in figure 2.

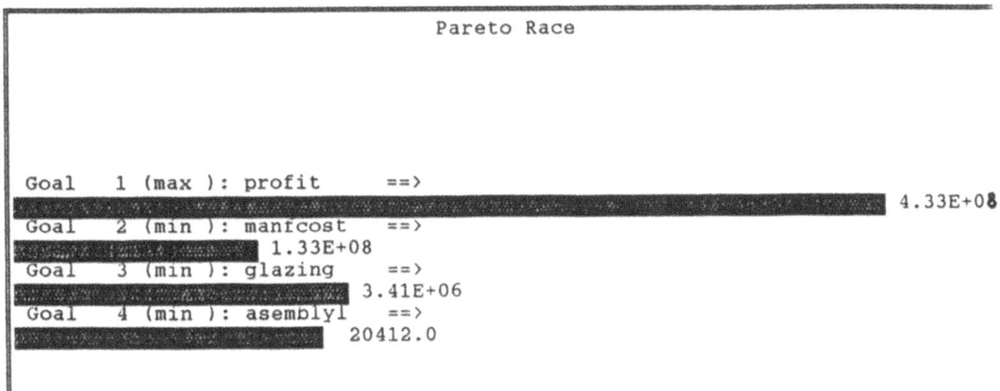

Figure 2. Optimum Solution Obtained with VIG

6. SUMMARY and DISCUSSION

We discussed the different features of The Belenson-Kapur method, the CCM and the ITM in our previous study. When we experimented with the ITM, each solution took approximately 5-7 minutes on IBM-XT for a problem with 4 objective functions 96 variables 137 constraints. Most of the constraints were upper bounds and/or lower bounds. We had to solve 9 linear programming problems to present a solution in multiple criteria sense to our decision maker. This was one iteration. We obtained our decision makers preference then reduced the weight space and solved another 9 linear programming problems. During our experimentation we thought it would have been much more efficient if one could obtain a solution; then dynamically see the impact of changing one objective function on others. In this respect the VIG seemed to be promising. The VIG is a versatile decision support system. Problem size is limited to 96 variables and 100 constraints. Constraints include a number of objective functions as well (which could be up to 10). We were lucky with VIG because of some of the features of our problem. Although we had 96 variables 141 constraints (4 of which were objective functions), some of the constraints were inactive. We could detect them in our previous study. We were able reduce our model into manageable size for VIG.

We found some inaccuracies in calculations when our problem size grew larger. Inaccuracy can be quite important for someone using the software to solve comparatively larger scale problems. Inaccuracy detection might not be so transparent for some problems. However we think it would not be difficult for the author to overcome this.

We thought problem size to be solved by the software could be enlarged easily if the author builds upper and lower bounds into the software. Most of our constraints were upper and lower bounds.

We were not able to get hard copies. We printed the results using a text editor(sidekick). We would like to see this feature embedded into the software.

We hope our readers will not get the impression that we did not appreciate the software. In fact we think that it is one of the best microcomputer based software for multiobjective linear programming.

The author wishes to thank Professor Ram. M.R. Kasuganti (Youngstown State University), for his constructive comments and suggestions.

REFERENCES

1. Belenson, S.M. & Kapur, K.C. 1973 "An algorithm for Solving Multicriterion Linear Programming Problems with Examples.",O.R. Quarterly., 24(1), 65-77

2. Cunningham, K. & Schrage L. 1987 VINO, The Scientific Press, Redwood City, California

3. Karpak, B. 1988a Multiple Criteria Decision Support Systems in Production Planning, A Micro Experience, Working Paper, Youngstown State University, Youngstown, Ohio, USA. (unabridged version of Karpak, 1988b)

4. Karpak, B. 1988b Multiple Criteria Decision Support Systems in Production Planning, A Micro Experience, forthcoming in: B. Karpak, S. Zionts (eds.) Multiple Criteria Decision Making and Risk Analysis Using Microcomputers Springer - Verlag

5. Karpak, B. 1978 Cok Amacli Optimizasyon Modellerinin Irdelenmesi Oyunlar kuramina Dayali Etkilesimli Bir Algoritma Onerisi ve Cok Amacli Bir Uretim Planlama Surecine uygulanmasi(in Turkish). Unpublished Thesis to qualify as an Associate professor,Istanbul University, faculty of Business Administration,Rumelihisariustu,Istanbul,Turkey

6. Karpak, B. and Kuruuzum, A. 1987 "Contracting Cone Method vs Modified Two Person Zero-Sum Game Approach: An Implementation in Production Planning" in Y.Sawaragi,K.Inoue and H.Nakayama, (Eds.), Toward Interactive and Intelligent Decision Support Systems,Springer Verlag, New York, pp. 343-352

7. Korhonen, P.J. 1987 VIG (Visual Interactive Goal Programming) User's Guide

8. Korhonen, P.J.,& Wallenius, J., 1987 A Pareto Race, Working Paper, Helsinki School of Economics.

9. Steuer, R.E. 1986. Multiple Criteria Optimization: Theory Computation and Application, Wiley, New York

10. Steuer, R.E. 1977. "Vector Maximum Gradient Cone Construction Techniques. Lecture Notes in Economics and mathematical Systems, in Zionts(ed) 155 Springer Verlag, New york.

12. Steuer, R.E. & Choo, E.U. 1983 "An Interactive Weighted Tchebycheff Procedure for Multiple Objective Programming ", Mathematical Programming 26(1), 326-344

A MULTIOBJECTIVE MODEL FOR LOCATING PUBLIC FACILITIES ON AN UNCONGESTED TRANSPORTATION NETWORK

Kostas G. Zografos
Department of Civil and Architectural Engineering
University of Miami

Herbert S. Levinson
Polytechnic University of New York

Robert Cromley
University of Connecticut

Aris Sissouras
University of Patras

ABSTRACT

A Multiobjective model for the location of public facilities on an uncongested transportation network is presented. The model combines the Mini-Max and Mini-Sum objectives and considers capacity constraints for the location of the facilities. The problem of locating health care centers in a rural area of Greece is used to display the capabilities of the proposed model.

1.0 INTRODUCTION

Location problems are encountered whenever public or private organizations seek to rationalize the distribution of their service facilities. Locational decisions of private and public sector organizations generally are governed by the same principle, that of fulfilling effectively the organizations goals under operational and resource constraints. The locational decisions of private sector generally seek to minimize costs or to maximize profits. Alternatively, public facility locational decisions involve a different set of values, and their objective function is difficult to be quantified in monetary terms [13]. Public facility location problems often must balance "spatial conflicts among the areas served or impacted by the facility" [2]. Therefore, public facility location problems are inherently multiobjective in nature. Multiobjective formulations have been used to locate a wide range of public facilities from fire stations [14], and emergency medical facilities

[1], to airports [3], and links of transportation networks [12].

This paper presents a Goal Programming model for the location of public facilities on an uncongested transportation network. An application of the proposed model for the location of primary health care centers is also presented.

2.0 PROBLEM STATEMENT

The proposed multiobjective public facility location problem can be stated as follows:

On an uncongested transportation network locate a given number of public facilities so as to:

1. Minimize the maximum travel time from demand centesr to their closest facility.
2. Minimize the total weighted travel time from demand centers to their closest facility locations.
3. The population served by each facility does not exceed an upper limit.
4. The population served by each facility exceeds a minimum threshold value.

The first objective of the proposed model corresponds to the minimax problem [6,7,8,9,10], and requires the minimization of the travel time of the most remotely located users of the public facility system. This objective is frequently used when the services of the facilities should be offered within a critical time threshold, i.e. emergency services [11].

The second objective corresponds to the mini-sum problem [6,8,11], and requires and minimization of the weighted travel time for all the users in system. The first two objectives of the proposed model cover two of the most important and usually conflicting objectives of the agencies providing public services that is efficiency and equity. While the first criterion is targeted to the equitable distribution of services, the second produces an efficient locational pattern. Criteria three and four concern the utilization of the facilities and hence they define threshold values for their efficient operation.

3.0 THE MATHEMATICAL MODEL

The mathematical expression of the multiobjective location model is as follows:

$$\text{MIN } F = W_1 \sum_{i=1}^{N} d_{1i}^{+} + W_2 d_2^{+} + W_3 \sum_{j=1}^{M} d_{3j}^{-} + W_4 \sum_{j=i}^{M} d_{4j}^{+}$$

Subject to:

$$\sum_{j=1}^{M} t_{ij} Y_{ij} - d_{1i}^{+} \leq T \qquad \forall\, i$$

$$\sum_{i-1}^{N} \sum_{j=1}^{M} P_i t_{ij} Y_{ij} - d_2^{+} \leq b_2$$

$$\sum_{i=1}^{N} Y_{ij} P_i + d_{3j}^{-} \geq C_{\ell} Y_{jj} \qquad \forall\, j$$

$$\sum_{i=1}^{N} Y_{ij} P_i - d_{4j}^{+} \leq C_u\, Y_{jj} \qquad \forall\, j$$

$$\sum_{j=1}^{M} Y_{ij} = 1 \qquad \forall\, i$$

$$\sum_{\substack{i=1 \\ i \neq j}}^{N} Y_{ij} \leq (N-1)\, Y_{jj} \qquad \forall\, j$$

$$\sum_{j=1}^{M} Y_{jj} = S$$

Where:

b_2 = The attainment level for the second goal.

C_ℓ = The minimum population served by any facility of the system.

C_u = The maximum population served by any facility of the system.

d_{1i}^+ = Positive deviational variable for goal one

d_2^+ = Positive deviational variable for goal two

d_{3j}^- = Negative deviational variable for goal three

d_{4j}^+ = Positive deviational variable for goal four.

M = The number of candidate facility locations.

N = The number of demand nodes (population centers).

P_i = Population of center (i)

S = The number of facilities to be located

T = Threshold value of travel time

t_{ij} = Minimum network travel time between population center (i) and facility location (j).

Y_{ij} = 1 If demand at (i) is assigned to a facility located at (j); 0 Otherwise

$W_1, \ldots W_4$ = Premptive priority factors for goals one through four.

4.0 APPLICATION OF THE MODEL

The proposed MLM (1)-(8) was applied to locate primary health care centers in a rural area of Greece. The transportation network of the region under consideration is shown in Figure 1. Villages represented by nodes one, five, eleven, fifteen, twenty-eight, and forty-nine are considered as candidate health center locations [15]. The number of health care centers to be located is equal to three, i.e. S=3. The travel time between the cadidate facility locations and the population centers (t_{ij}) was calculated using the Floyd's algorithm [5]. The study area consists of fifty nine towns and villages with a total population of 46,412 inhabitants. The distribution of the population for the fifty nine towns of the study area is given on Table 1.

An iterative goal programming was used to generate alternative scenarios for the location of the health care centers and the allocation of population centers to them. The SPERRY UNIVAC/FMPS mixed integer programming code was used to implement the iterative goal programming algorithm. Initially the upper and lower bounds of the goal attainment levels were generated using the first step of the GPSTEM method [4]. At the second step the goal attainment levels for the goals are decided by the decision maker and the problem is solved. The results of this step are presented in Table 2. At the third step the results of the solution are evaluated. If the decision maker is satisfied then the solution procedure terminates. If not, then the decision maker can set new goal attainment levels and the procedure is repeated.

Table 3 presents the results of three alternative scenarios examined for the purpose of this presentation by an hypothetical decision maker (D.M.). The first five columns of the table indicate the priority structure and the goal attainment levels for the three iterations, while the last five columns present the values of the deviational variables.

In the first iteration goal one has the highest priority and goals two and four are not satisfied. The location of the facilities and the allocation of population to them generated by the first iteration of the algorithm are presented graphically in Figure 1. Facilities are located at nodes one, eleven and fifteen. The catchment area of each facility is bounded by a bold faced line. The results of the second iteration are given in the second row of Table 3, and indicate that goal one is deviated by 1.9 minutes while goal four is deviated by 4096 persons.

Figure 2 presents the results of the second iteration of the

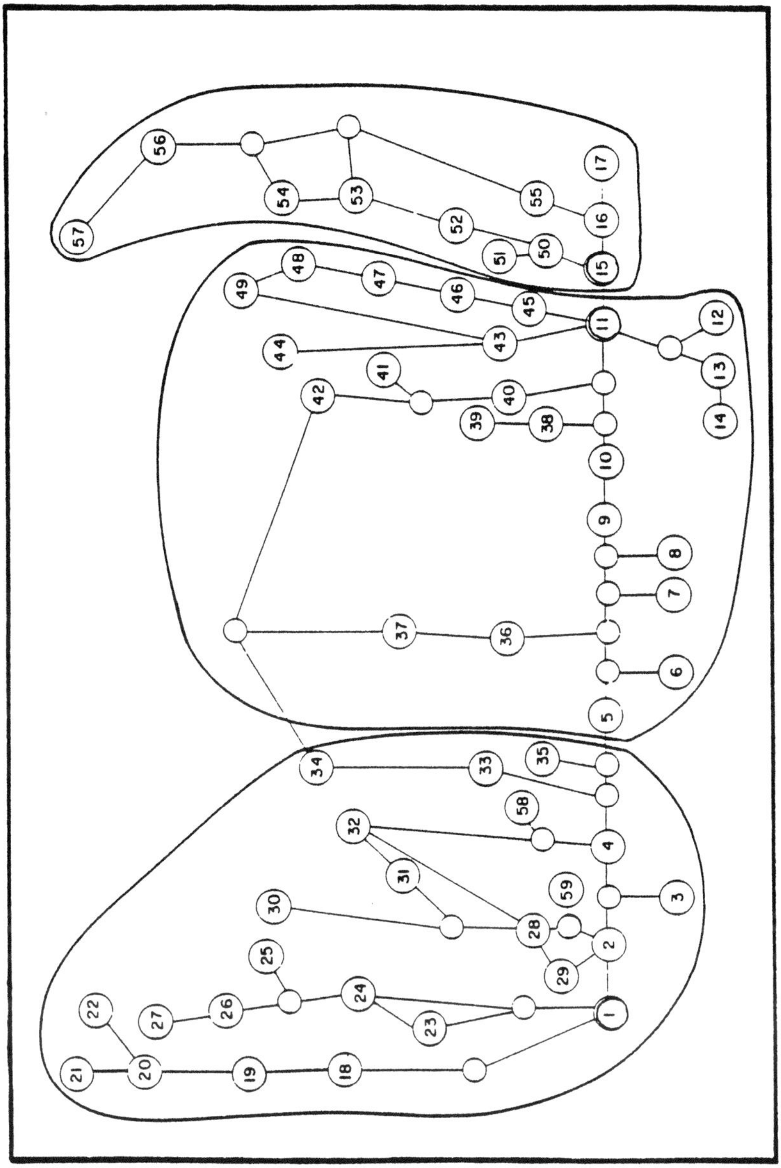

Figure 1: Results of the First Iteration

Table 1

Population of the Study Area

Site #(i)	Population Pi#	Site #(i)	Population Pi	Site #(i)	Population Pi
1	1320	21	96	41	476
2	450	22	60	42	305
3	151	23	312	43	608
4	361	24	288	44	492
5	1848	25	72	45	700
6	435	26	104	46	296
7	655	27	96	47	386
8	406	28	1624	48	282
9	453	29	191	49	655
10	381	30	73	50	394
11	20955	31	102	51	345
12	213	32	429	52	57
13	1113	33	85	53	91
14	526	34	280	54	171
15	1822	35	271	55	309
16	814	36	125	56	67
17	811	37	257	57	278
18	377	38	621	58	359
19	207	39	361	59	698
20	131	40	567		

Source: National Statistical Service of Greece, Population of Greece, 1981 census.

Table 2: Upper and Lower Bounds for the Goal Attainment Levels

Bound	Mini-Max (Minutes)	Weighted P-Median (Person Minutes)	Lower Pop. Limit (Persons)	Upper Pop. Limit (Persons)
Upper	32.90	168181.20	5000	34096
Lower	31.00	145741.70	5000	30000

Table 3: Results of the Multiobjective Location Model

Iteration Number	GOAL				GOAL DEVIATIONS			
	Mini-Max (Minutes)	Mini-Sum (Person-Minutes)	Lower Poulation Limit (Persons)	Upper Population Limit (Persons)	Σd_{1i}^{+}	d_2^{+}	Σd_{3j}^{-}	Σd_{4j}^{+}
1	W_1^{1*} $T = 31.0$	W_2^{2} $b_2 = 145741.7$	W_3^{4} $Cl = 5,000$	W_4^{3} $C_u = 30,000$	0.0	22439.9	0.0	3116
2	W_1^{1} $T = 32.9$	W_2^{2} $b_2 = 145741.7$	W_3^{4} $Cl = 5,000$	W_4^{3} $C_u = 30,000$	1.9	0.0	0.0	4096
3	W_1^{2} $T = 32.9$	W_2^{3} $b_2 = 150,000$	W_3^{4} $Cl = 5,000$	W_4^{1} $C_u = 30,000$	1.9	46.5	0.0	2166

(*) Superscripts Indicate Goal Attainment Priorities.

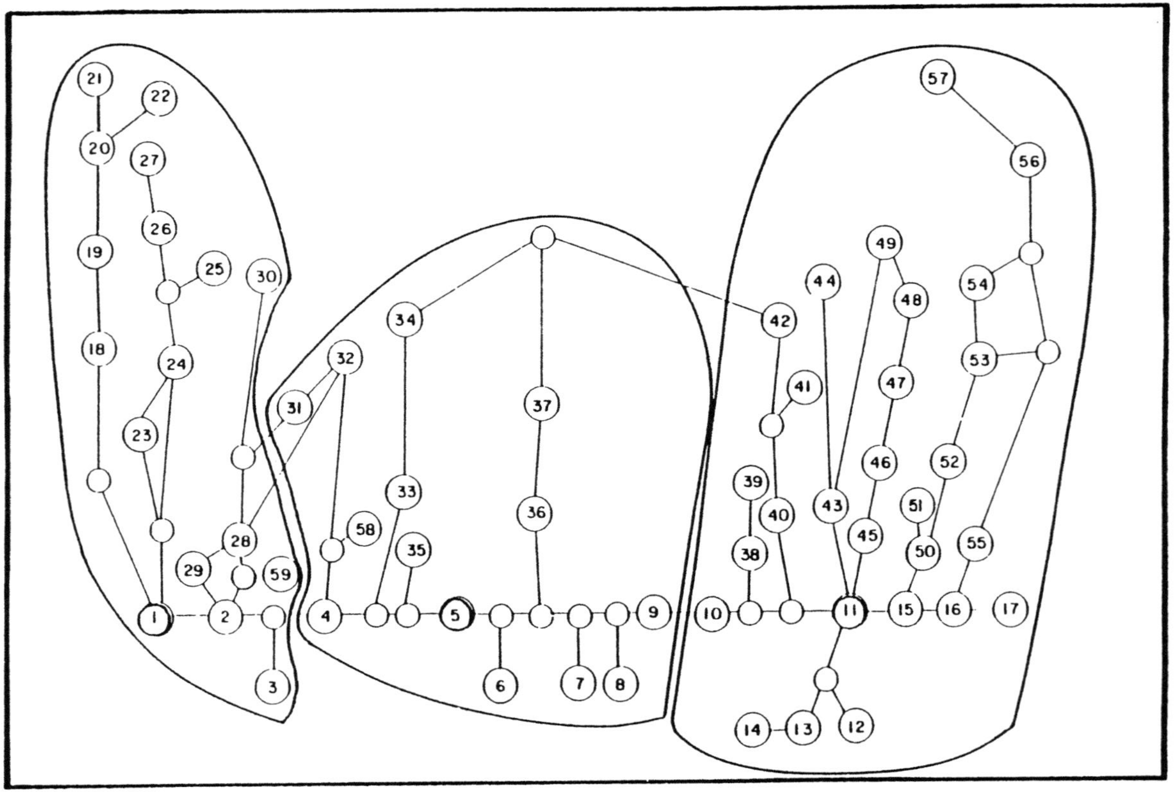

Figure 2: Results of the Second Iteration

algorithm. Note that the facilities now are located at nodes one, five, and eleven and that their catchment areas differ from the catchment areas of the previous iteration. Finally a third iteration is performed by allowing an increase in the total weighted travel time from 145741.7 to 150000 person minutes.

The results of the third iteration are presented in the third row of Table 3. The locations of the facilities remain the same as in the second iteration but the catchment areas of the facilities have been changed to order to reduce the overutilization of the facility located at node eleven by 1930 persons. As a result of the third iteration, the population of nodes ten, thirty-eight, thirty-nine and forty are assigned now to the facility located at node five. It is obvious that according to this iteration certain users of the system they are not assigned to their nearest facility.

5.0 CONCLUSIONS

A goal programming approach for the location of public facilities on an uncongested transportation network was presented. Although the model was discussed in conjunction to the location of health care centers in a rural area, it should be pointed out that the model with little or no modification can be used for the location of public facilities which require for their locationthe same criteria used in this formulation.

REFERENCES

1. Charnes and Storbeck, "A Goal Programming Model for the Siting of Multilevel EMS Systems," Soc. Econ. Plan. Sci. vol. 14, 1984, pp.155-161.

2. Cohon, Jared, Multiobjective Programming and Planning. Academic Press, New york, NY., 1978.

3. DeNeufville, R. and R. Keeney, "Use of Decision Analysis in Airport Development for Mexico City," in A. Drake, R. Keeney, and P. Morse (Eds.), Analysis of Public Systems, Cambridge, Mass. M.I.T. press, 1972.

4. Fichefet, J., "GPSTEM: An Interactive Multiobjective Optimization Method," in Progress in Operations Research Vol. 1, A. Prekopa, (Ed.), pp. 317-332. North Holland, Amsterdam, 1976.

5. Floyd, R. W., "Algorithm 97-Shortest Path," Comm. of ACM, Vol. 5, p. 34.

6. Goldman, A. J., "Optimal Location for Centers in a Network," Transportation Sciences, No. 3, pp. 352-360, 1969.

7. Goldman, A. J., "Optimal Center Location in Simple Networks for Optimal Facility Placement," Transportation Sciences, Vol. 5, 1971, pp. 212-221.

8. Goldman, A. J., "Minimax Location of a Facility in a Network," Transportation Science, Vol. 6, 1972, pp. 195-201.

9. Hakimi, S. L., "Optimal Locations of Switching Centers and the Absolute Centers and Medians of a Graph," Operations Research, Vol. 12, 1964, pp. 450-459.

10. Hakimi, S. L., E. F. Schmeichel, and J. G. Pierce, "On p-Centers in Networks," Transportation Science, Vol. 12, 1978, pp. 1-15.

11. Handler, G. Y. and P. B. Mirchandani, Location on Networks Theory and Algorithms, M.I.T. Press, Cambridge, Mass., 1984.

12. Leinbach, T. R. and R. G. Cromley, "A Goal Programming Approach to Public Investment Decisions: A Case Study of Rural Roads in Indonesia." Socio-Economic Planning Sciences, 17:1, 1983.

13. ReVelle, C., D. Marks, and J. Liebman, "An Analysis of Private and Public Sector Location Models." Management Science, Vol. 16, No. 11, 1970.

14. Schilling D.A., C. ReVelle, J. Cohon, and D. Elzinga, "Some Models for Fire Protection Locational Decisions," European Journal of Operational Research, Vol. 5, 1980, pp.1-7.

15. Zografos, K. G. and R. G. Cromley, "Modelling Large Scale Hierarchical Public Facility Location Problems." Proceedings at the 1987 Conference on Planning and Design in Urban and Regional Planning, E. Alexander (Ed.), American Society of Mechanical Engineers, 1987, pp. 97-103.

A MULTICRITERION OPTIMIZATION MODEL FOR INDUSTRIAL PROMOTION: THE CASE OF KOREA

M.T. Tabucanon and M.J. Shin
Division of Industrial Engineering and Management
Asian Institute of Technology
P.O. Box 2754, Bangkok 10501, Thailand

Abstract

Industrial policies form major parts in a country's national economic development plan. Countries like Korea, where manufacturing is a prime engine of economic growth, need sound industrial policies to promote balanced growth of industries. This paper develops a generalized goal programming model for selecting which industries need, and to what extent they are, to be promoted in line with national economic goals. The national economic objectives pursued in the model include the following: Foreign Exchange Earnings, Employment Opportunities, Domestic Content, Energy Consumption, Import Substitution, and Productivity. The model is applied to six industrial sectors, namely, Fabricated Metal Products, General Industrial Machinery and Equipment, Electrical Equipment and Apparatus, Electronic and Communication Equipment, Transportation Equipment, and Measuring, Medical and Optimal Instruments. The generality of the model makes it possible to be applied to any industrial sector or subsectors, and it can serve as a software laboratory for selecting types of industrial activities which need to be given promotion. It can also be applied to other countries with appropriate modifications.

1. Introduction

The National Economic and Social Development Plan is a prevailing instrument used by many developing countries to increase social welfare and promote economic development and hence, industrialization. Industrial development which is determined mainly by the economic, political and social conditions of the country cannot be achieved for all industries simultaneously due to many obstacles. Variety of conditions deter industrial growth in several ways. Industrial development planning is therefore essential to policy makers in order to, among other things, identify industries which need promotion to achieve the planned goals in a balanced fashion.

In the case of Korea, overall industries have grown up dramatically due to effective past economic development plans. Therefore, the economic situation in Korea is somewhat different from the other developing countries. Industrial structure re-adjustment is now needed to take full advantage of the increasing specialization of the international economic structure. The objective of this paper is to present a model which can be used for selection of the industrial sectors which need to be promoted.

2. Model Development

In this paper, industries are divided into large group (sector); each sector is divided into industry group (subsector); and each subsector is divided into individual industries. Six objectives are considered in the model namely, improvement of exports, increase of employment opportunity, increase of domestic content, lessening energy consumption and imports, and improvement of productivity.

The model is developed under the backdrop of the following scenarios - world and domestic economic circumstances will not change suddenly and world demand is very large (assumed infinite) relative to Korea's overall production.

The decision variable used in the model is X_{ijk} which represents the total amount of production of industry type k(1, 2, 3, , , e) which belong to j(1, 2, , ,n) subsector, and i(1, 2, , , m) sector.

Export Goal

Local production which can be considered as the dominant factor that affects exports should at least exceed the local demand and it should be maximized to achieve the export goal.

The export goal of sector i can be expressed as follows:

$$\sum_{j=1}^{n} \sum_{k=1}^{e} (p_{ijk}^{t-1} \alpha^t) X_{ijk}^t + (d_i^-)^t - (d_i^+)^t = Ex_i^t \tag{1}$$

where, t : Planned time horizon (t=1,2, ,...,5);

p_{ijk}^{t-1} : Expected export portion of products X_{ijk} at period t-1;

α^t : Expected increase (decrease) rate of export ratio at period t;

Ex_i^t : Export goal of sector i at period t;

X_{ijk}^t : Decision variable.

The decision variable represents the total amount of production of industry type k which belongs to j subsector, k sector at period t.

The objective is to minimize the under-achievement of the export goal, d_i^-.

Employment Goal

Assuming that employment is directly proportional to production capacity, a goal of employment can be expressed as;

$$\sum_{i=1}^{m} \sum_{j=1}^{n} \sum_{k=1}^{e} (w_{ijk}^{t-1} \beta^t) X_{ijk}^t + (d_{m+1}^-)^t - (d_{m+1}^+)^t = W^t \tag{2}$$

where, w_{ijk}^{t-1} : Average labor requirement to produce unit of products X_{ijk}^{t-1} at period t-1;

$$\left(= \frac{\text{Total number of employment in industry k}}{X_{ijk}} \right)$$

β^t : Expected increase (decrease) rate of employment at period t;

W^t : Minimum desirable number of employment level at period t.

The objective is to minimize the under-achievement of the employment goal, d^-_{m+1}.

Domestic Content Goal

The ratio of domestic content can be expressed in terms of ratio of the value of domestic produced parts to total production cost. Low domestic content gives rise to increase in corresponding import of parts, raw materials, etc. Thus, domestic content goal can be expressed as follows:

$$\sum_{i=1}^{m} \sum_{j=1}^{n} \sum_{k=1}^{e} (r^{t-1}_{ijk}\, \Gamma^t)\, X^t_{ijk} + (d^-_{m+2})^t - (d^+_{m+2})^t = D^t \tag{3}$$

where, r^{t-1}_{ijk} : The ratio of domestic content of products X_{ijk} at period t-1;

$$\left(= \frac{\text{The value of domestic products used in } X_{ijk}}{\text{Total manufacturing cost of } X_{ijk}} \right)$$

Γ^t : Expected increase (decrease) rate of the domestic content ratio at period t;

D^t : Maximum expected level of domestic supply in order to produce the total products at period t.

The objective is to minimize the under-achievement of the domestic supply goal, d^-_{m+2}.

Energy Consumption Goal

For energy deficient countries like Korea, it is preferable that industries which have low energy consumption need promotion.

$$\sum_{i=1}^{m} \sum_{j=1}^{n} \sum_{k=1}^{e} (t^{t-1}_{ijk}\, \delta^t)\, X^t_{ijk} + (d^-_{m+3})^t - (d^+_{m+3})^t = E^t \tag{4}$$

where, t^{t-1}_{ijk} : Energy consumption per unit production at period t-1,

$$\left(= \frac{\text{Total energy consumption in industry k}}{\text{Total production in industry k}} \right)$$

δ^t : Expected increase (decrease) rate of energy consumption ratio at period t;

E^t : Energy consumption goal at period t.

The objective is to minimize overachievement of the energy consumption goal, d^+_{m+3}.

Import Substitution Goal

A deliberate policy of import substitution aims at accelerating domestic production over- and above- process of natural import substitution, and it affects employment, net earning of foreign exchange, balance of payments, etc.

Import substitution can be measured by the ratio of self-sufficiency.

$$\sum_{i=1}^{m} \sum_{j=1}^{n} \sum_{k=1}^{e} (1 - s_{ijk}^{t-1} \pi^{t}) (d_{ijk}^{t-1} \sigma^{t}) X_{ijk}^{t} + (d_{m+4}^{-})^{t} - (d_{m+4}^{+})^{t} = I_{m}^{t} \qquad (5)$$

where, s_{ijk}^{t-1} : The ratio of self-supply in industry k at period t-1,

$$\left(= \frac{\text{Domestic market - Imports}}{\text{Domestic market}} \right)$$

π^t : Expected increase (decrease) rate of the self-supply ratio at period t;

d_{ijk}^{t-1} : The ratio of domestic market to total products at period t-1;

$$\left(= \frac{\text{Total amount of domestic market}}{\text{Total amount of production}} \right)$$

σ^t : Expected increase (decrease) rate of domestic market ratio at period t;

I_m^t : Maximum expected level of imports (finished products) at period t.

The objective is to minimize overachievement of the imports, d_{m+4}^{+}.

Productivity Goal

Generally, productivity can be measured in various ways. Among them, productivity of labor (= Gross value added/number of employment) and productivity of capital (= Gross value added/Total assets) are the basic indicators used most frequently, [3] and [12]. In both indicators, value added concept is used.

To measure productivity, value is employed.

$$\sum_{i=1}^{m} \sum_{j=1}^{n} \sum_{k=1}^{e} (v_{ijk}^{t-1} \mu^{t}) X_{ijk}^{t} + (d_{m+5}^{-})^{t} - (d_{m+5}^{+})^{t} = V^{t} \qquad (6)$$

where, v_{ijk}^{t-1} : The ratio of value added per unit products at period t-1;

$$\left(= \frac{\text{Total amount of value added in product k}}{\text{Total amount of product k}} \right)$$

μ^t : Expected increase (decrease) rate of value added ratio at period t;

V^t : Value added goal at period t.

The objective is to minimize underachievement of the value added goal, d_{m+5}^{-}.

Sum of Production

The sum of production of individual industries should be equal or greater than the production of the previous year.

$$\sum_{i=1}^{m} \sum_{j=1}^{n} \sum_{k=1}^{e} X_{ijk}^{t} \geq e^{t}.X_{ijk}^{t-1} \tag{7}$$

where, X_{ijk}^{t-1} : Production at period t-1;

e^{t} : Expected increase (decrease) rate of total sum of production at period t.

Growth Rate of Industry

A certain ratio of growth should be maintained in each industry.

$$\alpha_{l}^{t}.\ X_{bi}^{t-1} \leq X_{i}^{t} \leq \alpha_{u}^{t}.\ X_{bi}^{t-1} \tag{8}$$

where, X_{bi}^{t-1} : The production of sector i at period t-1;

α_{l}^{t}, α_{u}^{t} : Desired minimum and maximum growth rate of sector i at period t.

Investment

The concept of ratio of gross value added to tangible fixed assets is employed in efficiency investment goal.

$$\sum_{i=1}^{m} \sum_{j=1}^{n} \sum_{k=1}^{e} \frac{(v_{ijk}^{t-1}\ \mu^{t})}{(q_{ijk}^{t-1}\ \phi^{t})} (X_{ijk}^{t} - X_{ijk}^{t-1}) \leq I_{n}^{t} \tag{9}$$

where, q_{ijk}^{t-1} : The ratio of gross value added to tangible fixed assets for industry k at period t-1;

$$\left(= \frac{\text{Gross value added}}{\text{Tangible fixed assets - Construction in process}} \right)$$

ϕ^{t} : Expected increase (decrease) rate of ratio q_{ijk}^{t-1} at period t;

I_{n}^{t} : Net investment capacity for tangible fixed assets at period t.

Priority Structure

Since this paper is based on the Sixth Five-Year Plan of Korea, the priorities and weights are deduced from the plan. The priorities for the problem under consideration are taken to be the following:

Priority Level 1(P_1) Export Goal. The first priority is to minimize the under-achievement of the target of export goal, $P_1(d_1^-, d_2^-, d_3^-, d_4^-, d_5^-, d_6^-)$.

Priority Level 2(P_2) Employment Goal. The second priority is to minimize the under-achievement of the desired level of employment goal, $P_2(d_7^-)$.

Priority Level 3(P_3) Productivity Goal. The third priority is to minimize the under-achievement of the desired level of value-added, $P_3(d_{12}^+)$

Priority Level 4(P_4) Domestic Content Goal. The fourth priority is to minimize the under-achievement of the desired level of domestic contents, $P_4(d_8^-)$

Priority Level 5(P_5) Import Substitution. The fifth priority is to minimize the over-achievement of the expected level of import substitution, $P_5(d_{11}^+)$

Priority Level 6(P_6) Energy Consumption Goal. The sixth priority is to minimize the over-achievement of the target of energy consumption, $P_6(d_{10}^+)$.

Given these priorities, the objective function takes the form:

$$\text{Min. } Z = P_1(d_1^-, d_2^-, d_3^-, d_4^-, d_5^-, d_6^-) + P_2(d_7^-) + P_3(d_{12}^-) + P_4(d_8^-) + P_5(d_{11}^+) + P_6(d_{10}^+)$$

Subject to: All the constraints from eqs.(1) to (9).

If all the expected increase (decrease) ratios (α^t, β^t, , ϕ^t) are assumed 1, then this model becomes static.

3. Model Application

The Machinery and Electrical & Electronic industry of Korea is used as a case application of this model. These industry covers approximately 30% of the whole manufacturing sector. The sectors covered are Fabricated Metal Products, General Industrial Machinery and Equipment, Electrical Equipment and Apparatus, Electronic and Communication Equipment, Transportation Equipment, and Measuring, Medical and Optical Instruments.

The GP model formulated above consisting of 67 decision variables, 22 deviational variables, 11 goal constraints and 136 system constraints was solved using the SLGP approach through the MPSX package of IBM [4].

The "satisficing" solution of the basic run in terms of production in 1991 and goal achievement for each priority levels including goal levels and actual solution are summarized in Tables 1 and 2 respectively.

The deviation of the goal constraints for this solution are as follows:

Goal	Negative Deviation	Positive Deviation
Priority 1		
Sector 1	0	384.1
Sector 2	0	0
Sector 3	0	313.5
Sector 4	0	0
Sector 5	0	1265.6
Sector 6	0	0
Priority 2	0	211.7
Priority 3	0	484.4
Priority 4	0	0
Priority 5	0	3290.8
Priority 6	0	491.8

The results of the basic run shows that first to fourth priority goals, i.e., export, employment, productivity, and domestic content goals are achieved. However, the fifth and sixth goals, i.e., import substitution and energy consumption goals are not achieved. From this result, export for sectors 2,4,6 are completely achieved while export for sectors 1,3,5 are over-achieved, employment and productivity goals are over-achieved, and domestic content goal is fully achieved. On the other hand, investment for sectors 1-6 need 8242.3 billion won (Korean currency) which is less than the planned investment, and the total production for sectors 1-6 will reach 43847.5 billion won which is much more than the expected production. To achieve the import substitution and energy consumption goals, development of raw materials, parts industries and effective energy saving techniques are needed.

The basic run identifies the industries which need promotion during the 6th five-year plan period. The details are shown in Table 3. According to the National Economic and Social Development Plan, the average growth rate of the whole selected industries is 8.7% every year during the planned period. Table 3 also shows the industries which need to grow more than 8.7% per annum. Other results are shown in Tables 4 and 5.

4. Concluding Remarks

This paper develops a generalized goal programming model for selecting which industries need, and to what extent they are, to be promoted in line with the national economic goals. The model is also developed in order to simulate the optimal growth avenues of the selected industries. Approach followed needs the articulation of development goals and a statement of their relative priorities and an indication of desired level. The model is also flexible enough to allow testing on goal targets and priority structures in order to assess their influence on the solution. GP model formulated in this paper can aid policy makers in framing the industrial promotion policies, and it provides a powerful and flexible tool that is able to select the industries which meet the national planning objectives and to simulate the growth avenue to meet the national targets.

Policy makers can make adjustments when they apply this model. Moreover, the mathematical simplicity of the formulated GP model and the easy availability of linear computer code to obtain solutions make the model implementation rather easy.

Table 1 Goal Achievement for Basic Run

Priority	Goal Type	Goal Level	Actual Solution	Achievement
P_1	Maximize Export			
	for Sector 1	2460.0	2844.1	Achieved
	Sector 2	1209.5	1209.5	"
	Sector 3	1209.5	1523.0	"
	Sector 4	9676.0	9676.0	"
	Sector 5	4838.0	6103.6	"
	Sector 6	483.75	483.75	"
P_2	Maximize			
	Employment	1290.7	1502.4	"
P_3	Maximize			
	Productivity	13494.5	13978.6	"
P_4	Maximize			
	Domestic-content	35701.6	35701.6	"
P_5	Minimize			Not
	Imports	10668.0	13958.8	Achieved
	Minimize Energy			Not
	Consumption	1305.2	1797.0	Achieved

All units are Billion Won except P_2 (1000 persons)

Table 2 Industrial Structure Change under Basic Run

Unit: Billion Won ('80 Price)

Year / Sector	1986		1991	
	Production	Ratio	Production	Ratio
Sector 1	2,934.8	10.99	5,085.0	11.60
Sector 2	4,757.5	17.82	6,678.6	15.23
Sector 3	2,513.7	9.41	4,154.0	9.47
Sector 4	8,414.7	31.51	13,565.0	30.94
Sector 5	7,626.3	28.56	13,452.8	30.68
Sector 6	456.3	1.71	912.0	2.08
Total Production	26,703.4	100.00	43,847.4	100.00

Table 3 Model Solution for Basic Run (Unit: Billion Won)

X_{ijk}	Production	X_{ijk}	Production
X_{111}	286.0 *	X_{332}	104.0
X_{112}	808.0 *	X_{333}	119.0
X_{121}	262.0 *	X_{334}	235.0
X_{122}	926.0 *	X_{411}	1292.0 *
X_{123}	1086.0 *	X_{412}	2724.0 *
X_{124}	168.0	X_{413}	2300.0 *
X_{131}	84.0	X_{421}	1324.0
X_{132}	252.0 *	X_{422}	161.0
X_{133}	148.0	X_{431}	796.0 *
X_{134}	734.0 *	X_{432}	298.0
X_{135}	381.0	X_{433}	1890.1 *
X_{211}	1099.0	X_{434}	168.0
X_{221}	218.0	X_{435}	95.0
X_{222}	172.0 *	X_{436}	1746.0 *
X_{231}	706.0 *	X_{441}	514.0
X_{232}	76.0 *	X_{442}	262.0
X_{233}	496.0 *	X_{511}	5032.7 *
X_{234}	81.0	X_{512}	188.0 *
X_{235}	88.0 *	X_{513}	26.0
X_{236}	1884.0 *	X_{521}	380.0 *
X_{241}	50.0	X_{522}	140.0 *
X_{251}	54.0 *	X_{531}	3580.0 *
X_{252}	87.0	X_{532}	658.0 *
X_{253}	905.9	X_{533}	1324.0 *
X_{261}	760.0 *	X_{534}	1577.0
X_{311}	856.0 *	X_{541}	53.0
X_{312}	120.0 *	X_{551}	142.0 *
X_{313}	188.0 *	X_{552}	160.0 *
X_{314}	1062.0 *	X_{553}	188.0 *
X_{321}	81.0	X_{611}	96.0 *
X_{322}	107.0	X_{612}	178.0 *
X_{323}	134.0	X_{621}	240.0 *
X_{324}	94.0	X_{631}	398.0 *
X_{331}	1054.0 *		

*: Grow more than 8.7% per annum.

Table 4 Priority Structure

Run	1	2	3	4	5	6	7	8	9	10
Max. Export	1	2	2	3	3	6	1	1	2	1
Max. Employment	2	1	4	6	4	5	4	1	1	2
Max. Productivity	3	5	1	5	2	4	2	4	3	3
Max. Domestic content	4	3	5	1	6	3	6	2	1	3
Min. Import	5	4	5	2	5	1	3	2	1	2
Min. Energy	6	6	3	4	1	2	5	3	4	1

Table 5 Industrial Structure Change Under Different Priority Structure

(Unit: %)

	1986	1991								
		Basic	Run2	Run3	Run4	Run5	Run6	Run7	Run8	Run9	Run10
Sector 1	10.99	11.60	10.8	10.6	13.4	8.4	12.6	10.4	11.6	13.4	10.9
Sector 2	17.82	15.23	15.6	15.3	11.9	13.2	11.7	16.2	15.2	11.9	15.6
Sector 3	9.42	9.47	9.9	8.2	10.9	9.1	11.4	8.2	9.5	10.9	8.2
Sector 4	31.51	30.94	32.2	33.8	27.8	35.3	25.8	32.8	30.9	27.8	34.1
Sector 5	28.56	30.68	29.4	29.9	35.0	32.5	37.5	30.3	30.7	35.0	29.0
Sector 6	1.71	2.08	2.1	2.2	1.1	1.6	1.1	2.2	2.1	1.1	2.2
Total	100	100	100	100	100	100	100	100	100	100	100

References

[1] A Handbook of Korea (1983), 5th Edition, Ministry of Culture and Information.

[2] ADB (1982), Industrial Growth and Employment in Developing Asian Countries: Issues and Perspectives for the Coming Decade.

[3] Financial Statement Analysis (1986), The Bank of Korea.

[4] Ignizio, J.P. and Perlis, J.H. (1979), Sequential Linear Goal Programming: Implementation via MPSX, Computers and Operations Research, Vol. 6, pp.141-145.

[5] Ignizio, J.P. (1982), Linear Programming in Single and Multiple-Objective Systems, Prentice-Hall.

[6] Input-Output Table (1985), The Bank of Korea.

[7] Jose, V.D. and Tabucanon, M.T. (1986), Multiobjective Models for Selection of Priority Areas and Industrial Project for Investment Promotion, Engineering Costs and Production Economics. October 1986, pp.173-184.

[8] Tabucanon, M.T. (1988), Multiple Criteria Decision Making in Industry, Elsevier.

[9] The 5th Five-Year Economic & Social Development Plan (1982-1986), Government of The Republic of Korea.

[10] The Statistics of Exports (1986), Ministry of Trade & Industry, Republic of Korea.

[11] The Statistics of Imports (1986), Ministry of Trade & Industry, Republic of Korea.

[12] The Handbook of Machinery Industry (1987), Korean Machinery Industry Promotion Association.

[13] The 6th Five-Year Economic & Social Development Plan (1987-1991), Government of The Republic of Korea.

[14] Yearbook of Energy Statistics (1986), Ministry of Energy and Resources, and Korea Energy Economics Institute.

[15] Yearbook on Korean Economy & Business (1986), Business Korea, Co., Ltd.

V·I·S·A - VIM for MCDA

Valerie Belton
University of Strathclyde
Department of Management Science
Glasgow

Stephen P Vickers
Heriot-Watt University
Department of Accountancy and Finance
Edinburgh

Abstract

The strengths of visual interactive modelling in the development of decision support systems have been widely acclaimed in recent years and this is now being recognised within the field of multiple criteria decision aiding. The experience of one of the authors suggests that the most substantial part of the effort in multiple criteria modelling needs to be devoted to sensitivity analysis of the parameters, particularly of the criteria weights. However, our experience with commercially available software is that little attention is paid to this aspect of the analysis beyond a consideration of the effect of varying each criterion weight independently. The task of more in depth sensitivity analysis is generally left to ad-hoc investigations at the discretion of the analyst and the decision maker(s). V·I·S·A is a software package which makes effective use of simple interactive graphics to enable the user (decision maker or analyst) to investigate multi-dimensional changes in the vector of criteria weights. This paper discusses visual interactive modelling in general terms and considers its relevance to multiple criteria decision aid. We describe V·I·S·A and its application to a small problem. The problem of assessing the effectiveness of the approach is considered and some preliminary proposals for research are put forward.

Introduction

Although visual interactive modelling (VIM) has been part of the toolbag of the OR analyst for almost a decade it has not until recently been embraced by the MCDA community, despite a strong emphasis within that community on interactive methods for decision aid. To some extent this is not surprising as MCDA, itself a relatively newly emerged field of study, has been principally concerned with technical developments and, one might easily conclude from a quick survey of the literature, increasing mathematical sophistication, rather than with applications and interface with related fields of study such as psychology, management information systems and decision support systems. It is not our intention to discredit such work that does exist, merely to express concern that it is not an attitude which pervades all work in MCDA. However, if MCDA is to avoid becoming a sterile collection of tools, effort must be made to tailor those tools in a way which makes them attractive to the practitioner, whether it be in the field of operational research, MIS, DSS, or any other, and ultimately to the decision maker. The widely acclaimed success of visual interactive modelling indicates an important opportunity and one that is now being seized by a number of people active in MCDA [See 4,6,14,15]

In this paper we begin by discussing briefly the concepts of visual interactive modelling in general terms and their relevance to MCDA. The following section reviews the use of visual and interactive information displays in MCDA to date. We then describe V·I·S·A, a program to assist in the evaluation of a simple multi-attribute value function incorporating visual interactive sensitivity analysis, illustrating its use in the assessment of student presentations. In the final section we consider the problem of evaluating the effectiveness of the visual interactive component as a means of improving decision making and of increasing user acceptance. Previous work of a similar type is discussed briefly and a suggestion for a framework for investigation of effectiveness is put forward for discussion.

Visual Interactive Modelling

Visual interactive modelling is described by Bell [1] as "...the process of building and using a visual interactive model to investigate issues important to decision makers. The VI model has three essential components; a mathematical model, a visual display of the status of the model, and interactions that permit the status of the model to be changed".

Two types of visual model may be identified, the representational graphic model and the iconic graphic model. The former makes use of standard graphical representations to present information derived from a mathematical model, for example, bar charts, graphs or pie charts. Bell [1] comments that from the point of view of the OR community representational graphic models are no more than implementation vehicles for OR models; he adds that often the underlying model is of little interest, as a consequence of which such visual models are more widely recognised as the domain of the DSS or MIS specialist. In an iconic graphic display the real system is depicted in symbolic form. The most common application area is visual interactive simulation (VIS) which makes extensive use of pictures; others are project management, using Gantt charts and networks, and decision analysis using decision trees.

The frequently cited advantages of using VIM are:

Improved communications between the decision maker and analyst
Fewer errors in the construction of models
A more user-friendly model
Increased likelihood of implementation
Increased acceptance of the model by the decision maker
Increased ease of validation
Models and solutions are easier to explain

The important disadvantages appear to relate to the time taken to develop the model and the difficulty of designing effective displays, ultimately reflected in the cost of the system.

The available software is generally accessible to an analyst for the construction of VI models, of either the representational or iconic graphic form, rather than directly to the decision maker for the analysis of a particular problem. Such software ranges from general purpose integrated packages such as Symphony and Framework, which are well suited to the integration of representational graphics alongside simple models and used in a diversity of applications areas, to the specialist tools for the development of VIS models, for example, See Why, Hocus, Witness and Genetik, used almost exclusively by OR practitioners.

VIM for MCDA?

What are the implications of the above for the MCDA community? Could VIM become a part of the toolbag of the MCDA analyst? Are the potential benefits likely to outweigh the potential problems?

Taking the definition of VIM given above it is easy to incorporate MCDA; there is an underlying mathematical model in all approaches, albeit of varying degrees of complexity; interactions that permit the status of the model to be changed are central to most of MCDA; and the status of the model is often amenable to visual display. Further to this it seems that MCDA, with its emphasis on the incorporation of values and judgement, is well described by Bell [1] in the following statement about the applicability of VIM: "VIM may be of value for problems that have a clear optimal solution but VIM appears to be most useful for semi-structured problems where the objective of the analyst is to incorporate the knowledge or experience of the decision maker in the problem solving loop". If one is seeking to increase user acceptance of MCDA then the advantages described above should surely be welcomed.

What are the costs? It is our belief that VI models for MCDA will, at least initially, take the form of a shell - the general model structure and VI interface are prespecified, thus software development time is zero. The analyst needs to be expert in MCDA only, not in computer programming, her task is to structure the particular problem, to elicit appropriate information and to enter this into the program. The tangible costs are thus those of acquiring, or developing, the appropriate software. We discuss possible intangible costs in a later section.

We should ask ourselves whether VIM for MCDA is simply an implementation vehicle or more than this. The former viewpoint coincides with what Bell [1] describes as the passive VIM school of thought. The other extreme, the active VIM school, believes that VIM can improve the performance of the problem solver. We feel, although as yet cannot offer substantive evidence, that the incorporation of VIM in MCDA approaches will enhance the decision maker's learning about the problem and as a consequence lead to "improved" decision making. This issue is also discussed in greater depth in a later section.

Use of Visual and Interactive Methods in MCDA To Date

Of course, the use of computer graphics is not new to MCDA. Similarly, the principle of interaction with the decision maker is well established. However, with the exception of the work by Korhonen and colleagues [12,13,14] the use of graphical presentations in MCDA has fallen short of VIM. Below we briefly review the nature of interaction and the use of graphics to date.

Interaction with the decision maker generally occurs in one of the following ways:

Interactive input of information.
The decision maker may make use of a visual representation to scale inputs or may be periodically provided with feedback about the consistency of stated preferences and allowed the opportunity to make changes (as, for example, in the analytic hierarchy process [20]).

Progressive articulation of preferences.
This is common in approaches which gradually explore the set of efficient solutions. For example, the decision maker may be asked to respond to questions designed to yield information about her marginal substitution rates between objectives.

Feedback
The decision maker may be presented with a value function derived from holistic preference judgements and allowed the opportunity to suggest improvements.

Sensitivity analysis.
Once the results of an analysis are available the decision maker or analyst can select parameters over which to perform sensitivity analyses.

Standard graphical displays, occasionally allowing simple interaction, have been used in the following ways:

Input of information.
Bar charts are used widely to represent the relative performance of alternatives with respect to a criterion or the relative importance

of criteria. The decision maker can control the length of the bars.

Value paths.

Many authors have used line graphs, sometimes referred to as value paths, to illustrate the performance of a small set of alternatives. For example, see Schilling and Cohon [21] and Steuer [22]. This simple technique is incorporated in several computer packages implementing multi-attribute value functions. It is a useful visual tool for identifying dominating and dominated or near dominated alternatives.

Value functions.

Line graphs are used to represent value functions as a collection of univariate functions. For example, such an approach is implemented in Prefcalc (Jacquet Lagreze) and in POLICY, a program developed to implement the ideas of Social Judgement Theory [9].

Hierarchical structure

In methods incorporating a hierarchical value function the hierarchy of criteria is displayed as a tree on the screen. For example, see Hiview (Phillips et al) and Expert Choice (the computer program for the implementation of the analytic hierarchy process - Saaty and Forman).

Sensitivity analysis.

Line graphs are used to display one dimensional sensitivity analyses in the multi-attribute value function approach, for example Hiview and V·I·S·A (Belton and Vickers), and in Expert Choice.

Maps.

Using the statistical technique of multi-dimensional scaling multivariate data may be reduced to a two dimensional plot derived from information about indifferences (Rivett, [17]) or preferences (G Roy, [18]). Proximity in the map should imply similarity of preference.

Outranking graphs.

One of the outputs of outranking approaches such as Electre (B Roy, [18]) and Promethee (Brans and Mareschal, [6]) is the outranking graph. A visual display is the most effective way of presenting this, however, implementation in a computer system is not a simple matter.

As indicated earlier the only truly visual interactive method to be reported in the MCDA literature to date is that developed by Korhonen [13,14]. This is an implementation of multi-objective linear programming with an attractive dynamic graphic interface allowing the user to explore the set of efficient solutions and to obtain simultaneous feedback about the changes in the values of all objectives.

V·I·S·A - Visual Interactive Sensitivity Analysis for MCDA.

In this section we describe the use of visual interactive sensitivity analysis in a multiple criteria model for choice between discrete alternatives.

The Motivation to Develop V·I·S·A

The motivation for this work came from research into a means of extending sensitivity analysis within the simple multi-attribute value function model beyond the one dimensional analyses of the type described above [3]. Extensive experience in the use of simple multiple criteria models to advise decision makers [2] indicated a need for more sophisticated sensitivity analyses and an effective way of presenting the information gathered from such analyses to the decision maker. Even in problems involving only a few criteria, the amount of information yielded by a complete multi-dimensional sensitivity analysis can be overwhelming and the analyst is still faced with the problem of extracting what is useful to the decision maker. With the existing software it was not possible to do this analytically and to do so in an ad-hoc manner was a time consuming process which could lead to loss of interest by the decision maker and conceivably to a feeling that the analysis was no longer under her control. The method to be presented addresses a number of these problems by making use of interactive graphics.

The Model

The underlying model is a simple multi-attribute value function of the kind described in Belton [2]. It can be used for the evaluation of a number of well defined alternatives according to a hierarchy of criteria defined by the decision maker. For a hierarchy with a single level of criteria the model is as follows:

$$V_i = \Sigma w_j x_{ij}$$

Where V_i = the overall evaluation of alternative i

w_j = the weight assigned to criterion j

x_{ij} = the evaluation of alternative i according to criterion j

For hierarchies with more than one level of criteria the model extends naturally. A family of criteria is defined as the set of criteria having the same parent criterion. Each criterion is assigned a weight relative to other members of its family. The weight assigned to a criterion which has descendants relative to its own family members represents the aggregate weight of all descendants of that criterion. Alternatives are evaluated with respect to all end criteria, i.e. criteria having no descendants, and these scores are aggregated from the bottom up.

It is the responsibility of the analyst working with the decision maker to assess whether or not this is an appropriate model for the problem, to structure the problem and to elicit initial values for the x_{ij} and the w_j. The present version of the program offers no assistance in this respect. Future developments include interactive input routines incorporating graphical procedures and elicitation of values by pairwise comparisons. Input is via a menu system which allows addition of new criteria or alternatives, changes to current values or deletion of parts of the model with relative ease.

The system incorporates some of the static graphics described above, namely value paths and one dimensional sensitivity analyses. Examples of these are given for the application discussed below. The innovative part of the system is the interactive sensitivity analysis. This is very simple in concept. It allows the user to investigate interactively the effect on overall scores of any changes she may wish to make to the allocation of criteria weights within a particular family. A more detailed explanation is given in the context of the example in the following section.

Application of V·I·S·A

The system is very new and thus it is too early to assess the impact of the VI component. In the following section we discuss the question of effectiveness. It has been demonstrated to a wide range of potential users, both analysts and practising managers. The initial reaction is one of considerable interest and it is already in use in ICI for monitoring projects. In addition it has proved to be an effective and innovative way of teaching multiple criteria modelling to undergraduate students. Implemented on a DEC VAX computer at the University of Kent at Canterbury it has been used by a group of 60 students for assessment work. Almost all students made considerable use of the interactive analysis, some indiscriminately but a number in a very thoughtful manner. We hope that this is the sort of experiential learning that will prove useful to those who go on to pursue careers in management science and will have more impact on those following careers in general management than standard textbook learning.

The program will be illustrated by an application with which one of the authors was involved, in which presentations made by groups of students were evaluated. The presentations were of papers on a topic relating to combinatorial optimisation which have been published in recent years in OR journals or books. For the students this was the basis on which their performance for the course was to be assessed. In addition to the assessment by three examiners the students were required to assess each other using the same set of criteria. The hierarchy of criteria, formulated by the examiners and discussed with the students before acceptance, is presented in figure A.

Each participant was required to score each of the presentations on a 0 (Bad) to 10 (Excellent) scale for each of the bottom level criteria. The analysis presented for illustration is that of one of the examiners. Figure B displays the value path presentation of four of the presentations (there were seven actual presentations but only a selection are shown in order to maintain the clarity of the picture in black and white). It can be seen that there is no presentation which dominates all the others. Presentation 4 does well on all criteria and dominates Presentation 3; Presentations 1 and 2 have differing strengths and weaknesses.

Figure A **Hierarchy of Criteria for Evaluation of Presentations**

- Quality of presentation
  - Organisation
  - Timing
  - Sharing of work
  - Content of presentation
    - Is the message well conveyed?
    - Is the interest of the audience maintained?
    - Is it easy to understand?
  - Nature of presentation
    - Use of visual aids
    - Is the speed ok?
    - Is it clearly audible?
    - Is the attitude of the speaker good?
  - Handling of questions

Figure B **Value Path Graph - Top Level Criteria**

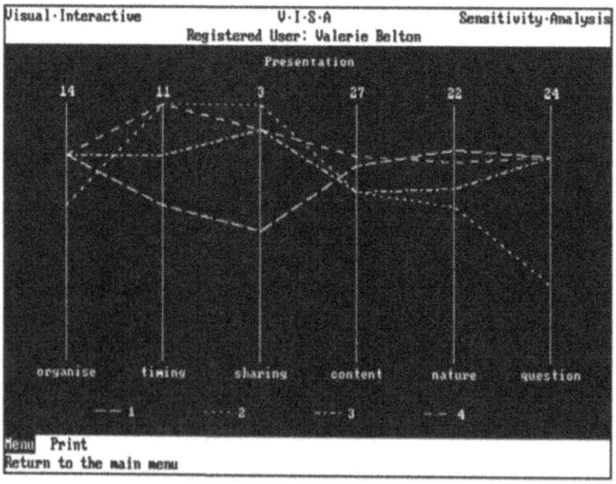

The overall scores for the four presentations are 76, 60, 74 and 82 respectively. A one dimensional sensitivity analysis of the weight assigned to "timing", illustrated in figure C, shows that the 4th presentation remains preferred irrespective of the weight assigned to this criterion, although there is some switching of preferences between the other alternatives.

Figure C Sensitivity Analysis of the Weight Assigned to Timing

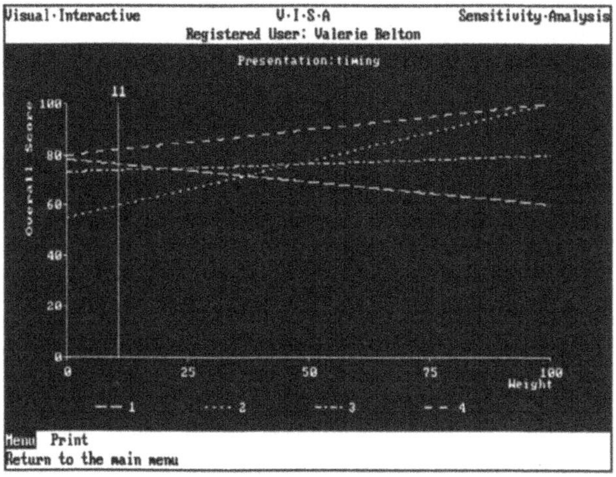

The visual interactive display is as shown in figure D. The top bar chart represents the overall scores of the alternatives being evaluated, given the current status of the model. The bottom bar chart represents the distribution of criteria weights within the selected family. By using the cursor control keys the user can select a particular criterion and change the weight allocated to it. All other weights change to preserve the normalisation, their relative values being retained. The effect of the change on the overall scores is seen immediately as the height of the bars changes. Figure E illustrates the effect of changing the emphasis on criteria weights to place greater value on how well the presentation is organised rather than the content. It can be seen that the fourth presentation remains preferred but the order of preference for the other three has changed. It is interesting to note that the dominated alternative, presentation 3 (the domination is not affected by changing criteria weights at this level), is now preferred to presentations 1 and 2, which are not dominated by presentation 4. The user can display the current values of the criteria weights and overall scores on the graph, obtain a printout of the current status of the model, opt to retain the current set of values or keep the initial set of values.

Figure D **Visual Interactive Sensitivity Analysis - 1**

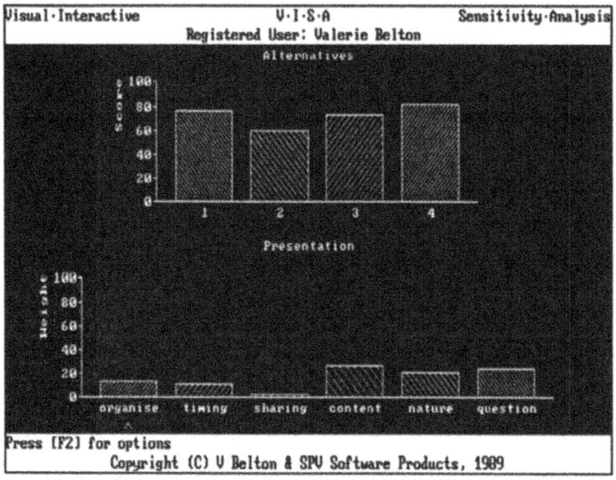

Figure E **Visual Interactive Sensitivity Analysis - 2**

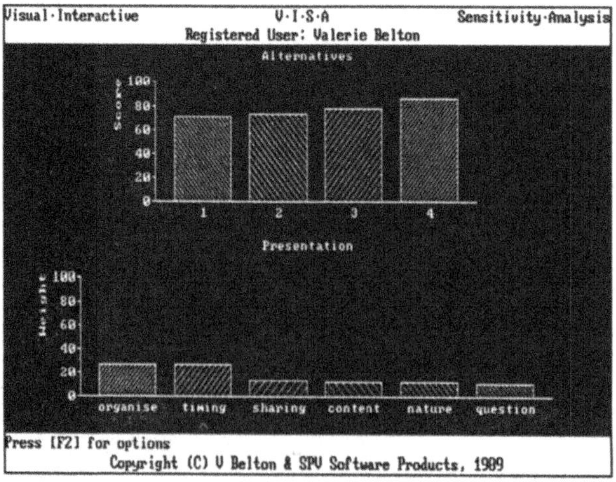

Is VIM an Effective Tool for MCDA?

How do we assess the effectiveness of such a tool? We must begin by defining effectiveness. The ultimate aim must be to improve decision making. Thus we can ask the question "Does V·I·S·A improve decision making?". But compared with what? Do we compare it to a multiple criteria model without visual interactive sensitivity analysis, but with static graphics? Do we compare it to a model without graphics? Or do we compare it to decisions taken without the use of a multiple criteria model?

All these questions are of interest, but to begin with we wish to address the first question, taking it as a matter of faith that the use of a model improves decision making, and feeling that the second comparison compounds the effect of using graphics at all and using interactive graphics. Thus the problem is one of evaluating the effectiveness of an additional facility which enhances the power of existing facilities. However, we see the additional facility as more than an alternative form of representational graphics for the presentation of information, but a move from passive to active VIM.

An academic assessment of effectiveness might stop there. However, we wish to consider effectiveness in a broader sense. To be truly effective in improving decision making the tool has to be used. In the responses to Bell's survey on the use of VIM [1] there was much enthusiasm about the success of VIM in facilitating communication with the decision maker and related benefits. If MCDA is to become an accepted tool of practising management scientists the same criteria for success must be applied. So we should also ask if the inclusion of a VI component in a MCDA model helps in moving towards success in this way.

These are clearly two very separate objectives although they may be achieved by the same vehicle. We do not believe that they are fundamentally conflicting, although we are concerned that a user can be seduced by effective graphics into accepting the results of a poor analysis or an ill-founded approach. However, subject to a belief in the underlying method, the increased likelihood of implementation through the use of the VI interface is not at the expense of reduced quality decisions and vice versa. We are fortunate in MCDA not to have to concern ourselves with some of the issues facing VI simulation in this respect. We do not have to worry that a visually attractive presentation can tempt the user into accepting a decision lacking in statistical validity. We can even brush aside concerns about sub-optimality; in a multiple criteria context the "optimal" choice is defined by the decision maker's preferences and thus is the decision maker's preferred choice. However, this does mean that we have no external criteria against which to validate a decision or assess its quality.

Thus, if conflict between the two objectives is not a matter for concern, we can concentrate on how to assess the extent to which they are achieved. A practitioner's viewpoint may be that "the proof of the pudding is in the eating", but we feel that one of the roles that academics can, and should, adopt in relation to innovation in the field of managerial decision aiding is analogous to that of an official taster. This can be achieved by a combination of experimental work and empirical observation. Below we discuss some of the problems of conducting useful experimental work in this area and propose a means of experimentation in the hope that this will stimulate discussion of the problem.

Towards the Design of an Experimental Programme

One can hypothesise an experimental situation in which decision makers use different tools to aid decision making in a similar, or possibly the same, situation giving rise to comparative measures of effectiveness - there is a host of literature on work of this kind in the field of MIS [5,7], and MCDA [10].

The task of finding a problem of sufficient reality and decision makers who will relate to it and are willing to participate is one which faces any work of this kind. The problem of determining measures of effectiveness is particularly knotty in this instance. As we commented above there are no external criteria against which we can assess the "quality" of a decision and hence we must use criteria which relate to the process of decision making. We agree with French [8] that a principal aim of a multiple criteria decision aid should be to facilitate the decision maker's learning about and understanding of the problem. We are still faced with the task of assessing learning and in particular the extent to which VI analysis enhances learning. This is a problem which is fundamental to the whole of management science, but particularly difficult in the case of MCDA and other softer approaches to OR where the system being modelled is largely one of values and beliefs rather than a physical reality. We believe that the lack of discussion of the issue in the literature is testimony to its difficulty rather than its perceived irrelevance.

The literature cited above on comparative studies in the field of MIS concentrates on assessing the effectiveness of alternative methods of display of information, for example, tabular or graphical, colour or monochrome, in situations where some external measure of performance is available, such as performance in monetary terms. Some recent work in the area of visual interactive simulation has looked at the quality of decisions made using the VI model in situations where an "optimal" solution is known from queuing theory [15]. The results are not encouraging for developers of VIS models! The work referred to in MCDA has followed a number of lines of investigation; comparison of the results of different MCDA models; comparison of the results of models with intuition, expert performance, or performance of decision makers who have been taught to behave in a particular way; and comparison of the results of interactive models with a pre-specified utility function. This work offers a number of criteria for judging the validity of MCDA models, but no suggestions for assessing the "value" of a model.

Work carried out at the Decision Analysis Unit, London School of Economics, on analysing decision conferences [23] and the handling of decision problems with differing degrees of structure [11] is more relevant to the type of investigation we would like to perform. This work has developed indices to capture the work a decision maker does in handling a problem in terms of structuring and analysis.

Having debated these issues at length we are of the opinion that the only way to proceed is through action research. In order to assess how decision makers use the tool we must observe them working on live problems. There are of course a number of different situations in which the approach might be used: an individual working with the assistance of a skilled analyst; an individual working alone; a group facilitated by a skilled analyst. The nature and scope of learning in the group process is of course more extensive, introducing the opportunity for interactive learning about the perspectives and priorities of others. The presence of an analyst introduces many further possibilities for bias; a skilled analyst should ensure more effective use of the tool, but will almost certainly exercise some control. Personalities and interpersonal interactions become significant in the presence of an analyst and in a group process.

We suggest that it is necessary to develop a means of logging a consultation and a means of using the log in a debriefing session. In the case of a single decision maker, with or without analyst support, the log would provide information on how much time was spent on different aspects of analysis and how the user moved between alternative displays of information and other analysis. The log of a group process would necessarily be more complex. The aim of the debriefing session would be to reconstruct the decision maker's thought processes, to identify points at which questions arose and how she sought to answer them and to identify points at which significant insights arose. The hypothesis to be examined is that the VI feature stimulates more questions to be asked although the answer may not necessarily come from the VI analysis. We are interested to seek your views on this as a means of experimentation.

Summary

In this paper we have looked at work on visual interactive modelling and the implications for multiple criteria decision aiding. The success of VIM in increasing user acceptance and understanding of the underlying model suggests that it could be effectively incorporated in MCDA. We look briefly at the use of graphics in MCDA to date and developments into use of VIM, going on to describe in more detail the use of V·I·S·A, a program for the development of a multi-attribute value function incorporating visual interactive sensitivity analysis. In conclusion we discuss the issues in assessing the effectiveness of such a tool in improving decision making, suggesting that in order to do so we must develop a means of measuring learning through careful monitoring of the use of the tool on live decisions followed by debriefing of the decision maker. The problem faced here is one which is central to MCDA and many other areas of soft OR and we hope to open up a debate on the need for work of this kind and on the means of conducting it.

Acknowledgement
We would like to acknowledge the support of Professor Jakob Krarup and DIKU, University of Copenhagen. Much of the work for this paper was carried out during the time the first author spent there as a guest professor.

References

[1] PC Bell, *Visual interactive modelling in 1986*, in V Belton and RM O'Keefe, *Recent developments in operational research*, Operational Research Society, 1986

[2] V Belton, *The use of a simple multi-criteria model to assist in selection from a shortlist*, Journal of the Operational Research Society, 1985

[3] V Belton, *A comparative study of methods for multiple criteria decision aid*, PhD Thesis, Cambridge, 1986

[4] V Belton and SP Vickers, *Use of a simple multi-attribute value function incorporating Visual Interactive Sensitivity Analysis for multiple criteria decision aid*, paper presented at the Third International Summer School on MCDA, 1988.

[5] I Benbasat and AS Dexter, *An experimental evaluation of graphical and color enhanced presentation*, Management Science, 1985

[6] JP Brans and B Mareschal, *Bankadvisor, a Promethee method for industrial evaluation*, paper presented at the Mini Euro Conference on VIM, 1988

[7] G DeSanctis, *Computer graphics as decision aids: directions for research*, Decision Sciences, 1984

[8] S French, *Decision Theory*, Wiley, 1986

[9] KR Hammond, J Rohrbaugh, J Mumpower and LA Adelman, *Social Judgement Theory: applications in policy formulation*, in MF Kaplan and S Schwartz, *Human judgement and decision processes in applied settings*, Academic Press, 1977

[10] BF Hobbs, *Experiments in MCDM and what we can learn from them*, in YY Haimes and V Chankong, *Decision making with multiple objectives*, Springer Verlag, 1985

[11] P Humphreys, AI Oldfield and J Allan, *Intuitive handling of decision problems: a five level empirical analysis*, Decision Analysis Unit, London School of Economics, 1987

[12] PJ Korhonen, *Computer graphics for MCDM*, in YY Haimes and V Chankong, *Decision making with multiple objectives*, Springer Verlag, 1985

[13] PJ Korhonen and J Laasko, *A visual interactive method for solving the multiple criteria problem*, European Journal of Operational Research, 1986

[14] PJ Korhonen and J Wallenius, *VIG - a visual and dynamic decision support system for multiple objective linear programming*, paper presented at the Third International Summer School on MCDA, 1988.

[15] RM O'Keefe and I Pitt, *Cognitive style and the use of visual interactive models*, paper presented at the Mini Euro Conference on VIM, 1988

[16] H Pastijn, *An interactive ordinal multiple criteria decision making tool*, paper presented at the Mini Euro Conference on VIM, 1988

[17] BHP Rivett, *Multidimensional scaling for multiobjective policies*, Omega, 1977

[18] B Roy, *The Electre methods*, paper presented at the Third International Summer School on MCDA, 1988.

[19] G Roy, *An application of multi-dimensional scaling to policy selection*, University of Western Australia, School of Architecture, Paper 6009

[20] TL Saaty, *The analytic hierarchy process*, McGraw Hill, 1980

[21] D Schilling and RC Cohen, *An approach to the display and analysis of multiobjective problems*, Socio Economic Planning Sciences, 1983

[22] RE Steuer, *Multiple criteria optimisation: theory, computation and application*, Wiley, 1986

[23] S Wooler, *Analysing decision conferences*, Decision Analysis Unit, London School of Economics, 1987

A HYBRID MODEL OF STRATEGIC PLANNING

Peter Naude
Graduate School of Business
University of Cape Town
Cape Town 7700
South Africa

Edwin Scholtz
Byrne Flemming Consultants
P O Box 783402
Sandton 2146
South Africa

INTRODUCTION

This case study shows the rationale behind and the benefits of using a hybrid model of strategy formulation, based on both the GE 9-box approach (see, for example, Hofer and Schendel, 1978) and the Analytic Hierarchy Process (Saaty, 1980). It shows how the two approaches can be combined into a useful strategic planning model.

The research was undertaken in 1987 by the Strategy Consulting Division of P-E Corporate Services, a management consulting firm in South Africa, for the Research and Development arm of a large local firm. The firm's R&D team had traditionally been funded by the firm, but the top management had recently decided that it should become a self supporting and self managed unit. The problem presented to us was to develop a workable strategic planning model, and to take the R&D team through the planning exercise.

BACKGROUND

Strategy involves fit: the fitting of an organisation's internal capabilities to the external environment in which it operates. This internal-external paradigm is the basis for many strategic planning models, with the underlying assumption being that the internal factors are controllable, and the external ones uncontrollable. By analysing a firm and its environment in terms of these internal-external or controllable- uncontrollable factors, an optimal fit between the two can be developed to guide management in its task of allocating scarce resources.

The simplest of these internal-external models is the traditional SWOT analysis that divides the internal forces into strengths and weaknesses, and the external environment into opportunities and threats. The model's advantage of operational simplicity is also its disadvantage, in that a 2x2 matrix often does not capture the complex reality facing managers.

A second model using the same format is the familiar 'farmyard' model of the Boston Consulting Group, where each product or SBU is analysed according to an external factor (market growth) and an internal factor (relative market share) (see, for example, Day 1977). Based on the original research done on Experience Curves and the PIMS database, this model has the same advantage of simplicity, and it is relatively easy to determine the logical course of action to follow depending on the relative position within the matrix. The major disadvantages of this model are those of scale and scope. A 2x2 matrix is still too restrictive, and decisions are based on only two criteria rather than on multiple criteria which reflect reality more accurately.

A more comprehensive and useful model, the GE 9-box, is illustrated in Figure 1 below. It is still based on the same two dimensional internal (business strengths) and external (market attractiveness) typology, but does offer significant advantages. It relaxes the 2x2 matrix into a 3x3 one, thereby permitting intermediate values to be separately analysed, and treats both axes as composite measures, permitting multiple criteria to be used. For details on the mathematical procedure used to map the different products or SBUs, see Smith (1985).

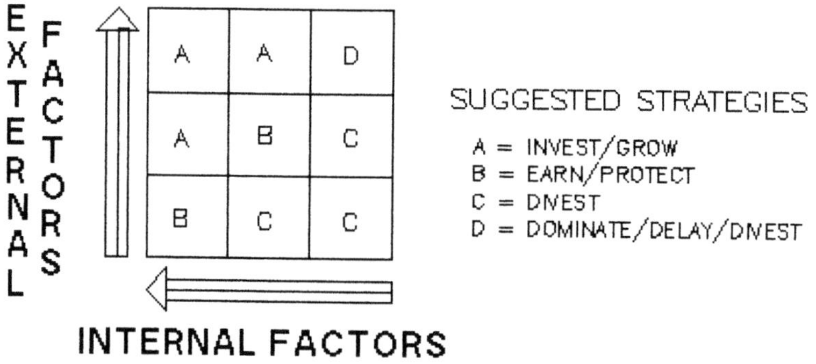

Figure 1

The G-E 9 Box

Again, this model has various shortcomings. First, the model does not incorporate time. There is thus no way to plan for changes in strategy as the stages in a product's life cycle varies. Second, if too many factors are incorporated in either axis the maximum relative contribution falls. For example, if four factors make up Industry Attractiveness, each could contribute up to 25%, whereas if ten were included the figure falls to 10%. Therefor, the more the model tries to capture reality by including more factors, the less sensitive it becomes.

THE APPROACH UTILISED

In order to try and overcome the concerns expressed above, we developed a hybrid model that utilised the basic GE 9-box model, but strengthened it by incorporating the AHP approach.

Our first step with the management of the R&D team was to get them to determine a set of four planning dates that were relevant to their product ranges and environment. The dates that were set were 1988 (i.e. the "here and now"), 1990, 1993, and 1998. A scenario was then developed that covered this time span (This was done using the familiar PEST - Political, Economic, Social, and Technological model). The objective of this was to make the team aware of the changing environment in which they operated, and the impact that this would have on the demand for their type of products.

The second stage was to get the team to construct a hierarchical tree to be used in the subsequent strategy development. The first level of the tree was to have only two nodes: internal (i.e. business strengths) and external (i.e. industry attractiveness) factors. The second (and subsequent, if desired) levels then had to identify those factors that determined the level of business strength or industry attractiveness.

Figure 2 indicates the structure of this tree. For both internal and external factors it so happened that five factors were chosen. Each of these ten factors was in turn made up of various sub-factors. These lower factors accounted for such a small percentage of overall importance that they were either combined into the definition of the host factor or excluded. The only exception to this was the division of external strategic considerations into software based technology and hardware based technology elements, each of which was sufficiently important to be retained. A brief definition of each of the lower level elements is given in the Appendix.

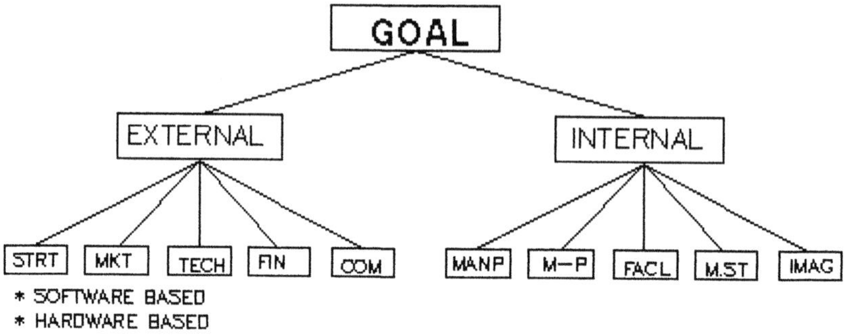

<u>Figure 2</u>

The Hierarchical Tree

These various factors were then rated using the pairwise methodology of the AHP as to their relative importance to the unit in the "here and now" time period. After this was completed, it was redone for the remaining three time periods, based on the assumptions that underpinned the scenarios. This was one of the most interesting exercises undertaken, since it forced the new management team to think through a range of factors in a structured way, focusing on how these compared in importance, how this relationship changed over time, and hence what the correct strategic response of the unit should be.

The results of this step appear in Figures 3a and 3b. What was immediately apparent was the overriding importance to the unit of strategic developments in the market place that were beyond their control. This importance itself was predicted to change quite radically, and their marketing effort had to place increasing emphasis on the software based technology applications. The second major determinant of success was seen as being the market place: the size, growth, and profitability of various product/markets. Again, this was an illuminating finding, since the unit in its current environment was allocated research by a coordinating head-office committee: it would now have to develop its own ability to assess external marketing opportunities, a completely new task to the traditional R&D team.

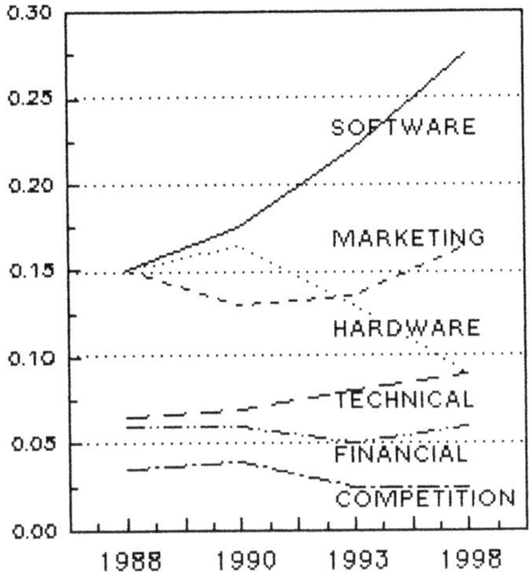

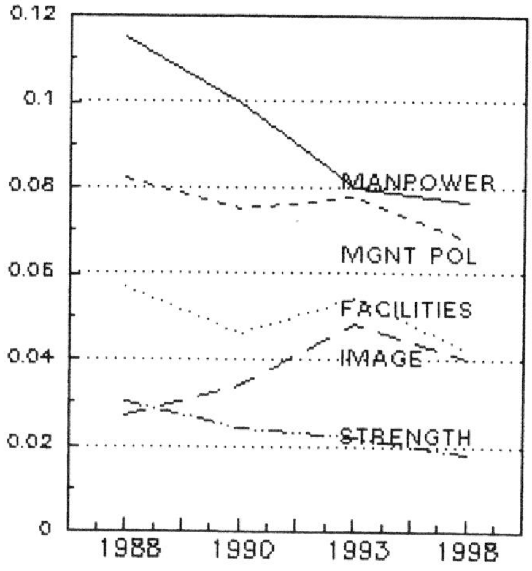

Figure 3a — External Factors Over Time

Figure 3b — Internal Factors Over Time

The next step was to take each broad product offering of the organisation and to rank it on a 5-point scale (5 = well above average, 1 = well below average) according to how well, comparatively, it featured on the eleven internal/external factors. This was done for each of the eight product groups, and repeated for each of the four time periods. The eight product groups were:

A: Intelligent Instrumentation
B: Communication Exchanges
C: Robotics
D: Ceramic Machining Components
E: Laser Cutting
F: Distributed Control
G: Network Management
H: Non-contact Measurement

By now taking an imaginary eleventh product as one that scored well above average on every factor, it was possible to calculate a datum point to compare all other points to. Figure 4 indicates the results. The advantages of this over the traditional GE 9-box are that it takes into consideration not only the relative positions between products, but also their variability over time. For example, non-contact measurement was seen as being past maturity and in the decline stage of its life cycle by 1998. A product such as ceramic machining components, however, was very much a 'dog' at present, but was worth continued investment, given its perceived growth.

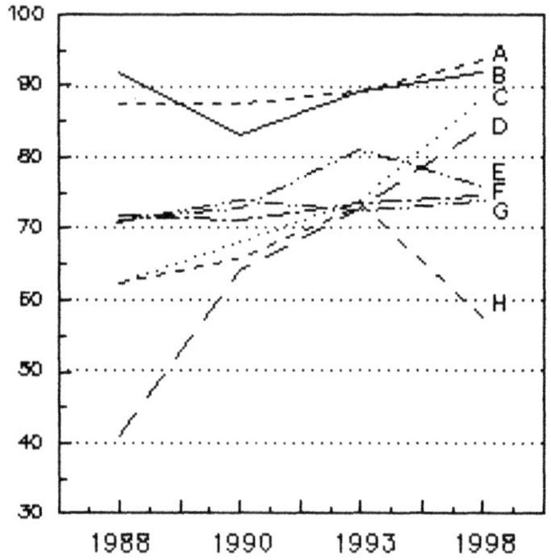

Figure 4

Results at the Product Level

The data in Figure 4 were calculated from the scores of the various products on both internal and external factors. If these are now kept separate, it is easy to reproduce the data in an original GE 9-box format, as shown in Figure 5. This is done by again taking the fictitious product's score as being 100% on each axis, and scaling the other products accordingly. We now see that products such as Intelligent Instrumentation and Communication Exchanges that scored so high in Figure 4, are still the best rated products. The spread of products across the portfolio is interesting in itself: whereas there is quite a spread on the external axis, the spread is not so great on the internal one. Apart from Communication Exchanges, no product rates particularly highly on the internal axis. This lack of perceived success on internal (i.e. controllable) factors is not atypical for an organisation that to date has never had to be market oriented, such as a R&D unit.

In addition, the element of time can now be incorporated onto the traditional GE 9-box diagramme. By doing similar calculations for each time period, the likely movement of products can be plotted. Thus we see Non-contact Measurement declining to the divest area of the chart, and Ceramic Machining Components becoming increasingly attractive.

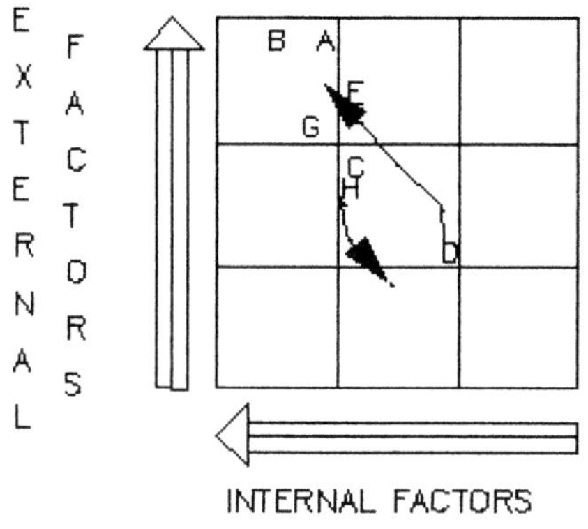

Figure 5

Results in G-E 9 Box Format

CONCLUSION

We believe that this methodology has particular advantages over the other methods reviewed here. These are:

* By drawing up a tree of internal/external factors, management is forced to structure their thinking about the company/environment fit.

* By analysing this tree via the AHP methodology the relative importance of the various elements is indicated. Those contributing little to the total are easily combined or excluded. The lack of sensitivity that may occur using the GE 9-box does not arise.

* By repeating the analysis for selected time periods, far more meaningful conclusions can be drawn. It has the advantage both of highlighting how internal\external factors vary in importance (and hence how management must plan to change their responses over time), and of how various products are likely to perform over time (and hence what portfolio should be held at any stage).

APPENDIX

DEFINITION OF THE LOWER LEVEL ELEMENTS

EXTERNAL FACTORS

Strategic:	Those factors dictating industry attractiveness that are outside the firm's control. This in turn was subdivided into hardware based technology (primarily mining) and software based engineering (primary electronics).
Market:	The size, growth, and profitability of the various market segments.
Technology:	The type and nature of the technology, its complexity, appropriateness, and acceptability.
Financial:	The financial viability of operating in that market, typified by entry and exit barriers.
Competition:	This examined particular types of competition, such as educational institutions, research institutions, and other private research organisations.

INTERNAL FACTORS

Manpower:	The availability, quantity, and quality thereof.
Management Policy:	Concerning especially the ability in project, financial, manpower, and marketing management
Facilities:	The technologies, R&D infrastructure, information, and equipment available.
Marketing Strength:	The operational efficiency of being able to exploit marketing opportunities.
Image:	Factors such as the level of credibility and speed of response of the organisation.

REFERENCES

Day, G.S. 1977
Diagnosing the Product Portfolio
Journal of Marketing, Vol 41, pp 29 - 38

Hofer, C.W., and Schendel, D. 1978
Strategy Formulation: Analytical Concepts
West Publishing Company

Saaty, T.L. 1980
The Analytic Hierarchy Process
McGraw-Hill

Smith, J.G. 1985
Business Strategy
Basil Blackwell

ON HEALTH CARE DISTRICTS PLANNING BY MULTIOBJECTIVE ANALYSIS WITH THE MPSX/370 PACKAGE

Włodzimierz Ogryczak[1] and Jacek Malczewski[2]

[1]Institute of Informatics
Warsaw University
Warsaw, Poland

[2]Institute of Geography and Spatial Organization
Polish Academy of Sciences
Warsaw, Poland

Abstract

In the paper we formulate a mathematical model and perform a multiobjective analysis of the real-life problem of health service areas reorganization.The model is formalized in terms of linear programming with three objective functions. The values of these functions vary significantly depending on assumed hierarchy of the objectives. Nevertheless, the multiobjective analysis allows us to find an efficient solution and this yields a significant improvement in the health care system performance. The multiobjective analysis is performed with the MPSX/370 package which can be regarded as a standard commercial tool for solving linear programming problems.

1. Introduction

A system of health care comprises of two elements: facilities offering services (supply) and a population using the health care facilities (demand). Relations between these elements are defined by the system organization. Depending on the organization, decisions where to place a treatment can be taken by patients themselves or they are prescribed by the health care authorities. In Poland the second type of organization is in use, i.e., the scope of functioning of various facilities is limited to districts assigned to them due to some administrative decisions. So, functionality of the system depends primarily on "correctness" of the decisions taken concerning delimitation of the districts. Moreover, a centralized administration system lacks mechanisms which would allow to correct dynamically the district pattern according to changes in the population. In effect, district patterns are frequently characterized by a low degree of functionality. Therefore, a selection of an optimal pattern of districts is a crucial problem in the sphere of health service organization. To state a proper objective function for such an optimization is hardly possible. For this reason, while solving the districting problem one has to use tools of multiobjective mathematical programming, which allows for evaluating functionality of the planned organization from many points of view. Such an approach is presented in this paper.

We formulate a mathematical model and perform a multiobjective analysis to delimit boundaries of districts serviced by general and dental clinics in the Wola quarter of Warsaw. The model is formalized in terms of linear programming with three objective functions. The multiobjective analysis is performed with the MPSX/370 package which can be regarded as a standard commercial tool for solving linear programming problems.

2. Statement of the model

Health care in Warsaw-Wola is governed by the two Health Service Complexes: Warsaw-Wola East and Warsaw-Wola West, denoted further by WWE and WWW, respectively. Within WWE there are six facilities offering both general and dental services. In the area of WWW there are four general and dental clinics. One should note that while the general and dental clinics share the same buildings, they differ in their service potentials.

The service potentials of clinics was determined in terms of the number of patients that can be served by a given clinic (see [6,7]). Further we will denote service potential of the j-th clinic for general or dental treatment by R^g or R^d, respectively.

The demand for services in health care was identified with the number of people above the age of 15, since the clinics under consideration offer services only to adults. In order to identify a spatial pattern of demand for health care, the area under study was partitioned into spatial units. Taking into account distribution of the population in Warsaw-Wola, a square sized 0.25×0.25 km was selected as the spatial unit. Further, the number of people above the age 15 living in the i-th spatial unit was identified as the number of patients P_i. There were 86 such units in WWE and 133 in WWW.

Partitioning of the area allowed us to discern quite precisely demands for health care services as well as to measure distances between several spatial units and the health care facilities. Taking into account the urban morphology and the transportation network, it was accepted that the city-block metric was the best approximation to real distances. So, the distance between the i-th spatial unit and the j-facility was defined by the formula

$$d_{ij} = |x_i - x_j| + |y_i - y_j|,$$

where (x_i, y_i) and (x_j, y_j) denote the co-ordinates of the centre of the

i-th spatial unit and the co-ordinates of the j-th facility, respectively. All the distances were calculated in kilometres. Certain connections between the spatial units and the facilities were eliminated as unacceptable. For instance, if reaching the given facility required use of at least two bus or tram lines.

Partitioning of the area into the spatial units transforms the districting problem into a generalized assignment problem (GAP) (see [9]). Each unit (identified with its centre) has to be assigned to a certain facility, with respect for service potentials of several facilities. While introducing decision variables

$$u_{ij} = \begin{cases} 1 & \text{if } i\text{-th unit is assigned to } j\text{-th facility} \\ 0 & \text{otherwise} \end{cases}$$

one can formulate constraints of the model as follows:

$$\sum_{j=1}^{n} u_{ij} = 1 \qquad \text{for } i=1,2,\ldots,m \qquad (1)$$

$$\sum_{i=1}^{m} P_i u_{ij} \le R_j \qquad \text{for } j=1,2,\ldots,n \qquad (2)$$

$$u_{ij} = 0 \text{ or } 1 \qquad (3)$$

where:

m denotes the number of spatial units,

n denotes the number of facilities,

R_j represents R_j^g or R_j^d depending on type of the clinic under consideration.

Connectivity of the districts and the other expected features of the district pattern have to be guaranteed by a proper definition of objective functions.

Due to the requirement (3), GAP is a problem of integer programming and needs rather complex methods for solving. However, significance of the requirement (3) in the model raises some doubts. This requirement means that no spatial unit can be split up between two or more facilities and it forces the district pattern to be fully consisted with rather artificial structure of square spatial units. On the other hand, it is known from the theory of linear programming, that solving a continuous model (when the condition (3) is neglected) by the simplex method leads to splitting up only a very small number of units (at most n-1). Furthermore, linear programming has developed efficient techniques for multiobjective analysis of large models. (see e.g. [2]). Therefore, we decided to neglect the requirement (3) from our model.

Due to omission of the integrality requirement, the model can be now expressed in the form of a typical transportation problem:

$$\sum_{j=1}^{n} t^{g}_{ij} = P_{i} \qquad \text{for } i=1,2,\ldots,m \qquad (4)$$

$$\sum_{i=1}^{m} t^{g}_{ij} \le R^{g}_{j} \qquad \text{for } j=1,2,\ldots,n \qquad (5)$$

$$t^{g}_{ij} \ge 0 \qquad \text{for } i=1,2,\ldots,m;\ j=1,2,\ldots,n \qquad (6)$$

and similarly

$$\sum_{j=1}^{n} t^{d}_{ij} = P_{i} \qquad \text{for } i=1,2,\ldots,m \qquad (7)$$

$$\sum_{i=1}^{m} t^{d}_{ij} \le R^{d}_{j} \qquad \text{for } j=1,2,\ldots,n \qquad (8)$$

$$t^{d}_{ij} \ge 0 \qquad \text{for } i=1,2,\ldots,m;\ j=1,2,\ldots,n \qquad (9)$$

where variables t^{g}_{ij} and t^{d}_{ij} represent the number of people from the i-th unit allocated to the j-th general clinic and the j-th dental clinic, respectively. Variables corresponding to the unacceptable allocations are omitted in the model (i.e., forced to zero).

In consultation with the decision makers, minimization of the average distance covered by the patients was selected as the main criterion which measures spatial efficiency of the district pattern. This criterion is represented by minimization of the following function:

$$\frac{1}{2P} \sum_{i=1}^{m} \sum_{j=1}^{n} d_{ij}(t^{g}_{ij}+t^{d}_{ij})$$

where P denotes the global number of patients.

As one can easily notice, minimization of the average distance leads to partitioning of the problem into two independent transportation problems related to general and dental clinics, respectively. Therefore, consideration of the average distance as only one criterion would lead to completely disregarding of the dependence between regions for various types of clinics. Moreover, solutions of the corresponding transportation problems could generate unconnected districts for certain facilities. In our model we avoid such weaknesses by introducing additional objective functions.

Minimization of the maximum distance travelled by the patients was originally proposed as the second criterion for optimization. This criterion can be regarded as a measure of spatial equality. It turned out, however, that the distribution of population in relation to the location of facilities makes this criterion passive. This means that the solutions minimizing the average distance minimize simultaneously the maximum distance. Such a situation was caused by existence of an isolated spatial unit located at a considerable distance from all the clinics. Minimization of the maximum distance covered by all the patients was then reduced to minimization of the maximum distance travelled by inhabitants of that single isolated spatial unit. Therefore, we modified the second criterion to make greater provision for the level of patients satisfactory in all the spatial units and simultaneously to force obtaining connected districts.

For each spatial unit we determine an extra distance travelled by patients as the difference between the distance to the clinic to which they were allocated and the distance to the nearest facility. Defined in such a way the extra distance measures unsatisfactory of patients from the spatial unit under consideration. In case when the spatial unit is split up among two or more clinics, the mean distance travelled by inhabitants of the spatial unit is considered.

As the second criterion we use minimization of the maximum extra distance. This criterion needs some additional constraints in the model. Therefore. we introduce the following inequalities:

$$v - \sum_{j=1}^{n} \frac{d_{ij}}{P_i} t_{ij}^{g} \geq -d_{i*} \quad \text{for } i=1,2,\ldots,m \tag{10}$$

$$v - \sum_{j=1}^{n} \frac{d_{ij}}{P_i} t_{ij}^{d} \geq -d_{i*} \quad \text{for } i=1,2,\ldots,m \tag{11}$$

where d_{i*} denotes the distance between the i-th spatial unit and the nearest clinic, i.e., $d_{i*} = \min_j d_{ij}$, and v is the maximum extra distance to be minimized as the second objective function.

As noted earlier, in the area under study the general and dental clinics are located at the buildings. Despite of this they may have different service areas due to different service potentials (capacities). For this reason, maximization of the multiple functionality of boundaries was selected as the third criterion for optimization. This means that we decided to maximize the number of patients using both the general and the dental clinic at the same building.

The third criterion can be easily formulated in terms of goal programming (see e.g. [4]). Therefore, we introduce into the model the following additional constraints:

$t^g_{ij}-t^d_{ij}+r^-_{ij}-r^+_{ij} = 0$ for $i=1,2,\ldots,m; j=1,2,\ldots,n$ (12)

$r^-_{ij} \geq 0$ for $i=1,2,\ldots,m; j=1,2,\ldots,n$ (13)

$r^+_{ij} \geq 0$ for $i=1,2,\ldots,m; j=1,2,\ldots,n$ (14)

The third criterion is then represented by minimization of the following function:

$$\frac{1}{2}\sum_{i=1}^{m}\sum_{j=1}^{n}(r^-_{ij} + r^+_{ij})$$

which counts the number of patients that use a general clinic located at the other site than the dental clinic.

Finally, we obtain the linear programming model including:

1° variables t^g_{ij}, t^d_{ij}, r^-_{ij}, r^+_{ij}, v;
2° constraints (4) to (14);
3° three objective functions to be minimized:

$$z_1 = \frac{1}{2P}\sum_{i=1}^{m}\sum_{j=1}^{n} d_{ij}(t^g_{ij} + t^d_{ij}), \quad (15)$$

$$z_2 = v\ , \quad (16)$$

$$z_3 = \frac{1}{2}\sum_{i=1}^{m}\sum_{j=1}^{n}(r^-_{ij} + r^+_{ij})\ . \quad (17)$$

The model formulated in this manner has 559 constraints and 813 variables while analyzing districts for WWE, and 726 constraints and 745 variables for WWW. Thus in both cases the corresponding (single-objective) linear programming problems could be easily processed by the MPSX/370 package (see [3,9]).

3. Pay-off matrices and lexicographic optimization

As the first step of the multiobjective analysis we performed minimization of each objective function separately, i.e., for each area we solved three single-objective linear programs:

$min\ \{z_k$ subject to (4)-(14)$\}$ for $k=1,2,3$ (18)

In effect, we get the so-called pay-off matrices presented in Table 1. The pay-off matrix is a well-known device in multiobjective programming. It gives values of all the objective functions (columns) obtained while solving several single-objective problems (rows) and thereby it helps to understand the conflicts between different objectives. Furthermore, the pay-off matrix contributes the basis for the calculation of frequently used reference vectors such as the utopia vector and the nadir vector. The utopia vector represents the best values of each objective considered separately, and the nadir vector represents the worst values of each objective observed during optimization of another objective function. These vectors, however, are not attainable, i.e., there are no feasible solution with such objective values.

Table 1. Pay-off matrices

Minimized function	Obj. values for WWW			Obj. values for WWE		
	z_1	z_2	z_3	z_1	z_2	z_3
z_1	2.227	1.840	3470	1.098	0.488	16346
z_2	2.230	1.633	3470	1.098	0.488	16346
z_3	2.228	1.840	3443	1.153	1.100	10185

Unfortunately, the separate optimization of the individual objectives usually does not result in unique optimal solutions and thereby the pay-off matrix is not uniquely defined. The utopia vector is the only unique information in the pay-off matrix. Moreover, optimal solutions to the several problems (18) can be nonefficient. Therefore we decided to use (similarly as in [5]) the lexicographic optimization to generate extended pay-off matrices which guarantee unique information. That is, we examined solutions obtained under various hierarchies of the objective functions. This process required to solve 3!=6 lexicographic problems for each area under consideration. We solved the lexicographic problems with the MPSX/370 package using a special control program.

While analyzing the results of lexicographic optimization for WWW (see Table 2) one finds that there are no significant changes of the objective values depending on the used hierarchy. Namely, z_1 varies from 2.227 to 2.230, z_2 takes value 1.633 or 1.840, and z_3 takes value 3443 or 3470. So, only the value z_3 (i.e., the maximum extra distance) depends significantly on the hierarchy of the objective functions. Furthermore, the solution corresponding to the hierarchy $z_2/z_3/z_1$ differs from the utopia vector only with respect to the value z_1 and this difference is of order 0.1%. Thus there is no reason to search for any better solution. Note that such a good solution does not happen while performing the standard single-objective analysis (see Table 1).

The results of lexicographic optimization for WWE (see Table 2) indicate a considerably higher, in comparison with WWW, variation of the objective values depending on the used hierarchy. Namely, z_1 varies from 1.098 to 1.152, z_2 varies from 0.488 to 1.088, and z_3 varies from 10185 to 16257. So, differences in the objective values are of order of 15% for z_1, on the level of 60% for z_3, and even overstep 100% for z_2.

Table 2. Extended Pay-off matrices

Minimized function	Obj. values for WWW			Obj. values for WWE		
	z_1	z_2	z_3	z_1	z_2	z_3
$z_1/z_2/z_3$	2.227	1.840	3470	1.098	0.488	16257
$z_1/z_3/z_2$	2.227	1.840	3470	1.098	0.488	16257
$z_2/z_1/z_3$	2.228	1.633	3470	1.098	0.488	16257
$z_2/z_3/z_1$	2.230	1.633	3443	1.136	0.488	11441
$z_3/z_1/z_2$	2.230	1.840	3443	1.150	1.088	10185
$z_3/z_2/z_1$	2.230	1.633	3443	1.152	0.765	10185

The results obtained from lexicographic optimization were consulted with the decision makers competent to delimit health care districts in WWE and WWW. It was decided that the optimal solution to the hierarchy $z_2/z_3/z_1$ could be accepted as a basic solution for the districts delimitation in WWW. On the other hand, similar consultations concerning WWE led to the conclusion that any lexicographic solution could not be regarded as a final solution, and it was necessary to search for some satisfying compromise solution.

4. Interactive and parametric analysis

The search for a satisfying solution was performed with the SIMOLP procedure (see [8,10]). SIMOLP is a simple tool for solving multicriteria linear programming problems with the MPSX/370 package. It is prepared in the MPSX Control Language using the MPS Report Generator (see [3]) for modification of the problem during computations. Essentially, SIMOLP does not solve the problem. It rather makes the user choosing the best solution.

At first the utopia vector is calculated. Further, the feasible solution nearest to the ideal vector is computed using the weighted Chebyshev norm. The computed solution is an efficient solution to the original multiobjective problem. It is presented to the user as a current solution. The user is asked whether he finds this solution satisfactory or not. Depending on the user's opinion, some new weights are generated. A new efficient solution is computed and presented as a current solution. The process is repeated as long as necessary. The sequence of generated solutions converges to some efficient solution due to a special technique used for calculating the weights. Thus after a few interactive steps a stabilization of the process is observed. There exists, however, a possibility to perform open search within the whole efficient set without stabilization.

The theoretical background of the SIMOLP procedure can be summarized as follows:

1° any current solution generated by SIMOLP is an efficient solution;

2° for any efficient solution there exist weights such that the solution minimizes the weighted Chebyshev distance to the ideal vector;

3° the set of weights which can be generated by SIMOLP is dense with respect to the set of all (positive) weights.

After a few runs of the SIMOLP procedure a solution with an objective vector

$$z_1 = 1.142, \ z_2 = 0.537, \ z_3 = 11204$$

was found as the most satisfactory. Similarly as in WWW, the proposed solution gave all characteristics significantly better than the present districts pattern (see Table 3). However, in comparison to the utopia vector, all the objective values of the proposed solution were worse by about 10%. Therefore, we decided to perform a parametric analysis (see e.g. [1]) as "an optimality proof".

Table 3. Final solutions

Pattern of disricts	Obj. values for WWW			Obj. values for WWE		
	z_1	z_2	z_3	z_1	z_2	z_3
Present	2.273	2.678	3754	1.193	0.810	13020
Proposed	2.228	1.633	3443	1.142	0.537	11204

The parametric analysis was performed as follows. Assuming that an objective z_{k_1} deviates at most by the defined percentage from the ideal value, we carried out parametric optimization moving from minimization of the objective z_{k_2} to minimization of z_{k_3}. Namely, we solved the following parametric problems:

$$min\{(1-\alpha)z_{k_2}+\alpha z_{k_3} \text{ subject to (4)-(14) and } z_{k_1}\leq(1+\delta)u_{k_1}\}$$

where δ means an acceptable level of the relative degradation of the objective value, and the parameter α varies from 0 to 1.

Using the parametric procedures of the MPSX/370 package, we performed calculations for three levels of the relative degradation δ: 1%, 5% and 10%. The obtained results illustrate the interdependence of objectives z_{k_2} and z_{k_3} at an accepted level of degradation of the objective z_{k_1}. Analysis of the results showed that it is not possible to obtain a satisfactory solution with δ equal to 1% or 5%. On the other hand, at the level of relative degradation δ equal to 10%, the value z_3 decreases to 11204 and further reduction of this value is not possible without overstepping of the 10% limit on the z_2 degradation. Moreover, it turned out that a relatively small decrease of the value z_3 led to relatively high increase of the value z_2. Therefore, the solution generated by the SIMOLP procedure was definitely accepted as the best compromise solution.

5. Concluding remarks

In this paper we formulated the mathematical model and performed the multiobjective analysis to delimit boundaries of districts serviced by general and dental clinics in the Wola quarter of Warsaw. In effect, we got the districts patterns which had all the characteristics significantly better than the present patterns of districts (see Table 3).

The objective functions used in this study takes into account specific circumstances of the health care organization in Warsaw-Wola. In this sense, they can be considered as a unique collection of objectives applicable to the particular problem. Nevertheless, these objectives should be also attractive while solving different districting or location-allocation problems. Minimization of the average distance is generally used in districting problems. Minimization of the maximum extra distance was defined for the specific case considered in this study, but probably in many models this objective would turn out to be an attractive alternative to the well-known and frequently used criterion of minimization of the maximum distance. The third objective was used in the study due to the specific organization of health care in Poland. However, the problem of multiple functions of boundaries frequently appears in relation to a spatial arrangement of administrative units. Therefore the proposed formalization of that problem in terms of linear goal programming could represent an interesting proposal.

Furthermore, the multiobjective analysis was performed with the MPSX/370 package. Thus only a standard linear programming tool was used.

References

1. Guddat J., Guerra-Vasquez F., Tammer K., Wendler K.:*Multiobjective and Stochastic Optimization Based on Parametric Optimization.* Akademie-Verlag Berlin, Berlin 1985.
2. Hwang C.L., Masud A.: *Multiple Objective Decision Making - Methods and Applications.* Springer-Verlag, Heidelberg 1979.
3. *IBM Mathematical Programming System Extended/370 - MPSX/370: Program Reference Manual.* IBM SH19-1095, 1979.
4. Ignizio J.P.: *Goal Programming and Extensions.* Heath, Lexington 1976.
5. Kok M., Spronk J.: A note on the pay-off matrix in multiple objective programming. Delft University of Technology, Report 85-06 (1985).
6. Malczewski J.: Spatial organization and efficiency of the primary health care system: a case study of Warsaw-Wola (in Polish) Ph.D. Thesis, Institute of Geography and Spatial Organization, Polish Academy of Sciences, Warsaw 1986.
7. Malczewski J., Ogryczak W.: A multi-objective approach to reorganization of health service areas: a case study. *Environment & Planning* A20 (1988) 1461-1470.

8. Ogryczak W.: SIMOLP - Simple Interactive Multiple Objective Linear Programming via MPSX/370: Short User Manual. Warsaw University, Technical Report (1984).
9. Ogryczak W.: *Mathematical Programming with the MPSX Package* (in Polish). Warsaw Univ. Press, to appear.
10.Ogryczak W., Malczewski J.: Health care districts planning by multiobjective analysis with the MPSX/370 package. *Archiwum Automatyki i Telemechaniki* 32 (1987) 369-381.
11.Ross G.T., Soland R.M.: A multicriteria approach to the location of public facilities. *EJOR* 4 (1980) 307-321.

An Interactive Support System for Bond Trading

Hirotaka Nakayama

Department of Applied Mathematics, Konan University
8-9-1 Okamoto, Higashinada, Kobe 658, JAPAN

ABSTRACT

In portfolio selection problems, many companies are now widely trying to use mathematical analysis for bond trading. In this paper, some bond trading problem is formulated by a kind of multi-objective fractional problem. An effective method for this kind of problem will be suggested along with an experimental result.

1. INTRODUCTION

Interactive multi-objective programming methods have made much progress in the past decades, and now are applied widely to real problems. Above all, blending problems, e.g., feed blending, raw material blending for industrial products, etc., can be very effectively solved by some mathematical programming approach. The author has been collaborating with several companies in Japan. In this report, an example for bond trading support system will be introduced on the basis of the author's experience.

Bond traders are facing almost every day a problem which bonds and what amount they should sell and/or buy in order to attain their customers' desires. The economic environment is changing day by day, and sometimes gives us a drastic change. Bond traders have to take into account many factors, and make their decisions very rapidly and flexibly according to these changes. The number of bonds to be considered is usually more than 500, and that of criteria is about ten as will be shown later. The amount of trading is usually more than 100 million yen, and hence even a slight difference of portfolio combination might cause a big difference to profit or loss. This situation requires some effective method which helps bond traders following faithfully their value-judgment on a real time basis not only mathematically but also in such a way that their intuition fostered by their experiences can be taken in.

Bond portfolio problems are a kind of blending problems. Therefore, mathemetical programming approach can be used very effectively. However, the traditional mathematical programming approach with a single objective function can not take in the value-judgment and intuition of bond traders so easily in a flexible manner for the changes of environment. We shall show that some interactive multi-objective programming method fits to this purpose.

2. MATHEMATICAL FORMULATION FOR BOND PORTFOLIO

We shall follow the mathematical model given by Konno and Innori (1987). Assume that an investor holds u_j units of bonds B_j, $j=1,..,N$. Associated with B_j, we have the following indices:

c_j: coupon to be paid at a fixed rate (yen/bond/year)

f_j: principal value to be refunded at maturity (yen/bond)

p_j: present price in the market (yen/bond)

t_j: maturity (number of years until its principal value is refunded)

Returns from bonds are the income from coupon and the capital gain due to price increase. Bond portfolio problems are to determine which bonds and what amount the investor should sell and/or buy taking into account many factors, say, expected returns and risk, the time needs money for another investment, and so on. Therefore, we have the following criteria for each bond B_j:

[I] returns

(i) direct yield (short term index of return)

$$\gamma_j=c_j/p_j$$

(ii) effectiv yield

$$\nu_j = \frac{\left[c_j\{(1+\alpha)^{t_j} - 1\}/\alpha + f_j\right]^{1/t_j}}{p_j} - 1$$

where α is the interest rate.

[II] risk

(iii) price variation

$$\pi_j = t_j/(1+\mu_j t_j), \quad j=1,\ldots,N$$

Let x_j $(j=1,\ldots,n_1)$ and X_k $(k=1,\ldots,n_2)$ denote the number of bonds to be sold and to be purchased, respectively. Then S_0 and S_1 represent respectively the total quantity of bonds and the total value of bonds after the transaction. Namely,

$$S_0 = \sum_{j=1}^{N} u_j - \sum_{j=1}^{n_1} x_j + \sum_{k=1}^{n_2} X_k$$

$$S_1 = \sum_{j=1}^{N} p_j u_j - \sum_{j=1}^{n_1} p_j x_j + \sum_{k=1}^{n_2} P_k X_k.$$

In addition, we set

$$S_2 = \sum_{j=1}^{N} p_j t_j u_j - \sum_{j=1}^{n_1} p_j t_j x_j + \sum_{k=1}^{n_2} P_k T_k X_k.$$

Then the average for each index is taken as an objective function:

(i') average direct yield

$$F_1 = (\sum_{j=1}^{N} \gamma_j p_j u_j - \sum_{j=1}^{n_1} \gamma_j p_j x_j + \sum_{k=1}^{n_2} \Gamma_k P_k X_k) \;/\; S_1$$

(ii') average effective yield

$$F_2 = (\sum_{j=1}^{N} \nu_j p_j t_j u_j - \sum_{j=1}^{n_1} \nu_j p_j t_j x_j + \sum_{k=1}^{n_2} \nu'_k P_k T_k X_k) \;/\; S_2$$

(iii') average price variation

$$F_3 = (\sum_{j=1}^{N} \pi_j u_j - \sum_{j=1}^{n_1} \pi_j x_j + \sum_{k=1}^{n_2} \pi'_k X_k) \;/\; S_0$$

Our constraints are divided into soft constraints and hard constraints:

[III] soft constraints

(iv) average unit price

$$F_4 = \left(\sum_{j=1}^{N} p_j u_j - \sum_{j=1}^{n_1} p_j x_j + \sum_{k=1}^{n_2} P_k X_k\right) / S_0$$

(v) average maturity

$$F_5 = \left(\sum_{j=1}^{N} t_j u_j - \sum_{j=1}^{n_1} t_j x_j + \sum_{k=1}^{n_2} T_k X_k\right) / S_0$$

(vi) specification of time of coupon to be paid

$$F_6 = \sum_{i \in I_1} x_i / S_0$$

$$F_7 = \sum_{i \in I_2} x_i / S_0$$

where I_m is the set of indices of bonds whose coupon are paid at the time t_m.

[IV] hard constraints

(vii) budget constraints:

$$\sum_{j=1}^{n_1} -p_j x_j + \sum_{k=1}^{n_2} P_k X_k \leqq C$$

$$\sum_{k=1}^{n_2} P_k X_k \geqq C$$

(viii) specification of bond

$$l_j \leqq x_j \leqq u_j, \quad j=1,\dots, n_1$$

$$L_k \leqq X_k \leqq U_k, \quad k=1,\dots, n_2$$

For this kind of problem, the satisficing trade-off method developed by the author (Nakayama 1984, see also Sawaragi-Nakayama-Tanino 1985) can be effectively applied. In the following section, we shall give a brief review of the method.

3. SATISFICING TRADE-OFF METHOD

The algorithm of the satisficing trade-off method is summarized as follows:

Step 1. (setting the ideal point) The ideal point $F^*=(F_1^*,\ldots,F_r^*)$ is set, where F_i^* is large enough, for example, $F_i^*=\text{Max}\ \{F_i(x) \mid x\varepsilon X\}$. This value is fixed throughout the following process.

Step 2. (setting the aspiration level) The aspiration level $\bar{F}_i^k$ of each objective function F_i at the k-th iteration is asked to the decision maker. Here $\bar{F}_i^k$ should be set in such a way that $\bar{F}_i^k < F_i^*$. Set k=1 for the first iteration.

Step 3. (weighting and finding a Pareto solution by the Min-Max method)
Set

$$w_i^k = \frac{1}{F^*_i - \bar{F}_i^k}, \tag{3.1}$$

and solve the Min-Max problem

$$\underset{x\varepsilon X}{\text{Min}}\ \underset{1\leq i\leq r}{\text{Max}}\ w_i^k|F^*_i - F_i(x)| \tag{3.2}$$

Let x^k be a solution of (3.2).

Step 4. (trade-off) Based on the value of $F(x^k)$, the decision maker classifies the criteria into three groups, namely,

(i) the class of criteria which he wants to improve more,

(ii) the class of criteria which he may agree to relaxing,

(iii) the class of criteria which he accepts as they are.

The index set of each class is represented by I_I^k, I_R^k, I_A^k, respectively. If $I_I^k=\emptyset$, then stop the procedure. Otherwise, the decision maker is asked his new acceptable level of criteria $\bar{F}_i^k$ for the class of I_I^k and I_R^k. For $i\varepsilon I_A^k$, set $\bar{F}_i^k=F_i(x^k)$. Go back to Step 3.

Remark 3.1. In LP type problems, we can use parametric analysis in trade-off very effectively. This device enables us to trace Pareto surface without solving any additional Min-Max problem (Nakayama 1987, Korhonen-Wallenius 1987). If we consider cases of only either sale or purchase, our problem is of LP type. In cases for bond-trade taking into account both sale and purchase, since the auxiliary min-max problem is of the linear fractional programming, we need some device for solving it.

4. AN ALGORITHM FOR LINEAR FRACTIONAL MIN-MAX POROBLEM

Let each objective function in our bond trading problem be of the form $F_i(x)=p_i(x)/q_i(x)$ $(i=1,\dots,r)$ where p_i and q_i are linear in x. Then since

$$F_i^* - F_i(x) = \frac{F_i^* q_i(x)-p_i(x)}{q_i(x)} := \frac{f_i(x)}{g_i(x)},$$

the auxiliary Min-Max problem (3.2) becomes a kind of linear fractional Min-Max problem. For this kind of problem, several methods have been developed: Here we shall use a Dinkelbach type algorithm (Borde-Crouzeix 1987, Ferland-Potvin 1985) as is stated in the following:

Step 1: Let $x^0 \varepsilon X$. Set $\theta^0 = \underset{1 \leq i \leq r}{\text{Max}} \; f_i(x^0)/g_i(x^0)$ and $k=0$.

Step 2: Solve the problem

$$(P_k) \qquad T_k(\theta^k) = \underset{x \varepsilon X}{\text{Min}} \; \underset{1 \leq i \leq r}{\text{Max}} \; (f_i(x)-\theta^k g_i(x))/g_i(x^k)$$

Let x^{k+1} be a solution to (P^k).

Step 3: If $T_k(\theta^k)=0$ then stop: θ^k is the optimal value of the given Min-Max Problem, and x^{k+1} is the optimal solution.

Step 4: If not, take $\theta^{k+1} = \underset{1 \leq i \leq r}{\text{Max}} \; f_i(x^{k+1})/g_i(x^{k+1})$.

Replace k by k+1 and go to Step 2.

Note that the problem (P_k) is the usual linear Min-Max problem. Therrefore, we can obtain its solution by solving the following equivalent problem in a usual manner:

$$(Q_k) \qquad \text{Minimize} \quad z$$

$$\text{subject to} \quad (f_i(x)-\theta^k g_i(x))/g_i(x^k) \leq z, \quad i=1,\dots,r$$

5. AN EXPERIMENTAL RESULT

A result of our experiments is shown below: The problem is to décide a bond portfolio among 37 bonds selected from the market. The holding bonds are x(1)=5000, x(9)=1000, x(13)=2500, x(17)=4500, x(19)=5500, x(21)=6000, x(23)=5200, x(25)=4200, x(27)=3200 and x(37)=3800. The experiment was performed by a worker of a security company in Japan who has a carreer of acting as a bond trader.

	Pareto sol.	Asp.Level (Target Range)	Lowest	Highest	Sensitivity
F1 (max)	5.8335	5.8000	5.4788	5.9655	0.0133
F2 (max)	6.8482	6.8000	6.7165	6.8698	0.0000
F3 (min)	0.1292	0.1300	0.1262	0.1325	1.0000
* F4	102.7157	$F4 \leq 103.00$			
* F5	4.0000	$4.00 \leq F5 \leq 5.00$			
* F6	0.2000	$0.20 \leq F6$			
* F7	0.2000	$0.20 \leq F7$			

X(1)=	1042.0643	X(2)=	400.0000
X(3)=	400.0000	X(4)=	200.0000
X(5)=	0.0000	X(6)=	400.0000
X(7)=	0.0000	X(8)=	0.0000
X(9)=	547.3548	X(10)=	0.0000
X(11)=	0.0000	X(12)=	0.0000
X(13)=	2500.0000	X(14)=	200.0000
X(15)=	0.0000	X(16)=	200.0000
X(17)=	4500.0000	X(18)=	0.0000
X(19)=	5321.9573	X(20)=	0.0000
X(21)=	6000.0000	X(22)=	0.0000
X(23)=	5200.0000	X(24)=	0.0000
X(25)=	4200.0000	X(26)=	0.0000
X(27)=	3200.0000	X(28)=	274.6025
X(29)=	400.0000	X(30)=	200.0000
X(31)=	0.0000	X(32)=	200.0000
X(33)=	400.0000	X(34)=	400.0000
X(35)=	0.0000	X(36)=	0.0000
X(37)=	3800.0000		

The asterisk of F_4 to F_7 implies soft constraints. In our system, we can change objective function into soft constraints and vice versa. Here, the bond trader changed F_2 (effective yield) into a soft constraint, and F_4 (unit price) into an objective function. Then under the modified aspiration level by trade-off, the obtained result is as follows:

	Pareto sol.	Asp.Level (Target Range)	Lowest	Highest	Sensitivity
F1 (max)	5.8629	5.9000	5.8193	5.9608	0.0000
* F2	6.8482	6.8482 ≦ F2			
F3 (min)	0.1302	0.1292	0.1291	0.1322	1.0000
F4 (min)	102.3555	102.1000	102.0676	102.7228	0.0043
* F5	4.0000	4.00 ≦ F5 ≦ 5.00			
* F6	0.2000	0.20 ≦ F6			
* F7	0.2000	0.20 ≦ F7			

X(1)=	0.0000	X(2)=	400.0000
X(3)=	400.0000	X(4)=	200.0000
X(5)=	0.0000	X(6)=	400.0000
X(7)=	0.0000	X(8)=	0.0000
X(9)=	139.4913	X(10)=	0.0000
X(11)=	0.0000	X(12)=	0.0000
X(13)=	2500.0000	X(14)=	381.5260
X(15)=	0.0000	X(16)=	200.0000
X(17)=	4500.0000	X(18)=	0.0000
X(19)=	5277.1577	X(20)=	0.0000
X(21)=	6000.0000	X(22)=	0.0000
X(23)=	5200.0000	X(24)=	0.0000
X(25)=	4200.0000	X(26)=	0.0000
X(27)=	3200.0000	X(28)=	14.8920
X(29)=	400.0000	X(30)=	200.0000
X(31)=	0.0000	X(32)=	200.0000
X(33)=	400.0000	X(34)=	400.0000
X(35)=	400.0000	X(36)=	222.7743
X(37)=	3800.0000		

The result was satisfactory to the testee. He recognized that this method (software) is very easy to use and flexible for variation of the desire of investers, and therefore will work well as a support system for the real bond trading. It has been decided that this system will be built in a total bond/stock trading system in a security company in Japan very soon after some modification of input/output systems onlined with the market and such that other criteria (e.g. transaction cost) may be taken into account too.

6. CONCLUDING REMARKS

Recently, much attention has been being paid to the intelligent information process by computers. Such intelligent information process as diagnosis with less value-judgment will be treated in an effective way by artificial intelligence or knowledge engineering. However, among intelligent information process, the value-judgment will remain hard to treat to the last. In portfolio problems such as bond trading, it is very important to get a solution reflecting faithfully the value-judgment of customers. In order to support bond traders in a flexible manner for the multiplicity of value-judgment and complex changes of economic environment, the cooperative system of man and computers are very attractive: above all, interactive multi-objective programming methods seem promising.

It should be noted that the traditional mathematical programming decides a solution once a mathematical model is formulated. Then it is difficult to get a solution faithful to decision makers' value judgment. On the other hand, the interactive multi-objective programming technique is an 'open system' to the value judgment of decision makers, and hence it is flexible to the multiplicity of value judgment.

REFERENCES

Borde, J. and Crouzeix, J.P. (1987), Convergence of a Dinkelbach-type Algorithm in Generalized Fractional Programming, Zeitscrift fur Operations Research 31, 31-54.

Ferland, A. and Potvin, J. (1985), Generalized Fractional Programming: Algorithms and Numerical Experimentation, European J. Operational Research 20, 92-101.

Konno, H. nad Inori, M. (1987), Bond Portfolio Optimization by Bilinear Fractional Programming, Research Paper IKSS 87-4, Institute of Human and Social Sciences, Tokyo Institute of Technology

Korhonen, P. and Wallenius, J.: A Pareto Race, Working Paper F-180, Helsinki School of Economics (1987)

Nakayama, H. (1984), Proposal of Satisficing Trade-off Method for Multiobjective Programming, Transact. SICE, 20: 29-35 (in Japanese)

Nakayama, H. and Sawaragi, Y. (1984). Satisficing Trade-off Method for Interactive Multiobjective Programming Methods, in M. Grauer and A.P. Wierzbicki (eds.) Interactive Decision Analysis, Proceeding of an International Workshop on Interactive Decision Analysis and Interpretative Computer Intelligence, Springer: 113-122.

Nakayama, H. (1987), Sensitivity and Trade-off Analysis in Linear Multi-objective Programming, presented in IIASA workshop in Albena/Bulgaria: also revised version is in preparation.

Sawaragi, Y., Nakayama, H. and Tanino, T. (1985). Theory of Multiobjecitve Optimization, Academic Press

A CAREFUL LOOK AT EFFICIENCY IN MULTIPLE OBJECTIVE LINEAR PROGRAMMING

Pekka Korhonen and Jyrki Wallenius
Helsinki School of Economics
Runeberginkatu 14-16,
00100 Helsinki, Finland

ABSTRACT

In this paper we discuss several common misunderstandings and pitfalls in multiple objective linear programming research and provide some suggestions for avoiding them. The discussion focuses on widely used concepts of efficiency. We also point out seemingly harmless misunderstandings pertaining to improper use of payoff tables. We illustrate the discussion using graphical and numerical examples.

Keywords: Multiple Criteria Decision-Making, Efficiency, Utility.

1. Introduction

Multiple criteria decision-making (MCDM) has been a popular research area for more than a decade. Over the years several procedures have been developed for solving optimization problems having multiple criteria. The types of problems addressed and the specifics of such procedures vary. A common goal is to attempt to help a decision-maker (DM) make good (efficient) decisions.

MCDM procedures usually make various assumptions concerning the DM's preference structure and behavior. In such procedures, the DM is asked to input preference information in the form of tradeoffs, pairwise comparisons, aspiration levels, etc. The DM's responses are used by the computer system to generate a, hopefully good (improved) solution or a set of such solutions. In interactive procedures, several iterations are usually required before a most preferred solution is found. The search is terminated using some stopping rule, or simply when the DM feels that the current solution is good enough. The underlying assumptions determine, which solutions are considered as possible compromise solutions. Unrealistic or incorrect assumptions in this respect may produce less than satisfactory results. Therefore, these issues have significant implications for the design and development of multiple criteria decision methods.

ı this paper we point out several common pitfalls in multiple criteria ecision-making, especially in the context of multiple objective linear rogramming (MOLP). We also make some suggestions for avoiding them. his paper is based on our own odysseys, conference presentations by olleagues, and published articles. We do hope that the discussion ıelps researchers avoid some of the most common pitfalls and fallacies .n their future work.

The rest of the paper is organized as follows. In the second section we discuss different definitions of efficiency, efficient tradeoffs, efficient extreme points, etc. In the third section, we discuss common fallacies regarding our perception of the efficient region. In the fourth section we point out seemingly harmless misunderstandings pertaining to an improper use of payoff tables. The fifth section concludes the paper.

2. Efficient Solutions and Efficient Extreme Point Solutions

Let us consider the following multiple objective linear programming problem:

$$\begin{aligned} &\text{"max" } \mathbf{q}, \\ &\text{s.t. } \mathbf{q} \in Q = \{\mathbf{f}(\mathbf{x}) \mid \mathbf{x} \in X\}, \end{aligned} \tag{2.1}$$

where $\mathbf{f}(\mathbf{x}) = (f_1(\mathbf{x}), f_2(\mathbf{x}), \ldots, f_k(\mathbf{x})) = C\mathbf{x}$, is a k-vector of objective functions, C is a (k x n)-coefficient matrix of the objective functions, $\mathbf{x}$ is a vector of decision variables, and $X = \{\mathbf{x} \mid A\mathbf{x} \le \mathbf{b}, \mathbf{x} \ge \mathbf{0}\}$ is the set of feasible decisions, where A is an (m x n)-coefficient matrix, and $\mathbf{b}$ a rhs-vector.

2.1. Efficiency vs. Weak Efficiency

Definition 1. A point $\mathbf{x}^o \in X$ is efficient iff there does not exist another $\mathbf{x} \in X$ such that $f_i(\mathbf{x}) \ge f_i(\mathbf{x}^o)$ for all $i=1,2,\ldots,k$ and $f_i(\mathbf{x}) \ne f_i(\mathbf{x}^o)$ for at least one $i \in K=\{1,2,\ldots,k\}$.

Definition 2. A point $\mathbf{x}^o \in X$ is weakly-efficient iff there does not exist another $\mathbf{x} \in X$ such that $f_i(\mathbf{x}) > f_i(\mathbf{x}^o)$ for all $i \in K$.

Criterion vectors corresponding to (weakly) efficient points are called (weakly) nondominated solutions. The set of nondominated solution vectors is denoted by Q^E and the set of weakly nondominated solution vectors by Q^W, respectively.

In randomly generated artificial examples, in which the coefficient matrix is relatively dense, weakly efficient solutions are not necessarily common. It may lead researchers to think that weakly efficient solutions are an exception rather than a rule. However, in practical problems, which contain "box constraints" for decision variables and structural (in)equality constraints, weakly efficient solutions appear quite often. A simple example in Table 1 demonstrates this.

	x_1	x_2	x_3	
q_1	1.0	1.0		-> max
q_2			1.0	-> min
Row 1:		1.0		≤ 1
Row 2:	1.5		-1.0	$= 0$
		x_1, x_2, x_3		≥ 0

Table 1. An Example of Weakly Efficient Solutions

From the example of Table 1 we see that Q^E is a subset of Q^W:

$$Q^E = \{(q_1,q_2) \mid q_2 = 1.5*(q_1-1),\ q_1 \geq 1\}, \text{ and}$$

$$Q^W = Q^E \cup \{(q_1,q_2) \mid q_2 = 0,\ q_1 \in [0,1)\}.$$

Since only nondominated solutions are relevant from the DM's point of view, it is important that MCDM-procedures take explicitly into account the possibility of generating weakly nondominated (but dominated) solutions.

In theory, all nondominated solutions can be generated as a solution to the following scalar maximum problem, as shown by Geoffrion (1966,1967):

$$\max \ \boldsymbol{\mu}'\mathbf{q} \qquad \text{s.t} \quad \mathbf{q} \in Q, \tag{2.2}$$

where $\boldsymbol{\mu} \geq 0$, $\boldsymbol{\mu} \neq \mathbf{0}$, and often normalized according to $\boldsymbol{\mu}'\mathbf{e} = 1$, where

'=(1,1,...,1). However, the solution to (2.2) may be weakly nondominated but dominated). To avoid this, we have to assume that $\mu>0$, as shown by eoffrion (1968). The use of formula (2.2) for generating all nondoinated solutions leads to two problems:

.. Generally, we cannot solve a scalar parametric problem in the open set $U = \{\mu \mid \mu > 0\}$, but in the closed set $U^c = \{\mu \mid \mu \geq \epsilon e, \epsilon>0\}$. In problem (2.1), all solutions are generated, if ϵ is "small enough", but "small enough" is difficult to specify in advance.

2. For all nondominated solutions, it is not possible to find a vector μ, which would generate that solution as a unique solution to the problem.

Similar problems arise, although some other composite rules for objective functions would be used (see, e.g., Wierzbicki, 1980 and 1986).

In other words, an incorrect choice of $\epsilon>0$ leads to excluding some nondominated solutions. Ideally, a method should enable a DM to consider all nondominated solutions and nothing but them, but this is easier said than done. Therefore, our recommendation is that, instead of generating a subset of all nondominated solutions, one should generate all nondominated solutions and explicitly inform the DM, when weakly nondominated (but dominated) solutions appear.

2.2 Efficient Extreme Point Solutions

For its simplicity, many authors have used the linear function $\mu'q$, $\mu\geq 0$, as an efficient solution generator (for example, Zionts and Wallenius, 1976). However, the linear function only produces extreme point solutions. Implicitly, its use is based on the assumption that the utility (value) function of the DM is, or more realistically, can be approximated by a linear function. In that case, efficient extreme points are sufficient to characterize the efficient frontier. Several techniques for generating efficient extreme point solutions have been developed that are based on pivoting from one efficient extreme point solution to another (see Steuer, 1986; and Yu and Zeleny, 1975).

There are, however, two problems in operating solely with efficient extreme point solutions:

1. In problems of realistic size, the number of efficient extreme point solutions is very large (see, e.g, Steuer, 1986).

2. The DM may not be satisfied with an extreme point solution as an approximation to the most preferred solution.

As a solution to the first problem, many authors have proposed effective methods to sample the set of efficient extreme point solutions (see, Steuer, 1986; Zionts and Wallenius, 1976). As far as the second problem concerns, a set of artificial extreme point solutions can be generated. However, the problem remains, how to specify such artificial extreme point solutions.

Despite the above difficulties, the theory developed for efficient extreme point solutions is relevant, because any MOLP-problem can be reformulated using the Chebychev-type of achievement scalarizing function proposed by Wierzbicki (1980) in a way which makes all efficient solutions extreme point solutions:

$$\begin{aligned} &\min \ y \\ &\text{s.t.} \\ &\mathbf{q} + \mathbf{w}y \geq \mathbf{q}^* \\ &\mathbf{q} \in Q, \end{aligned} \qquad (2.3)$$

where $\mathbf{w}>\mathbf{0}$, and $\mathbf{q}^*$ is an ideal vector (see, Wierzbicki, 1986). For a given $\mathbf{w}$, as a solution to problem (2.3), we may generate a weakly nondominated solution. In fact, all weakly nondominated solutions can be generated by varying $\mathbf{w}$. Because problem (2.3) is linear, all solutions are extreme point solutions. This formulation makes it possible to extend the ideas regarding efficient tradeoffs to non-extreme point solutions of the original constraint set.

2.3 Adjacent Efficient Tradeoffs

By performing a pivot, one can move from an extreme point solution to an adjacent extreme point solution. The edge in the criterion space connecting two extreme points is called an adjacent tradeoff. Tradeoffs connecting two nondominated solutions are of special interest in multiple objective linear programming.

It is a well-known result that a nondominated tradeoff leads from one ondominated extreme point to another (Steuer, 1986, p. 217). However, he reverse is not true. A tradeoff connecting two nondominated olutions is not necessarily efficient. See the example of Table 2.

	x_1	x_2	
q_1	1		-> max
q_2		1	-> max
Row 1:	1	3	$\leq$ 12
Row 2:	3	1	$\leq$ 12
Row 3:	1	1	$\geq$ 4
	x_1,	x_2	$\geq$ 0

Table 2. An Example of an Inefficient Tradeoff Connecting Two Nondominated Solutions

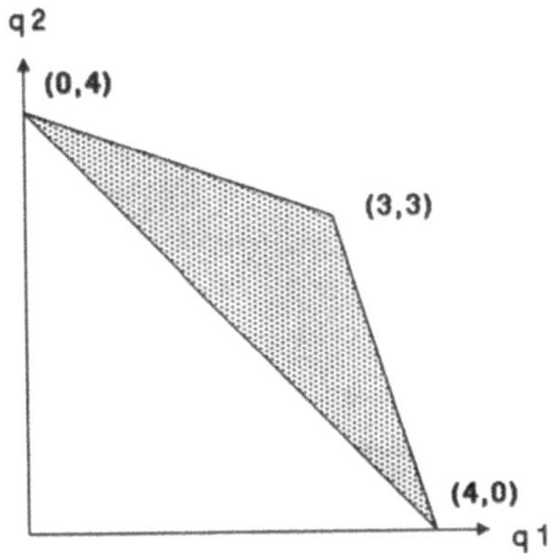

Figure 1. An Illustration of the Example in Table 2

The problem in Table 2 has three nondominated solutions: (4,0), (3,3), and (0,4). As we can see, points (4,0) and (0,4) are adjacent, but the corresponding tradeoff ((4,0),(0,4)) is dominated (see Fig. 1).

It is unnecessary to ask the DM to evaluate dominated tradeoffs. Therefore, it is important to diagnose which of the adjacent tradeoffs are nondominated (see, Zionts and Wallenius, 1980).

2.4 Efficient Solution Generators

As we have discussed, a simple way to generate efficient extreme point solutions is to maximize a weighted sum of the objectives: $\mu'q$, $\mu\geq 0$. By varying the parameter μ, all nondominated extreme point solutions can be found.

Actually, **μ'q** is a member of the more general class of functions called **achievement (scalarizing) functions** (see, Wierzbicki, 1980 and 1986). Such a function projects any given (feasible or infeasible) point onto the efficient frontier. An achievement function may be specified in different ways. A practical and simple form for this function is the following:

$$z(\mathbf{g},\mathbf{q},\mathbf{w}) = \max \{g_i - q_i / w_i,\ i=1,2,\ldots,k\}, \tag{2.4}$$

where **w** is a positive weighting vector and **g** an arbitrary k-vector. By minimizing $z(\mathbf{g},\mathbf{q},\mathbf{w},G)$ subject to $\mathbf{q} \in Q$, we find an (at least) weakly nondominated solution point $\mathbf{q}^o$. To generate only nondominated solutions, more complicated forms have to be used, or the dominance of each solution has to be checked. However, this simple form is adequate, if we only need a computationally efficient and handy efficient solution generator. If we replace $\mathbf{q}^*$ in problem (2.3) by **g**, we see how simple a formulation is adequate for applying the achievement function. The achievement function can also be used by varying **w** instead of **g**. Actually, our original problem formulation (2.3) was an example of this.

Several authors have used formulation (2.3) for projecting infeasible solutions onto the efficient frontier, especially in goal programming. But it is not well-known that this also works in projecting feasible (dominated) solutions onto the efficient frontier.

3. Visual Fallacies Regarding the Efficient Region

Usually, multiple criteria decision problems are visualized in two or three dimensional criterion spaces. This may cause misunderstandings. Let us consider a problem defined by means of inequality constraints. It seems natural to think that generally at an arbitrary nondominated solution point only one constraint is binding. However, this is not true.

The problem has two nondominated extreme point solutions (19.8, 0) and (0,10), which together with the connecting edge form the efficient frontier. We can easily see that both constraints are binding for all nondominated solutions.

	x_1	x_2	x_3	
q_1	1.0		1.0	-> max
q_2		1.0		-> max
Row 1:	1.0	1.0	.01	$\leq$ 10
Row 2:	.01	1.0	1.0	$\leq$ 10
			$x_1, x_2, x_3 \geq 0$	

Table 3. An Example Demonstrating the Number of Binding Constraints

Next, let us consider the following general multiple objective linear programming problem formulation (the parametric form) (see, e.g., Korhonen and Wallenius, 1987):

$$
\begin{aligned}
&\min \ y \\
&\text{s.t.} \\
&C\mathbf{x} + y\mathbf{w} \geq \mathbf{g}, \\
&A\mathbf{x} \qquad\ \ \leq \mathbf{b}, \\
&\quad \mathbf{x} \qquad \geq \mathbf{0}, \\
&\quad y \text{ unrestricted},
\end{aligned}
\tag{3.1}
$$

where C is a (k x n) - coefficient matrix of objectives of rank k, $\mathbf{w}>\mathbf{0}$ a k-vector of weights, $\mathbf{g}$ a k-vector of aspiration levels, A an (m x n) - matrix of coefficients of rank m, $\mathbf{b}$ an m-vector of constants, $\mathbf{x}$ an n-vector of decision variables, and y is a scalar variable. By varying vectors $\mathbf{w}$ or $\mathbf{g}$, any efficient solution can be generated (see also, Korhonen and Laakso, 1986a and 1986b; Wierzbicki, 1986).

The basic solution of problem (3.1) always consists of m+k variables. Let $\mathbf{z}^g$ be the k-vector of surplus variables that stands for the objective function rows and $\mathbf{z}^b$ the m-vector of slack variables that stands for the actual constraints. Obviously, y is always a basic variable. Because $\mathbf{w}$ and the unit vectors corresponding to $\mathbf{z}^g$-variables are linearly dependent, they cannot all be in the basis, simultaneously. This means that at most k-1 of the $\mathbf{z}^g$ variables can be in the basis. The rest of the variables are either $\mathbf{x}$- or $\mathbf{z}^b$-variables. Let us now consider the "original" constraint set $X = \{\mathbf{x} \mid A\mathbf{x} + \mathbf{z}^b = \mathbf{b},\ \mathbf{x}\geq\mathbf{0},\ \mathbf{z}^b\geq\mathbf{0}\}$. If y and all (k-1) $\mathbf{z}^g$-variables are in the basis, the number of $\mathbf{x}$- and $\mathbf{z}^b$-variables in the basis is m. This means that the solution is an extreme point of the polyhedron X, provided that no degeneracy occurs. On the other hand, if no $\mathbf{z}^g$ variables are in the basis, then the total number of **x**- and $\mathbf{z}^b$-variables in the basis is m+k-1. Thus, the solution lies on a k-1 dimensional facet of the polyhedron X, which does not mean that only one constraint would be binding.

4. Payoff Tables and Minimum Criterion Values

Minimum criterion values over the efficient set are of special interest to decision-makers, because they can be used to characterize the relevant ranges of the criterion values. Payoff tables are widely used in interactive multiple objective linear programming procedures to obtain estimates of the minimum criterion values (see, Isermann and Steuer, 1987). Many authors have discussed the fact that payoff tables only provide estimates (usually overestimates, sometimes underestimates) of the actual minimum criterion values (see, for example, Steuer, 1986; and Weistroffer, 1985). Yet, many authors have used these estimates, albeit with caution. The use is not dangerous, if the estimates are provided purely for information purposes. If these values are used to restrict the values of the objectives, it **may** cause problems, as shown below.

To demonstrate the consequences of misusing the minimum criterion values of a payoff table, consider the following simple example:

	x_1	x_2	x_3	
q_1	1.00			-> max
q_2		1.00		-> max
q_3			1.00	-> max
Row 1:	1.00	1.00	1.00	≤ 4
Row 2:	0.05	-0.05	1.00	≤ 2
Row 3:	1.00	-0.05	0.05	≤ 2
	x_1, x_2, x_3			≥ 0

Table 4. An Example Demonstrating the Misuse of a Payoff Table

Let us assume that the DM is interested in finding a solution max $\{\min(q_1,q_3)\}$ using some interactive procedure. If he/she assumes that the minimum criterion value for q_2 is 1.9048, as presented in the payoff table, then the solution is E: (1.0476, 1.9048, 1.0476). However, without any assumption concerning the minimum value of q_2, the solution is D: (1.9130, 0.1740, 1.9130). All nondominated solutions lie on the "diamond" ABCD. But if we restrict the values of q_2, only the triangle ABC is considered. Thus, about half of the nondominated solutions are excluded from further consideration.

The payoff table of the problem is given below:

	q_1	q_2	q_3
q^1	2.0952	1.9048	0.0000
q^2	0.0000	4.0000	0.0000
q^3	0.0000	1.9048	2.0952

Table 5. The Payoff Table of the Problem in Table 4

The problem is illustrated in Fig. 2.

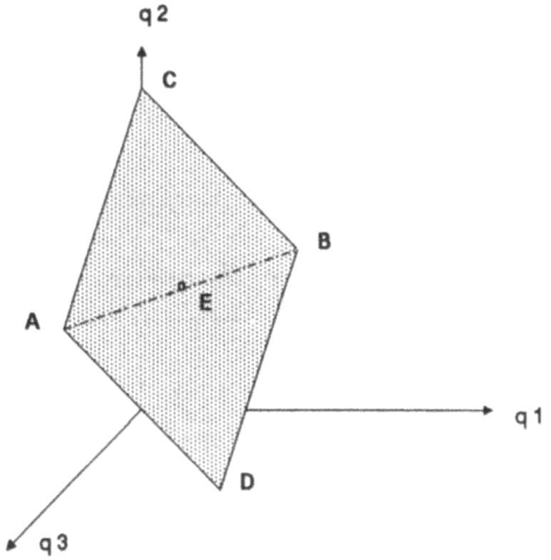

Figure 2. An Illustration of the Example in Table 4

5. Conclusion

As Steuer (1986) has expressed it so well, multiple criteria decision-making is an area that is exciting and challenging, but fraught with deceptions and pitfalls. In this tutorial, we have discussed several such pitfalls and fallacies in the context of multiple objective linear programming. In many instances, we have provided suggestions for avoiding them.

References

Geoffrion, A. M. (1966), "Strictly Concave Parametric Programming, Part I: Basic Theory", Mgmt Sci., 13, 244-253.

Geoffrion, A. M. (1967), "Strictly Concave Parametric Programming, Part II: Additional Theory and Computational Considerations", Mgmt Sci., 13, 359-370.

Geoffrion, A. M. (1968), "Proper Efficiency and the Theory of Vector Maximization", J. of Mathematical Analysis and Applications, 22, 618-630.

Isermann, H. and Steuer, R. E. (1987), "Computational Experience Concerning Payoff Tables and Minimum Criterion Values over the Efficient Set", Eur. J. Opl Res., 33, 91-97.

Korhonen, P. and Laakso, J. (1986a), "A Visual Interactive Method for Solving the Multiple Criteria Problem", Eur. J. Opl Res., 24, 277-287.

Korhonen, P. and Laakso, J. (1986b), "Solving Generalized Goal Programming Problems Using a Visual Interactive Approach", Eur. J. Opl Res., 26, 355-363.

Korhonen, P. and Wallenius, J. (1987), "A Pareto Race", Working Paper, DIS 85/86-13, Arizona State University, forthcoming in Naval Research Logistics.

Steuer, R. E. (1986), Multiple Criteria Optimization: Theory, Computation, and Application, Wiley, New York.

Weistroffer, H. R. (1985), "Careful Usage of Pessimistic Values is Needed in Multiple Objectives Optimization", Oper. Res. Letters, 4, 23-25.

Wierzbicki, A. (1980), "The Use of Reference Objectives in Multiobjective Optimization", in G. Fandel and T. Gal (eds.), Multiple Criteria Decision Making, Theory and Application, Springer-Verlag, 468-486.

Wierzbicki, A. (1986), "On the Completeness and Constructiveness of Parametric Characterizations to Vector Optimization Problems", OR-Spektrum, 8, 73-87.

Yu, P. L. and Zeleny, M. (1975), "The Set of All Nondominated Solutions in Linear Cases and a Multicriteria Simplex Method", J. of Mathematical Analysis and Applications, 49, 430-468.

Zionts, S. and Wallenius, J. (1976), "An Interactive Programming Method for Solving the Multiple Criteria Problem", Mgmt Sci., 29, 519-529.

Zionts, S. and Wallenius, J. (1980), "Identifying Efficient Vectors: Some Theory and Computational Results", Opns Res., 28, 785-793.

ALTERNATIVE SCALES IN AHP

Bertram Schoner and William C. Wedley
Faculty of Business Administration
Simon Fraser University, Burnaby, B.C., Canada, V5A 1S6

Abstract

The nine point scale within AHP has been criticised on various grounds. These include the fact that the scale is bounded, that semantic anchoring phrases are employed without validation, and that users are not informed that the numerical values generated are actually assumed to be ratio scaled. Nevertheless, applications of the scale to the measurement of weight, light intensity and distances have generated accurate estimates. Furthermore, it appears to be the case that decision makers do not experience difficulty in using the scale.

The method of Magnitude Estimation of S.S. Stevens has been successfully used in psychophysics for more than thirty years. It too has been designed to generate measurement on a ratio scale, but without the aid of the computer and without the support of consistency checks. This method could be easily substituted for the standard questioning procedure within AHP, with consistency checks added.

This paper reports some preliminary results from an experiment in which the semantically anchored version of the AHP scale and Magnitude Estimation are compared. Magnitude Estimation was found to be more accurate.

Key words: Multi-Criteria Decision Making, Decision Theory, Scaling Methods.

1. Introduction

The Analytic Hierarchy Process (AHP) was developed by Thomas L. Saaty (1977, 1980, 1982) as a decision analysis procedure for complex subjective problems. Unlike most multiattribute utility procedures which require the decision maker to use mathematical equations and explicit models, Saaty's technique de-emphasizes the underlying mathematics and the model complexity. The AHP input consists of direct pairwise comparisons of the decision maker's preferences under the guidance of a user-friendly computer program. This program converts the preferences to local priorities, which, in turn, are synthesized into overall decision priorities.

AHP employs a nine point scale to measure the strength of preference of one item over another. The choice of a particular value, say 7, indicates that one item is regarded as seven times more important than the other. Judgements may be made in either of two modes. The verbal comparison mode employs semantically anchored points, with 1 = equal preference; 3 = moderate preference; 5 = strong preference; 7 = very strong preference and 9 = absolute or extreme preference. The actual numbers are not shown in this version. Respondents are instructed to select a point on the scale that corresponds to their feelings about the magnitude of the comparison on the basis of the semantic cues, with values corresponding to 2, 4, 6, 8 selected for comprise between judgements. In the numerical mode respondents choose numbers directly, with decimal values permitted.

The resulting ratio values and their reciprocals are entered into a positive reciprocal matrix (i.e. $a_{ij} = 1/a_{ji}$ and $a_{ii} = 1$). Saaty has shown that the principal eigenvector of this matrix measures the relative priorities of the items. Moreover, the largest eigenvalue of this matrix can be used to calculate a measure of judgemental consistency. Since the calculations are complex, it is imperative that a computer be used. The complexity, however, is masked -- the user only has to input the paired comparisons and the computer generates the rest.

The inconsistency ratio (the ratio of the consistency index to the mean consistency index of random reciprocal matrices) provides the user with guidance on the degree of judgemental consistency. To be far enough away from the random situation (and therefore to be sufficiently consistent), Saaty (1980) suggests that this ratio should be less than .10 and definitely less than .20. If the comparisons are considered to be too inconsistent, the user can achieve greater consistency by re-evaluating and altering some of them. It should be noted that although consistency is a desirable attribute, it is possible to be consistent but inaccurate. In the end, the user must ensure that the priorities are logical and reasonable.

There are a number of reasons to suspect that the 1 to 9 scale for measuring strength of preference in AHP could be improved. First, the constraint of a range of unity to nine in the ratio may be inappropriate either because one or more of the comparisons lie outside that range, or because the comparisons all lie within a small portion of the range. Second, although Saaty indicates that it is possible to assign decimal values to responses in numerical mode, the philosophy of the scale mitigates against this. Five of the nine points are anchored by descriptive phrases with instructions to use points in between (the even numbered points) if uncertain. When responses are confined to integer values, the sensitivity of the scale is reduced. Finally, the particular descriptive phrases employed have not been validated by judging groups to demonstrate that they actually do correspond to the values to which they are attached. Such a demonstration would ideally demonstrate both the accuracy and the stability of the attached values of the phrases within and across groups.

In the study reported herein, the accuracy of the 9 point scale of AHP is compared to a simpler, non-computer method known as Magnitude Estimation (ME). The two methods are tested for the ability of subjects to use them in the prediction of colour proportions under different degrees of fuzziness. Since the true proportions in different pictures are known, we have correct answers from which to compare accuracy.

The method of Magnitude Estimation has been in existence much longer than AHP. Its popularization and development in psychophysics are credited to Stevens (1956, 1957). The method is simple to use, and like AHP, is claimed to yield measurement on a ratio scale. Although the method has its critics (e.g., Anderson, 1974), it is not our purpose to enter into the debate between its proponents and detractors. Rather, our intent is to provide a well recognized alternative to which the accuracy of standard AHP estimation procedure might be compared.

AHP and ME are similar in many respects. In addition to generating ratio scales, both are based on paired comparisons. ME, however, does not employ the entire set of possible pairs. Instead, only n-1 comparisons are made, which is in essence an interconnected spanning tree of the AHP matrix.

The method of ME is illustrated in the following example. Suppose we are judging the relative weight of five objects. One of the objects would be chosen as the standard. Although, in theory, this standard could be any of the five, Stevens (1975, p.26) reports more accurate results when the standard is one of the objects near the middle of the range. A given number of points, usually 10, is assigned to the standard. Subjects then assign points to each of the remaining objects so that the ratio of points assigned to the object relative to the points assigned to the standard are proportional to the perceived ratio of weights. If we were estimating proportions or percentages, the assigned numbers may be normalized to sum to unity or one hundred.

The task can be performed easily by pencil and paper, and it is quite flexible. Indeed, in the bulk of Stevens' work, he abandoned the practice of assigning a fixed number of points to the standard. Instead, he preferred to have the subject assign whatever number seemed appropriate to the first stimulus presented and then to employ that stimulus and number as the standard.

2. Experimental Method and Design

Magnitude Estimation was chosen as the technique to compare with AHP, because it is simple, flexible, and well documented in the literature. A variant of ME was used in which subjects first rank the objects. Then, in keeping with Stevens' (1975) finding, the middle ranked object is used as the standard for paired comparisons. The standard was given a value of 100.

In choosing a test exercise, it is important to conduct a test in which the correct answer is available but unknown to subjects. Saaty (1980) has written:

> Of considerable interest to us must be the issue of how closely the priority vector developed by our method matches the "real" priority vector. One way to ascertain this is to apply the method to situations which allow the determination of the actual numbers.

Several known-value stimuli were considered, and a decision was made to have subjects estimate the proportion of colors displayed on an overhead transparency under different degrees of fuzziness. The transparency was divided into 400 squares, with each square coloured by one of five distinct colours. The degree of fuzziness was controlled by employing two different spreads in the proportions and two different methods of distributing the colors throughout the picture. The two different spreads were:

(1) low spread - .10, .16, .20, .24, .30

(2) high spread - .05, .10, .20, .25, .40

The two distribution methods were:

(1) random - a random spacing of the colors over 400 squares.

(2) compact - adjacent squares forming irregular shapes.

The low spread random spacing should be the most fuzzy and difficult picture to estimate; the high spread compact, the easiest.

Sixteen different pictures were used, with colours assigned in four sets of spread-distribution combinations (i.e. low spread random, low spread compact, high spread random, high spread compact). Within each set, colors were assigned in different patterns to the spread proportions.

Each subject performed two evaluations -- the first under ME and then the other under AHP (or vice versa). Eight different presentation patterns were followed in order to give equal balance to method order (AHP or ME first), spread (high or low), and distribution (random or compact).

Subjects were 144 undergraduate business administration students who volunteered to participate in the exercise. As an inducement to participate and give accurate estimations, small monetary prizes ($30) were announced and awarded to the persons who got the most accurate estimations by ME and AHP. The measure of accuracy was not disclosed.

With ME, subjects were asked to rank the five colours in the order of the areas -- from No. 1 which covered the greatest area in the picture to No. 5 which covered the least area in the picture. Beside the ranking was a column for allocating points, with 100 points assigned to the middle rank (i.e. the third rank). Subjects were then asked to assign a number to each of the other four colours which represented the area relative to the standard of 100. They were told, for example, that "... if colour No. 5 covers half the area of No. 3 you will assign it a value of 50, or if colour 1 covers twice No. 3's area you will give it a 200." They could use any number they considered appropriate, and they were advised to not let a previous assignment influence a subsequent answer. Subjects were then advised to total the points and calculate the priorities, although this step was not necessary, because it could be done by the researchers. A calculator was nearby to help in the normalization process. Upon seeing their priorities, subjects had the opportunity to go redo their paired comparisons.

The AHP procedure was totally computerized. A variant of the numerical mode of AHP was employed in which subjects saw both semantic anchors and numerical values. Subjects were asked which element of a paired comparison was more important, where importance was defined as area covered, and by how much. Rather than rank the colours, the computer presented the colours in a fixed order (blue, black, yellow, purple, red) for the paired comparisons. The program included explanations of the pairwise comparison procedure and the rating scale. Entries into the bottom half of the matrix were displayed on the screen (the other half is unnecessary since it is the inverse), and after all entries were completed, the priorities and inconsistency ratio were displayed. At that point, users were allowed to test the matrix for ordinal transitivity or alter any of the entries. Through alterations, a person could hill climb to get a lower inconsistency ratio or more satisfying priorities. Once satisfied, the user could exit the program.

Accuracy in estimating the true proportions was used as the measure of effectiveness of the two methods. It was calculated for each individual as the mean absolute percent error (deviation) from the true weights, as given by the formula:

$$\text{MAPE} = \frac{1}{5} \cdot \sum_{i=1}^{5} \left(\frac{|x_i - t_i|}{t_i} \right) \times 100$$

where x_i is the estimate of the ith colour proportion and t_i the true proportion.

Although other measures of accuracy could be used, in our opinion MAPE is the most valid and interesting measure of error. Consider two colours, one which covers forty percent of the area and another which covers five percent of the area, as is actually the case in the high spread condition. An estimate of 0.44 for the first colour may be considered reasonably close, but an estimate of 0.09 for the second colour may be regarded as highly inaccurate. Although the actual absolute error is identical in both cases, the percentage error is 10% in the former case and 80% in the latter. Clearly, loss functions may be formulated for which MAPE would not be a valid measure of error, but in the absence of such a formulation we consider MAPE to be the measure of choice.

3. Results

Main Effects

It was anticipated that order would have an effect on the accuracy of estimates. In particular, since each subject was required to perform two estimates, with learning taking place from the first to the second, one might expect a tendency for the second estimate to be superior. Perhaps with less justification, and no priors with respect to the outcome, one might also expect that accuracy could depend on which colour was in the dominant position. In fact, analysis revealed that neither colour nor order had a noticeable effect on accuracy, and the miniscule effects observed were not statistically significant. These two factors will not be referred to further.

As may be seen from Table 1, ME was significantly more accurate than AHP for all conditions of speed and fuzziness.

The data were analyzed to see if there was a relationship between subjects' final consistency measures and their accuracy as measured by MAPE; that is, whether or not subjects with better consistency measures tended to have better accuracy as measured by their mean absolute proportion error. The correlation between the consistency measure and accuracy was 0.28, significant at the 0.02 level. This relationship, though weak, prompted us to reexamine the data with only those subjects who did not violate the ten percent inconsistency constraint. The original pattern of results was repeated.

TABLE 1

ANOVA Results

Mean Absolute Percent Error

	Means (1)	Means (2)			F	Significance
Main Effects:					41	***
Method	AHP 27.6	ME 15.9			102	***
Spread	High 20.2	Low 23.4			7.4	**
Fuzziness	Compact 19.7	Random 23.9			13.4	***
2-Way Interaction:						
Method-Spread	AHP-High 23.7	ME-High 16.7	AHP-Low 31.4	ME-Low 15.2	15.5	***
Method-Fuzziness	AHP-Comp 26.7	ME-Comp 12.7	AHP-Random 28.5	ME-Random 19.3	4.3	*

Significant levels: *** .001
** .01
* .05

Interaction Effects

Two significant interaction effects were found. First, AHP was much more sensitive to spread than was ME. The mean proportional accuracy of AHP in the high spread condition was roughly two-thirds that in the low-spread condition, whereas the results for ME on high spread and low spread were virtually identical. Although the interaction between method and fuzziness was significant at the five percent level, the magnitude of the effect was small.

Summary of Results

The results of this experiment were surprisingly clear and strong. The 1 to 9 AHP scale was markedly inferior to ME given the task of estimating colour proportions. The superiority of ME was most evident when the spread between the various proportions occupied by the colours was relatively small and when the colours were compact rather than randomly placed.

4. Discussion

AHP and ME Differences

In this experiment, the ME technique produced the more accurate priorities. In attempting to explain ME's superiority, the major differences between the two techniques can be given closer scrutiny. These differences are in consistency calculations, ranking procedures, reference points, and rating scales. But first, some caveats are in order.

The particular scale employed in this experiment did not permit decimal values, although this is possible in the numerical mode of Expert Choice (1983), the standard AHP program. This reduced sensitivity might be expected to affect the low-spread estimates more seriously than the high spread estimates, and account for the relative inferiority of AHP low spread estimates. It should be borne in mind, however, that many tests by Saaty (1977, 1980) have shown that the weights are insensitive to small perturbations in the comparisons. Second, subjects were informed that the AHP paired comparison value represented how much more important one colour was than another in the transparency, but not that this represented ratio-scale measurement. Finally, one may question the relevance of the assigned task.

The assigned task has a number of significant advantages. First, it is easily comprehended by subjects. Second, it can be expected to be independent of subjects' mathematical sophistication. Third, subjects are presented with all of the information necessary to complete the task without trivializing the exercise. Fourth, by providing a task which involves a single criterion one can focus on the scale itself, thus avoiding the problems associated with the generation of a composite measure or of missing criteria. Finally, both the spread of proportions and the "fuzziness" of presentation are easily manipulated in the design.

The major weakness in the design relates to its apparent lack of managerial relevance. Nevertheless, Saaty (1977a) has used known-value stimuli such as air distances, light intensities and weights to demonstrate the validity of AHP; tasks equally distant from managerial relevance. We would argue that it is very reasonable to insist that a scale designed to measure complex and subjective managerial constructs with unknown answers first be tested on simple and objective physical constructs with known answers.

Our preliminary thoughts on plausible explanations for the superiority of ME in this experiment are as follows:

1) Whereas AHP makes 1/2(n)(n-1) comparisons, ME makes only n-1. The extra comparisons in AHP are redundant and used to check for inconsistency. It is possible that this redundancy forces doubt into the minds of users, causing them to become less accurate. Alternatively, the requirement to perform all comparisons is a time-consuming and possibly tedious exercise which may cause users to lose their concentration. If so, later comparisons will be less accurate, and users will be less willing to try to improve their weights.

2) A second difference was the prior ranking of colours. In the version of ME used in the experiment, subjects first ranked the items before making the paired comparisons. This ranking allowed them to get a preliminary idea of each colour's ordinal importance, and this knowledge may have helped them to improve their ratio comparisons. In AHP, no such preliminary structuring took place.

3) A third possibility is that AHP requires a much longer training period than was provided in this experiment. Other work by the authors does suggest that inconsistency values improve with familiarity with the program.

Finally, the ME scale may really be superior to the AHP nine point scale in generating ratios.

References

Anderson, N.H., "Algebraic Models in Perception," in Handbook of Perception, Psychophysical Judgement and Measurement, E.C. Carterette and M.P. Friedman, (eds.), Academic Press, 2 (1974), 215-291.

Expert Choice, Decision Support Systems, McLean, Va, 1983.

Saaty, Thomas L., "A Scaling Method for Priorities in Hierarchical Structures", Journal of Mathematical Psychology, 15, No. 3 (1977), 234-81.

____________, The Analytic Hierarchy Process, New York, McGraw Hill, 1980.

____________, Decision Making for Leaders, Belmont, California: Wadsworth Inc., 1982.

Schoemaker, Paul J.H. and Carter C. Waid, "An Experimental Comparison of Different Approaches to Determining Weights in Additive Utility Models," Management Science, 28, No. 2 (1982), 182-96.

Stevens, S.S., "The Direct Estimation of Sensory Magnitudes -- Loudness," American Journal of Psychology, 69 (1956), 1-25.

____________, "On the Psychological Law," Psychological Review, 64 (1957), 153-81.

____________, Psychophysics, New York: John Wiley, 1975.

DESCRIBING AND FORMALIZING THE EVALUATION PROCESS OF PORTUGUESE NAVY OFFICERS

CARLOS A. BANA E COSTA
Center of Urban and Regional Systems
of the Technical University of Lisbon
IST/CESUR, Av. Rovisco Pais
1000 Lisboa - Portugal

CESAR F. DINIS DAS NEVES
Center for Naval Operations
of the Portuguese Navy
Min. Marinha, Pr. Município
1000 Lisboa - Portugal

ABSTRACT

During his career each officer of the Portuguese Navy is evaluated on average twice each year by his superior commander, always according to the same set of military attributes. This information will be taken into account when considering officers for promotion.

In this paper we try to formalize a methodology for aggregating the informations for each officer and for ranking officers according to their performances. It has the peculiarlity of combining MAUT and Outranking MCDA approaches and it is suitable for dealing with situations of incomplete inter-attribute preference information.

In section 1 we describe the problem. Section 2 deals with the analysis of the characteristics of the evaluation process, the identification of the main constraints for preference modeling and aggregation. The synthesis of the existing information is the subject of Section 3, where we apply the so-called "Outweigh Analysis" combined with the "distillation procedure" of the ELECTRE III method.

1. DESCRIBING THE PROBLEM

Let $A = \{a_1,...,a_i,...,a_n\}$ be a set of n candidates - officers with the same naval rank r, in competition for promotion to the above naval rank (r+1) - for example, a set of Lieutenant-Commanders considered to be candidates for promotion to Commander.

Let $X = \{X_1,...,X_j,...,X_m\}$ be a set of m military attributes. The m = 23 attributes are arranged in a hierarchical structure under 7 intermediate sub-sets (Y) of attributes, these falling under 3 major sub-sets (Z), as figure 1 shows.

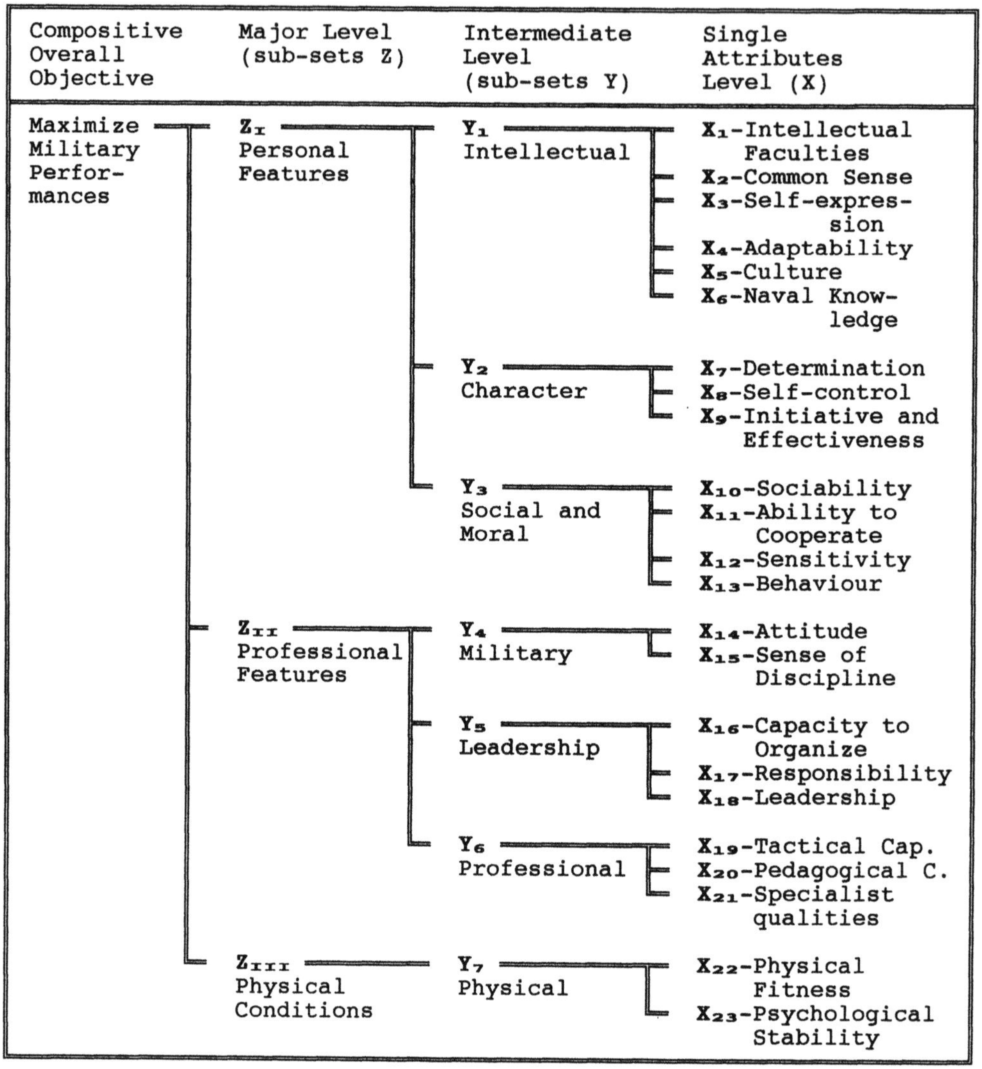

FIGURE 1 - Hierarchy of the Military Attributes

The overall objective of our decision-aid problem is to help the Navy to rank candidates for promotion, in decreasing order of preference under a family of military attributes. So, the output of the decision model must be a ranking procedure on A, to be systematically used everytime a set of officers is to be compared for promotion.

2. FORMALIZING THE PROBLEM

2.1. The Navy Evaluation System of Officers

Each candidate a_i has been evaluated, over time and since his last promotion, by a set of judges $J^k{}_i$ $(k=1,\ldots,p_i)$ - each judge $J^k{}_i$ being the superior commander of a_i in the sequence of the military units where a_i has been working. For this evaluation the Navy Law establishes a five point (ordinal) measurement scale E, common to the 23 attributes, to be used by each judge $(J^k{}_i)$ in establishing the level (e_{ij}) that officer a_i deserves - during the period of observation (usually, six months) - in each attribute X_j.

For the definition of the partial value of an officer in each attribute, the Navy follows a process which briefly consists of the following steps (we will not discuss its implicit theorectical assumptions):

Step 1: Using a specific procedure π, the levels assigned to an officer by each judge are corrected in order to take into account the personal interpretation each judge makes of the ordinal scale. π is applied within the set (J) of all the judges that have evaluated the candidates of a set A $(J = \{J^k{}_i: k = 1,\ldots,p_i \text{ and } i = 1,\ldots,m\})$. So:

$$e^k{}_{ij} \xrightarrow[\pi]{} c^k{}_{ij}$$

being:

$e^k{}_{ij}$ - the level of attribute X_j assigned by judge $J^k{}_i$ to a_i

$c^k{}_{ij}$ - the corrected value of $e^k{}_{ij}$ by procedure π.

Theoretically, π defines a mapping of E on R which provides a interval scale of measurement (see Roberts [1979, chapter 2]).

Step 2: Implicitly assuming a linear (Navy) preference function in each attribute X_j $(v(c_{ij}) = c_{ij})$, the partial value (v_{ij}) of each candidate a_i is computed as the arithmetic mean of the c_{ij} values:

$$v_{ij} = \sum_{k=1}^{p_i} c^k{}_{ij} \;/\; p_i \qquad (p_i \text{ - number of judges of } a_i).$$

The v_{ij} values define a pre-order on A for the attribute X_j, so that:

$$a_i \succsim_j a_{i'} \text{ if and only if } v_{ij} \geq v_{i'j}$$

where $\succsim$ denotes a preference relation on A^2).

The methodology developed in this paper will be illustrated in comparing 16 candidates for promotion. Their partial values for the 23 military attributes, computed by the above described Navy process, are shown in the decision matrix on Table 1, where we add two columns with the lowest and highest values on each attribute.

DECISION MATRIX																	RANGE	
F \ A	a_1	a_2	a_3	a_4	a_5	a_6	a_7	a_8	a_9	a_{10}	a_{11}	a_{12}	a_{13}	a_{14}	a_{15}	a_{16}	WORST VALUE	HIGHEST VALUE
X1	3.63	4.49	3.56	3.89	3.27	3.35	3.84	3.86	3.44	4.01	3.24	3.28	3.88	3.76	3.41	3.64	3.24	4.49
X2	3.82	2.71	4.46	4.05	3.78	3.73	4.23	4.34	4.10	3.15	3.60	3.56	3.66	3.40	4.21	3.90	2.71	4.46
X3	4.01	4.20	3.77	4.26	3.40	3.62	3.71	3.60	3.20	3.60	3.64	3.40	3.83	3.23	3.68	3.57	3.20	4.26
X4	3.97	3.46	4.24	4.22	4.40	4.60	4.37	3.96	3.92	3.87	4.09	3.44	4.30	3.82	3.98	4.02	3.44	4.60
X5	3.88	3.99	3.72	4.10	3.95	3.29	3.71	3.55	3.28	3.87	5.23	3.68	3.68	3.22	3.43	3.53	3.22	5.23
X6	3.76	3.78	4.62	4.27	3.71	3.75	4.02	4.10	3.75	3.96	4.12	3.49	3.89	3.50	3.97	3.87	3.49	4.62
X7	3.35	3.58	3.75	3.88	3.45	4.00	3.91	3.75	4.01	3.73	3.55	3.01	4.25	4.08	3.76	3.84	3.01	4.25
X8	4.11	4.65	4.10	3.73	3.74	3.27	3.97	3.96	3.90	3.66	3.58	4.16	3.77	3.70	3.82	3.73	3.27	4.65
X9	3.50	3.40	3.64	3.99	3.15	3.90	4.20	3.73	3.68	3.74	3.32	3.00	4.15	3.99	3.70	3.83	3.00	4.20
X10	3.61	3.88	3.77	4.69	3.90	3.37	4.31	3.56	3.86	4.29	3.20	3.66	3.66	3.44	4.34	3.68	3.20	4.69
X11	4.16	3.65	4.27	4.42	3.62	4.53	4.34	4.34	4.49	3.65	3.68	3.72	3.80	4.04	4.22	4.09	3.62	4.53
X12	3.79	3.65	3.82	4.04	3.61	3.79	4.15	3.90	4.05	3.77	3.42	3.60	3.76	3.67	3.77	3.91	3.42	4.15
X13	3.59	3.71	3.77	4.05	3.91	3.72	4.40	4.12	4.44	3.56	3.85	3.35	4.03	3.77	3.94	4.14	3.35	4.49
X14	3.87	3.67	4.07	4.18	3.75	3.48	3.12	4.25	4.21	3.79	3.60	3.47	4.35	3.67	3.92	3.88	3.12	4.35
X15	3.81	3.66	4.17	4.16	3.56	3.90	4.09	3.90	4.45	3.62	3.46	3.74	4.41	3.92	3.87	3.88	3.46	4.45
X16	3.68	3.45	3.45	3.90	3.73	4.07	3.56	3.82	3.13	3.43	3.59	3.29	3.97	3.82	3.40	3.94	3.13	4.07
X17	3.68	3.12	4.14	4.06	3.71	4.11	4.15	4.41	4.32	3.92	3.54	3.33	4.30	4.05	3.98	4.04	3.12	4.41
X18	3.57	3.13	3.92	3.96	3.26	3.24	4.80	3.49	3.81	3.35	3.38	3.28	3.81	3.46	3.96	3.72	3.13	4.80
X19	*	*	3.78	4.28	3.66	*	*	3.93	*	4.79	*	*	3.59	*	3.86	*	3.59	4.79
X20	3.91	4.18	3.38	4.48	3.89	2.93	4.17	3.59	*	3.28	3.32	3.87	3.65	3.97	4.11	3.12	2.93	4.48
X21	3.79	3.83	3.62	4.15	3.89	3.02	4.16	3.85	3.44	3.94	2.64	3.25	4.53	3.92	3.98	3.85	2.64	4.53
X22	3.78	3.74	4.38	4.09	3.72	3.91	3.95	3.93	3.84	3.35	3.27	3.33	3.36	4.20	4.02	3.57	3.27	4.38
X23	3.94	4.40	3.95	3.89	4.25	3.74	4.33	3.95	4.15	3.70	3.95	3.95	4.07	3.96	4.28	4.09	3.70	4.40

TABLE 1

2.2. The Overall Value of the Candidates

The overall value (V_i) of each candidate a_i, for the set X of attributes, is traditionally accepted in the Portuguese Navy to be calculated by an average sum aggregation of his partial values v_{ij}:

$$V_i = \sum_{j=1}^{23} w_j . v_{ij} \qquad (\text{with } w_j > 0 \text{ and } \Sigma_j w_j = 1) \tag{1}$$

where w_j is a scaling constant for the X_j attribute.

Consequently, any sub-set T of attributes of X ($T \subset X$) is assumed to be preferencially independent in X.

As pointed out by many authors (see, for instance, Chankong and Haimes [1983, pp. 80]), the validation of this assumption is not easy for a large number of real-life decision theory applications. This seems to be reinforced by the great lack of (at least, published) MCDM applications of additive value (or utility) functions giving satisfactory descriptions of the process followed for validating the mutually preference independence condition.

In the context of our case, these difficulties increase due to the fact that a well-defined Decision Maker does not exists. Nevertheless, as the additive model is well accepted in the Navy Staff, we will not question it in this paper. Constant value trade-offs between each pair of attributes are also acceptable (observe, in the two last columns of table 1, the small range of variation of the partial values in each attribute).

2.3. Non-Realism of Assigning Precise Numerical Values to the Scaling Constants

Unfortunatly, it is not possible to define a unique vector $\underline{w}$, $\underline{w} = (w_1,\ldots,w_j,\ldots,w_n)$, of precise numerical values for the scaling constants, acceptable beyond controversy to "all the Navy". Thus, no overall values of the candidates can be calculated from expression (1) and so it is impossible to compare the candidates on the basis of V_i numerical values.

But some (ordinal) relationships between scaling constants of sub-sets of attributes can be assessed, as section 2.4. describes.

In the following we will use the more generalized designation "weight" instead of "scaling constant" - we assume as well-known that the scaling constants, in the above described context, do not directly "characterise the intrinsic relative importance of the attributes" (Vansnick [1986]) (see, also, Belton [1986] and Vansnick [1984] for a discussion of the different meanings of the concept of "weight" in various MCDM approaches).

2.4. Modeling Inter-Attribute Preferences - The Available Information on the Weights

A top-down procedure has been adopted for the assessement of relationships between attribute weights. First, we consider the "Major Level" of the hierarchy of attributes (Figure 1). The three mutually exclusive sub-sets of attributes in this hierarchy level are:

$$Z_I = \{X_1,\ldots,X_{13}\}; \quad Z_{II} = \{X_{14},\ldots,X_{21}\}; \quad Z_{III} = \{X_{22},X_{23}\}$$

We asked a few officers of the Navy Staff to compare pairs of these sub-sets, by puting them questions of the type (Keeney and Raiffa [1976, pp. 121-122]):

Suppose a hypothetical officer whose evaluations on the 23 military attributes are all the worst possible evaluation level, and you have the option of improving some of the evaluations from the worst to the best evaluation level. Would you rather improve the levels of the attributes in the subset Z_j or in the sub-set $Z_{j'}$?

Answers to these paired comparisons revealed a consensus on:

$$w_I > w_{II} > w_{III}$$

being w_I, w_{II} and w_{III} the weight functions of Z_I, Z_{II} and Z_{III}, respectively.

Coming down the hierarchy, we have concluded that the conditional weights of the sub-sets of attributes under the same node of the upper hierarchical level are identical:

$$w(Y/Z) = w(Y'/Z) \qquad \left(\text{with } \sum_{Y \subset Z} w(Y/Z) = 1\right) \tag{2}$$

for all intermediate sub-sets Y, Y′ under the same major sub-set Z.

Similarly:

$$w(X_j/Y) = w(X_{j'}/Y) \qquad \left(\text{with } \sum_{X_j \in Y} w(X_j/Y) = 1\right) \tag{3}$$

for all attributes X_j, $X_{j'}$ under the same intermediate sub-set Y.

Furthermore, theorectically we have:

$$w_j = w(X_j/Y).w(Y/Z).w(Z) \quad (4)$$

Thus, based on expressions (2), (3) and (4) we get the following expressions of each weight w_j, as a function of one of the weights (w_I, w_{II} or w_{III}) on the major hierarchical level:

$$\begin{cases} w_1 \text{ to } w_6=1/6.1/3.w_I=w_I/18;\ w_7 \text{ to } w_9=w_I/9;\ w_{10} \text{ to } w_{13}=w_I/12 \\ w_{14}=w_{15}=1/2.1/3.w_{II}=w_{II}/6;\ w_{16} \text{ to } w_{18}=w_{II}/9;\ w_{19} \text{ to } w_{21}=w_{II}/9 \\ w_{22}=w_{23}=1/2.1.w_{III}=w_{III}/2 \end{cases} \quad (5)$$

with $w_I > w_{II} > w_{III}$.

2.5. Reducing the Dimension of the Problem

Under the conclusions of section 2.4., the overall value of each candidate (V_i) will be an additive function of the (un-known) weights w_I, w_{II}, w_{III} of the sub-sets of attributes Z_I, Z_{II}, Z_{III}:

$$V_i = \sum_{j=1}^{23} w_j.v_{ij} = f(w_I,w_{II},w_{III}) = \sum_{j=I}^{III} w_j.z_{ij} \quad (6)$$

where the values z_{ij} ($i = 1,\dots,n$ and j = I to III) are obtained by directly applying expressions (5) in expression (1) and using the data of the decision matrix of Table 1. Their values are shown in the (reduced) decision matrix of Table 2, where each column is the profile (z_I, z_{II}, z_{III}) of each candidate a_i.

Consequently, we have reduced the dimension of our problem from $\mathbb{R}^{23}$ to $\mathbb{R}^3$. This allows the graphical overview of the "Set of Feasible Weights" (SFW) defined on the space of the weights by the set of constraints:

$$w_I + w_{II} + w_{III} = 1 \text{ and } w_I, w_{II}, w_{III} > 0 \text{ and } w_I > w_{II} > w_{III} \quad (7)$$

The SFW is the triangle (a) in Figure 2, with extreme points (0,0,1), (0,1/2,1/2) and (1/3,1/3,1/3).

Due to the incomplete information on inter-attribute preferences, the weight functions w_I, w_{II} and w_{III} must be assumed as uniformly distributed in SFW.

REDUCED DECISION MATRIX				RANGE	
A \ Z	z_I	z_{II}	z_{III}	WORST VALUE	HIGHEST VALUE
a_1	3.76	3.74	3.85	3.74	3.85
a_2	3.79	3.64	4.09	3.64	4.09
a_3	3.93	3.85	4.17	3.85	4.17
a_4	4.10	4.15	3.99	3.99	4.15
a_5	3.63	3.69	4.01	3.63	4.01
a_6	3.77	3.49	3.82	3.49	3.82
a_7	4.11	3.94	4.24	3.94	4.24
a_8	3.90	3.94	3.94	3.90	3.94
a_9	3.90	3.85	4.01	3.85	4.01
a_{10}	3.75	3.67	3.59	3.59	3.75
a_{11}	3.53	3.36	3.68	3.36	3.68
a_{12}	3.48	3.47	3.64	3.47	3.64
a_{13}	3.92	4.06	3.93	3.92	4.06
a_{14}	3.71	3.85	4.06	3.71	4.06
a_{15}	3.88	3.88	4.15	3.88	4.15
a_{16}	3.85	3.84	3.88	3.84	3.88

TABLE 2

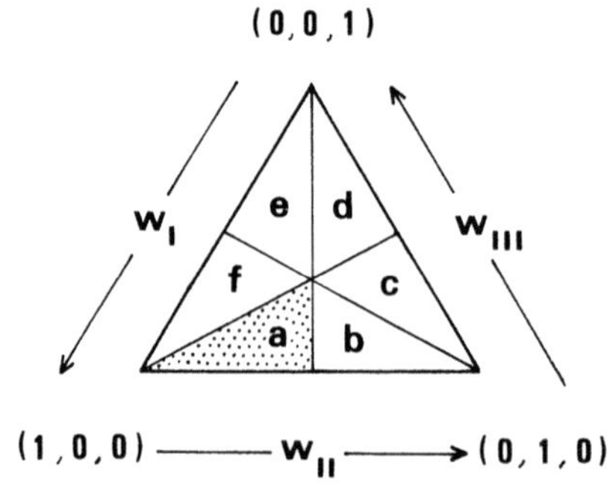

FIGURE 2

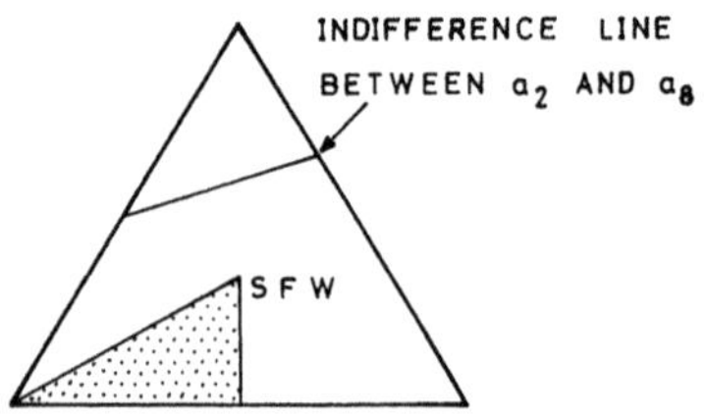

FIGURE 3

3. SEARCHING FOR A PROCEDURE TO DEFINE A PRE-ORDER IN THE SET OF CANDIDATES FOR PROMOTION

According to the conclusions of sections 1 and 2, the procedure for defining a pre-order on A, taking into account the information contained on Table 2, must be, on the one hand, based on the simple weighted value function aggregation model, and, on the other hand, be suitable for dealing with ordinal information on the weights.

To make the synthesis of the relative merits of the candidates, we will follow a pairwise comparison approach, the Outweigh Analysis, introduced in Bana e Costa [1988, section 3.2]. For a given SFW, it allows the construction of a fuzzy outranking relation (see Roy [1977] for theorectical definition) based on the relative dimensions of the sub-sets defined in SFW by its intersection with the indifference plan of each pair of candidates.

3.1. Dominance Analysis in SFW

For some pairs $(a_i, a_{i'})$, a situation of "dominance" in SFW can hold even if general dominance does not. In fact, the intersection of an indifference plan can be empty with SFW and at the same time not empty with the space of the weights, as Figure 3 shows.

For each vector of weights respecting the constraints (7), i.e., in each point P of SFW, the global preference situation between each pair of candidates, under the preference aggregation conditions of section 2, is given by the following theorem:

THEOREM - Let SFW be a set of feasible weights. For each point P ($\underline{w}^P{}_j$) in SFW, only two preference situations can arise for each pair $(a_i, a_{i'})$ of alternatives:

$$a_i \succ^P a_{i'} \quad \text{iff} \quad \Sigma_j\, w^P{}_j.(v_{ij} - v_{i'j}) > 0$$

or

$$a_i \sim^P a_{i'} \quad \text{iff} \quad \Sigma_j\, w^P{}_j.(v_{ij} - v_{i'j}) = 0$$

COROLLARY 1 - If $a_i \succ^P a_{i'}$ for all P then a_i dominates $a_{i'}$ in SFW.

Table 3 shows the dominance matrix in SFW, to which corresponds the (partial) dominance order of figure 4. Although "poor", this is the "most sure" information about global comparative merits of the candidates. Besides, any procedure for defining a richer order on A has to respect the dominance order.

	a_1	a_2	a_3	a_4	a_5	a_6	a_7	a_8	a_9	a_{10}	a_{11}	a_{12}	a_{13}	a_{14}	a_{15}	a_{16}
a_1	1	0	0	0	1	0	0	0	0	1	1	1	0	0	0	0
a_2	0	1	0	0	1	1	0	0	0	1	1	1	0	0	0	0
a_3	1	1	1	0	1	1	0	0	1	1	1	1	0	1	1	1
a_4	1	1	1	1	1	1	0	1	1	1	1	1	1	1	1	1
a_5	0	0	0	0	1	0	0	0	0	0	1	1	0	0	0	0
a_6	0	0	0	0	0	1	0	0	0	0	1	1	0	0	0	0
a_7	1	1	1	0	1	1	1	1	1	1	1	1	1	1	1	1
a_8	1	1	0	0	1	1	0	1	1	1	1	1	0	1	0	1
a_9	1	1	0	0	1	1	0	0	1	1	1	1	0	1	0	1
a_{10}	0	0	0	0	0	0	0	0	0	1	1	1	0	0	0	0
a_{11}	0	0	0	0	0	0	0	0	0	0	1	0	0	0	0	0
a_{12}	0	0	0	0	0	0	0	0	0	0	0	1	0	0	0	0
a_{13}	1	1	0	0	1	1	0	1	1	1	1	1	1	1	1	1
a_{14}	0	0	0	0	1	0	0	0	0	0	1	1	0	1	0	0
a_{15}	1	1	0	0	1	1	0	0	0	1	1	1	0	1	1	1
a_{16}	1	1	0	0	1	1	0	0	0	1	1	1	0	0	0	1

1 - a_i dominates $a_{i'}$ in SFW

TABLE 3 - Dominance Matrix

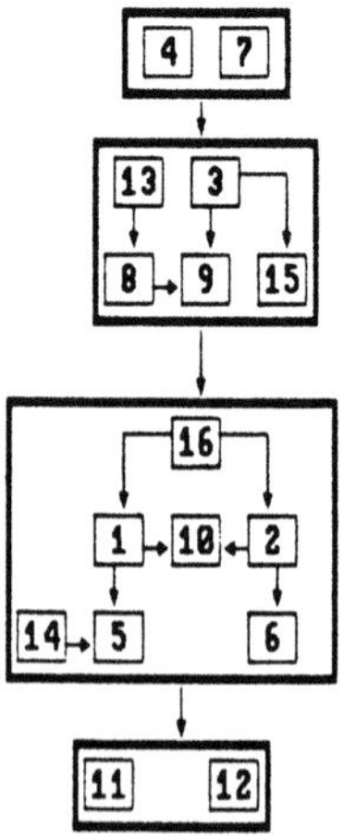

FIGURE 4 - Dominance Graph

3.2. Degrees of Credibility and Fuzzy Outranking Relation

In a situation of non-dominance in SFW between two alternatives a_i and $a_{i'}$, SFW is divided into two sub-sets SFW(i,i') and SFW(i,i') such that, according to the theorem:

COROLLARY 2 - Being P a point of SFW, two preference situations can hold:

$$a_i \succsim^P a_{i'} \quad \text{iff } P \in SFW(i,i')$$

or

$$a_{i'} \succsim^P a_i \quad \text{iff } P \in SFW(i',i)$$

Thus, the weights being uniformly distributed in SFW, the probability that P ∈ SFW(i,i') can be given by the relative size of the sub-set SFW(i,i') in SFW. We see this probability as the degree of credibility ($d_{ii'}$) to be assigned to the statement "a_i is at least as good as $a_{i'}$", in SFW, and one has:

$$\begin{cases} 0 \le d_{ii'} \le 1 \\ d_{ii'} = 1 - d_{i'i} \\ d_{ii'} = 1 \text{ (and } d_{i'i} = 0\text{) iff } a_i \text{ dominates } a_{i'} \text{ in SFW.} \end{cases} \qquad (8)$$

The values of $d_{ii'}$ define a fuzzy outranking relation (S^d) in A^2, which summarizes the results of the pairwise comparisons between the candidates (Roy et al. [1986, pp.327]).

Several ways can be chosen for calculating the credibility degrees. Both Hinloopen et al. [1983] and Bana e Costa [1988] propose to associate them with the relative size of the respective sub-sets of feasible weights. In our three attribute case, we will calculate $d_{ii'}$ as the area of SFW(i,i') divided by the total area of SFW. This process respects (8), takes into account the uniform distribution of the weights in SFW and, furthermore, it has the advantage that the fuzzy outranking relation thus defined is transitive, as proved by Bana e Costa and Vincke [1988].

Table 4 shows the numerical values of the credibility degrees for our Navy case.

	MATRIX OF CREDIBILITY DEGREES															
	a_1	a_2	a_3	a_4	a_5	a_6	a_7	a_8	a_9	a_{10}	a_{11}	a_{12}	a_{13}	a_{14}	a_{15}	a_{16}
a_1	1	.232	0	0	1	.922	0	0	0	1	1	1	0	.214	0	0
a_2	.768	1	0	0	1	1	0	0	0	1	1	1	0	.369	0	0
a_3	1	1	1	0	1	1	0	.840	1	1	1	1	.757	1	1	1
a_4	1	1	1	1	1	1	.834	1	1	1	1	1	1	1	1	1
a_5	0	0	0	0	1	.484	0	0	0	.292	1	1	0	0	0	0
a_6	.008	0	0	0	.516	1	0	0	0	.333	1	1	0	.066	0	0
a_7	1	1	1	.166	1	1	1	1	1	1	1	1	1	1	1	1
a_8	1	1	.160	0	1	1	0	1	1	1	1	1	0	1	.630	1
a_9	1	1	0	0	1	1	0	0	1	1	1	1	0	1	.138	1
a_{10}	0	0	0	0	.708	.667	0	0	0	1	1	1	0	.062	0	0
a_{11}	0	0	0	0	0	0	0	0	0	0	1	.599	0	0	0	0
a_{12}	0	0	0	0	0	0	0	0	0	0	.401	1	0	0	0	0
a_{13}	1	1	.243	0	1	1	0	1	1	1	1	1	1	1	.999	1
a_{14}	.786	.631	0	0	1	.934	0	0	0	.938	1	1	0	1	0	.023
a_{15}	1	1	0	0	1	1	0	.370	.862	1	1	1	.001	1	1	1
a_{16}	1	1	0	0	1	1	0	0	0	1	1	1	0	.977	0	1

TABLE 4 - Matrix of Credibility Degrees

3.3. Exploring the Fuzzy Relation for Ranking the Candidates

The information contained in the matrix of the credibility degrees must now be synthetised for a final ranking of the candidates. We will adopt the "distillation procedure" proposed by Roy [1978] (see also Ostanello [1985] and Roy et al. [1986]).

When $0 < d_{ii'} < 1$ (i.e., no dominance exists between a_i and $a_{i'}$) the remaining candidates are called upon to play a role in defining the final preference situation between a_i and $a_{i'}$. On this basis, the fuzzy outranking relation is explored by a sequential process. In each step a non-fuzzy (trivial) outranking relation is defined as follows:

For a given "separation level" δ, a_i outranks $a_{i'}$ iff:

$$d_{ii'} > \delta \quad \text{and} \quad d_{ii'} - d_{i'i} > s(d_{ii'}) \tag{9}$$

where the "discrimination threshold" $s(d_{ii'})$ has the linear form:

$$s(d_{ii'}) = a + b.d_{ii'}$$

Values for the parameters a and b must be fixed by the analyst, having been recommended that $a/b < 0$ (Skalka e al. [1983]). We will do $a = -2b$.

As, from (8), $d_{ii'} + d_{i'i} = 1$, expression (9) becomes:

$$d_{ii'} > 1/2 + 1/2.s(d_{ii'}) \qquad (10)$$

Futhermore, for our case it has been considered it too risky to accept values of $d_{ii'} < 2/3$. In this situation, expression (10) gives rise to $a = 1/2$ and $b = -1/4$.

We are now in position of applying the distillation procedure to the data of table 4, which results in the final ranking of the candidates of figure 5. A sensitivity analysis has shown this ranking significantly stable when other values are used for the parameters a and b (provided they respect the condition $d_{ii'} > 2/3$).

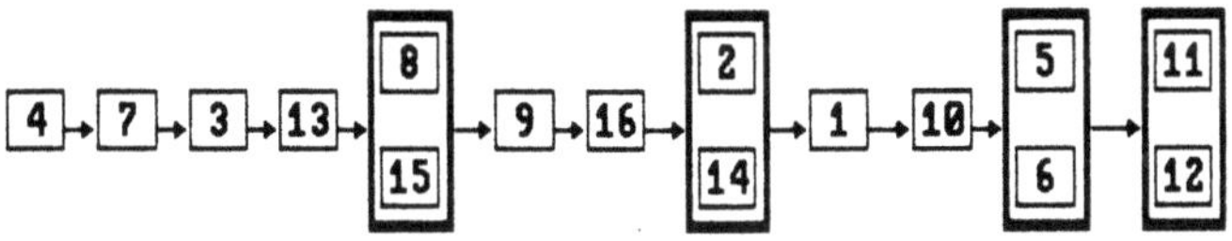

FIGURE 5

4. FINAL COMMENTS AND FUTURE RESEARCH

In this paper we propose a methodology to rank Portuguese Navy officers for promotion, using the so-called "Outweigh Analysis" and the "distillation procedure" of the ELECTRE III method. The acceptance of the methodology by a random sample of Navy Staff officers has been far better than one would antecipate for a new approach to a very sensitive problem.

The original data involved is the set of evaluations given to each officer by his superior commanders, since his last promotion. This evaluation process has a great degree of subjectivity, as shown by Neves et al. [1987], which we do not analyse here.

The methodology seems to be easily applicable to other similar cases, provided that particular attention be payed to the type of data and the links between the "outweigh" calculation of the credibility degrees and the values of the "discrimination threshold" parameters used in the "distillation". The absence of a study of the theorectical consistence of those links is a weak point of the paper, which we intend to surpass in future research. Futhermore, other improvements will be considered, such as the analysis of the stability of all possible rankings within the set of feasible weights (see, Hinloopen et al. [1983], Tavares [1986] and Bana e Costa [1988]) and the study of other approaches to the problem, directly taking into account the stochastic nature of the data (see, in the outranking context, Mareschal [1986]).

REFERENCES

Bana e Costa, C.A., "A Methodology for Sensitivity Analysis in Three-Criteria Problems: A Case Study in Municipal Management", European Journal of Operational Research, 33(2), 1988, pp. 159-173.

Bana e Costa, C.A. and Vincke, Ph., "Quelques Questions sur la Prise en Compte de l'Imprécision des Coefficients d'Importance des Critères", communication presented at the 27th Meeting of the European Working Group on Multiple Criteria Decision Aid, Mons, March, 1988 (not published).

Belton, V., "A Comparison of the Analytic Hierarchy Process and a Simple Multi-Attribute Value Function", European Journal of Operational Research, 26(1), 1986, pp. 7-21.

Chankong, V. and Haimes, Y.Y., Multiobjective Decision Making: Theory and Methodology, North-Holland, New York, 1983.

Hinloopen, E., Nijkamp, P. and Rietveld, P., "Qualitative Discrete Multiple Criteria Choice Models in Regional Planning", Regional Science and Urban Economics, 13, 1983, PP. 77-102.

Keeney, R. and Raiffa, H., Decisions with Multiple Objectives: Preferences and Value Tradeoffs, Wiley, New York, 1976.

Mareschal, B., "Stochastic Multicriteria Decision Making and Uncertainty", European Journal of Operational Research, 26 (1), 1986, pp. 58-64.

Neves, C., Nogueira, J., Santos, J. and Pereira, G., "Aloca ção de Pessoal", 1987 (to appear in Investigação Operacional).

Ostanello, A., "Outranking Methods", in Multiple Criteria Decision Methods and Applications, Eds. Fandel, G., Spronk, J. and Matarazzo, B., Springer, New York, 1985, pp. 41-60.

Roberts, F.S., Measurement Theory with Applications to Decision Making Utility and the Social Sciences, Addison-Wesley, London, 1979.

Roy, B., "Partial Preference Analysis and Decision Aid: The Fuzzy Outranking Relation Concept", in Conflicting Objectives in Decision, Eds. Bell, D., Keeney, R. and Raiffa, H., Wiley, New York, 1977, pp. 40-74.

Roy, B., "ElECTRE III: Un Algorithme de Classement Fondé sur une Représentation Floue des Préferences en Présence de Critères Multiples", Cahiers du Centre d'Etudes de Recherche Opérationnelle, 20 (1), 1978, pp. 3-24.

Roy, B., Méthodologie Multicritère d'Aide à la Décision, Economica, Paris, 1985.

Roy, B., Présent, M. and Silhol, D., A Programming Method for Determining which Paris Metro Statios should be Renovated", European Journal of Operational Research, 24 (2), 1986, pp. 318-334.

Skalka, J.M., Bouyssou, D. and Bernabeu, Y.A., "ELECTRE III et IV: Aspects Méthodologiques et Guide d'Utilisation", Document du LAMSADE nº 25, Université de Paris-Dauphine, Paris, 1984.

Tavares, L.V., "Multicriteria Scheduling of a Railway Renewal Program (Case Study), European Journal of Operational Research, 25 (3), 1986, pp. 395-405.

Vansnick, J.C., "Strengh of Preference. Theoretical and Practical Aspects", in Operational Research 84, Ed. Brans, J.P., North-Holland, Amsterdam, 1984, pp. 449-463.

Vansnick, J.C., "On the Problem of Weights in Multiple Criteria Decision Making (the Compensatory Approach)", European Journal of Operational Research, 24 (2), 1986, pp. 288-463.

INCONSISTENT LINEAR CONSTRAINTS AND OBJECTIVES IN MOLP PROBLEMS

David I. Soloveitchik
Policy and Planning Division
Ministry of Energy and Infrastructure
Jerusalem, Israel.

Abstract

In this paper we will develop a procedure for defining and exploring the set of right-hand side (RHS) vectors of linear programming (LP) and multiobjective linear programming (MOLP) problems with inconsistent RHS-vectors.

Introduction

It is well known that for the LP problem:

$$\max \{ Z = c \cdot x : Ax \leq b, \ x \geq 0 \} \qquad (1.1)$$

the optimal solution $\bar{x}(b)$ defines the level of the objectives $\bar{z} = c \cdot \bar{x}(b)$ which can be attained for a given RHS-vector b .But in some cases this level $\bar{z}$ may not satisfy the DM because the RHS-vector might be inconsistent or subject to error and, as a rule, has not been analyzed sufficiently.

One such problem is: how we must change the RHS-vector so as to improve the level of the objectives? There are many cases of decision making in practice when the initial data of model are contradictory. Similarly, in goal programming and MOLP, the aspiration levels of the constraints and/or the objectives themselves may be mutually contradictory or inconsistent.

As well as discussing inconsistent constraints or objectives in models of decision making, we will also consider inconsistent models. The exploration of such models can give the decision-maker (DM) the possibility of a more detailed consideration of underlying process itself. Some models and methods have been proposed to solve the problem of inconsistent linear inequalities: models for the correction of b vectors [1] , methods of finding all minimal and maximal consistent sub-systems of linear inequalities [2] , linear parametric programming models (LPP) [3] , flexible planning problem (FP) [4] , fuzzy decision models [5] , [14] , De Novo Programming [6] , the method of comities [7] , etc.

After solving such problems only one solution is obtained and it is assumed that this solution is acceptable to the DM.

In this paper we consider another approach to the problem. It is based on the description of the set of "feasible" RHS- vectors D_f:

$$D_f = \{f\colon \exists x \geq 0, \quad Ax = f, \; \tilde{f} \leq f \leq \hat{f}\} \tag{1.2}$$

The set of vectors f includes both the constraint vectors and the vector of objective functions $c_1 \cdot x, c_2 \cdot x, \ldots c_l \cdot x$.

It is assumed that in practice the DM starts his problem with RHS- vector $\bar{f} \notin D_f$, i.e. an infeasible formulation. We will consider the problem of finding f - "feasible vectors", which lie as closed to $D_{\bar{f}}$ as possible using a real-valued function $g(f-\bar{f})$ as the measure of the proximity between points $f \in D_f$ and $\bar{f}$ in E_m e.g. the Euclidean or any other norm.

The problem of finding the best alternative to $\bar{f}$ can be defined as:

$$\min\{g(f-\bar{f})\colon f \in D_f\} \tag{1.3}$$

The solution is the "feasible" RHS- vector f nearest to the vector $\bar{f}$. The main difficulty of solving problem (1.3) is finding an efficient description of the set D_f. It follows from definition (1.2) that D_f is polyhedron in E_m. For most practical problems the DM would not be able to use an analytical description, even if simplified, e.g. an algorithm which decreases the number of inequalities [4].

We will present an effective way to describe a set of feasible vectors f. To do so we will randomly generate the vectors from the interval $[\tilde{f}, \bar{f}]$. For each such vector we can solve the LP problem to check whether it is feasible or not. In such a way we can obtain a set of feasible vectors. In order to save computation time we propose special procedure to divide the whole set of vectors into feasible and infeasible ones.

Recall that D_f implicitly includes all the alternatives of the DM's RHS- vectors for the initial constraints. A very effective description of this set can be achieved by using decision rules

$$r_p = f \cdot y_p^* \quad (p = 1, \ldots, P)$$

where y_p^* is determined using Pattern Recognition or in another way.

On the one hand, this description sufficiently reduces the amount of redundant information. On the other hand, it is given in a form which is convenient for the decision-making process.

Three kinds of vectors are examined: "feasible" f-vectors, "infeasible" $\bar{f}$-vectors and "suspicious" s-vectors. If we do not know whether the vector being considered is an f or $\bar{f}$ -vector, it is denoted as an s-vector and these are the vectors which are used as an input to the iteration step of our algorithm.

To describe the set of f-vectors with the help of decision rules it is necessary to develop the procedure by which the initial data sets of the $f, \bar{f}$ and s -vectors are defined. Such a procedure significantly reduces the computing time required.

The first step of this method is the statistical simulation of possible constraints. It is suggested that for this purpose the DM should give intervals of permissible ranges for some of the constraints.

In the second step we define the set of random computer-generated vectors:

$$D_s = \{f_s: \tilde{f}_s \leq f_s \leq \hat{f}_s, \ s = 1, \dots, R\} \tag{1.4}$$

D_s consists of f or $\bar{f}$ -vectors ($D_s = D_f \cup D_{\bar{f}}$). For each random computer-generated vector we test whether it belongs to D_f or $D_{\bar{f}}$.

In this paper we will present a time-saving computer procedure of separating f and $\bar{f}$ -vector in D_s, which is made more efficient by using the matrix of reduced costs in the simplex method with special rule for introducing vectors into the basis.

Methodology

It is well known that one of the following alternatives is true: if inequality

$$Ax \leq f \tag{2.1}$$

has a nonnegative solution, then there are nonnegative solutions for the following system of inequalities

$$yA \geq 0, \ \ y \geq 0 \tag{2.2}$$

$$f \cdot y < 0 \tag{2.3}$$

The corresponding LP problem can be formulated:

$$\min f \cdot y < 0 \tag{2.4}$$

$$\text{s.t. } yA \geq 0, \quad y \geq 0 \tag{2.5}$$

$$y \cdot e = 1 \tag{2.6}$$

$$\text{or } yA' \geq l, \quad y \geq 0 \tag{2.7}$$

$$\text{where } A' = \begin{pmatrix} A \\ e \end{pmatrix}, \quad l = \begin{pmatrix} 0 \\ 1 \end{pmatrix}$$

Equation (2.6) eliminates the case when the minimum (2.4) is $-\infty$. If we find an optimal vector $y^* \geq 0$ for this problem, with $f_t \cdot y^* < 0$ then f_t is an $\bar{f}$-vector. For any given solution y^*, it is simple to calculate the set $D_{\bar{f}}(y^*)$ of $\bar{f}$-vectors by observing that these vectors must satisfy

$$r = f_t \cdot y^* < 0, \quad t = 1, \dots, \bar{f}_1 \tag{2.8}$$

$$D_{\bar{f}}(y_1^*) = \{\bar{f}_t : \bar{f}_t \cdot y^* < 0, \quad t = 1, \dots, \bar{f}_1\} \tag{2.9}$$

We note that every solution y^* is associated with a subset of $\bar{f}$-vectors $D_{\bar{f}}(y^*)$. Our aim is to construct such a set $\{y_k^*\}$ and corresponding $D_f(y_k^*)$. The inequality (2.8) is used as the first decision rule. At the end of the algorithm, p decision rules and the corresponding $D_f(y_k^*)$, $D_{\bar{f}}(y_k^*)$, $k = 1, \dots, p$ sets of f and $\bar{f}$-vectors are obtained.
Note that

$$D_f = \bigcup_{k=1}^{p} D_f(y_k^*) \tag{2.10}$$

$$\text{and } D_{\bar{f}} = \bigcup_{k=1}^{i} D_{\bar{f}}(y_k^*) \tag{2.11}$$

Let $A' = (B|N)$ be a partition of A' in (2.7). Assume that $i-1$ y_k vectors have been identified. As the i-th step of the procedure proposed for the identification of f-vectors we use the matrix of reduced costs (Δ^i).
It is well known that a column of Δ, say Δ_j, gives the marginal rates of change of the function $f \cdot y$ in (2.4) - (2.7) caused by the introduction of variable y_j into the basis.

The simplex method with special rule of introducing variable y_j into the basis is considered. The variable y_j is introduced into the basis when the number of negative components of the vector Δ_j is maximal. Solution y_2^* is obtained as a result.
The second decision rule is formed

$$r_2 = f \cdot y_2^*$$

It is well known that if $\Delta_{tj} \geq 0$ for all j, then the current LP solution for problem (2.4) - (2.7) is optimal for given f_t. If $f_t \cdot y_1^* > 0$ for any $f_t \in D_s$ and $\Delta_{tj} \geq 0$ for all j, then the vector f_t is classified as "feasible". The set of decision rules $\Delta^i \geq 0$ connected the with vector y_i^* is being examined. This procedure repeats the same steps. The algorithm is continued as long as $\Delta_{tj} < 0$ for some f_t. If $\Delta_{tj} \geq 0$ for all t, j it means that D_f and $D_{\bar{f}}$ have been determined finally.
The algorithm terminates if the number of f-vectors obtained at the previous steps is sufficient for the investigation. Interactive process is organized to determine the conclusion of our procedure.
Briefly our approach consists of the following steps:

Step 1. We simulate the set of random computer-generated RHS-vectors D_s according to the rule given by the DM. In the process of calculations we make the selection corresponding to the real process of modelling.

Step 2. The arithmetic mean $-f-$ of the vectors $f_t \in D_s$ is determined.

Step 3. The following LP problem

$$\begin{aligned} &\min f \cdot y \\ &\text{s.t. } yA \geq l, \quad y \geq 0 \end{aligned} \tag{2.12}$$

is being solved. Let y_1^* be an optimal solution of this problem.

Step 4. Calculate

$$r_1^t = f_t \cdot y_1^*, \quad t = 1, \dots, R \tag{2.13}$$

for each t.

Step 5. If $r_1^t < 0$ then f_t is contained in $D_{\bar{f}}(y_1^*)$, thus the subset $D_{\bar{f}}(y_1^*)$ is formed.

Step 6. Consider the vectors $f_t \in D_s \backslash D_{\bar{f}}(y_1^*)$. Any vectors f_t such that $\Delta_{tj} \geq 0$ for all j, is feasible, thus the subset $D_f(y_1^*)$ is formed.

Step 7. The subset

$$D_s^1 = D_s \setminus \left(D_f(y_1^*) \cup D_f(y_1^*) \right) \tag{2.14}$$

is used as input for the next step.

Step 8. Compute Δ_{ij}. Introduce the vector y_j into the basis (y_j is the vector with the greatest number of negative reduced cost components). At this step the solution y_2^* is obtained. Go to Step 4.

Steps 4 - 8 are repeated until satisfactory representation is obtained. At this point, the set of s -vectors is used as the input for Step 2.

Step 9. Methods of Pattern Recognition and Discriminant Analysis [11], [9] are used to form a piecewise linear decision function.

The step-by-step procedure of forming piecewise linear decision is:
(1) The sample of consistent and inconsistent vectors is formed by means of the procedure described above (Steps 1-8).
(2) This sample is divided into k clusters by using Cluster Analysis methods [10].
(3) Every cluster is being considered separately. In every cluster the linear discriminant function for separating consistent and inconsistent vectors is formed. The set of k such linear functions determines the required piecewise linear decision function.
(4) For every randomly generated vector (f_i) the the following problem is being solved: to which cluster does the vector f_i belong? Some methods for solving this problem have already been proposed (see, for example, [10], [11]).
If the vector f_i belongs to cluster j, we use the decision function of this cluster to check whether it is feasible or not. The exploration of the set of consistent vectors by the means of linear logical decision functions is also proposed [8].

Step 10. Methods of Multivariate Statistical Analysis [9], [12] are used to compress the information obtained and for presenting it to the DM for further consideration. At this step the DM chooses his most preferred vector (or vectors).
Such vectors give a good description of the RHS-vector and the objectives.. A limited number of LP problems (1.1) are solved to obtain the optimal solution x and the RHS-vectors. Methods of Regression Analysis are used to determine the inter-relations between the values of the RHS-vectors and the objectives.

Numerical Examples

Let us consider the numerical example used by S. Zionts [13]:

$$\begin{aligned} 3x_1 + x_2 + 2x_3 + x_4 &\geq b_1 \quad (66) \\ x_1 + x_2 + 2x_3 + 4x_4 &\geq b_2 \quad (80) \\ -x_1 + 5x_2 + x_3 + 2x_4 &\geq b_3 \quad (75) \\ 2x_1 + x_2 + 4x_3 + 3x_4 &\leq b_4 \quad (60) \\ 3x_1 + 4x_2 + x_3 + 2x_4 &\leq b_5 \quad (60) \end{aligned}$$

500 random vectors were generated using uniform distribution over the intervals (20-30, 30-80, 25-90, 40-70, 40-70).
The LP problem (2.12) was solved (y_1^* is the optimal solution). 39 f-vectors, 416 $\bar{f}$ -vectors and 45 s -vectors were obtained after the first iteration. These 45 s -vectors were used as input to the next iteration. The optimal solution y_2^* , and 16 $\bar{f}$ -vectors were identified. After solving four LP problems all 500 randomly generated vectors were classified: 55 f-vectors and 445 $\bar{f}$ -vectors were finally determined. The optimal solutions y_1^*, y_2^*, y_3^*, y_4^* were used as coefficients and four decision rules were formed (Table 1).
1000 random vectors were generated and the four decision ruled were used to determine the set of feasible vectors D_f. The vectors were classified as follows: 153 feasible vectors and 847 infeasible vectors. 10 infeasible vectors were classified as feasible.
Cluster Analysis was used to divide this set into two and three clusters. The typical vectors of the two and three clusters are given in Table 2.
The most interesting RHS-vectors are presented in Table 3. These vectors show that impossible to solve the LP problem with the RHS- vectors 66,60,75,60,60. All the components of of the RHS-vector must be changed, first component being the most important to change.
Another intervals (35-45, 60-75, 45-55, 60-70, 60-70) and the same distribution law were used to form 500 random RHS-vectors, of which 72 were consistent.

Let us consider the numerical example used in [5]:

$$\begin{aligned} x_1 + x_2 &\geq b_1 \quad (14) \\ -x_1 + 3x_2 &\geq b_2 \quad (21) \\ x_1 + 3x_2 &\geq b_3 \quad (27) \\ 4x_1 + 3x_2 &\leq b_4 \quad (45) \\ 3x_1 + x_2 &\leq b_5 \quad (30) \end{aligned}$$

After using the Fuzzy Approach [5] only one RHS-vector: (13.75,17.25,29.25,47.25,25.75) is determined.
In our approach the following intervals of RHS-vectors are consistent at the first step of investigation: (13.5-14.5,10-16,23-30,40-50,25-30). 100 random vectors were generated using uniform distribution. Only one consistent vector was obtained from this set. The interval of the second component of the RHS-vector was changed to 10-13 and after that only two vectors were obtained and they were not interesting to the DM.
At the next step the second component was changed again and the following intervals were examined: (13.5-14.5,17-25,23-30,40-50,25-30). 500 random RHS-vectors were generated using uniform distribution. 25 vectors from this set were consistent.
Analysis of the obtained vectors indicated that the 4-th component in the generated RHS-vector must be increased. The minimal level of b_4 was shown to be 47.2. In the original example this restriction is $4x_1 + 3x_2 \leq 45$.
Cluster Analysis was used to divide this set into two clusters. The typical vectors were: (14.29,15.73,29.3,49.66,27.36) and (13.81,17.29,29.36,47.46,25.38).

Conclusions

Some algorithms are used to solve the practical problem of inconsistent linear inequalities in multicriteria decision-analysis.
Further research in this direction must deal with developing methodology for analysing RHS-vectors in MOLP problems and their relations to the system of goals stated in the decision-making process.

Acknowledgement

The author wants to thank Professor Jonathan Kornbluth from Hebrew University for useful comments, support and suggestions.

References

1. Eremin I.I., Inconsistent problems of linear and convex programming. Nauka, Moscow (1983) - in Russian.

2. Chernikova N., Algorithm for finding the general formula of nonnegative solutions of system of linear inequalities. JVM and Math. Physics, 5, Moscow (1965) - in Russian.

3. T. Gal, Postoptimal analysis, parametric programming and related topics. McGraw-Hill, New York (1979).

4. C. Negoita, Management applications of system theory. Birkhouser Verlag, Basel and Stuttgart (1979).

5. Zimmerman H. J., Fuzzy programming and linear programming with several objective functions. Fuzzy Sets and Systems, 1, pp. 45-55 (1978).

6. M. Zeleny, Optimal system design with multiple criteria: De Novo Programming approach, Eng. Cost and Prod. Ec., 10, pp. 89-94 (1986).

7. Ablow C. and Kaylor D., Inconsistent homogeneous linear inequalities. Bull. Amer. Math. Soc., 71, 5, (1965).

8. Lbov G., Processing methods of diversity of types of experimental data. Nauka, Novosibirsk (1981) - in Russian.

9. Bolch B. and Huang C., Multivariate statistical methods for business and economics. Prentice-Hall Inc., Englewood (1971).

10. Duran B. and Odell P., Cluster analysis (a survey). Springer Verlag, New York (1974).

11. Sebestyen G., Pattern recognition by an adaptive process of sample set construction. IEEE Trans. on Information Theory, 8 (Sept. 1962).

12. Anderson T., An introduction to multivariate statistical analysis. Wiley, New York (1958).

13. S. Zionts, Multiple criteria decision making: an overview and several approaches. Working paper 454, SUNY at Buffalo (1980).

14. B. Werners, Interactive multiple objective programming subject to flexible constraints. EJOR, 31, pp. 342-349 (1987).

Table 1 - Optimal solutions

y_1^*	0.352	0.0947	0.146	0.212	0.194
y_2^*	0	0.0476	0.428	0	0.524
y_3^*	0.246	0.227	0.844	0.441	0
y_4^*	0.309	0.167	0.524	0.441	0

Table 2 - Typical RHS-vectors

N	b_1	b_2	b_3	b_4	b_5
I	47.59	65.4	29.59	69.34	64.94
II	25.1	44.31	29.74	42.51	43.19
III	13.61	24.75	23.21	44.37	25.34
I	25.1	44.31	29.74	42.51	43.19
II	39.06	45.92	25.83	62.33	44.79

Table 3 - RHS-vectors

N	b_1	b_2	b_3	b_4	b_5
1	42.83	60.94	47.29	68.58	68.29
2	36.12	74.9	47.85	69.38	62.45
3	38.75	60.75	45.55	62.07	66.56

MULTI FACTORIAL FINANCIAL PLANNING*

Jaap Spronk
Erasmus University
P.O.Box 1738, 3000 DR Rotterdam
The Netherlands

ABSTRACT

Financial planning has long been recognized as a problem in which uncertainty plays an important role. At the same time, the manager has to deal with a multiplicity of actors and, consequently, with a multiplicity of objectives. The framework described in this paper deals with both the uncertainty aspect and the multicriterial aspect of financial planning. It is based on two important developments in financial theory:

a. multi-factor representation of uncertainty
b. the revival of the concept of flexibility and its valuation through contingent claims analysis.

The approach is intended to support the planning process without assuming managers to be able to define exact weighting factors or to assess precise probability factors. Therefore, we believe the framework to have a good potential to be understood and used in practice.

1. INTRODUCTION

In this paper, financial plannning is seen as a structured process of identification and selection of present and future capital investment projects (including disinvestments) while taking account of the financing of these projects over time. This process can be visualised as follows:

---

* In the Department of Finance at Erasmus University, the multifactorial approach to financial modelling is quite frequently discussed. The author is grateful to those participating in these discussions, especially to Willem-Max van den Bergh, Marc Goedhart, Winfried Hallerbach and Hans Schaffers.

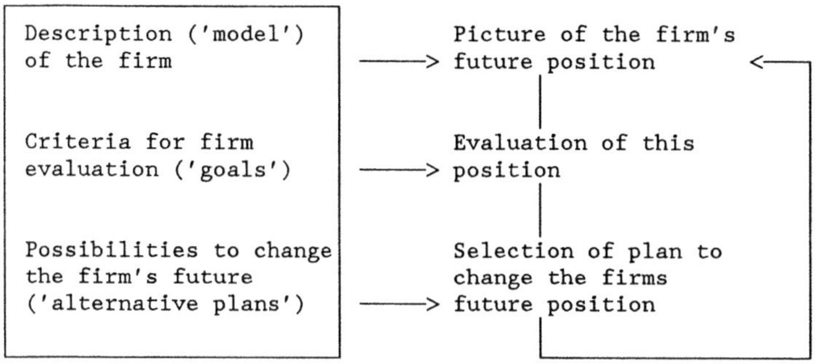

The firm has to derive its right to exist from the fact that it creates a surplus of value to society. In other words, the sum of contributions from the firm's participants should at least equal and preferrably be outweighted by the total of effects the firm has in return on these participants. Thus the firm combines inputs such as labour, money, goods and information into direct outputs such as goods and services. The output of the firm also includes indirect effects which may vary from positive effects like job satisfaction to negative effects like pollution. In this view, the firm is continuously managing a set of exchange relations with its participants. Clearly, there are different types of exchange relations. For instance, a relation can be a long-lasting one (e.g. with a part of the firm's employees) or be more volatile (e.g. a deal in a fully competitive market). Relations may also differ in the sanctionary power of the parties involved, which may vary from a client's power to refuse the purchase of the firm's products to the power to strike (employees) or to raid the firm and to strip its assets.[1)]
Obviously, the firm's exchange relations are dynamic: the firm can to a large extent choose its own participants who, in turn, generally have their alternatives outside the firm. This implies that the resultant effect of these relations is dynamic as well. In other words: the firm has to deal with a dynamic goal complex, a multiplicity of goals which vary over time.

In our opinion, any decision supportive approach to financial planning should take account of the reasonably well established fact that people are not very good in assessing reliable probabilities to the outcomes of future events. This may be partly explained by pointing at the limitations of the human mind. Another reason is of an even more fundamental nature: many future outcomes are contingent on decisions which are still to be made. In the case of capital investment projects, for instance, the

---

1) The kind of exchange relations a particular firm has to deal with depends on the type of its organisation. As compared with market deals, long lasting relations have both merits (e.g. a good possibility to combine the expertise of different participants, lower search and contracting costs, lower uncertainty through information exchange) and costs (e.g. bureaucracy, influence on the incentive structure). The size of these costs beyond doubt depends on the political-cultural environment of the firm.

future cash flows can generally not be well represented by some probability distribution alone. In addition it is often necessary to describe as well all the 'rights' and 'duties' connected with the project. For instance, the right to expand or to abandon the project or some legal duty to keep people on the payroll even in case it would be more economical to fire them. One might consider the use of decision trees to describe this type of situations. But in most practical cases this is impossible because of the number of possible outcomes being too large and/or because the timing of the potential outcomes (when something will happen) is largely unknown. The picture becomes even more complicated if one realises that the outcomes of a project are often not only contingent on the future decisions of the decision maker himself, but may also be contingent on the future decisions of the firm's participants, which are in their turn contingent on the firm's decisions. In other words, the way the firm manages its exchange relations influences the uncertainty surrounding the firm's cash flows.

From the above observations it is clear that the alternative plans to be considered by the decision maker are hard if not impossible to evaluate on the exclusive basis of the objective adopted by financial theory: the maximization of the shareholders' wealth. One problem is the contingincies, another is the game like nature of at least part of the firm's dynamic goal complex. However, cash flow remains one of the central concepts in developing a financial plan. The owners of the firm will evaluate the firm in terms of alternative investment opportunities. An obvious possibility is to invest in the stock market. This gives a clear reason to use the firm's market value as a benchmark for evaluating the firm and its investments. And to close the circle: the firm's market value directly depends on the firm's cash flows and the associated risk. If the firm's cashflows decrease and/or its riskiness increases, its market value will drop. As such, the shareholders may de facto be the most powerful group of participants of the firm. However, it is equally clear that the firm should not neglect the other participant's interests, both for its own sake and in the interest of the stockholders. In summary, the firm should for each financial plan try to assess all effects considered to be important by the most influential participants, taking account of the most important contingincies involved.

In this paper we describe a framework for financial planning in the firm which is based on two important developments in financial theory:

a. the use of multi-factor models to describe probability distributions of returns and
b. the revival of the concept of flexibility and its valuation through option pricing theory and contingent claims analysis.

We will show how these two ingredients can be used in a multiple criteria approach to financial planning in which the decision maker can systematically investigate the set of alternative financial plans, taking account of the most important parties, the most important uncertain factors and the most important contingincies, rights and duties involved. Preliminary experiences have given us the conviction that the framework has a good potential to be understood and used in practice.

The general ideas behind the proposed framework are outlined in the next section. In section 3 the framework itself is described after which section 4 is devoted to some aspects of the operationalization of the framework. Our conclusions are summarized in section 5.

2. THE MULTI FACTOR FRAMEWORK AND CONTINGENT CLAIMS

In the modelling of uncertainty one has to find a compromise between precision on the one hand and comprehensibility and managability by the model user on the other hand. Very often, uncertain outcomes of decision alternatives (projects, plans, etc.) are modelled as probability distributions defined on the range of possible outcomes. Decision makers are then required to assess the value of some parameters (e.g. mean and variance) of these distributions. In addition, decision makers often have to express their preferences (e.g. utility values) with respect to the uncertain outcomes. The thusly provided information is then used in the formulation of an objective function (e.g. the expected utility of the decision alternatives) or, alternatively, in formulating chance constraints. For the reasons given above we deviate from this approach by assuming that the uncertain phenomenae met in financial planning can generally not be modelled precisely. However, following Hallerbach and Spronk, 1986, we assume that for many of these phenomenae at least some structure can be found by using multifactor models. The results of a financial plan (to facilitate the exposition in this section, we take the firm's cash flow as the only relevant outcome variable) will depend on the one hand on the decisions made by the firm and on the other hand on the various forces and influences from its dynamic environment. We assume that, in general, it is very hard if not impossible to define a complete probability distribution over the value of the cash flows which may result. Instead, we assume that the firm is able to define its expectations concerning these cash flows and, in addition, that it is able to assess the sensitivity of these cash flows for unexpected changes in a number of factors which influence these cash flows. Consequently, the effect of a decision can be modelled as an expected level of the cash flows plus a series of sensitivites for unexpected changes in a number of factors influencing these cash flows. The firm does not necessarily know how these factors themselves will change in the future. Also, it may have found only some of the factors influencing its cash flows. Nevertheless, on basis of this way of modelling, a firm can estimate the firm's <u>aggregate</u> sensitivity (i.e. the sensitivity of all decisions combined) for the various factors it has found to be important. In this way, a given financial plan of the firm can be evaluated in terms of an expected cash flow development accompanied by a vector of sensitivities describing what deviations from this expectation could arise if one or more of the identified factors would get another value than expected. Clearly, the aggregate sensitivities can be treated by the decision makers as goal variables, which can be traded off against each other and against other goal variables, possibly within the framework of a formal multicriteria analysis. Such an analysis might show for instance that the firm's sensitivity for unexpected wage rate changes can be lowered, but at the price of increasing the firm's sensitivy for unexpected changes in the dollar rate.

It is conceivable that a firm is also evaluated in other terms than cash flows alone (e.g. employment level, labour relations, market power, etc.). Such a broadened evaluation can be done in a straightforward manner, be it that the amount of calculations increases and that one may want to investigate the interrelations between the evaluation criteria.

Many different factors will influence the future cash flows of the firm. Some of these factors will be well-identifiable and well-measurable. However, there will also be factors which are well-identifiable but less well measurable (e.g. because of limitations in the data or because the factor concerned can not easily be operationalized). In addition, there will be factors which are not well identifiable because of their too infrequent occurrence. The framework discussed here stresses the sensitivity of the cash flows for unexpected changes in well-measurable factors. However, in the evaluation of a plan, attention should also be paid to the identifiable but less-well measurable factors hidden in the specific risk component. The least to be done is to develop a list of this type of factors which can be used as an additional checklist for the evaluation of the plans. With respect to the factors which are less well identifiable one cannot do much more than to develop the reserves and the ability to handle unforeseen situations.

After the identification of the most important factors and the assessment of the firm's sensitivity for these factors, a number of questions remain. A first question is whether and, if so, how the firm is prepared for unexpected changes in the factors. Obviously, both the <u>defensive</u> and the <u>offensive</u> weapons of the firm should be considered. In both cases, the firm has two possibilities to armour itself. One possibility is to neutralize or to limit a risk beforehand by 'buying an insurance' with respect to this risk (where 'buying an insurance' should be understood in a broad sense). For instance, firms often insure themselves against the negative consequences of unexpected changes in the factors such as fire or exchange rate fluctuations (the latter can often be insured by buying valuta options). On the other hand, the firm can assure itself of the positive effects of unexpected changes in the factors (e.g. by acquiring the exclusive selling rights of a product in development). Another possibility to face risks, instead of buying an insurance, is to create sufficient elbow-room in the firm to be able to react adequately to an unexpected change in some factor if and at the moment it occurs. An example is not to insure against fire but instead to make a large enough reservation to be able to bear the negative consequences of fire. The importance of some elbow-room is illustrated by a comparison of the histories of PanAm and Braniff (example from Casey en Bartczak, 1985). At a certain moment both firms generated a negative operational cash flow. Braniff went bankrupt in 1982 while PanAm could survive longer by selling two important assets (Intercontinental Hotel and PanAm building). An evaluation of PanAm based on a cash flow analysis alone -i.e. without taking account of its 'elbow-room'- would clearly have given a wrong answer. A firm having a good and stable policy for recruitment and selection of personnel will produce less 'surprises' than a firm with a clumsy ad hoc policy. Of course, an evaluation backing an important decision will pay attention to such a factor, even if it is less well measurable.

Creating elbow-room can be viewed as buying an option. If an unexpected change of a factor materializes, the option holder has the right to use the elbow-room to react to this change. The possessor of the elbow-room is of course obliged to bear the consequences if the elbow-room is not sufficient. Elbow-room is often labeled as flexibility, where a distinction is made between operational and financial flexibility (cf. Kemna and Van Vliet, 1984, and Kemna, 1988). Examples of operational flexibility are the possibility to quickly adapt production (e.g. with respect to production volume or to product specification) according to the changing needs of the product's consumers. Already twenty years ago, the importance of financial flexibility was stressed by Donaldson, 1969, 1984, and in the Netherlands by Diepenhorst, 1962 (the latter made the useful distinction between defensive and offensive flexibility). Examples of financial flexibility are unused reserves (cash surpluses, unused credit facilities), unused debt capacity, the capability to reduce expenditures and the earlier mentioned possibility to sell assets. Not surprisingly, the creation and maintenance of elbow-room is not without costs: flexibility has its price (cf. Kemna en Van Vliet, op cit.).

3. MULTI FACTOR FINANCIAL PLANNING

In the multi-factorial approach to financial planning, the idea of replacing probability distributions by multi-factor representations is combined with the concept of contingent claims and with multiple criteria decision procedures. In broad lines, the resultant approach can be described as follows.

a. Identification of the firm's most important environmental factors. A distinction can be made between <u>operational</u> and <u>financial</u> factors. Operational factors relate to the firm's operational cash flows. Examples are the price of raw materials and the price of energy. Financial factors relate to the cash flows flowing from and to the suppliers of the firm's capital. Examples are interest rates and risk premiums required by the market.

b. Identification of projects, both real and financial. Real projects may vary from stopping a certain production line to the take-over of a competitor. Financial projects may vary from a modest one-year bank loan to a substantial emission of new stock.
c. Calculation expected cash flows of the firm as it is. (Obviously, the expected development of other output variables can be calculated as well. For ease of exposition hereafter only cash flow is assumed to be relevant). To calculate these expected cash flows, a 'base case' policy has to be defined, assuming that the environmental factors defined in (a) will adopt their expected values.
d. Estimation expected future cash flows of the projects. Here, the expected <u>incremental</u> cash flows are to be calculated. That is, the difference between the expected cash flows of the firm including the project and the expected cash flows of the firm not including the project (i.e. the expected values calculated in (c). As in (c), the expected incremental cash flows are calculated on the assumption that the environmental factors adopt their expected values.
e. Measurement of the firm's and the projects' sensitivities for unexpected changes in the environmental factors. The effect of the expected changes in the environmental factors has already be included in the expected cash flows calculated in (c) and (d).
f. Identification, specification and, if possible, valuation of the options ('rights and duties') available to the firm as it is and of those associated with new projects. The valuation of options is a relatively new area which is producing a stream of exciting results. Notwithstanding the large number of successes in that field (see Kemna,1988, for an overview) a lot of questions remain unanswered. Especially with respect to the valuation of real options one may expect more or less exact answers only in a limited amount of cases. Clearly, options may change the magnitude of the sensitivities of the cash flows. Alternatively, options may limit the effect of one or more unexpected factor movements without changing the sensitivities as such.
g. Both the firm as it is and the projects at hand are described by multidimensional profiles, consisting of the expected values calculated in (c) and (d) and of the sensitivities[2)] for unexpected changes in the environmental factors calculated in (e) and if necessary corrected for the effect of the options mentioned in (f).
h. On basis of these multidimensional profiles of the firm and the projects, the

---

2) The sensitivities estimated in the preceding steps relate to the cash flows streaming to the suppliers of capital. Given the firm's market value as a primary objective, one should try to find out how the financial market valuates different constellations of risk factors.

problem is to find the combination of projects which contributes best in terms of the firm's objectives: this is done by IMGP (see Spronk, 1981, 1985), in which the decision maker has the possibility to condition the set of possible project combinations by setting and systematically changing a series of goal constraints.

i. The result of (h) can be a single plan or a set of plans which is subjected to a secondary analysis. For instance, it is possible to investigate how a plan would perform given a constellation of unexpected factor changes (of course, several factor constellations can be investigated). Furthermore, one may want to evaluate the generated plans in terms of objectives which were not explicitly included in the model.

4. OPERATIONALISATION

With respect to the operationalisation of the multi factor framework, a series of problems arise. In this section we will address some of the problems associated with

a. the calculation of expected values and
b. the estimation of sensitivities.

Problems related to the identification and specification of projects, the specification and valuation of options and with the use of IMGP will not be dealt with here, because they are of a more general nature and because they are also dealt with elsewhere.

In order to calculate the expected cash flows, both of the firm 'as it is' and of the proposed projects a 'base case' or 'expected' scenario has to be defined. That is, a set of assumptions with respect to the expected values of the firm's most important environmental factors together with an overview of the decisions to be made by the firm in case the firm's environment would be as expected. The next step is to calculate the expected cash flows. This can be based on a more or less complicated model, relating the factor values and the decisions to the expected cash flows. A natural starting point for building such a model is a simple cash flow definition, for the firm as a whole and/or for the different divisions of the firm.

To estimate the sensitivities associated with the existing firm and with the projects, a wide variety of approaches is available. The degree of detail of the analyses may vary from an enumeration of the relevant factors on the one hand to complex statistical analyses on the other. Along this scale, the information obtained on the factors considered increases from a simple yes-or-no answer regarding their presence to.

Simple techniques to analyze factor-sensitivities are e.g. chartreading, checklists, qualitative interviews or even group-discussions. With these techniques, the relevance of factors is established in a qualitative manner. Some quantitative information is used in spreadsheet- or simulation-applications, in which not only the presence but also the estimated effect of a factor is incorporated. The effects may be obtained by rules of thumb, by extrapolation of historic data or by more complex estimation procedures. Information of a more quantitative nature may be obtained e.g. by time series analysis. Obviously, this is not without difficulties. For instance, factors influencing the cash flows are seldomly independent of each other. The influence of the factors on the cash flows may be described by means of linear models but is possibly better described through non-linear relations. How to separate expected from non-expected influences? Are the sensitivities to be measured on a monthly basis, on a quarterly basis or something in between? What about the potential instabilities in the sensitivities? And so on. A lot of questions remains to be answered. Nevertheless, the experiences acquired during the work following the paper by Hallerbach and Spronk,1986, suggest that the multi-factor road is not without difficulties but nevertheless seems to lead to valuable results.

5. CONCLUSIONS AND REMARKS

Although thusfar, no real application of the proposed approach was made (a study of some experiments with this approach will be finished soon), discussions with practitioners suggest that 'we are on the good way' and have lead to adaptations of the procedures used to evaluate and to select capital investment projects. Clearly, lots of problems are still to be solved, but we strongly believe it to be worthwhile to go through the process of solving them.

In this paper, no explicit attention was paid to the case in which financial planning involves two or more hierarchically distinct decision levels. However, in the preceding conference we presented an approach to two-level financial planning (see Schaffers and Spronk,1987) which can be extended to include the multifactorial approach adopted in the current paper (see Goedhart, Schaffers and Spronk, 1988).

REFERENCES

Casey, C.J. en N.J. Bartczak, 1985, Cash flow: voor buitenstaanders geen belangrijke informatie, Financieel Management.

Diepenhorst, A.I., 1962, Beschouwingen over de optimale financiele stuctuur van de onderneming, inaugurale rede, Bohn, Haarlem.

Donaldson, G., 1969, Strategy for Financial Mobility, Harvard University Press,

Boston.
Donaldson, G., 1984, Managing Corporate Wealth, Harvard University Press, Boston.
Goedhart, M., Schaffers, J. and J. Spronk, 1988, An Interactive Multi-Factor Procedure for Two-Level Financial Planning with Conflicting Goals, Report 87xx/f,Centre of Research in Business Economics, Erasmus University, Rotterdam.
Hallerbach, W. en J. Spronk, 1986, An Interactive Multi-Factor Portfolio Model, Report 8610/F,Centre for Research in Business Economics, Erasmus University, Rotterdam.
Kemna, A.G.Z., 1988, Options in Real and Financial Markets, PhD thesis, Erasmus University Rotterdam, Donner Boeken, Rotterdam.
Schaffers, J. and J. Spronk, 1987, Two-Level Financial Planning with Conflicting Goals, An Interactive Procedural Approach, in Y.Sawaragi, K. Ihoue and H. Nakayama (eds), Towards Interactive and Intelligent Decision Support Systems, Volume 1, Springer, Heidelberg, pp. 270-279.
Spronk, J., 1981, Interactive Multiple Goal Programming : Applications to Financial Planning, Martinus Nijhoff, Boston.
Spronk, J., 1985, Financial Planning with Conflicting Objectives, in G. Fandel and J. Spronk (eds), Multiple Criteria Decision Methods and Applications, Springer, Berlin, pp. 269-288.

THE USE OF PREFERENCE INFORMATION IN MULTIPLE CRITERIA INTERACTIVE PROCEDURES

Daniel VANDERPOOTEN
LAMSADE, Université de Paris-Dauphine
75775 Paris cedex 16, FRANCE.

ABSTRACT

The management of preference information in multiple criteria interactive procedures is shown to be highly related to different conceptions of the interactive approach in decision aid. Original ways of using preference information which allow to support learning of preferences while guiding the search for best compromises are introduced in this paper.

1. **Introduction**

Multiple criteria interactive procedures aim at deriving a final prescription which corresponds to a satisfactory compromise by integrating preference information progressively specified by a decision maker (DM). Such procedures are based on the alternation of calculation phases for constructing proposals with dialogue phases actively involving the DM. The management and use of preference information clearly play a crucial part in the design of these procedures. Classical and original strategies of treatment of preference information as well as their influence on the behaviour of the procedure are presented in this paper.

A formal description of a multiple criteria interactive procedure is briefly introduced in section 2. Despite a unified formal structure, these procedures may be conceived according to quite distinct conceptions. A basic distinction, presented in section 3, is between methods which conduct a search process and methods which aim at supporting a learning process. By considering the underlying assumptions made within each conception, we show that the way of taking preference information into account is quite different. A third conception of the interactive approach which involves an original treatment of preference information is then introduced.

Several concepts allowing to control the management of preference information are presented and illustrated in section 4. The contribution of these concepts is discussed in the final section.

2. Formal description of a multiple criteria interactive procedure

Let us define some preliminary notations to characterize a multiple criteria problem:

- A the set of potential alternatives (or actions),
- Z ($\subset \mathbb{R}^p$) the criterion space with its natural partial preorder $\geq$ such that:

$$\forall\, \mathbf{z} \in Z, \forall\, \mathbf{z}' \in Z, \quad \mathbf{z} \geq \mathbf{z}' \text{ iff } z_j \geq z'_j \ (j = 1, \ldots, p),$$

- a set of p criteria g_j ($j = 1, \ldots, p$),
- $Z_A = \{\mathbf{z} \in Z \ / \ z_j = g_j(a) \ (j = 1, \ldots, p), \ a \in A\}$ the set of potential criterion vectors.

Although many multiple criteria interactive procedures with distinctive features were proposed (see *e.g.* Steuer (1986), Vanderpooten and Vincke (1989) for a survey), it is possible to bring out a common formal structure.

A multiple criteria interactive procedure, when perceived from a purely formal point of view, is an algorithm for constructing a proposal on the basis of preference information supplied by the DM which is iterated until he is satisfied with the current proposal. Proposals usually consist of one or several alternatives which represent potential compromise solutions.

Even if preference information can be obtained in quite different ways (which is a critical point in the success of a procedure), only a few main types of elementary parameters are used to render it:

- Intercriteria parameters representing weighting coefficients or marginal rates of substitution associated with each criterion which reflect the relative importance of each criterion from the DM's point of view. They will be denoted by $(\lambda_1, \ldots, \lambda_p) = \boldsymbol{\lambda}$.
- Aspiration levels $(\bar{z}_1, \ldots, \bar{z}_p) = \bar{\mathbf{z}}$ which indicate the ideal criterion values from the DM's point of view.
- Reservation levels $(\bar{\bar{z}}_1, \ldots, \bar{\bar{z}}_p) = \bar{\bar{\mathbf{z}}}$ corresponding to requirements on each criterion such that any alternative which does not meet these requirements is rejected by the DM as unsatisfactory.

Let Ω^h be the set of preference parameters derived at iteration h from preference information progressively supplied by the DM. We have the following general relation:

$$\Omega^h = f(I^h, I^{h-1}, \ldots, I^0) \qquad (2.1)$$

where I^h represents preference information elicited from the DM at iteration h
and I^0 represents prior preference information.

A preference parameter may be composed with several elementary parameters. In particular, Ω^h may include parameters for delimiting a subregion of interest $Z_A^h \subset Z_A$ (Ω_1^h -parameters) and parameters for generating the proposal (Ω_2^h -parameters).

The generation of a proposal is performed using a scalarizing function s which, considering a preference parameter, gives an estimate of the interest of any criterion vector **z** from Z_A. Classical scalarizing functions are the weighted-sum, various weighted-norms and achievement functions (see *e.g.* Wierzbicki (1986) for more details). Proposal p^h is then constructed by optimizing s over Z_A (or Z_A^h if $\Omega_1^h \neq \emptyset$) for each parameter from Ω_2^h (with Ω_2^h possibly reduced to one element). Let Γ represent this construction process, we have:

$$p^h = \Gamma(Z_A, \Omega^h, s) \qquad (2.2)$$

It should be clear from this brief introduction (see Vanderpooten (1989) for more details) that the construction of the proposal p^h is highly dependent on the way of inferring Ω^h from preference information. This paper deals with the way of managing preference information so as to determine Ω^h (a critical point is to decide which pieces of information $I^h, I^{h-1}, \ldots, I^0$ are to be taken into account) and its influence on the general behaviour of an interactive procedure.

3. Conceptions of decision aid and management of preference information

Two classes of multiple criteria interactive procedures, related to different conceptions of the interactive approach in decision aid, may be distinguished:

- procedures conducting a search process which should finally produce a best compromise solution,
- procedures primarily aiming at supporting a learning of preferences.

Search-oriented procedures present the DM with a converging sequence of proposals. Within this approach, the final proposal is to be considered as an optimal prescription. However, such a mathematical convergence can only be achieved if some prior assumptions are made about the stability of the DM's preference structure. Moreover, the DM is supposed to provide preference information which is consistent with his stable preference structure. In particular, he is generally not allowed to give information which contradicts previous judgments.

By refuting these assumptions, some researchers have been arguing that the main objective of an interactive procedure should be to allow the DM to strengthen, or even to create, his preference structure. This can be achieved through a free exploration of the set of potential compromise solutions which allows trial and error. However, no formal guarantee can be given with respect to the convergence of these learning-oriented procedures. It is just expected that a final prescription will naturally emerge from the learning process induced by the procedure.

Each conception, because of its underlying assumptions with respect to the DM's preference structure, involves a specific treatment of preference information.

In a search-oriented conception, the preference information supplied by the DM at any iteration is supposed to remain valid throughout the whole interactive process. The acquisition of new pieces of information only serves to give more details about the stable DM's preference structure. The strategy for the management of preference information is then immediate: it consists in retaining every piece of information $I^h, I^{h-1}, \ldots, I^0$. Preference parameters used at iteration h are elaborated through a synthesis of previous pieces of information. Operationally, this synthesis is performed by simply updating parameters used at iteration h-1. We have then: $\Omega^h = f(I^h, \Omega^{h-1})$.

In a learning-oriented conception, the preference structure is not assumed to remain stable. Moreover, the DM is even encouraged to provide information in a trial and error fashion. Consequently, preference information cannot be considered as valid throughout the whole interactive process. The problem is then to know which pieces of information may be taken into account to construct the proposal p^h. Because there is no objective way of identifying information which remains valid, the usual strategy consists in only considering information which has just been supplied. Preference parameters used at iteration h are only inferred from this information. We have then: $\Omega^h = f(I^h)$.

We believe the second conception to be more adequate when considering the essence of decision making. However, we recognize that methods based on the first conception bring a support to the DM, by guiding his exploration of the set of potential compromise solutions (the problem is that this guiding results from a rigid converging process based on restrictive assumptions). On the contrary, a pure learning-oriented conception, which promotes a free exploration of the set of potential compromise solutions through trial and error, provides no guiding and can be suspected to reflect nothing but the wanderings of the DM.

We are then led to propose a third conception aiming at supporting learning of preferences while including guiding mechanisms. Such a learning-and-search-oriented conception clearly needs *flexible* guiding mechanisms. More precisely, an interactive procedure designed according to this conception should respect the three following principles:

(P.1) the exploration of potential compromise solutions must be guided by the procedure,
(P.2) the DM must be free to control the exploration process,
(P.3) the exploration process should not be perturbed by preference information no longer valid.

We should notice that if the DM identifies preference information that is no longer valid, he may react insofar as (P.2) is respected. However, he may not be able to identify this type of information *e.g.* because his attention is currently focused on other aspects. The identification of preference information no longer valid should then be partly taken in charge by the procedure, which justifies the introduction of (P.3). An interesting way of combining the three above principles is by imposing a limited form of consistency with respect to preference parameters.

4. **Consistency of preference parameters**

It is important to notice beforehand that we are not dealing here with consistency related to a preference structure. Such a consistency would amount to assuming the pre-existence of a stable preference structure as in a pure search-oriented conception. Consistency with respect to a preference structure clearly involves strong forms of consistency with respect to preference parameters. However, consistency of preference parameters does not necessarily involve any assumption on the preference structure.

The consistency we are interested in only refers to preference parameters which convey information provided by the DM. More precisely, we shall say that consistency is respected when preference parameters inferred from information obtained at iteration h are not in contradiction with parameters inferred from information obtained at iteration h-i. Attention should be paid to the fact that consistency concepts should be viewed as mere technical devices for supporting the DM in his exploration of the set of compromise solutions according to the above principles.

Different forms of consistency and their impact on the exploration process - considering principles (P.1), (P.2) and (P.3) - are now introduced.

We propose to identify different forms of consistency by defining specific conditions adapted to the type of preference parameter. These conditions may be modulated in order to define *weak* forms of consistency. Principles (P.2) and (P.3) clearly induce limited forms of consistency respectively termed as *restricted consistency* and *k-consistency*.

We illustrate the definition of such conditions in the case of reservation levels. Similar conditions could be defined for aspiration levels or weights.

Reservation levels usually serve for delimiting a subregion of interest $Z_A^h \subset Z_A$ from which p^h will be generated. The delimitation is simply performed as follows:

$$Z_A^h = \{\mathbf{z} \in Z_A \ / \ \mathbf{z} \geq \bar{\bar{\mathbf{z}}}^h, \ \bar{\bar{\mathbf{z}}}^h \in Z\} \quad (4.1)$$

where $\bar{\bar{\mathbf{z}}}^h$ is a vector of reservation levels defined at iteration h (see *e.g.* Benayoun et al. (1971)).

The set Ω^h of preference parameters derived at iteration h includes reservation levels $\bar{\bar{\mathbf{z}}}^h$ as Ω_1 -parameters. Preference information at iteration h - represented by I^h - consists of some values v_j^h more or less directly expressed by the DM which indicate his requirements concerning some criteria. We shall denote $J^h \subset J = \{1, \ldots, p\}$ the corresponding subset of criterion indices. For commodity and without any restriction, we shall set $J^0 = J$ and $v^0 = (v_1^0, \ldots, v_p^0) = \bar{\bar{\mathbf{z}}}^0$ where v_j^0 is small enough to guarantee $v_j^h \geq v_j^0$ for all h and for all $j \in J^h$ (*e.g.* v_j^0 less than the minimal value of z_j over Z_A). The way of determining $\bar{\bar{\mathbf{z}}}^h$ from current and previous pieces of information corresponds to the definition of f in (2.1). The complete definition of f will depend on the consistency condition which is retained. It is natural to define f such that:

if the consistency condition allows it:

- $\bar{\bar{z}}_j^h = v_j^h \quad \forall j \in J^h$ (4.2)
- $\bar{\bar{z}}_j^h = v_j^0 \quad \forall j \in J \setminus J^h$ (4.3)

otherwise:

- set $\bar{\bar{z}}_j^h$ to the minimal value compatible with the consistency condition. (4.4)

(4.3) and (4.4) impose a prudent use of reservation levels which may strictly increase only through (4.2). Moreover, we shall assume that each iteration is *informative* in the sense that $|J^h| \geq 1$ for all h and there exists at least one value $v_j^h \neq \bar{\bar{z}}_j^{h-1}$.

Before considering various consistency conditions, we must define a basic consistency condition between two parameters $\bar{\bar{z}}^h$ and $\bar{\bar{z}}^{h-i}$. This basic condition may be defined as:

$$\bar{\bar{z}}_j^h \geq \bar{\bar{z}}_j^{h-i} \qquad \forall j \in J^{h-i} \tag{4.5}$$

A weak version of this basic condition would be:

$$\bar{\bar{z}}_j^h \geq \bar{\bar{z}}_j^{h-i} - \Delta_j^{h-i} \qquad \forall j \in J^{h-i} \tag{4.6}$$

where Δ_j^{h-i} can be considered as a tolerance threshold.

In the following, we shall explicit conditions using (4.5). Similar conditions could be derived from (4.6). Conditions will be defined at a specific iteration h *without imposing any form of consistency on previous iterations.*

Complete consistency at iteration h may be defined as:

$$\forall j \in J \qquad \bar{\bar{z}}_j^h \geq \bar{\bar{z}}_j^{h-i(j)} \tag{4.7}$$

where $i(j) = \text{Min } \{t \in \{1, \ldots, h\} \; / \; j \in J^{h-t}\}$

This condition states that the current reservation level respect to a criterion g_j is forced to be greater than or equal to the most recent reservation level derived (through (4.2)) from preference information supplied by the DM. We have then the following results.

Result 1: If complete consistency is imposed at iteration h, then we have $\bar{\bar{z}}^h \geq \bar{\bar{z}}^{h-1}$.

Proof: Considering (4.7), if j is such that $i(j) = 1$, the result is clearly true. If $i(j) > 1$, we necessarily have $\bar{\bar{z}}_j^{h-i(j)} \geq \bar{\bar{z}}_j^{h-1}$ ($\bar{\bar{z}}_j^{h-1} > \bar{\bar{z}}_j^{h-i(j)}$ would be possible only through (4.2) which would imply $j \in J^{h-1}$ and $i(j) = 1$). Result 1 is then proved by considering again (4.7).

Result 2: If complete consistency is imposed at each iteration then the exploration process is converging.

Proof: Because of result 1, we have $\bar{\bar{z}}^h \geq \bar{\bar{z}}^{h-1}$ for all h. Thus, we get from (4.1) $Z_A^h \subset Z_A^{h-1}$ for all h and $Z_A^h \neq Z_A^{h-1}$ for all h because of our assumption about the informative nature of each iteration, which generates a strictly monotonous reduction of the subregion of interest. The final prescription is $p^k = \Gamma(Z_A, \Omega^k, s)$ - see (2.2) - where k is the smallest index such that $Z_A^{k+1} = \emptyset$.

Result 1 shows that the way of setting $\bar{\bar{z}}^h$ only depends on $\bar{\bar{z}}^{h-1}$ and on current information. We have then: $\Omega_1^h = f(I^h, \Omega_1^{h-1})$. Result 2 shows that complete consistency respects (P.1) in a very strong way. (P.2) and (P.3) are clearly not satisfied. In order to satisfy (P.2), we propose a concept of restricted consistency.

Restricted consistency at iteration h may be defined as:

$$\forall\, j \in J \setminus J^h \qquad \bar{\bar{z}}_j^{\,h} \geq \bar{\bar{z}}_j^{\,h-i(j)} \tag{4.8}$$

where $i(j) = \mathrm{Min}\ \{t \in \{1, \ldots, h\}\ /\ j \in J^{h-t}\}$

We have the following result.

Result 3: If restricted consistency is imposed at iteration h then we have $\bar{\bar{z}}_j^{\,h} \geq \bar{\bar{z}}_j^{\,h-1}\ \ \forall\, j \in J \setminus J^h$.

Proof : As in Result 1, we have $\bar{\bar{z}}_j^{\,h-i(j)} \geq \bar{\bar{z}}_j^{\,h-1}$ for all j. Because of (4.8), we get the result.

This result clearly shows that the DM may freely provoke converging or diverging stages in his exploration process. Here again we have: $\Omega_1^h = f(I^h, \Omega_1^{h-1})$. In order to satisfy (P.3), we propose a concept of **k**-consistency, where $\mathbf{k} = (k_1, \ldots, k_p)$ is such that each k_j represents a range of validity for the reservation level corresponding to criterion g_j.

k-consistency at iteration h may be defined as:

$$\forall\, j \in J \qquad (\, j \in \bigcup_{i=1}^{k_j} J^{h-i} \ \Rightarrow\ \bar{\bar{z}}_j^{\,h} \geq \bar{\bar{z}}_j^{\,h-i(j)}\,) \tag{4.9}$$

where $i(j) = \mathrm{Min}\ \{t \in \{1, \ldots, k_j\}\ /\ j \in J^{h-t}\}$

We have the following result.

Result 4: If **k**-consistency is imposed at iteration h then we have $(\, j \in \bigcup_{i=1}^{k_j} J^{h-i} \ \Rightarrow\ \bar{\bar{z}}_j^{\,h} \geq \bar{\bar{z}}_j^{\,h-1}\,)$.

Proof: similar to Result 3.

The exploration process includes converging stages only if $J \setminus \bigcup_{i=1}^{k_j} J^{h-i} = \emptyset$ or, more importantly, if the reservation levels on the corresponding criteria may still be considered as valid, which can be partly checked in an original way using contradiction (to be presented in section 5). Moreover, we have: $\Omega_1^h = f(I^h, I^{h-1}, \ldots, I^{h-k}, \Omega_1^{h-1})$ with $k = \mathrm{Max}\ \{k_j\ \ (j = 1, \ldots, p)\}$, where the only needed information in I^{h-i} $(i = 1, \ldots, k)$ is J^{h-i} (and not v_j^{h-i} for $j \in J^{h-i}$).

We are finally led to propose the concept of restricted **k**-consistency.

Restricted k-consistency at iteration h may be defined as:

$$\forall\, j \in J \qquad (\, j \in (\bigcup_{i=1}^{k_j} J^{h-i}) \setminus J^h \ \Rightarrow\ \bar{\bar{z}}_j^{\,h} \geq \bar{\bar{z}}_j^{\,h-i(j)}\,) \tag{4.10}$$

where $i(j) = \mathrm{Min}\ \{t \in \{1, \ldots, k_j\}\ /\ j \in J^{h-t}\}$

We have the following expected result.

Result 5: If restricted **k**-consistency is imposed at iteration h then

$$\text{we have } \left(j \in \left(\bigcup_{i=1}^{k_j} J^{h-i} \right) \setminus J^h \;\Rightarrow\; \bar{\bar{z}}_j^{\,h} \geq \bar{\bar{z}}_j^{\,h-1} \right)$$

Proof: obvious from the above results.

Here again, we have: $\Omega_1^h = f(I^h, I^{h-1}, \ldots, I^{h-k}, \Omega_1^{h-1})$ with $k = \text{Max}\,\{k_j\ (j = 1, \ldots, p)\}$, where the only needed information in I^{h-i} $(i = 1, \ldots, k)$ is J^{h-i}. Imposing restricted **k**-consistency on preference parameters allows to respect (P.2) and (P.3). It is clear that converging stages may not be frequent. The point we emphasize is that they occur under satisfying conditions *i.e.* when the DM wishes it ((P.2) being respected) and not through an artificial process of maintaining unrelevant information ((P.3) being respected).

5. **Conclusion**

In our opinion, multiple criteria interactive procedures must allow the DM to change his mind. This amounts to saying that preference information progressively supplied by the DM cannot be considered as valid throughout the whole process. Imposing a complete consistency annihilates one of the major contributions of the interactive approach. However, interactive procedures should also help the DM to perform an exploration of the set of compromise solutions. We briefly indicate the reasons why we believe that limited forms of consistency on preference parameters promote a guided exploration as well as learning of preferences.

First, it should be clear that any form of consistency guides the exploration by ensuring a continuity on preference parameters which involves a continuity on subregions of interest and thus on proposals. Insofar as p^h and p^{h+1} are derived from partly common information, they seem to be more related. Exploration is then perceived by the DM as a logical sequence. On the contrary, if previous pieces of information are systematically omitted, exploration is started again from scratch at each iteration.

As in a strictly learning-oriented conception, we emphasize the necessity of allowing a free exploration of the set of potential compromise solutions. Restricted consistency seems to be the most obvious way of allowing the DM to control his exploration process, while preserving non-contradictory previous information.

Learning-oriented procedures aim at stimulating a learning process based on trial and error. The introduction of **k**-consistency may generate another type of learning which could be termed as learning by contradiction. **k**-consistency involves indeed that no contradiction may be observed between proposals and preference information within the range of consistency. However, contradictions may clearly arise out of this range (e.g. p^h may not satisfy a reservation level expressed before iteration h-k_j). The way such a contradiction is perceived by the DM is very instructive with respect to his current preference structure. If the DM does not react, this contradiction is unimportant to him, which indicates that his preference structure has evolved. If he does, this may reveal strong aspects of his preference structure, in which case k_j could be given a greater value in subsequent iterations. From a practical viewpoint, it is well-known that the number of iterations performed by a DM during a session is rather small (see *e.g.* experiments by Wallenius (1975) and Brockhoff (1985) who respectively report median numbers of iterations of about 10 and 20). Thus, we suggest to set each k_j to a low value (1 or 2) at the beginning of the session; this does not really alter the significance of **k**-consistency especially when the procedure induces the DM to provide preference information regarding several criteria at each iteration.

We recently proposed a procedure based on these principles (briefly described in Vanderpooten and Vincke (1989)).

References

Benayoun, R., de Montgolfier, J., Tergny, J., and Larichev, O., (1971), "Linear Programming with Multiple Objective Functions: STEP Method (STEM)", *Mathematical Programming* **1**, 366-375.

Brockhoff, K., (1985), "Experimental Test of MCDM Algorithms in a Modular Approach", *European Journal of Operational Research* **22**, 2, 159-166.

Steuer, R.E., (1986), *Multiple Criteria Optimization: Theory, Computation and Application*, Wiley, New York.

Vanderpooten, D., (1989), "The Interactive Approach in MCDA: a Technical Framework and some Basic Conceptions", *Mathematical and Computer Modelling* - special issue on MCDM (to appear).

Vanderpooten, D., and Vincke, Ph., (1989), "Description and Analysis of some Representative Interactive Multicriteria Procedures", *Mathematical and Computer Modelling* - special issue on MCDM (to appear).

Wallenius, J., (1975), "Comparative Evaluation of some Interactive Approaches to Multicriterion Optimization", *Management Science* **21**, 12, 1387-1396.

Wierzbicki, A.P., (1986), "On the Completeness and Constructiveness of Parametric Characterizations to Vector Optimization Problems", *OR Spektrum* **8**, 2, 73-87.

CONSTRUCTION OF A MULTIATTRIBUTE RISK FUNCTION AND ITS PROPERTIES FOR ASSESSING HAZARDOUS EVENTS

Fumiko Seo

Kyoto Institute of Economic Research, Kyoto University
Kyoto 606, JAPAN

1. Introduction

This paper discusses the construction of the multiattribute risk function in a way that is consistent with preference analysis in an entire domain.

Regional or urban development programs in modern high-technology society often include some risk; the development programs usually embody adverse as well as beneficial effects on the local inhabitants. In this sense, the development programs are regarded as a risky choice problem. This paper concerns a methodological consideration on the risk assessment for a multiattribute risky choice problem. First, effectiveness of the expected utility hypothesis as a quantitative method for measuring the value of risk is discussed. It is examined whether the representation form of the utility function having a multiattribute risk attitude in measuring the gains from the development programs can be used without any contradiction for evaluation of the value of losses. Second, properties of the multiattribute risk function consistent with a preference function assessed in an entire domain are considered. Third, some examples of the risk evaluation are discussed.

2. Discussions on the expected utility hypothesis

2.1 Concepts of risky choice

Decision for the risky choice problem is a comparative evaluation of alternative policy programs which are described in terms of the probability distribution for their future return (gains and losses). The decision includes the conflicting goals between high return and low risk, where the risk is defined alternatively as (a) dispersion of

---

Acknowledgment. This research is supported by the grant from the Japan Ministry of Education, Science and Culture, for 1987-88 academic years.

outcomes or (b) the probability of losses, or the failure to obtain a certain level of return. It has been observed that, in managerial decisions, the dispersion of outcomes is not so much important as the considerations on the below-target return, or the probability of losses. In particular, the probability of ruinous losses receives the major concern. The decision maker usually assesses the losses with more weights than those on gains, and usually his responses to the losses are more extreme than to the gains (Libby and Fishburn 1977, Fishburn & Kochenberger 1979). We will call this situation the hazardous effect, and treat the managerial decisions in risky choice problems including the hazardous effects as the serious game, in contrast with the hobby game for personal choice in daily life under uncertainty. In the serious game, the catastrophic effects with ruinous losses are the matter of major concern.

2.2 Axiomatic representation of the risky choice

Alternative decision models for the risky choice have been developed: (1) parametric models such as the mean-variance trade-off model, which is based on comparing distribution parameters and is appropriate to treat the risky choice defined by (a) (Markowitz 1959, 1987); (2) the expected utility (EU) model, developed by von Neumann and Morgensterm (1944) and Savage (1954), which can more properly treat the serious game. Savage's axiom which presents the postulates on individual choice between gambles is composed of (i) weak order: connectivity and transitivity; (ii) the sure-thing principle (the substitution axiom), which asserts that a choice among actions is not affected by the payoff of an event producing the same payoff for both actions; (iii) admissibility that means inadmissibility of dominated actions; (iv) independence, which asserts that probability is assessed independently of the payoff; and (v) nontriviality: there is at least one pair of consequences such that a $\succ$ b in an event. The subjective expected utility (SEU) model is derived under Savage's axiom (i)-(v).

$$\begin{aligned} u(\hat{x}) &= u(\sum_{j=1}^{n} p_j x_j) \\ &= \sum_{j=1}^{n} p_j u(x_j) \end{aligned} \tag{2.1}$$

where the prospect $\hat{x} \triangleq (x_1, p_1; \ldots; x_n, p_n)$ shows an n-chance fork gamble.

The SEU model is based on three components.

(1) Representation of subjective probability, where psychological values of probability distributions should be articulated.

(2) Properties of utility function $u(x_i)$, where the linearity is not assured in determining the difference between the level of an attribute x_i and the psychological value $u(x_i)$ to be attached to the attribute level. The utility function $u(x_i)$ can be risk averse, or risk prone (Friedman & Savage 1948, Markowitz 1952, Arrow 1963, Pratt 1964). The asymmetric property in the domains of winning and losing has also been examined (Kahneman & Tversky 1979, Fishburn & Kochenberger 1979).

(3) Representation of mathematical expectation as a psychological value, which is based on the above components (1) and (2) and also holds the particular linear representation form (2.1).

The questions to the SEU hypothesis have been proposed. Allais's paradox (1952) mainly results from the violations of the consistency in holding the axiomatic properties. First is the violation of the substitution axiom; it shows a preference reversal due to shift in probability evaluation. Second is from the asymmetric properties in assessing preference functions, which include some violation of the independence axiom (Ellsberg 1961). These violations of the axiomatic properties represent the certainty effects, including the "more probability" effect and the "pseudo certainty" effect. As a result, the linearity in the representation form of the SEU model has also been confronted with contradiction.

2.3 Extension of the SEU model

In extension of the SEU model for coping with Allais's paradox, the following directions are proposed in accordance with the above components of the SEU model.

(1) Introduction of nonlinearity in probability evaluation. This means using the weighted estimation of probability distribution that can explain the certainty effects (Ellsberg 1961, Kahneman & Tversky 1979, Karmarkar 1978, Handa 1977).

(2) Introduction of a symmetric property of the risk attitudes in the preference function. A counterexample has been considered in terms of the reflection effect with skewed symmetry (Kahneman & Tversky 1979).

(3) Examination of compatibility of the multiattribute risk attitude with the EU model based on the multiattribute utility function. This examination leads to the reconsideration of relationship, or consistency, between the multiattribute utility function and the multiattribute risk function.

In the following discussion, the symmetric property of the preference function in both domains, winning and losing, is examined.

2.4 Properties of the preference function.

The prospect theory (PT) by Kahneman and Tversky (1979) proposed the reflection effect, which asserted that the preference function is concave in the domain for gains and convex for losses, putting the present situation as the reference point, and the response to losses is steeper than that to gains. This finding has been followed by the experiments for the multiattribute risk attitude (Payne, Laughhunn & Crum 1984): multiattribute risk aversion holds for gains, and multiattribute risk prone holds for losses. This finding is inconsistent with representation forms (additive or multiplicative) of the multiattribute utility function because the function should not include the reversal of parameters within an entire domain if it can consistently be assessed with the constant parameters.

However, in the serious games with the hazardous events, differing from the cases of the hobby games, the phenomenon of the preference reversal in the domains for gains and losses will seldom occur. The experiments that have supported the PT hypothesis have been performed only for the hobby games. The serious game, having the probability of catastrophic consequences, includes the following characteristics: (i) possible occurrence of irreversible losses, (ii) existence of a threshold level in acceptability of losses, and (iii) a more risk averse attitude to the ruinous losses than to the gains. The representation form of a preference function that is compatible with the properties (i)-(iii) in an entire domain brings it back to Bernoulli's classical shape. It means that the evaluation will be throughly risk averse in the entire domain of preference (Figure 1). The multiattribute preference function that includes the properties (i)-(iii) also embodies multiattribute risk aversion. It is known that the multiattribute risk averse attitude is implied by the multiplicative form of multiattribute utility function with the negative coefficient parameter (Richard 1975). Thus representation of multiattribute preference function embodying the multiattribute risk aversion is consistent with the multiattribute utility function in the multiplicative form when the reference point is moved to the worst situation of the attributes.

2.5 Effectiveness of the SEU model.

For examining the effectiveness of the SEU model in the extended form of the preference function, independence assumptions for the representation of SEU model are considered.

(i) Event preference independence (the sure thing principle for the multiattribute outcome $x \triangleq (x_1, \ldots, x_m)$. Individual preference between multiattribute gambles $\hat{x}$ and $\hat{y}$ is not affected by replacing a multiattribute outcome x_j in a chance fork j, j = 1,...,n, with x_j', where x_j and x_j' are common to both gambles.

(ii) Attribute preference independence. Individual preference between single attribute gambles $\hat{x}$ and $\hat{y}$ is not affected by replacing other attribute $\bar{x}$ with $\bar{x}'$, where $\bar{x}$ and $\bar{x}'$ are common to both gambles.

(iii) Utility independence (Keeney and Raiffa 1976). Attribute preference independence when the other attributes $\bar{x}$ and $\bar{x}'$ in both gambles are sure outcomes.

(iv) Marginality independence (Raiffa 1969). Individual choice is indifferent between gambles with identical marginal probability distributions on all attributes.

Experiments by Winterfeldt (1971), and Payne, Laughhunn and Crum (1984) presented the systematic violation of the marginality and attribute independence assumptions due to the multiattribute risk averse attitude. It means that the additive utility model should be confronted with a contradiction. However, it is also reported that the event independence and utility independence assumptions still hold some support from their experiments. As Winterfeldt (1980) stressed, the multiattribute risk aversion is compatible with the event and utility independence assumptions, and thus with the nonadditive expected utility model. It is asserted that the multiattribute risk aversion contradicts the additive part in the additive EU model, but is still consistent with the expected part of the EU model. This recognition leads to the support for the multiplicative form of multiattribute utility functions that can be nested in the SEU model.

3. Construction of the multiattribute risk function

Here we consider the construction of the multiattribute risk function (MRF) that is consistent with the preference function that embodies the multiattribute risk averse attitude in an entire domain.

Consider a policy program A with multiple impacts on a region. Possible consequences of the program execution are under uncertainty, and assessed in terms of multiple attributes that are selected as the measure for consideration of the multiple impacts and treated as uncertain quantities. Denote a set of m attributes as $X \triangleq \{X_1,...,X_m\}$ and a set of their quantities as $x \triangleq \{x_1,...,x_m\}$. A probability $p_\ell(x_\ell)$, $\ell = 1,...,m$, is assessed subjectively for each attribute x_ℓ. A set of probability function $p_j \triangleq \{p_1,...,p_m\}$ defines a risk profile of the program A when an event j occurs.

A multidimensional risk function (MRF) of the program A is

defined on the multiattribute risk profile p and represented as

$$R(p) \triangleq R(p_1,\ldots,p_m). \quad (3.1)$$

Properties of MRF are as follows.

(i) MRF represents a multidimensional adverse utility (aversion) function that represents negative human preference. A higher order of magnitude of the MRF-value indicates a higher degree of the evasion, or displeasure, for consequences of the program A.

(ii) Assessment for MRF is subjectively performed on each element p_ℓ in the risk profile p, whose value varies between 0 and 1, $p_{\ell j} > 0$, $\sum_{j=1}^{n} p_{\ell j} = 1$. $R(p) = 0$ when $p = 0$ and $R(p) = 1$ when $p = 1$.

(iii) Assessment for MRF includes the value trade-offs between the attributes in terms of their probability.

Now we must consider representation theorems of MRF.

Assumption 1 (Event independence).

The occurrences of attributes as possible consequences of an event A are independent of each other for the assessor as the subject of probability evaluation.

Assumption 2 (Aversion independence).

Preference (or aversion) for an attribute is independent of that of the other attributes.

Assumption 3 (Trade-off independence).

A trade-off between a pair of attributes is assessed independently of trade-offs for the other pairs of attributes.

Theorem 1 (Representation of MRF).

Under the Assumptions 1-3, the multidimensional (multiattribute) risk function for a state s_j is represented in an additive form,

$$R(p_1,\ldots,p_m) = \sum_{\ell=1}^{m} k_\ell r_\ell(p_\ell), \quad \text{when} \quad \sum_{\ell=1}^{m} k_\ell = 1, \quad (3.2)$$

or, in a multiplicative form,

$$R(p_1,\ldots,p_m) = \frac{1}{K}\left[\prod_{\ell=1}^{m} (Kk_\ell r_\ell(p_\ell) + 1) - 1\right], \quad (3.3)$$

when $\sum_{\ell=1}^{m} k_\ell \neq 1$, where $0 < k_\ell < 1$. K is a solution of $1 + K = \prod_{\ell=1}^{m} (1 + Kk_\ell)$, where $K \neq 0$ and $K > -1$. $0 \leqslant r_\ell \leqslant 1$ and $0 \leqslant R \leqslant 1$.

Proof of Theorem 1 is straightforward on the same line as that for the multiattribute utility function (MUF) by Keeney and Raiffa

(1976).

$r_\ell(p_\ell)$, is a single attribute risk function (SRF) for a state s_j and is assessed on the probability function $p_\ell(x_\ell)$, and $\partial r_\ell/\partial p_\ell > 0$. $r_\ell(p_\ell) = 0$ when $p_\ell = 0$, and $r_\ell(p_\ell) = 1$ when $p_\ell = 1$.

The existence of a critical level and asymptotic approach to it in assessing a risky attribute corresponds to the existence of a saturation level and asymptotic approach to it in utility assessment for gains. In the assessment of the preference function, the negative utility assessment for losses will rapidly approach -1.0 beyond a critical level for an attribute, although the utility assessment is bounded to 1.0. The mapping of this preference assessment onto SRF $r_\ell(p_\ell)$ derives the risk proneness property. In other words, SRF generally has a steeply increasing property corresponding to an increase of the probability for a risky attribute. This reflects the critical effect in the risk evaluation (Figure 2).

Assessment of scaling constants k_ℓ, $\ell = 1,\ldots,m$, on SRF can be performed as the result of trade-off experiments between probabilities for attributes to occur. The trade-off experiment is performed on the following questions.

Question I.

Consider two probability functions $p_r(x_r)$ and $P_h(x_h)$ for unwanted attributes x_r and x_h, where x_h is the most important attribute to be taken into consideration. Suppose that a probability p_r for an occurrence of an unwanted attribute x_r will change from 0 (the best situation) to 1 (the worst situation). Then, how much degradation can the decision maker (DM) allow from the present situation for the most important attribute x_h in terms of its probability ?

Question II.

Consider a probability function p_h for the most important attribute x_h. On the one hand, consider a certain outcome in which the probability p_h takes 1 (the worst situation) and all the other probabilities p_r, $r = 1,\ldots,m$, $r \neq h$, take 0 (the best situation). On the other hand, consider a probability mixture of uncertain consequences (lottery) in which all the probabilities p_ℓ, $\ell = 1,\ldots,m$, take 1 with a probability π or all the probabilities p_ℓ take 0 with a probability $1-\pi$. Then how much is the π-value which makes those two, the certain outcome and the lottery, indifferent for DM ?

Then, for introducing the degree of belief for occurrence of a multiattribute risk profile, the possible biases of the probability evaluation should be taken into consideration. For this purpose, we will use the weighted distribution in the representation form of the expected model for MRF.

$$ER \triangleq \sum_{j=1}^{n} \pi(s_j)R(p_j(x_j)), \tag{3.4}$$

where $R(p_j(x_j))$ is an MRF for an event j as a possible outcome of a risky program, and $\pi(s_j)$ is assessed as the degree of belief for the multiattribute risk profile $p_j(x_j)$ when a state s_j of the event j occurs. $\pi(s_j)$ can be a nonlinear function defined on s_j, which can be assessed on the risk profile p_j and thus $\pi(p_j)$. The representation (3.4) of MRF in the expected form is referred as the expected multiattribute risk function (EMRF). The weighted distribution function $\pi(s_j)$ can be constructed as a decision weight, and takes a hypothetical form (Figure 3). The weight also can reflect a diversification of evaluation for the risk profile.

EMRF can be consistent with the multiplicative form of MRF, which embodies the multiattribute risk averse attitude when $0 > K > -1$.

4. Examples for assessing the hazardous events

(1) Evidence shows the consistent risk aversion for hazardous events. It has been suggested that the preference reversal in assessing risky events be interpreted as representing the relative strength of attractiveness for sure gains and relative strength of aversion for sure losses. This means that preference is risk averse in the positive domain and risk prone in the negative domain. In result, the combined effect will lead to a contradiction (Figure 4). However, if the adverse outcomes include the hazardous events, the sure losses C will be preferred to the gamble D. Then preferences for both domains will be all risk averse, and thus the combined effects can be indifferent. The preference reversal will not occur and thus no contradiction will happen.

(2) The policy implication of constructing the risk function is the feasibility for using it to measure the beneficial effects of executing compensatory policy programs to counteract the occurrence of the hazardous losses. An example is to use the decremental values of MRF as the value of the characteristic function v(S) for a coalition S in the n-person cooperative game.

$$v(S) = \sum_{i \in S} \frac{1}{s} R_i(p_i) - R_S(p_S) \tag{4.1}$$

(3) The dispersion of evaluation for the risk profile can be

assessed in terms of probability distributions, and thus a group decision problem can be introduced as the decision weights $\pi(p_j)$. In this case, the weighted function is assessed as a cumulative distribution function obtained from an observation such as in a brainstorming process.

References.

Allais, M., The foundations of a positive theory of choice involving risk and a criticism of the poetulates and axioms of the American School (1952), in ed. by Allais, M. and O. Hagen, Expected Utility Hypothesis and the Allais Paradox, D. Reidel, 1979.

Arrow, K.J., The theory of risk aversion (1963), in Essays in the Theory of Risk-Bearing, North-Holland, 1971.

Ellsberg, D., 'Risk, ambiguity, and the savage axioms,' Quarterly Journal of Economics, 75, 1961.

Fishburn, P.C. and G.A. Kochenberger, Two-piece von Neumann-Morgenstern utility functions, Decision Sciences, **10**, 1979, 503-518.

Fisher, G.W., M.S. Kamlet, S.E. Fienburg, and D. Schkade, 'Risk preferences for gains and losses in multiple objective decision making,' Management Science, **32(9)**, 1986, 1065-1086.

Friedman, M. and L.J. Savage, The utility analysis of choices involving risk, Journal of Political Economy, **56**, 1948, 279-304.

Handa, J., Risk, probabilities, and a new theory of cardinal utility, Journal of Political Economy, **85(1)**, 1977, 97-122.

Kahneman, D. and A. Tversky, 'Prospect theory: an analysis of decision under risk,' Econometrica, **47(2)**, 263-291, 1979.

Karmarkar, U.S., Subjectively weighted utility:a descriptive extension of the expected utility model, Organizational Behavior and Human Performance, **21**, 1978, 61-72.

Keeney, R.L. and H. Raiffa, Decision with Multiple Objectives: Preference and Value Tradeoffs, John Wiley & Sons, 1976.

Libby, R. and P.G. Fishburn, 'Behavioral models of risk taking in business decisions: a survey and evaluation,' Journal of Accounting Reséarch, **15**, 1977, 272-292.

Markowitz, H.M., The utility of welfare, Journal of Political Economy, **60**, 1952, 151-158.

Markowitz, H.M., Portfolio Selection, Yale University Press, 1959.

Markowitz, H.M., Mean-Variance Analysis in Portfolio Choice and Capital Markets, Blackwell, 1987.

Payne, J.W., D.J. Laughhunn, and R. Crum, 'Multiattribute risky choice behavior: the editing of complex prospects,' Management Science, **30(11)**, 1350-1361, 1984.

Pratt, J.W., Risk aversion in the small and in the large, Econometrica, **32(1-2)**, 1964, 122-136.

Raiffa, H., Preferences for multiattributed alternatives, RM-5868-DOT/RC, The Rand Corporation, Calif., 1969.

Richard, S.F., 'Multivariate risk aversion, utility independence and separable utility functions,' Management Science, **22(1)**, 12-21, 1975.

Savage, L.J., The Foundation of Statistics, Wiley, 1954.

Seo, F. and M. Sakawa, Multiple Criteria Decision Analysis in Regional Planning: Concepts, Methods and Applications, D. Reidel Publishing Co., 1988.

Tversky, A. and D. Kahneman, 'The framing of decisions and the psychology of choice,' Science, **211**, 453-458, 1981.

von Neumann, J. and O. Morgenstern, Theory of Games and Economic Behavior, Wiley, 1944, 3rd ed. 1953.

von Winterfeldt, D., 'Additivity and expected utility in risky multiattribute preference,' Journal of Mathematical Psychology, 21, 66-82, 1980.

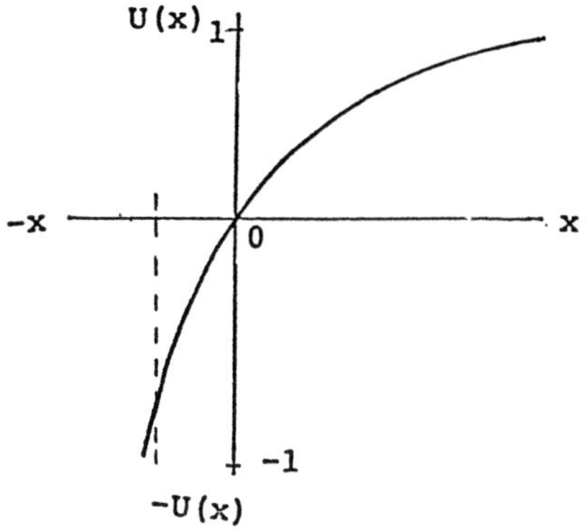

Figure 1

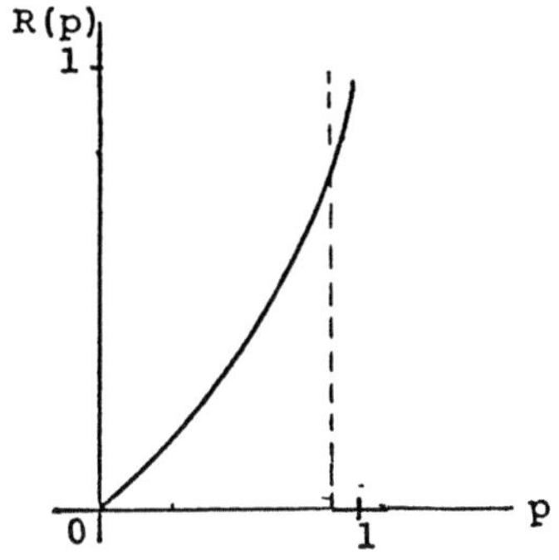

Figure 2

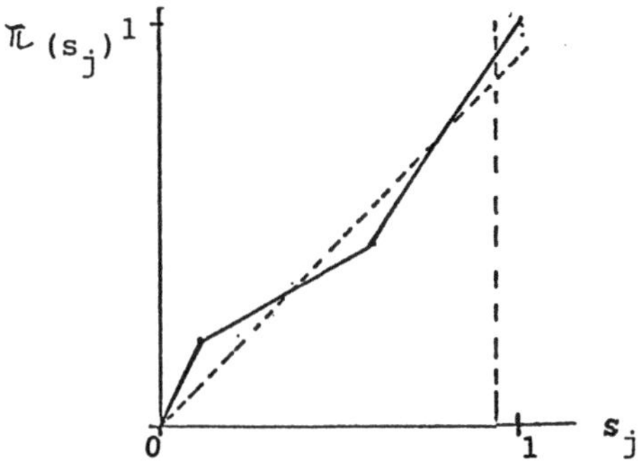

Figure 3

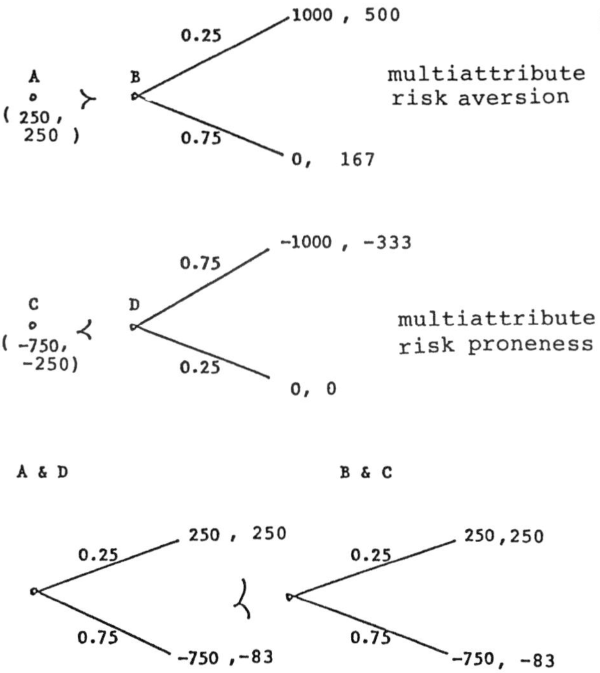

Figure 4

NONDOMINATED TRADEOFFS AND TERMINATION IN INTERACTIVE MULTIPLE OBJECTIVE LINEAR PROGRAMMING

Merja Halme and Pekka Korhonen

Helsinki School of Economics,
Runeberginkatu 14-16, 00100 Helsinki, Finland

Abstract

This paper deals with the problem of finding, for a given nondominated solution, a set of nondominated tradeoffs. Such tradeoffs play a central role in interactive multiple objective linear programming procedures when direction-finding and termination are considered. If the current solution is preferred to all of the tradeoffs, it is optimal, provided that the decision maker's value (utility) function is assumed to be (globally) pseudo-concave at the moment of consideration. If any of the tradeoffs is a direction of improvemet, a new search direction can be found and the procedure continues as earlier.

Finally, we point out that the stopping rule proposed above for using in interactive multiple objective linear programming procedures is analogous to verifying the Kuhn-Tucker conditions for optimality in a single objective problem.

Keywords: Multiple Objective, Linear, Interactive, Efficiency, Termination.

1. Introduction

Several interactive procedures have been developed for solving optimization problems having multiple criteria. In interactive procedures an exploration over the feasible or efficient set is conducted for locating the most preferred solution. As Steuer (1986) notes, interactive procedures are characterized by phases of decision-making alternating with phases of computation. At each iteration, one nondominated solution or a subset of nondominated solutions is generated and presented for the decision-maker's (DM) evaluation. As a result of the examination, the DM inputs information in the form of tradeoffs, pairwise comparisons, aspiration levels, marginal rates of substitution, etc. His/her responses are used to generate a presumably improved solution. The procedures terminate in a satisfactory or a satisfactorily near-optimal solution, or simply when the DM so chooses.

The nature and type of information requested from the DM differs from one procedure to another. Also, the mathematical assumptions upon which the procedures are based vary. Which solutions are considered as possible compromise solutions is heavily dependent on these assumptions. Unrealistic assumptions about the DM's behavior may force him/her to accept unsatisfactory solutions. On the other hand, in "free search" type of procedures, such as PARETO RACE (Korhonen and Wallenius, 1988), the DM may have difficulties in finding directions of improvement, particularly in the case of a large number of objectives.

Another major problem in interactive procedures is termination. Unfortunately, we cannot apply the theory developed in mathematical progamming as such, because the character of interactive procedures is different. Interactive procedures stop, when the DM is getting tired, the system terminates the search using some stopping rule, or the DM feels that the current solution is "good enough". However, to model the concept "good enough" is problematic.

In this paper, we study direction-finding and termination. Our purpose is to develop an approach, which requires a minimum amount of information from the DM at each iteration for checking optimality conditions, and provides the DM with a new search direction, if optimality is not reached. The approach is based on the assumption that the DM's value (utility) function is pseudo-concave at the moment of consideration. During the search process, no assumptions whatsoever concerning the value function are needed. The approach is based on considering nondominated tradeoffs at the current solution.

The paper consists of five sections. In the first section, we have described the purpose of the study. In the second section, some preliminary considerations are presented. The third section provides the theory, and in the fourth section, we show that our approach is analogous to verifying the Kuhn-Tucker conditions for optimality in the single objective problem. Concluding remarks are given in the last section.

2. Preliminary Considerations

We use the following notation. The n-dimensional Euclidean space is denoted R^n. The notation $\mathbf{x}$ represents a column vector, whereas $\mathbf{x}'$ represents a row vector. Matrices are denoted by capital letters, such as A, B, and C. The entry in row i and column j of a matrix A is denoted by a_{ij} and the jth column of A is denoted by $\mathbf{a}_j$. The zero vector is denoted by $\mathbf{0}$ and the vector with each component equal to one is denoted by $\mathbf{1}$. The ith unit vector is denoted by $\mathbf{e}_i$, and consists of zeros except for a 1 at the ith position. The (n x n)-identity matrix is denoted by I and sometimes by I_n to highlight its dimension.

Let us consider the following multiple objective linear programming problem:

$$\text{"max" } \mathbf{q} = \mathbf{f}(\mathbf{x}) = (f_1(\mathbf{x}), f_2(\mathbf{x}), \ldots, f_k(\mathbf{x}))' = C\mathbf{x},$$

$$\text{s.t.} \qquad \mathbf{x} \in X = \{\mathbf{x} \mid A\mathbf{x} = \mathbf{b},\ \mathbf{x} \geq \mathbf{0}\}, \tag{2.1}$$

where C is a (k x n)-coefficient matrix of the objective functions, $\mathbf{x}$ is a vector of decision variables, A is an (m x n)-coefficient matrix with rank m, and $\mathbf{b}$ is a non-zero rhs-vector. In addition, we assume that the matrix $[C',A']$ is of rank k+m. To simplify the notation, we assume that X is bounded. The set of feasible solutions in the criterion space R^k, is called the feasible region and denoted by Q, $Q = \{C\mathbf{x} \mid \mathbf{x} \in X\}$.

<u>Definition 2.1.</u> A point $\mathbf{x}^o \in X$ is efficient iff there does not exist another $\mathbf{x} \in X$ such that $f_i(\mathbf{x}) \geq f_i(\mathbf{x}^o)$ for all $i \in K = \{1,2,..,k\}$ and $f_i(\mathbf{x}) \neq f_i(\mathbf{x}^o)$ for at least one $i \in K$.

<u>Definition 2.2.</u> A point $\mathbf{x}^o \in X$ is weakly-efficient iff there does not exist another $\mathbf{x} \in X$ such that $f_i(\mathbf{x}) > f_i(\mathbf{x}^o)$ for all $i \in K$.

The criterion vectors $\mathbf{q} \in Q$ corresponding to (weakly) efficient points are called (weakly) nondominated solutions.

Definition 2.3. A nonempty set G_x defined in an n-dimensional Euclidean space R^n is called a **cone** with vertex **x** if $\mathbf{x} + \mathbf{y} \in G_x \Rightarrow \mathbf{x} + \mu\mathbf{y} \in G_x$ for all $\mu \geq 0$. The cone with vertex zero is denoted by G.

Definition 2.4. Let $\mathbf{x} \in X$. The cone of (weakly) efficient directions of X at **x**, denoted by D(**x**), is given by

$$D(\mathbf{x}) = \{\mathbf{d} \mid \mathbf{d} \neq \mathbf{0},\ \mathbf{x} + \mu\mathbf{d} \in X, \text{ and } \mathbf{x} + \mu\mathbf{d} \text{ is (weakly) efficient for all } \mu \in (0,\delta) \text{ for some } \delta > 0\}.$$

The corresponding cone $H(\mathbf{q}) = \{\mathbf{h} \mid \mathbf{h} = C\mathbf{d},\ \mathbf{d} \in D(\mathbf{x}),\ \mathbf{q} = C\mathbf{x}\}$ in the criterion space is called the cone of (weakly) nondominated tradeoffs at **q**.

It is a well-known result that a point **x** is an extreme point of X iff A can be decomposed into [B,N] such that

$$\mathbf{x} = \begin{bmatrix} \mathbf{x}_B \\ \mathbf{x}_N \end{bmatrix} = \begin{bmatrix} B^{-1}\mathbf{b} \\ \mathbf{0} \end{bmatrix}, \qquad (2.2)$$

where B is an m x m invertible matrix satisfying $B^{-1}\mathbf{b} \geq \mathbf{0}$ (see, e.g., Bazaraa and Shetty 1979, p. 56).

If X is nonempty and $\mathbf{x} \in X$, we can present any $\mathbf{x} \in X$ in the form corresponding to (2.2):

$$\mathbf{x} = \begin{bmatrix} \mathbf{x}_B \\ \mathbf{x}_N \end{bmatrix} = \begin{bmatrix} B^{-1}\mathbf{b} - B^{-1}N\mathbf{x}_N \\ \mathbf{x}_N \end{bmatrix}$$

where B is an (m x m) invertible matrix satisfying $B^{-1}\mathbf{b} - B^{-1}N\mathbf{x}_N \geq \mathbf{0}$, $\mathbf{x}_N \geq \mathbf{0}$, and $\mathbf{x}_B > \mathbf{0}$, if **x** is not a degenerate solution.

Definition 2.5. The cone G(**x**) in R^n is called a covering cone for X at point **x** if G(**x**) is a cone and if for all $\mathbf{y} \in X \Rightarrow \mathbf{y} - \mathbf{x} \in G(\mathbf{x})$. Further, G(**x**) is minimal, if $\mathbf{d} \in G(\mathbf{x}) \Rightarrow \mathbf{d}$ is a feasible direction.

Next we show, how the minimal covering cone can be found for any nondegenerate $\mathbf{x} \in X$.

Theorem 2.1. Let $\mathbf{x} \in X$ with

$$\mathbf{x} = \begin{bmatrix} \mathbf{x}_B \\ \mathbf{x}_N \end{bmatrix} = \begin{bmatrix} B^{-1}\mathbf{b} - B^{-1}N\mathbf{x}_N \\ \mathbf{x}_N \end{bmatrix} \geq \mathbf{0}$$

and $\mathbf{x}_B > \mathbf{0}$, then a minimal covering cone $G(\mathbf{x})$ at $\mathbf{x}$ can be found as follows:

$$G(\mathbf{x}) = \{\mathbf{d} \mid \mathbf{d} = \sum_{\substack{j=1 \\ (\mathbf{x}_N)_j > 0}}^{n-m} \mu_j \mathbf{t}_j + \sum_{k=1}^{n-m} \mu_k \mathbf{s}_k,\ \mu_j,\ \mu_k \geq 0\},$$

where

$$\mathbf{s}_k = \begin{bmatrix} -B^{-1}N \\ I_{n-m} \end{bmatrix}_k,\ k=1,2,..,n-m,$$

$\mathbf{t}_j = -\mathbf{s}_j$ for $j=1,2,\ldots,n-m$, if $(\mathbf{x}_N)_j > 0$.

Proof. First we prove that $G(\mathbf{x})$ is a covering cone. Clearly, $G(\mathbf{x})$ is a cone. Suppose that $\mathbf{y} \in X$. We show that $\mathbf{y} - \mathbf{x} \in G(\mathbf{x})$. Decompose $\mathbf{x}$ and $\mathbf{y}$ into $(\mathbf{x}_B,\mathbf{x}_N)'$ and $(\mathbf{y}_B,\mathbf{y}_N)'$, respectively. Now we can write

$$\mathbf{y} - \mathbf{x} = \begin{bmatrix} -B^{-1}N(\mathbf{y}_N - \mathbf{x}_N) \\ \mathbf{y}_N - \mathbf{x}_N \end{bmatrix} = \begin{bmatrix} -B^{-1}N \\ I_{n-m} \end{bmatrix} (\mathbf{y}_N - \mathbf{x}_N).$$

Hence, it follows that

$$\mathbf{y} - \mathbf{x} = \sum_{k=1}^{n-m} \mathbf{s}_k\ (\mathbf{y}_N - \mathbf{x}_N)_k \qquad (2.3)$$

by the definition of $\mathbf{s}_k$. If $(\mathbf{y}_N-\mathbf{x}_N)_k \geq 0$, then we can choose $\mu_k = (\mathbf{y}_N-\mathbf{x}_N)_k$. If $(\mathbf{y}_N-\mathbf{x}_N)_k < 0 \Rightarrow (\mathbf{x}_N)_k > 0$, because $\mathbf{y} \in X \Rightarrow \mathbf{y} \geq \mathbf{0}$. Now we denote $\mu_k = (\mathbf{x}_N-\mathbf{y}_N)_k > 0$, and set $\mathbf{t}_k = -\mathbf{s}_k \in G(\mathbf{x})$ by definition.

Assume $\mathbf{d} \in G(\mathbf{x})$. Since $\mathbf{x}_B > \mathbf{0}$, we can choose $\delta \in (0,\delta_1]$ such that

$$\delta_1 = \max_{\delta}\ \mathbf{x}_B + \delta \mathbf{d}_B \geq \mathbf{0},$$

where $\mathbf{d}$ is decomposed into $(\mathbf{d}_B, \mathbf{d}_N)'$. On the other hand, $\mathbf{x}_N + \delta \mathbf{d}_N \geq 0$ for sufficiently small δ, because $(\mathbf{d}_N)_k < 0$ only when $(\mathbf{x}_N)_k > 0$ according to the definition of $G(\mathbf{x})$. This proves that $\mathbf{d}$ is a feasible direction. The proof is complete.

When the point $\mathbf{x} \in X$ is (weakly) efficient, we can identify a cone of (weakly) efficient directions as a subset of the covering cone and use it for finding a new search direction or proving optimality .

3. Theory

For dealing with efficient solutions, we can use an achievement (scalarizing) function proposed by Wierzbicki (1980). Such a function projects any given (feasible or infeasible) point $\mathbf{g} \in R^k$ onto the set of (weakly) nondominated solutions. The function is suitable both for generating nondominated solutions and checking the dominance of a given solution.

An achievement function may be specified in different ways (see, e.g., Wierzbicki, 1980 and 1986). A practical and simple form for this function is the following:

$$z(\mathbf{g},\mathbf{q},\mathbf{w}) = \max \{g_i - q_i / w_i,\ i=1,2,...,k\}, \tag{3.1}$$

where $\mathbf{w}>\mathbf{0}$ is a k-vector of weights, and $\mathbf{g} \in R^k$ an arbitrary vector, the components of which are called aspiration levels. By minimizing $z(\mathbf{g},\mathbf{q},\mathbf{w})$ subject to $\mathbf{q} \in Q$, we find a weakly nondominated solution vector $\mathbf{q}^o \in Q^W$ (see, e.g., Wierzbicki, 1980 and 1986). If the given solution $\mathbf{g} \in R^k$ is feasible, then $\mathbf{q}^o \in Q$, $\mathbf{q}^o \geq \mathbf{g}$. To generate only nondominated (instead of weakly nondominated) solutions, more complicated forms have to be used. However, this simple form is computationally efficient.

Using the achievement function (3.1), the multiple objective linear programming problem can be presented in the following parametric form (see, Korhonen and Wallenius, 1988), which makes all efficient solutions extreme point solutions:

$$\begin{aligned} &\min\ y \\ &\text{s.t.} \\ &A\mathbf{x} = \mathbf{b} \\ &C\mathbf{x} + \mathbf{w}y - I\mathbf{z} = \mathbf{g} \\ &\mathbf{x},\mathbf{z} \geq 0, \end{aligned} \tag{3.2}$$

where y is a scalar variable to be minimized and **z** is the k-vector of surplus variables for objective rows. Given $\mathbf{g} \in R^k$, as a solution we will find a weakly efficient point **x**. To move on the efficient frontier, we can vary the vector **w** or the aspiration levels **g**. We use the latter way by formulating the problem in a parametric form:

$$
\begin{aligned}
&\min\ y \\
&\text{s.t.} \qquad\qquad\qquad\qquad\qquad\qquad\qquad\qquad (3.3) \\
&A\mathbf{x} = \mathbf{b} \\
&C\mathbf{x} + \mathbf{w}y - I\mathbf{z} = \mathbf{g} + t\mathbf{d} \\
&\mathbf{x}, \mathbf{z} \geq \mathbf{0},
\end{aligned}
$$

where $\mathbf{d} \in R^k$ is a so-called reference direction (Korhonen and Laakso, 1986a), and $t \geq 0$ a parameter. There are several ways to specify a reference direction. Korhonen and Laakso (1986a) proposed the use of aspiration levels for this purpose. After seeing the current solution $C\mathbf{x}^*$, where $\mathbf{x}^*$ is an efficient point, the DM is asked to give new aspiration levels $\mathbf{g}^1$, and a reference direction is computed $\mathbf{d} = \mathbf{g}^1 - C\mathbf{x}^*$. In PARETO RACE developed by Korhonen and Wallenius (1988) the reference direction is specified dynamically by the system. The system generates reference directions bearing a relationship to the DM's desire to make relative changes in the values of the objectives.

Neither in the original version of Korhonen and Laakso (1986a) nor in PARETO RACE (Korhonen and Wallenius, 1988), is the reference direction feasible. In this paper, we will consider feasible and nondominated reference directions with an aim to provide the DM with the maximal amount of relevant information about the environment of the current solution.

First, we prove the following Lemma.

Lemma 3.1. The efficient set of problem (2.1) can be presented as a subset of the union of the at most (k-1)-dimensional facets of the polyhedron $X=\{x \mid A\mathbf{x} = \mathbf{b},\ \mathbf{x} \geq 0\}$, where k is the number of objective functions.

Proof. Consider problem (3.2) and let $\mathbf{x}^*$ be an arbitrary efficient point corresponding to the optimum of y with a given **w** and **g**. The basic solution of problem (3.2) always consists of m+k variables. It is easy to see that y is always one of the basic variables. Because **w** and the unit vectors $\mathbf{e}_i, i=1,2,\dots,k$, corresponding to the surplus variables z are linearly dependent, they cannot all be in the basis, simultaneously. Thus, the maximum number of the z variables in the basis is k-1. The other basic variables are x-variables. Let us now consider the polyhedron $X = \{\mathbf{x} \mid \mathbf{Ax} = \mathbf{b},\ \mathbf{x} \geq 0\}$. If y and all (k-1) z-variables are in the basis, the number of x-variables in the basis is m. This means that the solution is an extreme point of the polyhedron X, provided that no degeneracy occurs. If no z-variables are in the basis, then the total number of x-variables in the basis is m+k-1, implying that the maximal facet, on which an efficient solution lies is a (k-1)-dimensional facet of the polyhedron X. This completes the proof.

Definition 3.1. An efficient point $\mathbf{x}^*$ is nondegenerate, if each optimal basic x-variable of problem (3.2) has a strictly positive value.

Definition 3.2. An efficient point $\mathbf{x}^*$ is said to be on an r-dimensional facet, if it is a relative interior point (see, Rockafellar, 1970, p.43) of an r-dimensional facet.

Corollary 3.1. Assume that $\mathbf{x}^*$ is a nondegenerate efficent point on an r-dimensional facet, then the whole r-dimensional facet is efficient.

Proof. The same basis is valid on the whole facet. Thus, if the optimality conditions are fulfilled at $\mathbf{x}^*$, they are fulfilled on the whole facet.

In the sequel, we will consider nondegenerate efficient solutions on a k-1 dimensional facet. Our aim is to find the vectors which span the covering cone at an efficient point $\mathbf{x}^*$, and identify the efficient directions.

The variables of problem (3.2) are denoted by $\mathbf{u}' = [\mathbf{x}', y, \mathbf{z}']$ and decomposed into the basic and nonbasic variables $\mathbf{u}' = [\mathbf{u}_B', \mathbf{u}_N']$. When necessary, we write $\mathbf{u}_B' = [\mathbf{x}_B', y, \mathbf{z}_B']$ and $\mathbf{u}_N' = [\mathbf{x}_N', \mathbf{z}_N']$. The coefficient matrix is decomposed into two parts, respectively:

$$\begin{bmatrix} A & 0 & 0 \\ C & \mathbf{w} & -I_k \end{bmatrix} = [B, N], \tag{3.4}$$

where an (m+k) x (m+k) matrix B is the basis and the (m+k) x (n+1-m) matrix N corresponds to nonbasic variables. Now we can write

$$Bu_B + Nu_N = \mathbf{r},$$

and furthermore,

$$\mathbf{u} = \begin{bmatrix} u_B \\ u_N \end{bmatrix} = \begin{bmatrix} B^{-1}\mathbf{r} - B^{-1}Nu_N \\ u_N \end{bmatrix}, \qquad (3.5)$$

where $\mathbf{r}' = [\mathbf{b}', \mathbf{g}']$, $u_N = \mathbf{0}$. Note that all weakly efficient solutions of problem (3.2) are extreme points of the extended linear system, but not of the original set of X.

Define the cone $G(\mathbf{u})$:

$$G(\mathbf{u}) = \{ \mathbf{d} \mid \mathbf{d} = \sum_{k=1}^{n+1-m} t_k \begin{bmatrix} -B^{-1}\mathbf{n}_k \\ \mathbf{e}_k \end{bmatrix}, \ t_k \geq 0, \ k=1,\ldots,n+1-m\}, \quad (3.6)$$

where $\mathbf{n}_k$ is the kth column of matrix N. We easily see that $G(\mathbf{u})$ is a covering cone for X at $\mathbf{u}$. In the case of a nondegenerate basic solution, all vectors of G are also feasible directions of problem (3.2) and thus G is minimal.

Denote by

$$\mathbf{d}_k = \begin{bmatrix} -B^{-1}\mathbf{n}_k \\ \mathbf{e}_k \end{bmatrix}, \quad k = 1,\ldots,n+1-m \qquad (3.7)$$

the spanning directions of $G(\mathbf{u})$.

At the optimum of problem (3.2), the optimality condition is $\mathbf{e}_j' B^{-1}N \leq \mathbf{0}$, where y is assumed to be the jth variable in the basis. In the case of a unique solution, the strict inequality holds. To compute the values of the objective functions at point $\mathbf{x}^*$, we can simply use the direct way $\mathbf{q}^* = C\mathbf{x}^*$ or the indirect way:

$$\mathbf{q}^* = \mathbf{g} - \mathbf{w}y^* + I\mathbf{z}^*, \qquad (3.8)$$

where y^* and z^* are the values of y and $\mathbf{z}$ at the optimum. In the sequel, we will use the indirect way (3.8), because we only consider the values of y and z-variables.

Except for the boundary points of the efficient frontier, the efficient solutions lying on lower than (k-1)-dimensional facets are rare, because the (k-1)-dimensional measure of those facets is zero. Thus their likelihood to appear is zero.

Consider an efficient point $\mathbf{x}^*$ on a (k-1)-dimensional facet. On a (k-1)-dimensional facet the basis consists of the variable y and (m+k-1) x-variables. Thus, all z-variables are nonbasic. We are especially interested in the efficient directions of the covering cone. In the next section, we will show that dealing with such directions is sufficient for checking optimality.

<u>Lemma 3.2.</u> At a nondegenerate efficient solution $\mathbf{x}^*$ on a (k-1)-dimensional facet no direction of the covering cone corresponding to a nonbasic x-variable is efficient.

<u>Proof.</u> On a (k-1)-dimensional facet, no z-variable is in the basis. The basic variables are y and (m+k-1) x-variables. On the other hand, we assumed that the solution is nondegenerate, implying that all basic x-variables have a strictly positive value. Thus, all directions (3.7) of the covering cone (3.6) are feasible. Corresponding to a unit change in the nonbasic x-variable, the changes in the values of the objective functions can be presented as follows:

$$\Delta \mathbf{q}_k = -\mathbf{w}\Delta y_k,$$

where $\Delta y_k = -\mathbf{e}_j' B^{-1}\mathbf{n}_k$, and $\mathbf{n}_k$ is a vector corresponding to one of the nonbasic x-variables. Note that $\Delta \mathbf{z}_k = 0$, because no z-variables are in the basis. From the optimality of y, it follows that $-\mathbf{w}\Delta y_k = -\mathbf{w}(-\mathbf{e}_j' B^{-1}\mathbf{n}_k) \leq \mathbf{0}$, indicating that all objective functions are decreasing.

<u>Lemma 3.3.</u> At a nondegenerate efficient solution $\mathbf{x}^*$ on a (k-1) dimensional facet, any direction of the covering cone corresponding to a nonbasic z-variable is efficient.

<u>Proof.</u> Let $\mathbf{x}^*$ be a nondegenerate efficient point on a (k-1) dimensional facet. The direction corresponding to a nonbasic z-variable is a feasible direction on the (k-1) dimensional facet. According to Corollary (3.1) the direction must be efficient.

<u>Theorem 3.1.</u> At a nondegenerate efficient solution $\mathbf{x}^*$ on a (k-1)-dimensional facet of the polyhedron $X=\{\mathbf{x} \mid A\mathbf{x} = \mathbf{b},\ \mathbf{x} \geq 0\}$, the

efficient part of the covering cone $G(\mathbf{x}^*)$ is spanned by the directions corresponding to the nonbasic z variables.

Proof. The cone $G(\mathbf{x}^*)$ can be presented as a nonnegative linear combination of the spanning directions corresponding to the nonbasic z- and x-variables. Each vector corresponding to an x-variable is not efficient, and thus it is not on the same (k-1)-dimensional facet as $\mathbf{x}^*$. On the other hand, the directions corresponding to the z-variables are efficient and span the efficient part of $G(\mathbf{x}^*)$.

4. Termination

In mathematical programming the convergence and the verification of optimality conditions play a central role. These considerations are based on the behavior of an objective function in some constraint set. In interactive procedures, no such explicit objective function is available. Instead, our problem is to optimize some implicit function, which may even change during the search process for behavioral reasons. In interactive procedures the convergence is no problem - the procedure always finds a solution in a few steps. The difficulty is to evaluate how good the final solution is - at the moment of termination. Therefore in interactive procedures, we have to concentrate on termination, and try to introduce some optimality conditions under certain assumptions about the objective function. However, instead of dealing with an objective function and its gradient we have to base the termination conditions on the use of feasible directions, which the DM can evaluate.

Korhonen and Laakso (1986a) introduced an optimality checking procedure based on the nondominated tradeoffs at the nondominated solution $\mathbf{q}^*$, which is assumed to be the best solution found by the DM, so far. They proved (Korhonen and Laakso, 1986b) that if none of these tradeoffs is a direction of improvement, the current solution $\mathbf{q}^*$ can be considered optimal, provided the DM's unknown value (utility) function is a pseudoconcave function of the decision variables at the moment of consideration. If any of these directions is preferable to the DM, a new improved search direction can be found.

In this section, we show that Korhonen and Laakso's procedure is analogous to verifying the Kuhn-Tucker optimality conditions in single objective optimization.

Theorem 4.1. Assume that $\mathbf{f}: \mathbf{R}^n \to \mathbf{R}^k$ and that f_i is linear for all $i=1,\ldots,k$. Moreover, assume $v:\mathbf{R}^k \to \mathbf{R}$ is a pseudo-concave function and increasing in all its arguments. Then function $u: \mathbf{R}^n \to R$, $u(\mathbf{x}) = v(\mathbf{f}(\mathbf{x}))$ is also pseudo-concave.

Proof. See Korhonen and Laakso (1986b).

Theorem 4.2. Assume that

$$G(\mathbf{x}^*) = \{\mathbf{d} \mid \mathbf{d} = \Sigma_i a_i \mathbf{s}_i \ , a_i \geq 0 \} \quad (4.1)$$

is a covering cone for X at $\mathbf{x}^*$, $X \subset F_{\mathbf{x}^*} = \{\mathbf{y} \mid \mathbf{y} = \mathbf{x}^* + \mathbf{d},\ \mathbf{d} \in G(\mathbf{x}^*)\}$, and $u(\mathbf{x})$ is a pseudo-concave function of $\mathbf{x}$ on $F_{\mathbf{x}^*}$.If $u(\mathbf{x}^*) \geq u(\mathbf{x}^* + a_i \mathbf{s}_i)$ for all $a_i \geq 0$ and $i=1,\ldots,r$, then $u(\mathbf{x}^*)$ is the global maximum of u in X.

Proof. See Korhonen and Laakso (1986b).

Theorem 4.2. says that we can stop the search procedure, if the DM does not like any of the directions spanning the covering cone $G(\mathbf{x}^*)$. Actually, we have to ask the DM to evaluate only nondominated tradeoffs to ensure the optimality of the current solution $\mathbf{x}^*$ (see, Korhonen and Laakso, 1986b, Theorem 3).

Now, we are ready to show that the above termination condition is analogous to the Kuhn-Tucker optimality conditions.

Theorem 4.3. Assume $u(\mathbf{x})$ is a pseudo-concave (value) function of $\mathbf{x}$ on $F_{\mathbf{x}^*}$, and the DM likes none of the directions $\mathbf{s}_i$, $i=1,2,\ldots,r$, spanning the covering cone $G(\mathbf{x}^*)$ (4.1), then the Kuhn-Tucker optimality conditions are fulfilled.

Proof. Because the DM does not like any of the directions $\mathbf{s}_i$, $i=1,2,\ldots,r$, it means that $u(\mathbf{x}^*) \geq u(\mathbf{x}^* + a_i \mathbf{s}_i)$ for all $a_i \geq 0$, $i=1,\ldots,r$, implying

$$\nabla u'(\mathbf{x}^*)\mathbf{s}_i \leq \mathbf{0}, \quad \text{for all } i=1,\ldots r.$$

Thus

$$\nabla u'(\mathbf{x}^*)(\mathbf{x} - \mathbf{x}^*) \leq \mathbf{0}, \text{ for all } \mathbf{x} - \mathbf{x}^* \in G(\mathbf{x}^*).$$

By definition, the covering cone consists of all solutions, $\mathbf{x} - \mathbf{x}^* \in G(\mathbf{x}^*)$, for which $A(\mathbf{x} - \mathbf{x}^*) = \mathbf{0}$. Thus

$$\forall(\mathbf{x} - \mathbf{x}^*) \ (A(\mathbf{x} - \mathbf{x}^*) = 0 \ \Rightarrow \nabla u'(\mathbf{x}^*) \ (\mathbf{x} - \mathbf{x}^*) \le 0 \ ).$$

By an alternative formulation of Farkas' Lemma (see, e.g., Bazaraa and Shetty, 1979, p. 47) the above is equivalent to

$$\exists \mathbf{y} \text{ such that } \nabla u(\mathbf{x}^*) = A'\mathbf{y},$$

which is the Kuhn-Tucker condition for optimality.

In section 3 we have shown that at most k efficient directions at a nondegenerate efficient point on a (k-1)-dimensional facet are needed. It means that for checking the optimality, we need to ask the DM only k questions.

5. Concluding Remarks

In this paper, we have shown how the efficient directions can be found for any nondegenerate efficient solution on a (k-1)-dimensional facet. We have pointed out that at most k such directions are efficient. These directions can be used for finding a new search direction in interactive multiple criteria procedures and making judgements about termination. If none of these efficient directions is preferable to the DM, then the current solution is optimal, provided that the DM's value function may be assumed to be pseudo-concave at the moment of termination.

In this study, we have considered only nondegenerate efficient solutions on (k-1)-dimensional facets. Degenerate solutions and solutions on the boundary of the efficient frontier are a subject for future research.

References

Bazaraa, M. and Shetty, C. , Nonlinear Programming: Theory and Algorithms, John Wiley & Sons, New York, 1979.

Korhonen, P., and Laakso, J., "A Visual Interactive Method for Solving the Multiple Criteria Problem", European Journal of Operational Research, 24(2), 277-287, 1986a.

Korhonen, P. and Laakso, J., "Solving Generalized Goal Programming Problems Using a Visual Interactive Approach", European Journal of Operational Research, 26(3), 355-363, 1986b.

Korhonen, P. and Wallenius, J., "A Pareto Race", Naval Research Logistics, 6, 615-623, 1988.

Rockafellar, R.T., Convex Analysis, Princeton, N.J., 1970.

Steuer, R., Multiple Criteria Optimization: Theory, Computation, and Application, Wiley, New York, 1986.

Wierzbicki, A., "The Use of Reference Objectives in Multiobjective Optimization", in Multiple Criteria Decision Making, Theory and Application, G. Fandel and T. Gal, Eds., Springer-Verlag, Berlin, pp. 468-486, 1980.

Wierzbicki, A., "On the Completeness and Constructiveness of Parametric Characterizations to Vector Optimization Problems", OR-Spectrum, 8, 73-87, 1986.

Partial Information and Sensitivity Analysis in Multi-objective Decision Making

Simon French and David Rios Insua
School of Computer Studies
University of Leeds
Leeds, LS2 9JT, UK

Abstract

Many decision aids have been designed to help decision makers choose between several alternatives. Such aids are typically intended to deal with some or all of the problems arising from conflicting objectives, uncertainty, groups of decision makers, etc.

In all quantitative analysis is essential to check the sensitivity of the conclusions to the data. In designing decision aids this is even more true as the decision makers become aware of the implications and possible inconsistencies of their judgements. Sensitivity analysis can focus on those judgemental inputs which are more important and, therefore, need to be revised the most carefully.

In this paper we review some of the previous approaches to sensitiviy analysis. With a few exceptions, they consider sensitivity to one or, at most, two data inputs at a time, the remaining data being taken as fixed. With recent advances in optimisation theory and with the advent of cheap computer power, it seems possible to provide a more general approach to sensitivity analysis, allowing for simultaneous variation in all the data. We briefly introduce some tools to help in this task.

1 The Problem

Our set of alternatives will be $A = \{a_1, \ldots, a_m\}$. The problem is introduced in the decision table below. The multi-attributed consequence $a_{ij} = (a^1_{ij}, \ldots, a^n_{ij})$ arises when alternative a_i is selected and θ_j is the state.

		States $\theta_1 \ldots\ldots \theta_k$
Probabilities		$p_1 \ldots\ldots p_k$
Alternatives	a_1	$a_{11} \ldots\ldots a_{1k}$
	a_2	$a_{21} \ldots\ldots a_{2k}$
	.	...
	.	...
	a_m	$a_{m1} \ldots\ldots a_{mk}$

If $k = 1$, the problem is one under certainty and we shall use a value function v to rank the alternatives. If $k > 1$, there is uncertainty present, modelled by the subjective probabilities p_j, and we shall use a utility function u whose expected value ranks them. Both cases will be represented by means of a function Ψ which ranks the alternatives. By building u, v, p we model the decision maker's (DM) preferences and beliefs, encoding her judgemental inputs in some parameters w: thus $\Psi = \Psi(a, w)$, the space of parameters being W. Although we know the uniqueness properties of u, v, p (French, 1986), it may be difficult to elicit precise information necessary to build them exactly. Hence two, not necessarily mutually exclusive, cases might arise:

- The DM is not able to locate w precisely, but she may give some constraints on it, which we shall represent by $w \in S \subseteq W$. This is a case of *decision-making under partial information.*

- The DM gives some estimate ω of w, but feeling doubtful about ω, she asks for help to identify which of her judgements are the critical ones in order to think about them more closely. In this case, *sensitivity analysis* comes to the fore.

Example

Suppose that $k > 1$, that u is additive and that the DM is comfortable with the assessments of the component utility functions $u_2, \ldots, u_n$, but not of u_1, the weights λ nor the probabilities p. Then:

$$\Psi(a_i, p, \lambda, u_1) = \sum_{j=1}^{k} p_j \sum_{t=1}^{n} \lambda_t u_t(a_{ij}^t).$$

Her assessments might lead to some estimates $\omega = (p^0, \lambda^0, u_1^0)$ of her probabilities, weights and component utility u_1 together with some constraints defining
$S \subseteq W$:

$$\alpha_i \leq p_i \leq \beta_i, \; p_i \leq p_j + p_k, \; \ldots, \lambda_1 \leq \ldots \leq \lambda_k, \; u_1 \text{ monotonic.}$$

It has been stated (Goicoechea et al 1981) that *'the absence of an efficient procedure to conduct sensitivity analysis'* is a shortcoming for the SEU approach. Our methods begin to address this. In fact, all the following applies to any aid based on a ranking function known up to several parameters. Further details and technical results appear in a companion paper (Rios Insua and French, 1989).

We shall assume that the mathematical programming problems proposed in the following have solutions.

2 Some previous approaches

Earlier work in this area may be found, *inter alia*, in Isaacs (1963), Fishburn (1965), Fishburn et al (1968), Giron and Rios (1980), Evans (1984), Schneller and Sphicas (1985), and Berger and O'Hagan (1987). Their common feature is that they assume precise utilities, but some kind of uncertainty about the probabilities.

There is a corresponding problem when uncertainty arises about weights in additive, or linear, value functions. Von Winterfeldt and Edwards(1987) treat both problems in parallel. They also observe that in practice the conclusions of a decision analysis seem remarkably robust to 'reasonable' variations in judgemental inputs, suggesting a *Flat Maxima Principle.* We believe sensitivity issues arise more often than they suggest. We also believe that the DM's understanding is so enhanced by sensitivity analyses that the empirical robustness is a side-issue.

Some authors suggest that sensitivity analysis is a craft. Edwards and Newman (1982) propose varying one parameter in discrete steps and observing the change in the output of the model. This is done in much of the available software: e. g. HIVIEW.

3 Some insights into the problem

3.1 Partial information in multiobjective decision-making

The form of the ranking function Ψ is known up to the parameters $w \in W$. Further, the DM has located w to lie in S. For shortness, $\Psi(a_j, w)$ will be written $\Psi_j(w)$. The solutions suggested by the model are those which maximise $\Psi_j(w)$ over j for some $w \in S$. These solutions have been called *potentially optimal* (p. o.) (Hazen,1986) or *quasi-Bayesian* (Rios, 1975). An analytic way of determining whether a solution a_j is p. o. is to solve the problem:

$$[Pj] \qquad \min_{w \in S, z_j} \quad z_j$$

$$s.\ t. \quad \Psi_k(w) - \Psi_j(w) - z_j \leq 0, \ k \neq j$$

If the optimal $z_j \leq 0, a_j$ is p. o. The relation with several other dominance and order concepts may be found, together with some algorithms in Rios Insua and French(1989). Other references in this area are White (1983) and Weber (1987).

3.2 Sensitivity analysis in multi-objective decision-making

Here we have an estimate ω of w and usually some further information, $w \in S$. We may use $\Psi_{.}(\omega)$ to rank the alternatives, but should be ready to criticise that ranking. From now on, a_* will be an optimal solution when $w = \omega$ and $\Psi_*(w)$ its evaluation. A useful concept is that of an *adjacent potentially optimal* (a. p. o.) alternative to a_*. Let $S_j \subseteq S$ be the set of parameters for which a_j is optimal: a_j is a. p. o. to a_* if $S_j \bigcap S_* \neq \emptyset$.

Example

Let $A = \{a_1, a_2, a_3, a_4\}$ as in the Fig. 1. Assume a linear value function with positive weights and that a_4 is the current optimal solution. Then a_2 is a. p. o. to a_4; a_1 is p. o. but not a. p. o. to a_4; a_3 is not p. o.

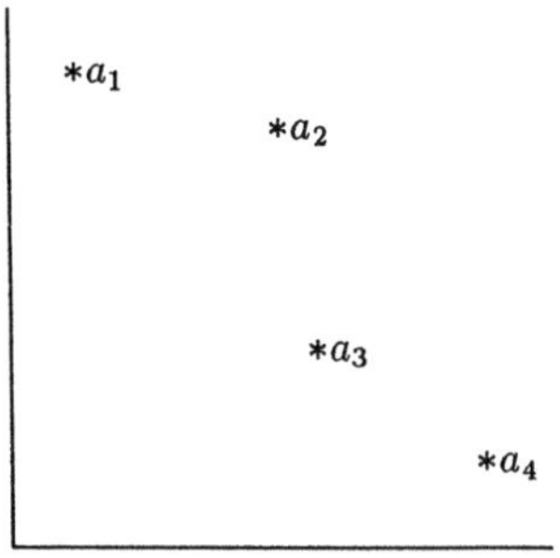

Figure 1: Potential optimality and adjacent potential optimality

To find whether a_j is a. p. o. to a_*, we may solve the problem:

$$[P0j] \qquad \min_{w \in S} \ (\Psi_j(w) - \Psi_*(w))^2$$

$$s.\ t. \quad \Psi_i(w) - \Psi_*(w) \leq 0, \forall i \neq j$$

If the optimal value of $[P0j]$ is 0, a_j is a. p. o. to a_*. The solution of this problem gives a w for which both a_j and a_* are optimal.

Solving these optimisation problems may be quite time-consuming. A less demanding approach is to determine the alternatives that outrank a_* for some $w \in S$. For example, we may try to solve, for each a_j, the problem:

$$[P1j] \qquad \min_{w \in S} \; \Psi_*(w) - \Psi_j(w)$$

If the optimal value of $[P1j]$ is negative, a_j outranks a_*. Alternatively we might solve the problem:

$$[P2j] \qquad \min_{w \in S} \; (\Psi_j(w) - \Psi_*(w))^2$$

If its optimal value is 0, a_j may outrank a_*.

These tools allow us, essentially, to determine the possible competitors of a_*. Having reduced the set of alternatives, the next task is to analyse how changes in ω modify the ranking. In this paper, we shall base our analysis on distances. Isaacs(1963), Fishburn et al (1968), Evans (1983) and Schneller and Sphicas (1984) have used Euclidean distances. A strictly Bayesian approach, would introduce a distribution over the parameters and a second order utility function. But this would imply that the DM would need to provide further information. We prefer to use this additional effort to try to reduce S in her primary problem.

Assume that S is convex, that the Ψ_j's are continuous and that a_j is a possible competitor of a_*. We would like to know how much we could vary ω before a_j outranks a_*. Thus we could solve the problem:

$$[Pdj] \qquad \min_{w \in S} \; d_j = d(w, \omega)$$

$$s.\ t. \quad \Psi_j(w) - \Psi_*(w) = 0$$

where $d(.\,,.)$ is some continuous distance (metric). This gives us the w for which a_j and a_* are indifferent and which minimises the distance to ω. Let $d = \min d_j$. Then, $d(w, \omega) \leq d$ implies $\Psi_*(w) \geq \Psi_j(w), \forall j$. Thus, d is an absolute measure of the sensitivity of the solution a_j. If we solve the problem:

$$[Pd] \qquad \max_{w \in S} \; d(w, \omega)$$

and call its optimal value δ, then $r = d/\delta$ gives us a relative measure of the sensitivity of a_* to changes in w:

- If $r = 1$, then $\Psi_*(w) \geq \Psi_j(w), \forall w \in S$ and $\forall j$; thus, a_* is completely insensitive.
- If $r = 0$, a_j is completely sensitive.
- For intermediate values, r monotonically increases with d.

Examples

- When $d(.\,,.)$ is the Euclidean distance, we can formulate the equivalent problem of $[Pdj]$:

$$[P3j] \qquad \min_{w \in S} .5 \sum w_i^2 - \sum \omega_i w_i$$

$$s.\ t. \quad \Psi_*(w) - \Psi_j(w) = 0.$$

- When $d(.\ ,.)$ is the Tchebycheff distance, $[Pdj]$ may be formulated:

$$[P4j] \qquad \min_{t,\, w \in S} \ t$$

$$s.\ t. \quad w_i - t \leq \omega_i$$
$$w_i + t \geq \omega_i$$
$$\Psi_*(w) - \Psi_j(w) = 0$$

Remarks

1. The a_j identified by finding d may be distance dependent.

2. When an L_p distance is used, a. p. o. alternatives to a_* are found.

3. Not every a. p. o. alternative is a minimiser for an L_p distance.

Proofs and counterexamples may be found in Rios Insua and French(1989), together with several other properties, which support the use of distance-based tools for sensitivity analysis. They lead to interesting numerical problems, which we are currently attacking.

3.3 Interface with the DM

All this information needs to be displayed in an informative way to the DM: what Von Winterfeldt and Edwards (1987) call 'sensitivity analysis for the client'.

3.3.1 Graphical representation

Above ω appears to be a natural origin for any graphical representation. We may represent, instead of the w's, the $(w - \omega)$'s. Cases which we might be interested to represent include the set of minimisers for different distances, the set of alternatives for some distances, etc. To do this, we may use several graphical tools from exploratory multivariate data analysis. For example, if we have used the Euclidean distance, we may apply a principal components analysis to the $(w - \omega)$'s and represent $d(w, \omega)$ in the directions of $(w - \omega)$ projected in the first two principal components. For other distances, it might be preferable to use principal coordinates analysis(see Everitt (1977)). Other possibilities which we are investigating are the use of bar charts or modified star graphs (Chambers et al, 1983).

3.3.2 Semantic representations

We have already mentioned the process of modelling the judgemental inputs of the DM, which consists of an iterative protocol of questioning/answering/encoding, and which leads to ω. From sensitivity analyses we get possible competitors of ω which lead to different alternatives. Thus, we could think of inverting the protocol, finding the questions and the answers which could lead to these competitors, pose them to the DM and use her comments to modify, if necessary, the estimate ω and/or the set S.

4 Computational experience

The approach has been preliminary tested at UMRCC using the NAG FORTRAN library with promising results, though some limitations are imposed to its use due to its current lack of flexibility.

Example

We use here an example adapted from Chankong and Haimes(1982). Consider a decision problem with three states and three bi-attributed alternatives. Assume that the utility function is additive, but that neither the weights nor the probabilities are known. The data of the problem are:

		States		
		θ_1	θ_2	θ_3
Alternatives	a_1	(-400,16)	(80,14)	(200,12)
	a_2	(10,20)	(20,18)	(50,16)
	a_3	(-100,20)	(0,8)	(100,8)

If p's designate probabilities and λ's designate weights, we have:

$$\begin{aligned}\Psi_1(p,\lambda) &= \lambda_1(-400p_1+80p_2+200p_3)+\lambda_2(16p_1+14p_2+12p_3)\\ \Psi_2(p,\lambda) &= \lambda_1(10p_1+20p_2+50p_3)+\lambda_2(20p_1+18p_2+16p_3)\\ \Psi_3(p,\lambda) &= \lambda_1(-100p_1+100p_3)+\lambda_2(20p_1+8p_2+8p_3)\end{aligned}$$

Assume that we know $\lambda_1 \geq \lambda_2$, $p_2 \geq p_3 \geq p_1$ and that ω has been found as : (p^0,λ^0)=(.25, .5, .25, .75, .25). In this case,

$$\Psi_1 = -4,\ \Psi_2 = 93/4,\ \Psi_3 = 11/4.$$

So a_2 is the 'optimal' solution. We have:

$$\begin{aligned}\Psi_2(p,\lambda)-\Psi_1(p,\lambda) &= \lambda_1(410p_1-60p_2-150p_3)+\lambda_2(4p_1+4p_2+4p_3)\\ \Psi_2(p,\lambda)-\Psi_3(p,\lambda) &= \lambda_1(110p_1+20p_2-50p_3)+\lambda_2(10p_2+8p_3).\end{aligned}$$

The table below gives the solutions to problems [Pnj]:

	p_1	p_2	p_3	λ_1	λ_2	Opt. value
ω	.250	.500	.250	.750	.250	—
P11	.000	.500	.500	1.000	.000	-105.000
P13	.000	.500	.500	1.000	.000	-15.000
P21	.198	.405	.396	.641	.358	.000
P23	.038	.532	.429	.571	.428	.000
P31	.181	.525	.293	.750	.249	.007
P33	.072	.511	.415	.765	.234	.059
P41	.185	.500	.314	.814	.185	.064
P43	.088	.500	.411	.911	.088	.161

Observe that both a_1, a_3 may outrank a_2. We represent this information graphically by calculating $(w-\omega)$, as in the table:

	$p_1 - p_1^0$	$p_2 - p_2^0$	$p_3 - p_3^0$	$\lambda_1 - \lambda_1^0$	$\lambda_2 - \lambda_2^0$
P31	-.069	.025	.043	.000	-.001
P33	-.178	.011	.165	.015	-.016
P41	-.065	.000	.064	.064	-.065
P43	-.162	.000	.161	.161	-.162

We may represent it by bar charts, as in figure 2, or star graphs, as in figure 3.

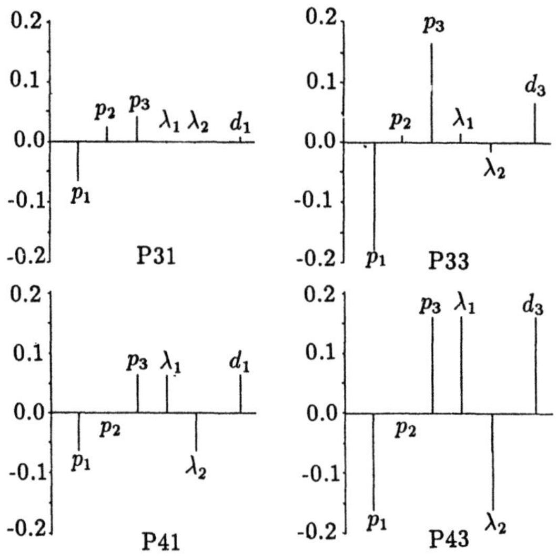

Figure 2: Bar charts showing least changes necessary in the parameters to cause an outranking of a_2 and minimal distance.

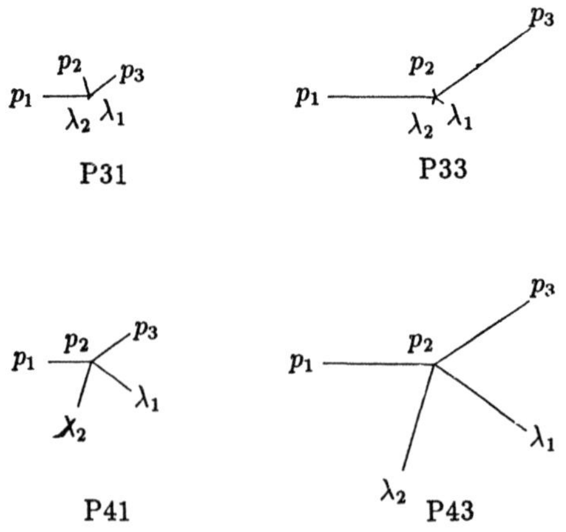

Figure 3: Star graphs showing least changes necessary in the parameters to cause an outranking of a_2.

As the dimensionality of the problem is low, we may represent it in the parameter space: see figure 4. The usual reductions to $P = \{p_2, p_3 \geq 0, p_2 + p_3 \leq 1\}$ for the probabilities and to L=[0,1] for the weights have been made. S_P and S_L will designate the restriction of S to P and L. A superscript * indicates initial value. A superscript i indicates values for which a_i outranks a_*. We shall only represent data relative to Euclidean distances.

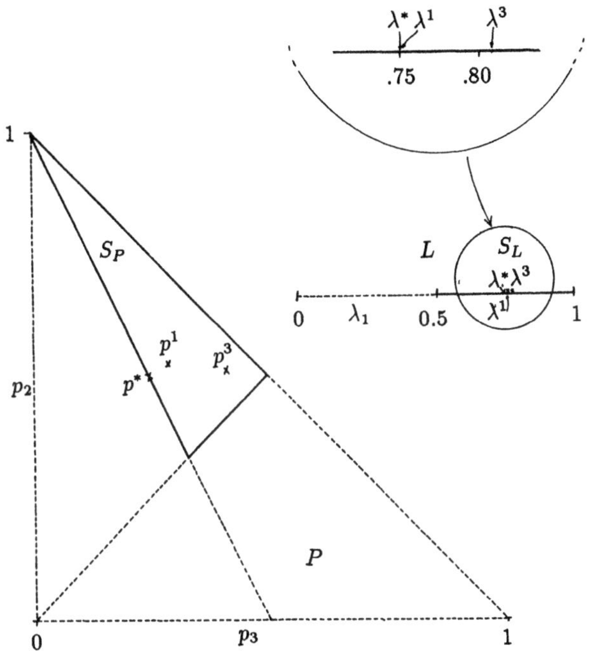

Figure 4: Nearest points to ω at which a_2 is outranked.

These graphs are very descriptive. They indicate the need for precision if we are to be sure that a_2 is optimal. The alternative which is its clearest competitor is a_1. Moreover, whatever distance is used, changes occur when p_1 is decreased, p_3 increased, λ_1 increased and λ_2 decreased: there need be little change in p_2. This information would clarify in the DM's mind the least changes in her judgements that would lead to a change in the suggested course of action.

5 Conclusion

We have briefly introduced some tools for sensitivity analysis and handling partial information in decision-making. Further details may be found in Rios Insua and French (1989). Preliminary results seem promising, although much more computational experience is required and there is a clear need to develop special algorithms and a flexible, user-friendly interface.

Acknowledgments

An initial version of this paper was prepared while the authors were at the Statistical Laboratory, Department of Mathematics, University of Manchester. Computing facilities were provided by the University of Manchester Regional Computing Centre.

David Rios Insua is supported by a grant of the Bank of Spain.

References

[1] Berger, J. and O'Hagan, A.(1987) *Ranges of Posterior Probabilities for Unimodal Priors with Specified Quantiles*, presented at the Third Valencia International Meeting of Bayesian Statisticians.

[2] Chambers, J.M., Cleveland, W.S., Kleiner, B. and Tukey,P.A. (1983) *Graphical Methods for Data Analysis*, Duxbury Press.

[3] Chankong, V. and Haimes, Y. (1982) *Multiobjective Decision-Making*, North Holland Pub. Co.

[4] Decision Analysis Unit, London School of Economics HIVIEW.

[5] Edwards,W. and Newman, J. R. (1982) *Multi-attribute evaluation*, Sage.

[6] Evans, J.R. (1984) Sensitivity Analysis in Decision Theory, *Decision Sciences*, 15, 239-247.

[7] Everitt, B.S. (1978) *Graphical Techniques for Multivariate Data*, Heinemann.

[8] Fishburn,P. (1965) Analysis of Decisions with Incomplete Knowledge of Probabilities,*Opns. Res.*,13, 217-237.

[9] Fishburn, P.,Murphy, A. and Isaacs, H. (1968) Sensitivity of Decisions to Probability Estimation Errors: A Reexamination, *Opns. Res.*,16, 254-267.

[10] French, S. (1986) *Decision Theory*, Ellis Horwood.

[11] Giron, F. and Rios, S. (1980) Quasi-Bayesian Behaviour: A More Realistic Approach to Decision-Making?, in Bernardo, De Groot, Lindley and Smith(eds), *Bayesian Statistics*, Valencia U.P.

[12] Goicoechea, A., Hansen, D. and Duckstein, L. (1982) *Multi-objective Decision-Analysis*, Wiley.

[13] Hazen, G.B. (1986) Partial Information, Dominance and Potential Optimality in Multiattribute Utility Theory, *Operations Research*, 34, 296-310.

[14] Isaacs, H.(1963) Sensitivity of Decisions to Probability Estimation Errors,*Opns. Res.*, 11, 536-552.

[15] Rios, S.(1975) Ordre Quasi-bayesien des Regles de Decision, *Proc. 40th Session of ISI*, Warsaw, 694-698.

[16] Rios Insua, D. and French, S.(1989) *A Framework for Sensitivity Analysis in Discrete Multi-objective Decision-Making*, Research Report, School of Computer Studies, University of Leeds.

[17] Schneller,G.O. and Sphicas, G.P.(1985) On Sensitivity Analysis in Decision Theory, *Decision Sciences*,16,399-408.

[18] Von Winterfeldt, D. and Edwards, W.(1987) *Decision Analysis and Behavioural Research*, Cambridge U.P.

[19] Weber, M. (1987) Decision Making with Incomplete Information *EJOR*, 28, 44-57.

[20] White, D.J.(1980)Optimality and Efficiency I, *EJOR*, 4,346-346.

ALGORITHMS FOR MULTI-OBJECTIVE NONLINEAR PROGRAMMING PROBLEMS: AN OVERVIEW

Subhash C. Narula
H. Roland Weistroffer
School of Business, Virginia Commonwealth University
Richmond, Virginia, USA

ABSTRACT

A number of real life problems have multiple objectives and constraints that are truly nonlinear in nature. Many efficient interactive algorithms have been proposed to solve these problems. These procedures differ in the philosophy/approach they follow. Even within a given philosophy/approach, the algorithms differ in several details. In this paper, our objective is to give an overview of some of the available algorithms.

1. INTRODUCTION

Since its development in the late 1940's, linear programming has enjoyed great popularity, and the number and variety of its applications has been remarkable. However, in the past two decades multiple objective methods have emerged as a major tool for decision makers. This may be largely due to the following two reasons. First, it is recognition of the fact that in today's complex organizational environment, a decision maker (DM) is regarded as someone who must attempt to achieve a set of objectives to their fullest possible extent in an environment of conflicting interests, incomplete information, and limited resources, Simon (1960). And second, the power of the microcomputer has grown, and it has found increasing use in support of business decision making. Microcomputers have made it possible to use interactive algorithms, making it easy to input the information and understand the output from the model, thus, facilitating the use of even complex models. Moreover, if the computer-user interface is friendly, the decision makers may find that information and the use of information more closely conform to their individual decision making or cognitive style. In fact, a wide variety of interactive approaches has been proposed to assist decision makers in finding preferred solutions to multiple criteria decision making problems.

It was natural that starting with (single objective) linear programming problems the first burst of research activity would concentrate on multi-objective linear programming (MOLP) problems. A number of very efficient interactive algorithms and computer programs have been developed to solve the MOLP problems. Further, many successful applications of these formulations and techniques have been reported in the literature. Clearly, the MOLP problems provide a better approximation to the real-world problems than the (single objective) linear programming formulations. However, most practical problems have multiple objectives and constraints that are truly nonlinear in nature. Although not studied as extensively as the MOLP problem, a number of efficient interactive algorithms have been proposed to solve a multi-objective nonlinear programming (MONLP) problem to the satisfaction of a DM.

In this paper, our objective is to give an overview of some of the interactive algorithms that are available to solve the MONLP problem. These algorithms differ in terms of the philosophy/approach they follow. Furthermore, within a given approach, they differ in terms of the procedure they use to find an initial and subsequent solutions, the computational algorithm used, the preference order preserving scalarizing function (or surrogate objective function) used, as well as the type of interaction with and the demands placed on the DM. The interested reader may also refer to White (1982) and Yu (1985) for some of these aspects of MONLP algorithms.

The rest of the paper is organized as follows: In Section 2, we present the problem formulation and basic philosophies/approaches the algorithms follow. In Section 3 we give an overview of the algorithms and conclude the paper with a few remarks in Section 4.

2. PHILOSOPHIES/APPROACHES

Consider a multi-objective nonlinear programming problem of the form

(MONLP) Maximize $f_j(\underline{x})$ $j = 1, \ldots, k$

Subject to

$$\underline{x} \in R = \{\underline{x} \mid r_i \leq g_i(\underline{x}) \leq d_i,\ i = 1, \ldots, p;\ \underline{x} \geq \underline{0}\}$$

where $\underline{x}^T = (x_1, \ldots, x_n)$ is an nx1 vector of decision variables. We assume that $f_j(\underline{x})$, $j = 1, \ldots, k$ and $g_i(\underline{x})$, $i = 1, \ldots, p$ are continuous functions of $\underline{x}$, and all objective functions have to be maximized.

The basic steps of many currently available interactive algorithms to solve problem (MONLP) can be succinctly stated as follows:

1. Find an initial feasible solution
2. Interact with the decision maker
3. Obtain a new solution. If this or some previous solution is acceptable to the DM, stop; otherwise, go to step 2.

Although the proposed interactive algorithms to solve the MONLP problem follow the same basic structure, they differ in terms of their philosophies/ approaches. Many of these procedures follow one of the following four (not mutually exclusive) philosophies/approaches, viz., the reference point, the decreasing expectations, the increasing requirements, and the bracketing approach.

The Reference Point Approach: The DM specifies a reference value for each of the objective functions (i.e., a reference point or reference solution), Wierzbicki (1982). These reference values may or may not be achievable. For this reason, instead of maximizing the overachievement or minimizing the underachievement of the objective function values with respect to (w.r.t.) these values, an order preserving scalarizing function (or surrogate objective function) is used to simultaneously maximize the overachievements and minimize the possible underachievements w.r.t. the given reference values. This solution is shown to the DM who then revises the reference values. A new solution is found and the process repeated until an acceptable compromise solution is found.

This approach is intuitively appealing. However, it seems to be a little arbitrary and lacks a clear strategy to find the best compromise solution.

The Decreasing Expectations (Optimistic) Approach: At the start of an algorithm the DM may specify the most optimistic objective function values as the desired values knowing full well that these values are not achievable simultaneously. Based on this information, an analyst finds a solution that minimizes the underachievement of the objective function values w.r.t. the specified desired values. This solution is presented to the DM who is asked to revise his/her expectations. The DM may lower the aspiration level for one or some of the objective functions. This process is repeated until an acceptable compromise solution (to DM) is found.

If the DM has any difficulty in specifying the initial desired values (aspiration levels) for the objective functions, one may use the ideal values f_j^*, $j = 1, \ldots, k$, as the starting values, where

$$f_j^* := \max_{\underline{x} \in R} f_j(\underline{x}) , \qquad j = 1, \ldots, k.$$

Note that the ideal values represent upper bounds on the objective function values on the solution set.

The optimistic approach can be viewed as a special case of the reference point approach in which all the reference values consistently exceed the objective function values at all the intermediate and compromise solutions. It has a well defined strategy to reach a desired compromise solution and does not place excessive demand on the DM. This approach is like a Dutch auction where the asking price starts very high and is gradually lowered until an offer is accepted. Lewandowski and Wierzbicki (1985) analyze the theoretical and methodological backgrounds of aspiration based techniques.

The Increasing Requirements (Pessimistic) Approach: The DM purposely specifies low required values for the objective functions which can be achieved with certainty. Therefore, the analyst tries to maximize the overachievement of the objective function values w.r.t. the specified required values. After reviewing the solution, the DM may revise some or all the required values upwards. A new solution is calculated and the process repeated until a satisfactory compromise solution is obtained.

If the DM has a problem in specifying the initial required values for the objective functions, an algorithm may start with the pessimistic values, f_{j*}, $j = 1, \ldots, k$, where

$$f_{j*} := \min_{i \in \{1,\ldots,k\}} f_j(\underline{x}^{(i)}) , \quad j = 1,\ldots,k ,$$

where $\underline{x}^{(i)} \in R$ are the ideal solutions (i.e., $f_j^* = f_j(\underline{x}^{(j)})$). It has been mentioned however by several authors, (e.g., Weistroffer (1985)), that pessimistic values f_{j*} are not lower bounds for the objective functions on R or on the subset of efficient solutions.

The pessimistic approach can also be viewed as a special case of the reference point approach in which all the reference values are always less than or equal to the objective function values at the intermediate and the final solution. Like the optimistic approach, it also has a well defined strategy to reach a satisfactory compromise solution and does not place excessive demand on the DM. This approach is like a regular auction where the bidding starts very low and the bid is gradually raised until a bid is accepted.

The Bracketing Approach: Instead of specifying a single optimistic, pessimistic or reference value for each objective function, the DM specifies two values, viz., a required value which must be achieved before a solution is acceptable, and a desired value beyond which any solution is acceptable. That is, the set of required and desired values form a bracket (or box) which contains an acceptable compromise solution. A solution is obtained such that it simultaneously minimizes underachievements w.r.t. the desired values and maximizes overachievements w.r.t. the required values. Based on this solution, the DM may lower some or all of the desired values, or may raise some or all of the required values, thus narrowing the size of the box that contains the most preferred solution. This process is repeated until all solutions within the box are sufficiently close to the best compromise solution. In other words, at this stage all the solutions in the box are acceptable.

If the DM has any difficulty in specifying the initial desired and required values for the objective functions, one may use f_j^* and f_{j*} as the starting values.

This approach is theoretically, intuitively and psychologically appealing. A DM may feel more comfortable specifying both a required and a desired value for each of the objective function values rather than just one value as in all the other approaches. As in the pessimistic approach, each intermediate solution satisfies all the specified required values.

3. SOME AVAILABLE ALGORITHMS

For each philosophy/approach we will discuss briefly some of the available algorithms. The interested reader should refer to the specific paper for more details of a particular algorithm.

The algorithms by Wierzbicki (1982), Lewandowski and Grauer (1982), Grauer (1983), and Masud and Hwang (1981) basically follow the reference point approach. Wierzbicki proposed the approach and suggested that once the DM has specified a reference point then a solution can be obtained by using some achievement scalarizing function; this function simultaneously maximizes the overachievement and minimizes the possible underachievements of the objective function values w.r.t. the reference values. Two such functions are:

$$\sum_{j=1}^{k} \{\max [0, r_j - f_j(\underline{x})]\}^2 - \sum_{j=1}^{k} \{\max [0, f_j(\underline{x}) - r_j]\}^2$$

and

$$\sum_{j=1}^{k} \log\left(\frac{f_j^* - f_j(\underline{x})}{f_j^* - r_j}\right)$$

where r_j's represent reference values and f_j^*'s represent the ideal values.

Lewandowski and Grauer (1982) developed a decision support system that operationalizes the reference point approach. In their system, the DM interactively changes the reference values on the basis of the knowledge gained by examining the intermediate solutions. They used

$$\frac{1}{\alpha} \ln \left[\frac{1}{k} \sum_{j=1}^{k} \left\{ \frac{f_j^* - f_j(\underline{x})}{f_j^* - r_j} \right\}^{\alpha} \right]$$

as the surrogate objective function. They obtained the initial and intermediate solutions by optimizing the surrogate objective function using a constrained nonlinear optimization procedure.

In the sequential goal programming method of Masud and Hwang (1981), the DM specifies the initial reference values within the limits of the pessimistic and the optimistic values of the objective functions. Then a principal solution is found by first minimizing the underachievements of the objective function values w.r.t. the reference values, and then maximizing the overachievements. They generate k alternate solutions by minimizing the underachievements of the objective function values w.r.t. the reference values such that one of the k objective functions (a different one for each of the k alternate solutions) satisfies its reference value and maximizes the overachievement. The DM may then either pick one of the k+1 solutions as the final compromise solution, or specify new reference values based on the insight gained from the k+1 solutions. The subsequent iterations differ from the first iteration in that all currently satisfied reference values are always satisfied. Clearly, the algorithm is computationally intense. Furthermore, at each iteration, the DM is offered a set of k+1 solutions and asked to either select one as the compromise solution, or provide new reference values.

The interactive goal programming algorithm of Weistroffer (1983) follows the optimistic approach. The DM specifies the initial optimistic desired values $\delta^{(0)}$ which are used to find the starting solution. The starting and the subsequent solutions are found by minimizing the underachievements of the objective functions w.r.t. the desired values by using

$$\sum_{j=1}^{k} [\max (0, \delta_j^{(\nu)} - f_j(\underline{x}^{(\nu)}))]^2$$

as the surrogate objective function, where $\underline{\delta}^{(\nu)}$ is the desired value at the νth iteration. The intermediate solutions are presented to the DM to help him/her decide which desired values to lower and by how much. The DM may lower one or more desired values; however, if he/she feels that some desired values have been lowered too far, he/she may raise these. Since a constraint can be formulated as an objective function with a fixed goal, all intermediate solutions can be obtained by using a program for unconstrained nonlinear programming problems.

The interactive overachievement programming algorithm of Weistroffer (1982), Benson's (1975) algorithm, the surrogate worth trade off method of Haimes and Hall (1974) and the SPOT algorithm of Sakawa (1982) follow the pessimistic approach, i.e., each algorithm starts with low required values for the objective function values and then the required values are gradually increased to arrive at an acceptable compromise solution.

In Weistroffer's (1982) algorithm the maximization of the over-achievements of the objective functions w.r.t. the specified required values is achieved via a barrier (or interior penalty) function technique, i.e.,

$$\max \sum_{j=1}^{k} \phi(f_j(\underline{x}) - \rho_j^{(\nu)}),$$

where ϕ is a barrier function and $\rho_j^{(\nu)}$ is the required value at an iteration ν. The constraints are formulated as objective functions with fixed required values. Thus, the maximization of the barrier function, at each iteration, can be accomplished via an unconstrained nonlinear programming code.

In Benson's (1975) algorithm the DM specifies the initial required values (called satisfactory goals). At each iteration, the DM is shown the solution and asked to identify the least satisfactory objective function; then this objective function is maximized subject to the constraints and the other objective functions satisfying their required values. The DM reviews the solution and identifies the least satisfactory objective function value and the process is continued until a compromise solution is found. Each intermediate solution is found using a constrained nonlinear programming code. Since at each iteration, one objective function is maximized, this information may not be as good as that from an algorithm where all objectives are maximized simultaneously.

The surrogate worth trade off method of Haimes and Hall (1974) and SPOT of Sakawa (1982) are similar in that at each iteration one objective function is optimized subject to ε - constraints (i.e., the required values) for all other objective functions over R. The DM is asked for a "marginal rate of substitution" between the optimized objective function and each of the other objective functions. The marginal rate of substitution is compared to the vector of Lagrange multipliers; if the two are sufficiently close, stop; otherwise, a proxy preference function is optimized in the direction of the difference. For the next iteration, the required values are changed and the process is repeated until a compromise solution is found.

The combined over- and under-achievement programming technique of Weistroffer (1984), the flexible modelling approach of Weistroffer (1987) and Narula and Weistroffer (1989), the interactive multi-objective decision making method using fuzzy satisficing intervals of Sakawa (1986) and the multiple goal programming algorithm of Spronk (1981) follow the bracketing approach.

In Weistroffer's (1984) algorithm, the simultaneous minimization of the underachievement w.r.t. to the desired values and the overachievement w.r.t. to the required values is accomplished by a penalty and a barrier function. Each intermediate solution is found by an unconstrained nonlinear programming code. In Sakawa's algorithm, instead of specifying the desired and the required values, the DM has to specify the fuzzy membership functions and update these at each of the intermediate solutions. The algorithm due to Narula and Weistroffer (1989) differs from the preceding two in that the DM does not have to distinguish between the objective functions and the constraints. The surrogate objective function of the form

$$\sum_{j=1}^{k} w_j \log \frac{\min [f_j(\underline{x}), d_j] - r_j}{d_j - r_j}$$

where $w_j \in [0, 1]$ specifies the degree to which a criterion function is considered an objective function.

In the multiple goal programming algorithm of Spronk (1981), each objective function is maximized subject to some required values for the remaining objective functions. The obtained values are called the restricted ideal values. The worst possible value for the restricted ideal solution, called the restricted pessimistic values, are also computed. The restricted ideal and pessimistic values, called the potency matrix, form a box which contains the most preferred solution. The DM initially specifies a finite number of required levels for each objective function such that

$$f_{j*} \leq \rho_j^1 \leq \dots \leq \rho_j^{m_j} \leq f_j^* .$$

The required values are set to the lowest levels and an initial solution is computed. The DM is then asked to identify the least satisfactory value (in this solution) which is replaced by the next required value and a new potency matrix is generated. If the DM considers the new potency matrix an improvement over the previous one, the procedure continues by finding a new solution until an acceptable compromise solution is found. It may be worth noting that in this procedure the initial and intermediate solutions are obtained by using the increasing requirements (pessimistic) approach; however, at each iteration the DM is also shown the box (or the potency matrix) that contains the compromise solution.

4. SOME REMARKS

Clearly a number of algorithms have been proposed to solve the multiple objective nonlinear programming problems. The procedures differ in terms of their philosophies/approaches, the methods used to find the initial, the intermediate and the final compromise solutions, the surrogate objective functions used, and the interaction with and the demands placed on the DM.

The algorithms discussed in this paper are by design interactive. However, the software implementing them is not always available in the public domain. Further, very little, if any, information on the computational effort required, or the reaction of and acceptability by the decision makers, and other details has been reported in the literature. This makes it very difficult to choose the "best" algorithm for a given problem.

Many more efficient algorithms are available for the multiple objective linear programming problems than for the corresponding nonlinear programming problems. With some effort, it may be possible to adapt some of these algorithms to solve the nonlinear programming problems. Roy and Wallenius (1988) have proposed one such algorithm. It is based on the Zionts and Wallenius (1976, 1983) interactive multiple objective linear programming method and the generalized reduced gradient techniques for nonlinear optimization of Abadie and Carpentier (1969) and Lasdon, Fox, and Ratner (1974). Its basic approach is to maximize a linear composite of the objective functions, compute the reduced gradients at the solution and check for efficient tradeoffs. If the tradeoffs indicate that the current solution is not optimum, a new linear composite function consistent with all previous responses of the decision maker is found and the process is repeated.

REFERENCES

Abadie, J. and Carpentier, J. (1969), "Generalization of the Wolfe reduced gradient method to the case of nonlinear constraints", in R. Fletcher (ed.): Optimization, Academic Press, New York.

Benson, R. G. (1975), "Interactive multiple criteria optimization using satisfactory goals", Doctoral thesis, University of Iowa, Iowa City.

Grauer, M. (1983), "Reference point optimization - the nonlinear case", in P. Hansen (ed.): Essays and Surveys on Multiple Criteria Decision Making, Springer-Verlag, Berlin.

Haimes, Y. Y. and Hall, W. A. (1974), "Multiobjectives in water resources systems analysis: The surrogate worth trade off method", Water Resources Research 10(4), 615-623.

Lasdon, L., Fox, R. L., and Ratner, M. (1974), "Nonlinear optimization using the generalized reduced gradient method," R.A.I.R.O. 3, 73-104.

Lewandowski, A. and Grauer, M. (1982), "The reference point optimization approach - methods of efficient implementation", Working paper WP-82-019, IIASA, Laxenburg, Austria.

Lewandowski, A. and Wierzbicki, A. (1988), "Aspiration based decision analysis and support, Part I: Theoretical and methodological backgrounds", Working paper WP-88-03,IIASA, Laxenburg, Austria.

Masud, A. S. and Hwang, C. L. (1981), "Interactive sequential goal programming", Journal of the Operational Research Society 32, 391-400.

Narula, S. C. and Weistroffer, H. R. (1989), "A flexible method for nonlinear multicriteria decision making problems", IEEE Transactions on Systems, Man, and Cybernetics SMC-19(4).

Roy, A. and Wallenius, J. (1988), "An algorithm and the supporting theory for nonlinear multiple objective optimization", Working paper, Arizona State University and University of Jyvaskyla, Finland.

Sakawa, M. (1982), "Interactive multiobjective decision making by the sequential proxy optimization technique: SPOT", European Journal of Operational Research 9(4), 386-396.

Sakawa, M. (1986), "Interactive multiobjective decision-making using fuzzy satisficing intervals and its application to an urban water resources system", Large Scale Systems 10, 203-213.

Simon, H. A. (1960). The New Science of Management Decision, Harper and Brothers, New York.

Spronk, J. (1981), Interactive Multiple Goal Programming, Martinus Nijhoff, Boston.

Weistroffer, H. R. (1982), "Multiple criteria decision making with interactive overachievement programming", Operations Research Letters 1(6), 241-245.

Weistroffer, H. R. (1983), "An interactive goal programming method for nonlinear multiple-criteria decision-making problems", Computers and Operations Research 10(4), 311-320.

Weistroffer, H. R. (1984), "A combined over- and under-achievement programming approach to multiple objectives decision making", Large Scale Systems 7, 47-58.

Weistroffer, H. R. (1985). "Careful usage of pessimistic values is needed in multiple objectives optimization", Operations Research Letters 4(1), 23-25.

Weistroffer, H. R. (1987), "A flexible model for multi-objective optimization", in J. Jahn and W. Krabs (eds): Recent Advances and Historical Development of Vector Optimization, Springer-Verlag, Berlin.

White, D. J. (1982), Optimality and Efficiency, John Wiley & Sons, Chichester.

Wierzbicki, A. P. (1982), "A mathematical basis for satisficing decision-making", Mathematical Modelling 3, 391-405.

Yu, P. -L. (1985), Multiple-Criteria Decision Making, Plenum Press, New York.

Zionts, S. and Wallenius, J. (1976), "An interactive programming method for solving the multiple criteria problem", Management Science 22, 652-663.

Zionts, S. and Wallenius, J. (1983), "An interactive multiple objective linear programming method for a class of underlying nonlinear utility functions", Management Science 29, 519-529.

DYNAMIC INTERACTIVE NETWORK ANALYSIS SYSTEM*

W. Ogryczak, K. Studziński and K. Zorychta
Institute of Informatics
Warsaw University
Warsaw, Poland

ABSTRACT

This paper describes the Dynamic Interactive Network Analysis System (DINAS) which enables the solution of various multiobjective transshipment problems with facility location using IBM-PC XT/AT microcomputers. DINAS utilizes an extension of the classical reference point approach to handling multiple objectives. In this approach the decision-maker forms his requirements in terms of aspiration and reservation levels, i.e., he specifies acceptable and required values for given objectives. A special TRANSLOC solver was developed to provide DINAS with solutions to single-objective problems. It is based on the branch and bound scheme with a pioneering implementation of the simplex special ordered network (SON) algorithm with implicit representation of the simple and variable upper bounds (VUB & SUB). DINAS is prepared as a menu-driven and easy in usage system armed with a special network editor which reduces to minimum effort associated with input a real-life problem.

1. THE DINAS SYSTEM

DINAS is a decision support system designed for solving multiobjective transshipment problems with facility location on IBM-PC XT/AT or compatibles (see Ogryczak et al. 1987, 1988, 1988a, 1988b, 1989). It requires 640K RAM and a hard disk or at least one floppy disk. DINAS can process problems consisted of:

- up to seven objective functions,
- a transportation network with up to one hundred of nodes and three hundred of arcs,
- up to fifteen potential locations.

A mathematical model of the problem is described in Section 2.

DINAS consists of three programs prepared in the C programming language:

1° an interactive procedure for efficient solutions generation,

2° a solver for single-objective problems,

3° a network editor for input data and results examination.

---

* This work was supported in part by the International Institute for Applied Systems Analysis under the Contracted Study: "Theory, Software and Testing Examples for Decision Support Systems" (see Ogryczak et al. 1988, 1988a, 1988b).

The basic concept of the interactive scheme for efficient solutions generation is as follows:

- the DM works with the system in an interactive way so that he can change his aspiration and reservation levels in any direction;
- after editing the aspiration and reservation levels, the system computes a new efficient solution by solving a corresponding single-objective problem;
- each computed efficient solution is put into a special solution base and presented to the DM as the current solution in the form of tables and bars which allow him to analyze performances of the current solution in comparison with the previous solutions.

Operations available in the DINAS interactive procedure are partitioned into three groups and corresponding three branches of the main menu: PROCESS, SOLUTION and ANALYSIS. The PROCESS branch contains basic operations connected with processing the multiobjective problem and generation of several efficient solutions. There are included operations such as editing and converting the problem, computation of the pay-off matrix, and finally, generation of a sequence of efficient solutions depending on the edited aspiration and reservation levels.

The SOLUTION branch contains additional operations connected with the current solution. The DM can examine in details the current solution using the network editor or analyze only short characteristics such as objective values and selected locations. Values of the objective functions are presented in three ways: as a standard table, as bars in the aspiration/reservation scale and as bars in the utopia/nadir scale. The bars show percentage level of each objective value with respect to the corresponding scale. The DM may also print the current solution or save it for using in next runs of the system with the same problem. There is also available a special command to delete the current solution from the solution base if the DM finds it as quite useless.

The ANALYSIS branch collects commands connected with operations on the solution base. The main command COMPARE allows the DM to perform comparison of all the efficient solutions from the solution base or of some subset of them. In the comparison only the short characteristics of the solutions are used, i.e., objective values in the form of tables and bars as well as tables of selected locations. Moreover, some commands which allow the DM to select various efficient solutions from the solution base as the current solution are included in this branch. There exist also an opportunity to restore some (saved earlier) efficient solution to the solution base.

A special TRANSLOC solver has been prepared to provide the multiobjective analysis procedure with solutions to single-objective problems. The solver is hidden from the user but it is the most important part of the DINAS system. It is a numerical kernel of the system which generates efficient solutions. According to the interactive procedure described in Section 3 the TRANSLOC solver has to be able to solve two kinds of single-objective problems: the first one associated with calculation of the pay-off matrix and the second one associated with minimization of the scalarizing achievement function. Both kinds of the problems have, however, the same main constraints which represent the feasible set of the generalized network model. Moreover, the other constraints of both the kinds of problems can be expressed in a very similar way (see Ogryczak et al. 1988 or 1989).

Both the single-objective problems are typical mixed integer linear programs, i.e., they are typical linear programs with some integer variables. Mixed integer linear programs are usually solved by branch and bound approach with utilization of the simplex method. The TRANSLOC solver also uses this approach. Fortunately, only a very small group of decision variables is required to be integer in our model. Therefore we can use a simple branch and bound scheme in the solver.

Even for a small transshipment problem with facility location the corresponding linear program has a rather large size. For this reason it cannot be solved directly with the standard simplex algorithm. In order to solve the program on IBM-PC XT/AT microcomputers it is necessary to take advantage of its special structure.

Note that the main group of the problem constraints represents typical network relations. All these rows can be handled in the simplex method as the so-called special ordered network (SON) structure (Glover and Klingman 1981,1985).

The standard simple upper bounds (SUB) are usually processed outside the linear programming matrix (Orchard-Hays 1968). Similarly, slightly more complex inequalities can be considered as the so-called variable upper bounds (VUB) and processed outside the matrix due to a special technique (Schrage 1975, Todd 1982).

Thus only a small number of inequalities has to be considered as typical rows of a linear program. While taking advantage of this fact, the TRANSLOC solver can process transshipment problems of quite a large dimension.

DINAS is armed with the built-in network editor EDINET. EDINET is a full-screen editor specifically designed for input and edit data of the generalized network model defined in Section 2. The essence of the EDINET concept is a dynamic movement from some current node to its neighbouring nodes, and vice versa, according to the network structure. The input data are inserted by a special mechanism of windows while visiting several nodes. Independently, a list of the nodes in the alphabetic order and a graphic scheme of the network is available at any time. Some special windows are also used for defining objective functions.

2. THE GENERALIZED NETWORK MODEL

A network model of the problem consists of nodes that are connected by a set of direct flow arcs. The set of nodes is partitioned into two subsets: the set of fixed nodes and the set of potential nodes. The fixed nodes represent "fixed points" of the transportation network, i.e., points which cannot be changed. Each fixed node is characterized by two quantities: supply and demand. The potential nodes are introduced to represent possible locations of new points in the network. Some groups of the potential nodes represent different versions of the same facility to be located (e.g., different sizes of a warehouse). For this reason, potential nodes are organized in the so-called selections, i.e., sets of nodes with the multiple choice requirement. Each selection is defined by the list of included potential nodes as well as by lower and upper numbers of nodes which have to be selected (located). Each potential node is characterized by a capacity which bounds maximal flow through the node. The capacities are also given for all arcs but not for the fixed nodes.

A several linear objective functions are considered in the problem. The objective functions are introduced into the model by given coefficients associated with several arcs and potential nodes. They will be called cost coefficients independently of their real character in the objective functions. The cost coefficients for potential nodes are, however, understood in a different way than for arcs. The cost coefficient connected to an arc is treated as the unit cost of the flow along the arc whereas the cost coefficient connected to a potential node is considered as the fixed cost associated with the use (location) of the node rather than as the unit cost.

For simplicity of the model and the solution procedure, we transform the potential nodes into artificial arcs. The transformation is performed by duplication of all potential nodes. After the duplication is done all the nodes can be considered as fixed and each potential node is replaced by an artificial arc which leads from the node to its copy. Due to the transformation we get a network with the fixed structure since all the nodes are fixed. Potentiality of artificial arcs does not imply any complication because each arc in the network represents a potential flow. Moreover, all the bounds on flows (i.e., capacities) are connected to arcs after this transformation. Additional nonstandard discrete constraints on the flow are generated only by the multiple choice requirements associated with the selections. Cost coefficients are connected only to arcs, but the coefficients connected to artificial arcs represent fixed costs.

A mathematical statement of this transformed problem takes the form of the following generalized network model:

minimize

$$\sum_{(i,j)\in A\setminus A_a} f_{ij}^{p} x_{ij} + \sum_{(i,j)\in A_a} f_{ij}^{p} y_{ij} \qquad p=1,2,\ldots,n_o \qquad (2.1)$$

subject to

$$\sum_{(i,j)\in A} x_{ij} - \sum_{(j,i)\in A} x_{ji} = b_i \qquad i\in N \qquad (2.2)$$

$$0 \leq x_{ij} \leq c_{ij} \qquad (i,j)\in A\setminus A_a \qquad (2.3)$$

$$0 \leq x_{ij} \leq c_{ij} y_{ij} \qquad (i,j)\in A_a \qquad (2.4)$$

$$g_k \leq \sum_{(i,j)\in S_k} y_{ij} \leq h_k \qquad k=1,2,\ldots,n_s \qquad (2.5)$$

$$y_{ij} = 0 \text{ or } 1 \qquad (i,j)\in A_a \qquad (2.6)$$

where

- n_o - number of objective functions,
- N - set of nodes (including copies of potential nodes),
- n_s - number of selections,
- A - set of arcs (including artificial arcs),
- A_a - set of artificial arcs,
- f_{ij}^{p} - cost coefficient of the p-th objective associated with the arc (i,j),

b_i - supply-demand balance at the node i,

c_{ij} - capacity of the arc (i,j)

g_k, h_k - lower and upper number of (artificial) arcs to be selected in the k-th selection,

S_k - set of (artificial) arcs that belong to the k-th selection,

x_{ij} - decision variable that represents flow along the arc (i,j),

y_{ij} - decision variable equal 1 for selected arc and 0 otherwise.

The generalized network model of this form includes typical network constraints (2.2) with simple upper bounds (2.3) as well as a special discrete structure (2.5)-(2.6) connected to the network structure by variable upper bounds (2.4). While analyzing the model we have to take advantage of all these features.

3. INTERACTIVE PROCEDURE FOR HANDLING MULTIPLE OBJECTIVES

There are many different concepts for handling multiple objectives in mathematical programming. We decided to use the so-called reference point approach which was introduced by Wierzbicki (1982). This concept was further developed in many papers and was used as a basis for construction of the software package DIDAS (Dynamic Interactive Decision Analysis and Support system). The DIDAS package proved to be useful in analyzing conflicts and assisting in decision making situations (Grauer et al. 1984).

The basic concept of the reference point approach is as follows:

(1) the decision-maker (DM) forms his requirements in terms of aspiration levels, i.e., he specifies acceptable values for given objectives;

(2) the DM works with the computer in an interactive way so that he can change his aspiration levels during sessions of the analysis.

In our system, we extend the DIDAS approach. The extension relies on additional use of reservation levels which allow the DM to specify necessary values for given objectives (Wierzbicki 1986).

Consider the multiobjective program associated with the generalized network model:

minimize q

subject to

$q = F(x,y)$

$(x,y) \in Q$

where

q represents the objective vector,

F is the linear vector-function defined by (2.1),

Q denotes the feasible set of the generalized network model, i.e., the set defined by conditions (2.2)-(2.6).

The reference point technique works in two stages. In the first stage the DM is provided with some initial information which gives him an overview of the problem. The initial information is generated by minimization of all the objectives separately. More precisely, the following single objective programs are solved:

$$\min \left\{ f^p(x,y) + (r_o/n_o) \sum_{i=1}^{n_o} f^i(x,y) \ : \ (x,y) \in Q \right\}$$

$$p=1,2,\ldots,n_o \qquad (3.1)$$

where f^p denotes the p-th objective function and r_o is an arbitrarily small number.

The so-called pay-off matrix

$$D = (q_{pj})_{p=1,\ldots,n_o;j=1,\ldots,n_o}$$

which yields information on the range of numerical values of each objective is then constructed. The p-th row of the matrix D corresponds to the vector (x^p,y^p) which solves the p-th program (3.1). Each quantity q_{pj} represents a value of the j-th objective at this solution (i.e., $q_{pj}=f^j(x^p,y^p)$). The vector with elements q_{pp}, i.e., the diagonal of D, defines the utopia (ideal) point. This point, denoted further by q^u, is usually not attainable but it is presented to the DM as a lower limit to the numerical values of the objectives.

Taking into consideration the j-th column of the matrix D we notice that the minimal value in that column is $q_{pp} = q^u_p$.

Let q^n_j be the maximal value, i.e.,

$$q^n_j = \max_{1 \le p \le n_o} q_{pj}$$

The point q^n is called the nadir point and may be presented to the DM as an upper guideline to the values of the objectives. Thus, for each objective f^p a reasonable but not necessarily tight upper bound q^n and a lower bound q^u are known after the first stage of the analysis.

In the second stage an interactive selection of efficient solutions is performed. The DM controls the selection by two vector-parameters: his aspiration level q^a and his reservation level q^r, where

$$q^u \le q^a < q^r \le q^n$$

The support system searches for the satisfying solution while using an achievement scalarizing function as a criterion in the single-objective optimization. Namely, the support system computes the optimal solution to the following problem:

minimize

$$\max_{1\le p\le n_o} u_p(q,q^a,q^r) + (r_o/n_o)\sum_{p=1}^{n_o} u_p(q,q^a,q^r) \qquad (3.2)$$

subject to

$$q = F(x,y), \qquad (x,y)\in Q$$

where r_o is an arbitrarily small number and u_p is a function which measures the deviation of results from the DM's expectations with respect to the p-th objective, depending on a given aspiration level q^a and a reservation level q^r.

The computed solution is an efficient (Pareto-optimal) solution to the original multiobjective model. It is presented to the DM as a current solution. The DM is asked whether he finds this solution satisfactory or not. If the DM does not accept the current solution he has to enter new aspiration and/or reservation levels for some objectives. Depending on this new information supplied by the DM a new efficient solution is computed and presented as a current solution. The process is repeated as long as the DM needs.

The function $u_p(q,q^a,q^r)$ is a strictly monotone function of the objective vector q with value $u_p=0$ if $q=q^a$ and $u_p=1$ if $q=q^r$. In our system, we use (similarly as in Wierzbicki 1986) a piece-wise linear function u_p defined as follows:

$$u_p(q,q^a,q^r)=\begin{cases} a_p(q_p-q_p^a)/(q_p^r-q_p^a)\;, & \text{if } q_p < q_p^a \\ (q_p-q_p^a)/(q_p^r-q_p^a)\;, & \text{if } q_p^a \le q_p \le q_p^r \\ b_p(q_p-q_p^r)/(q_p^r-q_p^a)+1, & \text{if } q_p^r < q_p \end{cases}$$

where a_p and b_p $(p=1,2,\dots,n_o)$ are given positive parameters. In the DINAS system, the parameters a_p and b_p are defined according to the formulae

$$a_p = a(q_p^r-q_p^a)/(q_p^a-q_p^u)$$

$$b_p = b(q_p^r-q_p^a)$$

where a and b are positive parameters computed as follows

$$a = 0.1 \min_{1\le j\le n_o} (q_j^a-q_j^u)/(q_j^r-q_j^a)^2$$

$$b = 10 \max_{1\le j\le n_o} 1/(q_j^r-q_j^a)$$

The parameters a_p and b_p satisfy inequalities: $a_p<1$ and $b_p>1$, and thereby the achievement functions u_p are convex. Minimization of the function u_p is then equivalent to minimization of a variable u_p defined as follows:

$$u_p = v_p + b_p v_p^- - a_p v_p^+ \qquad (3.3)$$

$$v_p - v_p^+ + v_p^- = (q_p - q_p^a)/(q_p^r - q_p^a) \qquad (3.4)$$

$$0 \leq v_p \leq 1 \qquad (3.5)$$

$$v_p^+ \geq 0, \ v_p^- \geq 0 \qquad (3.6)$$

4. CONCLUDING REMARKS

Initial experiences with the DINAS system on small testing examples confirm appropriateness of the used methodology for solving multiobjective transshipment problems with facility location. The interactive scheme is very easy to understand and it provides the DM with compressed the most important characteristics of generated efficient solutions. Moreover, the DM controls the system with unsophisticated parameters: aspiration and reservation levels. In effect, one easily reaches a satisfactory solution in a few interactive steps.

On the other hand, we have noticed that introducing of multiple objectives into the transshipment problem with facility location transformed this easy discrete problem into a complex one. Namely, it has turned out that the single-objective problem connected with minimization of the achievement function is far more complex than the original single-objective problems connected with minimization of several objective functions. The original single-objective problem is solved with the branch and bound method after examination only a few number of subproblems while minimization of the achievement function requires to analyze many branches of the tree. Thus for solving large real-life problems, rather a more advanced hardware than the standard IBM-PC XT should be used.

5. REFERENCES

Glover, F., Klingman, D. (1981). The simplex SON method for LP/embedded network problems. Mathematical Programming Study 15:148-176.

Glover, F., Klingman, D. (1985). Basis exchange characterization for the simplex SON algorithm for LP/embedded networks. Mathematical Programming Study 24:141-157.

Grauer, M., Lewandowski, A., Wierzbicki, A. (1984). DIDASS - theory, implementation and experiences. In M. Grauer, A.P. Wierzbicki (eds), Interactive Decision Analysis. Springer, Berlin 1984.

Ogryczak, W., Studziński, K., Zorychta, K. (1987). General concept of the Dynamic Interactive Network Analysis System (DINAS). Archiwum Automatyki i Telemechaniki 32:279-289.

Ogryczak, W., Studziński, K., Zorychta, K. (1988). A generalized reference point approach to multiobjective transshipment problem with facility location. In A.Lewandowski and A.P.Wierzbicki (eds), Theory, Software and Testing Examples in Decision Support Systems. IIASA, WP-88-071.

Ogryczak, W., Studziński, K., Zorychta, K. (1988a). Solving multi-objective distribution-location problems with the DINAS system. In A.Lewandowski and A.P.Wierzbicki (eds), Theory, Software and Testing Examples in Decision Support Systems. IIASA, WP-88-071.

Ogryczak, W., Studziński, K., Zorychta, K. (1988b). Dynamic Interactive Network Analysis System - DINAS version 2.1 (1988): user's manual. IIASA, WP-88-114.

Ogryczak, W., Studziński, K., Zorychta, K. (1989). A solver for the multiobjective transshipment problem with facility location. EJOR, in print.

Orchard-Hays, W. (1968). Advanced Linear-Programming Techniques. McGraw-Hill, New York.

Schrage, L. (1975). Implicit representation of variable upper bounds in linear programming. Mathematical Programming Study 4:118-132.

Todd, M.J. (1982). An implementation of the simplex method for linear programming problems with variable upper bounds. Mathematical Programming 23:34-49.

Wierzbicki, A.P. (1982). A mathematical basis for satisficing decision making. Math. Modelling 3:391-405.

Wierzbicki, A.P. (1986). On the completeness and constructiveness of parametric characterizations to vector optimization problems. OR Spectrum 8:73-87.

A SAFE CHOICE PROCEDURE IN INTERACTIVE PROGRAMMING

W. Michalowski* and T. Szapiro**
*Decision Analysis Laboratory
School of Business, Carleton University
Ottawa, Ontario K1S 5B6 Canada
**Institute of Econometrics
Central School of Planning and Statistics
Warsaw, Poland

Introduction

In this paper we consider a multiple criteria decision problem framed as a vector optimization program (**vop**). Within this framework we propose an approach to problem resolution which allows for the selection of a compromise decision. This decision is sought by a decision maker (**DM**) through an iterative process, with subsequent proposals of the decision in question generated at different stages of this process. At each stage, a **DM** is asked to compare the decision being considered with the worst compromise, represented by the least acceptable levels of performance, and to express his opinion about the compared outcomes. This information is later used to generate a new decision, and the process is repeated. Following changes in a **DM**'s preferences the worst compromise is subsequently displaced.

We distinguish two stages in a decision process modelled in the way described above. In a *predecision stage*, the decision problem is structured, and a **DM**'s tasks are kept to a minimum. In an *evaluation stage*, a **DM** evaluates the generated decision, and stop rules are activated if necessary. The evaluation is accomplished by asking a **DM** to classify the set of indices of outcomes into three nonvoid sets comprising the indices of outcomes to be changed, to be worsened and to be unchanged.

The paper is composed of three more sections. In the next section we discuss the framing of a decision problem as a **vop** and outline some issues related to this program resolution. Section 3 is devoted to the discussion of our proposal of a safe choice procedure, and in the last section we offer some conclusions.

Vector Optimization Program and its Resolution

Through the paper the following notation will be used:

- vector $\mathbf{x} \in \Re^n$ denotes a *decision* from the compact set X ($X \neq \emptyset, X \subset \Re^n$) which represents all admissible decisions;
- vector $\mathbf{y} = F(\mathbf{x})$, $F(\mathbf{x}) = [F_1(\mathbf{x}),...F_p(\mathbf{x})]$ denotes the *outcome* of a decision $\mathbf{x}$ such that $\mathbf{y} \in \Re^p$ ($\mathbf{y} = [y_1,..., y_p]$); the components F_j of F are called *outcome functions* which we assume are continuous and quasi-concave (Bazaraa and Shelly, 1979);
- set $Y = F(X) = \{\mathbf{y} \in \Re^p;\ \exists\, \mathbf{x} \in X\ \ F(\mathbf{x}) = \mathbf{y}\}$ denotes the *set of outcomes* associated with the set X.

The **vop** is described through the identification of the set X and the vector outcome function $F(\mathbf{x})$ for $\mathbf{x} \in X$ (Sawaragi et al., 1985). Resolution of **vop** requires the imposition of some additional preference structure on a problem. The search for possible structures may be performed in a decision space $\Re^n$ or in an outcome space $\Re^p$. It seems that the most natural structure is achieved through the introduction of a *relation of preference* $\geqq$ in an outcome space. This relation allows to classify the elements of the set Y into (Yu, 1985):

$$Y_{ND} = \{\mathbf{y} \in Y; \nexists\, \mathbf{y}' \in Y \quad \mathbf{y}' \neq \mathbf{y} \wedge \ \mathbf{y}' \geqq \mathbf{y}\} \text{ and } Y_D = Y - Y_{ND}.$$

The set Y_D is called a set of *dominated outcomes*, while the set Y_{ND} is the set of *nondominated outcomes.* Observe that $Y_D \cap Y_{ND} = \emptyset$ and $Y_D \cup Y_{ND} = Y$. Consequently, by taking inverse images of the sets Y_{ND} and Y_D, one arrives at the partition of the set X into:

$$X_E = \{\mathbf{x} \in X; \nexists\, \mathbf{x}' \in X \ \ F(\mathbf{x}') \neq F(\mathbf{x}) \wedge F(\mathbf{x}') \ \geqq \ F(\mathbf{x})\} \text{ and } X_{NE} = X - X_E$$

which are the sets of *efficient* and *nonefficient* decisions, respectively. Similarly, observe that $X_E \cap X_{NE} = \emptyset$ and $X_E \cup X_{NE} = X$. Moreover, $\mathbf{x} \in X_E$ iff $F(\mathbf{x}) \in Y_{ND}$ (Zeleny, 1974).

Using the structure imposed by the relation of preference, the resolution of **vop** may be considered as: for a given 3-tuple (X, F, $\geqq$) find X_E. If there exists a unique solution $\mathbf{x}'$ of **vop**, then $X_E = \{\mathbf{x}'\}$. However, this is not the case in most of the situations, so the concept of resolution of **vop** should be modified as: select a unique element $\mathbf{x}'$ from the set X_E of the **vop**.

There is a variety of methods designed to find $\mathbf{x}'$. Some of them operate through the imposition of ordering on X_E, for example, by assigning weights to individual outcome functions (Chankong and Haimes, 1983), or by assuming their lexicographic ordering (Ecker and Kouada, 1975). The other group of methods operates in a different manner, by the iterative imposition of a supplementary structure on a set of nondominated outcomes. These are the *interactive programming* methods (Steuer, 1986), where additional information about a DM's preferences is obtained through the interactive process of the analysis of local preferences, i.e. the preferences associated with certain decision (Nakayama, 1985). In a group of interactive programming methods we concentrate our attention on the improvement direction methods which encompass the selection of a decision point considered as a reference, together with the rules of the direction of search for some better decision (Korhonen and Lantto, 1986). Changes in the reference point used in an interactive procedure may be attributed to the determination if a given decision alternative is perceived as gain or as a loss (Tversky and Kahnemann, 1985), and as such may result in obtaining different solutions of **vop** when different methods are used.

A *safe choice* procedure proposed in this paper is an extension of the interactive programming method based on the worst compromise displacement. This was designed to solve multiple criteria linear programming problems (Michalowski, 1988), and was found to conform to the requirements of sufficiently flexible interactive methods (Laritchev and Nikiforov, 1987).

A safe choice procedure discussed here, takes its name from its "safety" feature meaning simultaneous use of the worst compromise and the ideal point as the points of reference. The final decision $\mathbf{x}'$ emerges as the result of avoidance of undesirable consequences (represented by the worst outcome levels), and is implicitly guided by the search for an ideal.

The Procedure

Let us discuss a safe choice procedure by first making a clear distinction between the preparatory activities and the activities comprising different stages of the decision process. The complete discussion of the mathematical background of this procedure is given in Michalowski and Szapiro (1988).

Preparatory activities

The operation of a safe choice procedure starts with the definition (by a DM or through some simulation exercise) of the *initial worst outcome levels* denoted by $\mathbf{y}^0 = [y_1{}^0,\dots, y_p{}^0]$. Each component $y_j{}^0$ of $\mathbf{y}^0$ corresponds to an outcome which initially fulfils the worst aspirations of a DM with respect to the j-th outcome function. Using the knowledge of $\mathbf{y}^0$ we represent the y - acceptable set as:

$$X(0) = \{\mathbf{x} \in X;\ F_j(\mathbf{x}) \geq y_j{}^0,\ j = 1,\dots,\ p\}$$

The y -acceptable set contains all decisions which have outcomes not worse than the initial worst outcome levels, and it may not contain all decisions from X_E. This set is normally not empty, and our goal is to find "the best" allowable improvement for the admissible decisions in this set. Determination of $\mathbf{y}^0$ and identification of $X(0)$ completes the preparatory activities.

The stages of the decision process composed of the activities which take place in the r-th iteration are discussed below.

Predecision stage

As the result of the activities which take place in this stage, the decision problem is further structured, and the worst outcome levels, the outcomes to be analyzed, and the corresponding worst compromise decision are generated. These activities are discussed below in greater detail.

Step PS1: Definition of the allowable improvement.

The predecision stage starts with generation of "the best" allowable improvement it is possible to achieve in relation to worst outcome levels considered in r-1 iteration. Denoting by $\mathbf{d} = [d_1, \ldots, d_p]$ the *vector of diversity coefficients* such that $\mathbf{d} = |\,\mathbf{a} - \mathbf{y}^0|$, (where $\mathbf{a}$ is an ideal point), we identify the selection of the "the best" allowable improvement as the solution of the problem:

P_r: $\max w$

s.t. $w \in W_r$

where $W_r = \{\, w \in \Re_+; \exists\, \mathbf{x} \in X(0, w_{r-1}^*)\ \ F_j(\mathbf{x}) \geq y_j^r + d_j w, j = 1, \ldots, p\}$ and $X(0, w_{r-1}^*) = \{\mathbf{x} \in X; F_j(\mathbf{x}) \geq y_j^{r-1} + d_j w_{r-1}^* \text{ for } j = 1, \ldots, p\}$ is the *worst compromise set* in r-th iteration.

The solution of problem P_r is given as $w_r^* = \sup W_r$, and it exists because W_r is compact and nonempty. If the set $X(0, w_r^*) = \{\, \mathbf{x}_r^E \,\}$, then this decision is considered to be a worst compromise, and is further analyzed. However, if $X(0, w_r^*)$ is a multi-element set, there is a necessity to select such a decision.

Step PS2: Generation of the worst compromise

We propose to look for a worst compromise decision in reference to the ideal point. The worst compromise, $\mathbf{x}_r^E \in X(0, w_r^*)$ is obtained as a solution of the following goal programming problem:

G_r: $\min \|\, F(\mathbf{x}) - \mathbf{a}\, \|$

s.t. $\mathbf{x} \in X(0, w_r^*)$

where symbol $\|...\|$ denotes the euclidean norm in $\Re^p$.

G_r is solved to find the worst compromise decision to be presented to a DM. This implies a process of evaluation by a DM of each of the generated decisions, and if the decision in question is perceived as unsatisfactory, the other one has to be generated. Therefore, the process is iterative, with each of the iterations divided into predecision and evaluation stage.

Evaluation stage

This is the stage when consecutive worst outcome levels (corresponding to $\mathbf{x}_r^E$) are evaluated. A DM controls the activities of the evaluation stage by being able to interrupt the process when satisfactory solution is obtained. Moreover, if the differences between outcomes of the worst compromise decisions generated in consecutive sessions become *e-accurate*, computational process is also interrupted (e-accuracy is defined by a DM). The possibilities and conditions of interrupting the procedure will be discussed in greater detail in a separate section.

Step ES1: Preference elicitation

A DM is confronted with the analysis of worst outcome vector $F(\mathbf{x}_r^E)$ associated with the worst compromise $\mathbf{x}_r^E$. The analysis of the components of $F(\mathbf{x}_r^E)$ results in the classification of the set of indices of outcome functions into three nonvoid subsets such as:

- $I^+(r)$ which is the subset of indices of those outcome functions whose values are to be changed at r-th iteration;
- $I^0(r)$ which is the subset of indices of those outcome functions whose values do not have to be changed at r-th iteration;
- $I^-(r)$ which is the subset of indices of those outcome functions whose values can be relaxed at r-th iteration.

The following properties hold for these three subsets: $I = I^+(r) \cup I^0(r) \cup I^-(r)$, and $I^+(r) \cap I^0(r) \cap I^-(r) = \emptyset$. The classification of the outcome functions into appropriate subsets allows the modification of the structure imposed on **vop**. This modification which operates on the elements of the worst outcome levels $\mathbf{y}^r$ is accomplished in the next step.

Step ES2: Worst outcome levels displacement

The information about a DM's preferences allows to displace the worst outcome levels according to changes in these preferences. Thus, the displacement of the worst outcome levels is defined as:

$$y_j^r = \begin{cases} y_j^{r-1} + d_j w_{r-1}^* & \text{for } j \in I^+(r) \\ y_j^{r-1} & \text{for } j \in I^0(r) \\ y_j^0 & \text{for } j \in I^-(r) \end{cases}$$

The modification of the worst outcome levels terminates the activities of an evaluation stage. Control is again transfered to the predecision stage, where new worst outcome vector $\mathbf{y}^r$ is used in step PS1 and new worst compromise is generated in step PS2.

Stop rules

The process of displacement/evaluation outlined above is not convergent in a mathematical sense. Therefore, it is necessary to introduce a mechanism which will control the transition between iterations. This mechanism should be under the control of a **DM**, but should also allow a procedure to come to a halt if certain pre-defined conditions take place. We assume, as it is a case for interactive programming methods, that the procedure is interrupted when an acceptable worst compromise decision is generated in the predecision stage. This stop rule is activated by a **DM**. However, we found that automatic interruption of the procedure is also necessary. This is when differences between values of the outcome functions generated in two consecutive iterations become insignificant. To measure this indifference, we propose to use the *vector of disparities* observed in r-th iteration $\Delta_r = [\Delta_r^1,..., \Delta_r^p]$, such that: $\Delta^j_r = |F_j(\mathbf{x}_r^E) - F_j(\mathbf{x}_{r-1}^E)|$.

At the beginning of the search process (or prior to the application of the procedure), a **DM** is asked to define a vector $\mathbf{e} = [e_1,..., e_p]$. The vector **e** is required for the identification when the disparities in observed outcomes become insignificant. Following this rule, the procedure is halted automatically when outcomes become e-accurate, i.e. $\forall j\ \Delta^j_r \leq e_j$.

The following algorithm gives a summary of a safe choice procedure:

Preparatory activities

1. calculate the initial worst outcome levels;
2. identify the y - acceptable set;

Predecision stage

3. solve the problem P_r in order to find "the best" allowable improvement;
4. find the worst compromise decision in a worst compromise set;

Evaluation stage

5. check if the stop rules may be activated => solution **x'** of **vop** is found; otherwise, classify the outcome functions according to a **DM**'s preferences;
6. displace the worst outcome levels to reflect change in a **DM**'s preferences, and return to 3.

The justification for the "safety" of the procedure can be found in interpretation of the steps 4, 5, and 6. The search for the solution of **vop** is guided by the willingness to avoid unsatisfactory outcomes. The least attractive levels of these outcomes were defined in step 1. The quest to avoid undesirable decision leads to the identification of the worst compromise set $X(0, w_r^*)$. In a moment when this set is identified, the perspective of search changes. Instead of avoiding the "bad" (this was achieved with the generation of $X(0, w_r^*)$), the search for a "good" worst compromise starts. The level of satisfaction is measured in term of a distance from the ideal point. The "good" worst compromise is generated in step 4, and further evaluated by a DM in step 5. The results of this evaluation may be twofold: the worst compromise is accepted and **vop** is solved, or further search is necessary.

During the sampling of the efficient set, a DM learns about the problem in hand through the analysis of worst compromise decisions. This "learning phenomenon" affects a DM's perception of the least acceptable levels and may result in changes of the preferences. These changes are modelled in the procedure by defining the displacement of the worst outcome levels (step 6), and if necessary, by allowing for the decrease of the worst outcome levels up to their initial values.

We argue that a safe choice procedure differs from most of the existing interactive programming methods as it combines the advantages of achievement function methods (Wierzbicki, 1980), displaced ideal methods (Zeleny, 1982) and classical reference point methods (Benayoun et al., 1971). Because of this property, the procedure is flexible enough to allow for the preference shifting and preference reversal effects (Reilly, 1982).

Conclusions

A safe choice procedure proposed in this paper is based on the earlier work of Michalowski (1988), and generalizes the concept of displaced worst compromise to a wider class of decision problems. We consider that strength of our approach lies in its ability to deal with the decisional framework of avoiding undesirable outcomes. In the course of the search for compromise decision, following the shifts in preferences, the way in which a DM perceives undesirable outcomes may change. Incorporation of this change is reflected in a manner how safe choice procedure is sampling the efficient set of **vop**. The final solution **x'** emerges as a result of incremental learning by a DM about his/her preferences.

References

Bazaraa M. S. and C. M. Shelly (1979) *Nonlinear Programming: Theory and Algorithms*. J. Wiley, New York.

Benayoun R., J. deMontgolfier, J. Tergny and O. Larichev (1971) "Linear Programming with Multiple Objective Functions: Step-method (STEM)", *Mathematical Programming*, vol. 1, no. 3.

Chankong V. and Y. Y. Haimes (1983) *Multiobjective Decision Making: Theory and Methodology*. North Holland, New York.

Ecker J. G. and I. A. Kouada (1975) "Finding Efficient Points for Linear Multiple Objective Programs", *Mathematical Programming*, vol. 8, no. 3.

Korhonen P. and O. Lantto (1986) "An Experimental Comparison of Some Reference Direction Techniques for MCDM Problems", Helsinki School of Economics.

Larichev O. and A. D. Nikiforov (1987) "Analytical Survey of Procedures For Solving Multicriteria Mathematical Programming Problems", in Y. Sawaragi, K. Inoue and H. Nakayama (eds) *Toward Interactive and Intelligent Decision Support Systems*. Springer Verlag, Tokyo.

Michalowski W. (1988) "Use of the Displaced Worst Compromise in Interactive Multiobjective Programming", *IEEE Transactions on Systems, Man, and Cybernetics*, vol. 18, no. 3.

Michalowski W. and T. Szapiro (1988) "A Procedure For Worst Outcomes Displacement in Multiple Criteria Decision Making", *Computers and Operations Research*, (forthcoming).

Nakayama H. (1985) "On the Components In Interactive Multiobjective Programming Methods", in M. Grauer, M. Thompson and A. P. Wierzbicki (eds) *Plural Rationality and Interactive Decision Processes*. Springer Verlag, Berlin.

Reilly R. J. (1982) "Preference Reversal: Further Evidence and Some Suggested Modifications in Experimental Design", *American Economic Review*, no. 72.

Sawaragi Y., H. Nakayama and T. Tanino (1985) *Theory of Multiobjective Optimization*. Academic Press, Orlando.

Steuer R. E. (1986) *Multiple Criteria Optimization: Theory, Computation and Application*. J. Wiley, New York.

Tversky A. and D. Kahneman (1985) "The Framing of Decisions and the Psychology of Choice", in G. Wright (ed) *Behavioral Decision Making*. Plenum Press, New York.

Wierzbicki A. P. (1980) "The Use Of Reference Objectives In Multiobjective Optimization", in G. Fandel and T. Gal (eds) *Multiple Criteria Decision Making, Theory and Application*. Springer Verlag, Berlin.

Yu P. L. (1985) *Multiple-Criteria Decision Making: Concepts, Techniques and Extensions*. Plenum Press, New York.

Zeleny M. (1974) *Linear Multiobjective Programming*. Springer Verlag, Berlin.

Zeleny M. (1982) *Multiple Criteria Decision Making*. McGraw-Hill, New York.

PERSONAL COMPUTER-AIDED INTERACTIVE DECISION MAKING FOR MULTIOBJECTIVE LINEAR PROGRAMMING PROBLEMS WITH FUZZY COEFFICIENTS AND ITS APPLICATIONS

Masatoshi SAKAWA
Department of Computer Science, Faculty of Engineering
Iwate University, Morioka 020, JAPAN
Hitoshi YANO
Department of Information Science, College of Economics
Kagawa University, Takamatsu 760, JAPAN

Abstract

In this paper, we focus on multiobjective linear programming problems with fuzzy coefficients and present a personal computer-aided interactive decision making method for obtaining the satisficing solution of the decision maker (DM) on the basis of the linear programming method. The fuzzy coefficients in the description of the objective functions and the constraints, which reflect the experts' ambiguous understanding of the nature of the coefficients in the problem-formulation process, are characterized by fuzzy numbers. The concept of (α,β)-multiobjective linear programming and (α,β)-Pareto optimality is introduced based on the α-level and β-level sets of fuzzy numbers. In our interactive decision making method, in order to generate a candidate for the satisficing solution of the DM which is also (α,β)-Pareto optimal, if the DM specifies the degree α and β of the α-level and β-level sets and the reference objective values, the minimax problem is solved by making use of the linear programming method, and the DM is supplied with the corresponding (α,β)-Pareto optimal solution together with the trade-off rates among the values of the objective functions and the degree α and β. Then by considering the current values of the objective functions, α and β as well as the trade-off rates, the DM acts on this solution by updating his/her reference objective values and/or the degree α,β. In this way the satisficing solution for the DM can be derived efficiently from among an (α,β)-Pareto optimal solution set. On the basis of the proposed method, interactive FORTRAN programs, which run on N.E.C. PC-9801 series personal computers, are developed to implement man-machine interactive procedures. An application to an optimal operation problem in a package system in automated warehouses is demonstrated together with the corresponding personal computer outputs.

1. INTRODUCTION

In multiobjective programming problems, multiple objectives are usually noncommensurable and cannot be combined into a single objective. Moreover, the objectives often conflict with each other in that any improvement of one objective can be achieved only at the expense of another. For most such multiobjective programming problems, in addition to the decision analyst's role, value-judgement-based analysis of a decision maker plays an essential role. To be more explicit, it is particularly important how to combine the roles of a decision maker and a decision analyst in order to find a compromise or satisficing solution of a decision maker which is also Pareto optimal. However, when formulating the multiobjective programming problem which closely describes and represents the real-world decision situation, various factors of the real-world system should be reflected in the description of the objective functions and the constraints. Naturally these objective functions and the constraints involve many coefficients whose possible values may be assigned by the experts. In the conventional approaches, such coefficients are fixed at some values in an experimental and/or subjective manner through the experts' understanding of the nature of the coefficients.

In most real-world situations, however, it may be reasonable to assume that the possible values of these coefficients are often only imprecisely or ambiguously known to the experts. In this case, it would certainly be more appropriate to interpret the experts' understanding of the coefficients as fuzzy numerical data which can be represented by means of fuzzy subsets of the real line known as fuzzy numbers [1]. The resulting multiobjective programming problem involving fuzzy coefficients would be viewed as the more realistic version of the conventional one [6,10,11].

In this paper, we focus on the multiobjective linear programming problems with fuzzy coefficients characterized by fuzzy numbers. To cope with the fuzzy coefficients, the concept of (α,β)-Pareto optimality is introduced by extending the ordinary Pareto optimality on the basis of the α-level and β-level sets of the fuzzy numbers. Then a new interactive decision making method is presented in order to derive the satisficing solution of the decision maker efficiently from among an (α,β)-Pareto optimal solution set on the basis of the linear programming method as a generalization of our previous results [7-10].

2. (α,β)-PARETO OPTIMALITY

Consider the following multiobjective linear programming problem involving fuzzy coefficients (MOLP-FC) :

$$\min \quad (\tilde{c}_1 x \ , \ \tilde{c}_2 x \ , \ldots , \ \tilde{c}_k x)$$

$$\text{subject to} \quad x \in X(\tilde{a},\tilde{b}) \triangleq \{ \ x \in E^n \mid \tilde{a}_j x \leqq \tilde{b}_j \ , j=1,\ldots,m \ ; \ x \geqq 0 \ \}. \qquad (1)$$

where $\tilde{c}_i = (\tilde{c}_{i1},\ldots,\tilde{c}_{in})$, $\tilde{a}_j = (\tilde{a}_{j1},\ldots,\tilde{a}_{jn})$ and $\tilde{b}_j$ represent fuzzy coefficients which reflect the experts' ambiguous understanding of the nature of the coefficients in the problem-formulation process.

We assume that $\tilde{c}_{i1},\ldots,\tilde{c}_{in}$, $\tilde{a}_{j1},\ldots,\tilde{a}_{jn}$ and $\tilde{b}_j$ in the MOLP-FC are fuzzy numbers [1] whose membership functions are $\mu_{\tilde{c}_{i1}}(c_{i1}),\ldots,\mu_{\tilde{c}_{in}}(c_{in})$, $\mu_{\tilde{a}_{j1}}(a_{j1}),\ldots,$ $\mu_{\tilde{a}_{jn}}(a_{jn})$ and $\mu_{\tilde{b}_j}(b_j)$ respectively. For simplicity of notation, define the following vectors:

$$c = (c_1,\ldots,c_k),\quad \tilde{c} = (\tilde{c}_1,\ldots,\tilde{c}_k),\quad a = (a_1,\ldots,a_m),$$

$$\tilde{a} = (\tilde{a}_1,\ldots,\tilde{a}_m),\quad b = (b_1,\ldots,b_m),\quad \tilde{b} = (\tilde{b}_1,\ldots,\tilde{b}_m).$$

Then we can introduce the following α-level and β-level sets [1] of the fuzzy numbers $\tilde{a}_{jr}$, $\tilde{b}_j$ and $\tilde{c}_{ir}$.

Definition 1. (level set)

(1) The α-level set of the fuzzy numbers $\tilde{a}_{jr},\tilde{b}_j$ is defined as the ordinary set $L_\alpha(\tilde{a},\tilde{b})$ for which the degree of their membership functions exceeds the level α:

$$L_\alpha(\tilde{a},\tilde{b}) = \{(a,b) \mid \mu_{\tilde{a}_{jr}}(a_{jr}) \geqq \alpha,\ \mu_{\tilde{b}_j}(b_j) \geqq \alpha,\ j=1,\ldots,m,\ r=1,\ldots,n\}. \quad (3)$$

(2) The β-level set of the fuzzy numbers $\tilde{c}_{ir}$ is defined as the ordinary set $L_\beta(\tilde{c})$ for which the degree of their membership functions exceeds the level β:

$$L_\beta(\tilde{c}) = \{ c \mid \mu_{\tilde{c}_{ir}}(c_{ir}) \geqq \beta,\ i=1,\ldots,k,\ r=1,\ldots,n \}. \quad (4)$$

For certain degrees α and β, the MOLP-FC can be understood as the following nonfuzzy (α,β)-multiobjective linear programming ((α,β)-MOLP) problem.

$$\min\ (c_1x,\ c_2x,\ \ldots,\ c_kx)$$

$$\text{subject to}\quad x \in X(a,b) \triangleq \{ x \in E^n \mid a_j x \leqq b_j,\ j=1,\ldots,m\ ;\ x \geqq 0 \}, \quad (5)$$

$$(a,b) \in L_\alpha(\tilde{a},\tilde{b}),\quad c \in L_\beta(\tilde{c}).$$

It should be emphasized here that in the (α,β)-MOLP the coefficients (a,b,c) are treated as decision variables rather than constants.

On the basis of the level sets of the fuzzy numbers, we introduce the concept of (α,β)-Pareto optimal solutions to the (α,β)-MOLP.

Definition 3. ((α,β)-Pareto optimal solution)

$x^* \in X(a,b)$ is said to be an (α,β)-Pareto optimal solution to the (α,β)-MOLP, if and only if there does not exist another $x \in X(a,b)$, $(a,b) \in L_\alpha(\tilde{a},\tilde{b})$, $c \in L_\beta(\tilde{c})$ such that $c_i x \leqq c_i^* x^*$, $i=1,\ldots,k$, with strict inequality holding for at least one i, where the corresponding values of coefficients $(a*,b*)$ and $c*$ are called α-level and β-level optimal coefficients.

As can be immediately understood from the definition, usually (α,β)-Pareto optimal solutions consist of an infinite number of points, and some kinds of subjective judgement should be added to the quantitative analyses by the decision maker (DM). The DM must select his/her satisficing or compromise solution from among (α,β)-Pareto optimal solutions.

3. MINIMAX PROBLEMS AND INTERACTIVE ALGORITHM

For generating a candidate for the satisficing solution which is also (α,β)-Pareto optimal, in our interactive decision making method, the DM is asked to specify the degree α, β and the reference levels of achievement of the objective functions, called reference levels. Observe that the idea of the reference levels or the reference point was first appeared in Wierzbicki [12]. For the DM's degree α, β and reference levels $\bar{z}_i$, $i=1,\dots,k$, the corresponding (α,β)-Pareto optimal solution, which is in a sense close to his/her requirement or better than that if the reference levels are attainable, is obtained by solving the following minimax problem.

$$\min_{x,v,a,b,c} \quad v \tag{6a}$$

$$\text{subject to} \quad c_i x - \bar{z}_i \leqq v, \quad i=1,\dots,k, \tag{6b}$$

$$a_j x \leqq b_j, \quad j=1,\dots,m, \quad x \geqq 0, \tag{6c}$$

$$(a,b) \in L_\alpha(\tilde{a},\tilde{b}), \quad c \in L_\beta(\tilde{c}). \tag{6d}$$

In this formulation, however, constraints (6b) and (6c) are nonlinear because the coefficients a,b and c are treated as decision variables. To cope with such nonlinearlities, we introduce the following set-valued functions $S_i(.)$ and $T_j(.,.)$.

$$S_i(c_i) = \{(x,v) \mid c_i x - \bar{z}_i \leqq v\}, \quad T_j(a_j,b_j) = \{x \mid a_j x \leqq b_j\}. \tag{7}$$

Then it can be easily verified that the following relations hold for $S_i(.)$ and $T_j(.,.)$, when $x \geqq 0$.

Proposition 1.

(1) If $c_i^1 \leqq c_i^2$, then $S_i(c_i^1) \supset S(c_i^2)$.

(2) If $a_j^1 \leqq a_j^2$, then $T_j(a_j^1,b_j) \supset T_j(a_j^2,b_j)$.

(3) If $b_j^1 \leqq b_j^2$, then $T_j(a_j,b_j^1) \subset T_j(a_j,b_j^2)$.

Now from the properties of the α-level and β-level sets for the vectors of the fuzzy numbers $\tilde{c}_i, \tilde{a}_j$ and the fuzzy numbers $\tilde{b}_j$, it should be noted here that the feasible regions for a_j, b_j and c_i can be denoted respectively by the intervals $[a^L_{j\alpha}, a^R_{j\alpha}]$, $[b^L_{j\alpha}, b^R_{j\alpha}]$ and $[c^L_{i\beta}, c^R_{i\beta}]$.

Therefore, through the use of Proposition 1, we can obtain an optimal solution to (6) by solving the following linear programming problem.

$$\begin{aligned} \min \quad & v \\ \text{subject to} \quad & c^L_{i\beta}x - \bar{z}_i \leqq v, \; i=1,\dots,k, \\ & a^L_{j\alpha}x \leqq b^R_{j\alpha}, \; j=1,\dots,m, \quad x \geqq 0 . \end{aligned} \tag{9}$$

It is important to realize here that this problem is no more nonlinear but an ordinary linear programming problem, and consequently, it is easy to solve it by making use of the simplex method of linear programming.

It should be emphasized here that the minimax problem is simply used as a means of generating an (α,β)-Pareto optimal solution, and if the DM is not satisfied with the current (α,β)-Pareto optimal solution, it is possible for him/her to improve the solution by updating his/her reference levels and/or the degree α,β.

The relationships between the optimal solutions to (9) and the (α,β)-Pareto optimal concept of the (α,β)-MOLP can be characterized by the following theorem.

Theorem 1.

(1) If x^* is a unique optimal solution to (9), then x^* is an (α,β)-Pareto optimal solution to the (α,β)-MOLP.

(2) If x^* is an (α,β)-Pareto optimal solution, and (a^*,b^*), c^* are α-level and β-level optimal parameter to the (α,β)-MOLP, then x^* is an optimal solution to (9) for some $\bar{z} = (\bar{z}_1,\dots,\bar{z}_k)$.

Now given the (α,β)-Pareto optimal solution for the degree α,β and the reference levels specified by the DM by solving the corresponding minimax problem, the DM must either be satisfied with the current (α,β)-Pareto optimal solution, or update his/her reference levels and/or the degree α, β. In order to help the DM express his/her degree of preference, trade-off information between a standing objective function and each of the other objective functions as well as between the degree α,β and the objective functions is very useful. Such trade-off information is easily obtainable since it is closely related to the simplex multipliers of the problem (9).

To derive the trade-off information, define the Lagrangian function L for the problem (9) as follows :

$$L = v + \sum_{i=1}^{k} \pi_i (c^L_{i\beta}x - \bar{z}_i - v) + \sum_{j=1}^{m} \lambda_j (a^L_{j\alpha}x - b^R_{j\alpha}) \tag{10}$$

If all $\pi_i > 0$ for each i, then by extending the results in Haimes and Chankong [3], it can be proved that the following expression holds [13]

$$-\frac{\partial(c^L_{1\beta}x)}{\partial(c^L_{i\beta}x)} = \frac{\pi_i}{\pi_1}, \quad i=2,\ldots,k. \tag{11}$$

Regarding a trade-off rate between $c_i x$ and α, the following relation holds based on the sensitivity theorem [2].

$$\frac{\partial(c^L_{i\beta}x)}{\partial\beta} = \sum_{i=1}^{k} \pi_i \frac{\partial(c^L_{i\beta})}{\partial\beta} x, \quad i=1,\ldots,k \tag{12}$$

$$\frac{\partial(c^L_{i\beta}x)}{\partial\alpha} = \sum_{j=1}^{m} \lambda_j \left\{ \frac{\partial(a^L_{j\alpha})}{\partial\alpha} x - \frac{\partial(b^R_{j\alpha})}{\partial\alpha} \right\}, \tag{13}$$

It should be noted here that in order to obtain the trade-off information from (11), (12) or (13), all the constraints $c^L_{i\beta}x - \bar{z}_i \leq v$, i=1,...,k of the problem (9) must be active for the current optimal solution. Therefore, if there are inactive constraints, it is necessary to replace $\bar{z}_i$ for inactive constraints by $c^L_{i\beta}x^* - v^*$ and solve the corresponding minimax problem (9) for obtaining the simplex multipliers.

We can now construct the interactive algorithm in order to derive the satisficing solution for the DM from among the (α,β)-Pareto optimal solution set. The steps marked with an asterisk involve interaction with the DM.

Step 0. Calculate the individual minimum and maximum of each objective function under the given constraints for $\alpha,\beta = 0$ and $\alpha,\beta = 1$.

Step 1* Ask the DM to select the initial value of α, β and the initial reference levels $\bar{z}_i$, i=1,...,k.

Step 2. For the degree α, β and the reference levels specified by the DM, solve the minimax problem.

Step 3*. The DM is supplied with the corresponding (α,β)-Pareto optimal solution and the trade-off rates between the objective functions and the degree α, β. If the DM is satisfied with the current objective function values of the (α,β)-Pareto optimal solution, stop. Otherwise, the DM must update his/her reference levels and/or the degree α, β by considering the current values of the objective functions and α, β together with the trade-off rates between the objective functions and the degree α, β and return to step 2.

Here it should be stressed for the DM that (1) any improvement of one objective function can be achieved only at the expense of at least one of the other objective functions for some fixed degree α, β, and (2) the greater values of the degree α, β gives worse values of the objective functions for some fixed reference levels.

4. AN INTERACTIVE COMPUTER PROGRAM AND ITS APPLICATION

Based on the methods described above, we have developed an interactive computer program for solving multiobjective linear programming problems with fuzzy coefficients. The entire program is written in FORTRAN, because FORTRAN language is very popular among many scientists and also almost all computer facilities are available for loading the programs with minor changes. Our program consists, roughly, of 1200 FORTRAN statements. The current personal computer version of our program is capable of solving problems with 50-100 variables, 25-50 constraints and 20 objectives under N.E.C. PC-9801 series personal computers of 384KB-640KB memory size with at least one diskette drive and a black and white or color monitor.

Our computer program is composed of one main program and several subroutines. The main program calls in and runs the subprograms with commands indicated by the user (DM). Here we give a brief explanation of the major commands prepared in our computer program.

(1) MINMAX: Displays the calculated individual minimum and maximum of each of the objective functions under the given constraints for α,β = 0 and α,β = 1.

(2) GO: Derives the satisficing solution for the DM from among the (α,β)-Pareto optimal solution set by updating the reference levels and/or the degree α,β.

(3) STOP: Exists from the program.

(4) SAVE: Saves all the necessary information, which has been put in, in a file.

(5) READ: Restores the information which was saved in the file.

In our computer program, the fuzzy coefficients, which reflect the experts' ambiguous understanding of the nature of coefficients in the problem-formulation process, are assumed to be characterized by the fuzzy numbers whose membership functions are either linear or exponential. (α,β)-Pareto optimal solutions for multiobjective linear programming problems are calculated by solving the minimax problems on the basis of the simplex method of linear programming.

Now consider an application of our interactive computer program under N.E.C. PC-9801 personal computers to an optimal operation of a packaging system in automated warehouses, where the DM (manager of warehouses) is supposed to decide packaging volume and the items for shipping next day by considering the total volume to be packed, the priority for packaging , and the balance of the inventory, etc. The problem formulation was originally given by Hamada [4], and the three objective linear-programming problem formulations were developed by Nakayama et al.[5] and the numerical example was solved by their interactive decision making method. Recently, the same problem was successfully solved by the interactive fuzzy satisficing method of Sakawa et al.[9]

In this paper, for incorporating the experts' ambiguity or fuzziness in the problem-formulation process, we modify the original three objective linear programming problem as the following one involving fuzzy coefficients.

$$\max \quad z_1(x) = \tilde{c}_1x_1 + \tilde{c}_1x_2 + 5\tilde{c}_2x_3 + 5\tilde{c}_2x_4 + 4\tilde{c}_3x_5 + 4\tilde{c}_3x_6 + 2\tilde{c}_4x_7 + 2\tilde{c}_4x_8$$

$$\min \quad z_2(x) = x_3 + x_4 + x_5 + x_6$$

$$\min \quad z_3(x) = 5.6x_1 - 5.6x_2 - 2.4x_3 + 2.4x_4 - 5.25x_5 + 5.25x_6 - 8.4x_7 + 8.4x_8$$

subject to

$$x_1 + x_2 \leqq 1000\tilde{c}_1,\quad x_3 + x_4 \leqq 200\tilde{c}_2,\quad x_5 + x_6 \leqq 250\tilde{c}_3,\quad x_7 + x_8 \leqq 500\tilde{c}_4\ ,$$

$$-x_1 + x_2 \leqq 300,\quad x_3 - x_4 \leqq 100,\quad x_5 - x_6 \leqq 200,\quad x_7 - x_8 \leqq 200,$$

$$x_1 + x_3 + x_5 + x_7 \leqq 1250(1 + \tilde{\epsilon}_w),\quad x_2 + x_4 + x_6 + x_8 \leqq 1250(1 + \tilde{\epsilon}_w),$$

$$1500\tilde{\alpha}_1 \leqq x_1 + x_2 + x_5 + x_6 \leqq 1500,\quad 1000\tilde{\alpha}_2 \leqq x_3 + x_4 + x_7 + x_8 \leqq 1000,$$

$$2x_1 + 2x_2 + x_3 + x_4 \leqq 1500(3 - \tilde{b}_1),\quad x_1 + x_2 + 6x_3 + 6x_4 \leqq 500(5 - 3\tilde{b}_1),$$

$$3x_5 + 3x_6 + x_7 + x_8 \leqq 200(7 - 3\tilde{b}_2),\quad x_5 + x_6 + 2x_7 + 2x_8 \leqq 200(10 - 3\tilde{b}_2),$$

$$x_j \geqq 0,\ j=1,\dots,8.$$

where $\tilde{c}_1,\dots,\tilde{c}_4$, $\tilde{\epsilon}_W$, $\tilde{\alpha}_1$, $\tilde{\alpha}_2$, $\tilde{b}_1$, $\tilde{b}_2$ are fuzzy numbers reflecting the experts' ambiguity.

In the following, some of the interaction processes using our computer program under N.E.C. PC-9801 personal computer are explained through the aid of some of the computer outputs.

Illustration 1

Using the MINMAX command, the calculated individual minimum and maximum of each of the objective functions z_1, z_2 and z_3 for α, β = 1 and α, β = 0 are displayed.

Illustration 2

Using the GO command, the minimax problem is solved for the initial reference levels and the degree α, β, and the DM is supplied with the corresponding (α,β)-Pareto optimal solution and the trade-off rates between the objective functions and the degree α, β. Since the DM is not satisfied with the current values of the (α,β)-Pareto optimal solution, the DM updates his/her reference levels.

The same procedure continues in this manner, until the DM is satisfied with the current values of the (α,β)-Pareto optimal solution. In this example, at the third iteration, the satisficing solution of the DM is obtained.

5. CONCLUSION

In this paper, we have proposed a personal computer-aided interactive decision making method in order to deal with the multiobjective linear programming problem with fuzzy coefficients characterized by fuzzy numbers. Through the use of the concept of the level sets of the fuzzy numbers, a new solution concept called the (α,β)-Pareto optimality has been introduced. In our interactive scheme, the satisficing solution of the DM can be derived from among an (α,β)-Pareto optimal solution set by updating the reference levels and/or the degree α,β together with the trade-off rates between the objective functions and the degree α,β. An application to an optimal operation problem in packaging systems in automated warehouses demonstrates the feasibility and efficiency of both the proposed method and the interactive computer program on N.E.C. PC-9801 series personal computers.

REFERENCES

[1] Dubois, D. and Prade, H., Fuzzy Sets and Systems : Theory and Applications (Academic Press, 1980).

[2] Fiacco, A. V., Introduction to Sensitivity and Stability Analysis in Nonlinear Programming (Academic Press, 1983).

[3] Haimes, Y. Y. and Chankong, V., Kuhn-Tucker multipliers as trade-offs in multiobjective decision-making analysis, Automatica, 15 (1979) 59-72.

[4] Hamada, T., Studies on optimal operations in automated warehouses, doctorial dissertation, Osaka Univ., Japan (1976).

[5] Nakayama, H., Karasawa, Y. and Dohi, S., Subjective programming applied to optimal operatioin in automated warehouses, Int. J. Systems Science, 11, (1980) 513-525.

[6] Orlovski, S. A., Multiobjective programming problems with fuzzy parameters, Control and Cybernetics, 13 (1984) 175-183.

[7] Sakawa, M., Interactive computer programs for fuzzy linear programming with multiple objectives, Int. J. Man-Machine Studies, 18 (1983) 489-503.

[8] Sakawa, M., Interactive fuzzy decision making for multiobjective linear programming and its application, Proc. IFAC Symposium on Fuzzy Information, Knowledge Representation and Decision Analysis (E. Sanchez, ed.) (1983) 295-300.

[9] Sakawa, M., Yano, H. and Yumine, T., An interactive fuzzy satisficing method for multiobjective linear-programming problems and its application, IEEE Trans. Systems, Man, and Cybernetics, SMC-17 (1987) 654-661.

[10] Sakawa, M. and Yano, H., Interactive decision making for multiobjective linear programming problems with fuzzy parameters, in Large-Scale Modelling and Interactive Decision Analysis (Ed. by G.Fandel, M.Grauer, A.Kurzhanski and A.P.Wierzbicki) (1985) 88-96.

[11] Tanaka, H. and Asai, K., Fuzzy linear programming problems with fuzzy numbers, Fuzzy Sets and Systems, 13 (1984) 1-10.

[12] Wierzbicki, A. P., The use of reference objectives in multiobjective optimization - Theoretical implications and practical experiences, WP-79-66, IIASA, Austria (1979).

[13] Yano, H. and Sakawa, M., Trade-off rates in the weighted Tchebycheff norm method, Large Scale Systems (in the press).

Illustration 1

```
COMMAND:
MINMAX

          INDIVIDUAL MINIMUM AND MAXIMUM FOR  α = β =1

          I          MINIMUM          I          MAXIMUM
---------+-------------------------+-----------------------
 Z( 1)  I              -2.4000    I              -1.0950
 Z( 2)  I                .0000    I             324.7059
 Z( 3)  I           -4650.0000    I           10325.0000

          INDIVIDUAL MINIMUM AND MAXIMUM FOR  α = β =0

          I          MINIMUM          I          MAXIMUM
---------+-------------------------+-----------------------
 Z( 1)  I              -5.0400    I                .0000
 Z( 2)  I                .0000    I             433.3333
 Z( 3)  I           -4650.0000    I           11522.0000
```

Illustration 2

```
COMMAND:
GO

---------------------< ITERATION 1 >---------------------

INPUT INITIAL REFERENCE VALUES Z(I) (I=1,3):
-5.0  0.0  -4650.0

INPUT THE DEGREE α  OF THE α  LEVEL SETS
FOR THE FUZZY PARAMETERS:
0.9

INPUT THE DEGREE β  OF THE β  LEVEL SETS
FOR THE FUZZY PARAMETERS:
0.9

APPROXIMATE SOLUTION TO THE MINIMAX PROBLEM
FOR YOUR REFERENCE VALUES
-----------------------------
     OBJECTIVE FUNCTION
-----------------------------
Z(1) =          -1.7772
Z(2) -          221.8541
Z(3) =        -4428.1459
------------------------------------------------------------
   X( 1) =              .0000     X( 2) =          300.0000
   X( 3) =            33.8906     X( 4) =             .0000
   X( 5) =           187.9635     X( 6) =             .0000
   X( 7) =           218.0547     X( 8) =           18.0547
------------------------------------------------------------

α , β -PARETO OPTIMALITY TEST
  EPS( 1)=      .5227963526
  EPS( 2)=      .0000000000
  EPS( 3)=      .0000000000
-----------------------------
      α , β -PARETO OPTIMUM
-----------------------------
Z(1) =          -2.3000
Z(2) =          221.8541
Z(3) =        -4428.1459
------------------------------------------------------------
   X( 1) =           248.3283     X( 2) =          548.3283
   X( 3) =            33.8906     X( 4) =             .0000
   X( 5) =           187.9635     X( 6) =             .0000
   X( 7) =           218.0547     X( 8) =           18.0547
------------------------------------------------------------

------------------------------------------------------------
TRADE-OFFS AMONG OBJECTIVE FUNCTIONS
 -Δ Z(2)/Δ Z(1) =               .1903
 -Δ Z(3)/Δ Z(1) =           2166.5327
------------------------------------------------------------
TRADE-OFF BETWEEN α  AND OBJECTIVE FUNCTIONS
 Δ Z/Δ α  =            3.8754
------------------------------------------------------------
TRADE-OFF BETWEEN β  AND OBJECTIVE FUNCTIONS
 Δ Z/Δ β  =       481984.2189
------------------------------------------------------------

ARE YOU SATISFIED WITH THE CURRENT OBJECTIVE VALUES OF
THE  α , β -PARETO OPTIMAL SOLUTION ?
NO
```

PERFORMANCE CHARACTERISTICS OF THREE INTERACTIVE SOLUTION STRATEGIES FOR BICRITERIA INTEGER PROGRAMMING

R. Ramesh +
Mark H. Karwan ++
Stanley Zionts +

+ School of Management
State University of New York at Buffalo
Buffalo, New York 14260

++ Department of Industrial Engineering
State University of New York at Buffalo
Buffalo, New York 14260

ABSTRACT

In this paper, we develop an efficient solution framework for bicriteria integer programming, and present three algorithms developed using this framework. The decision-maker's underlying utility function is assumed to be pseudoconcave and nondecreasing. The problem is solved interactively using a branch-and-bound approach. We present the performance characteristics of three algorithms developed using this framework with an illustration and computational experience with problems having upto 80 variables and 40 constraints. The results show that the proposed approach is viable and effective in solving practical bicriteria integer programming problems.

1. Introduction

This paper presents an efficient interactive framework for solving bicriteria integer programming problems, and the performance characteristics of three algorithms developed using this framework. A decision maker is assumed to have only an implicit utility function of the two objectives he wishes to maximize. The problem is solved by an interactive branch-and-bound method in which the decision-maker's preference structure is explored and assessed using pairwise comparison questions.

Several methods have been proposed for solving multicriteria integer programming problems in general. A classification of these methods is given in Zionts [15]. The multicriteria programming problem has been addressed using three approaches: vector optimization, maximization of an explicit utility function using multiattribute utility theory, and the maximization of an implicit utility function using interactive methods. Some of the important references on multicriteria integer programming using vector optimization are [2] [7] [9] [13]. The multiattribute utility theory has been used to solve several multicriteria decision problems [3] [6]. In this research, we address the bicriteria integer programming problem using the implicit utility maximization approach. The problem of concern is formulated as follows.

$$\text{Maximize} \quad g(Cx) \qquad \text{(BCIP)}$$

$$\text{Subject to:} \quad Ax \leq b,\ x \geq 0,\ \text{integer}$$

where C and A are (2 x n) and (m x n) matrices of criteria and constraint coefficients respectively, $b \in R^m$ is the vector of resources available, $x \in R^n$ denotes the vector of decision variables, and $g(\cdot)$ is the implicit utility function of the decision maker, which is assumed to be pseudoconcave and nondecreasing. The linear relaxation of problem (BCIP) is denoted as (BCLP).

The organization of this paper is as follows. Section 2 presents the proposed methodology, and Section 3 presents an illustration of the algorithms developed using this framework. Section 4 summarizes our computational experience with the algorithms, and the conclusions are presented in section 5.

2. The Methodology

The problem (BCIP) is solved using an interactive branch-and-bound framework. The framework employs a modified version of the method of Zionts and Wallenius [17] for multicriteria linear programming to solve the linear relaxations of the candidate problems generated in the branch-and-bound search. The initial step consists of relaxing the integrality constraints and solving problem (BCLP). If the optimal solution to (BCLP) is integer, then it is also optimal to problem (BCIP), and the search is terminated. Otherwise, a branch-and-bound search for the integer optimum is conducted. The feasible set is partitioned into subsets that either dominate the incumbent or are efficient with respect to it. The search is conducted separately in each set. This strategy is based on our initial computational experience which showed that the total number of questions can be significantly reduced using this approach. Using a linear composite objective, a candidate problem is partitioned by branching on a fractional variable. A branching strategy such as LIFO can be used for this purpose. A candidate problem can be fathomed if the optimal solution to its bicriteria linear relaxation is less preferred than the incumbent. We develop a preference structure representation mechanism using certain convex cones that can be used to avoid having to ask the decision maker questions in the following section.

2.1. Preference Structure Representation

We first introduce the notation and definitions used in the following analysis.

Let X^c and X^I denote the set of feasible solutions to problems (BCIP) and (BCLP) respectively, in the decision variable space. Hence $X^I \subseteq X^c$. An element of X^c is denoted as x. The objective function vector is defined for all $x \in X^c$ and is denoted by $f(x) = (c^1x, c^2x)^t$. The set $Z^c = \{f(x) \mid x \in X^c\}$ denotes the set of feasible solutions to (BCLP) in the objective function space. An element of Z^c is denoted as z.

Assuming a pseudoconcave utility function, a convex cone is defined as follows. Consider two distinct solutions z^1 and z^2. Let the decision maker prefer z^1 to z^2. Then the set of solutions falling on or dominated by the ray $\{z \mid z = z^2 + \mu\,(z^2 - z^1),\ \mu \geq 0\}$ is less preferred than z^1, and no more preferred than z^2, and hence can be eliminated from consideration. This ray is a convex cone, and is denoted as $c[z^1 \succ z^2]$. z^1 is called its 'upper generator' and z^2 its 'lower generator.' The theory underlying convex cones is developed in Korhonen et al. [8].

We use convex cones to represent the decision-maker's preference structure while solving problem (BCIP) as follows. Consider a candidate problem, whose solution to its bicriteria linear relaxation is z. Let z^I denote the current incumbent. If z is preferred to z^I, then we can eliminate the solutions dominated by $c[z \succ z^I]$ in searching for the integer optimum. Similarly, if z^I is preferred to z, then the solutions dominated by $c[z^I \succ z]$ can be eliminated. Such cones are termed 'global cones,' since they can be used throughout the branch-and-bound search in eliminating solutions from consideration. The set of global cones are ordered, and are used at every candidate problem, and whenever new global cones are generated in the procedure, they are added to this set. We develop the search strategies in the following discussion.

2.2. Search Strategies

The solution space is partitioned into two subsets S_1 and S_2 as follows.

$S_1 = \{z \mid z \quad Z^c, z_1 \geq z_1^I\}$ and $S_2 = \{z \mid z \in Z^c, z_1 \leq z_1^I - \varepsilon, z_2 \geq z_2^I\}$ where z^I is the initial incumbent, and ε is a small positive quantity. The search is conducted separately in S_1 and S_2. We employ three strategies in selecting a candidate problem for preference structure assessment during the branch-and-bound search as follows. In the strategy called BTI (Branching Till Integrality), the bicriteria linear relaxation of a candidate problem is solved only if the solution obtained from branching is integer. Otherwise, branching is continued. This strategy generates several candidate problems, but eliminates many continuous solutions from consideration. In the strategy called ZWA (perform Zionts and Wallenius procedure at All candidate problems), the bicriteria linear relaxation of every candidate problem is solved. This strategy generates fewer candidate problems, but the number of questions could increase since many continuous solutions may be considered. The third strategy which is denoted as GLT (Global Cones Test) is a combination of the other two strategies. In this strategy, the bicriteria linear relaxation of a candidate problem is solved if the solution obtained from branching is integer. Otherwise, the solution is tested for dominance by any of the global cones. If dominated, then the bicriteria linear relaxation is solved. Otherwise, branching is continued. In this strategy, we avoid the excessive branching as in BTI through the global cones test, and also avoid asking too many questions as in ZWA by selecting only those problems that have a high potential for being fathomed. This intuitive observation has been confirmed by our computational results. We present an illustration in the following section, that shows the behavior of the three strategies in solving problem (BCIP).

3. An Illustration

Consider the following problem:

$$\text{Maximize } z_1: -52x_1 +8x_2\ 11x_3 +38x_4 +39x_5 +23x_6 +19x_7 +62x_8 +45x_9 +28x_{10}$$

$$z_2: 92x_1 +46x_2 -46x_3 +99x_4 +59x_5 -59x_6 +38x_7 -38x_8 +81x_9 +0x_{10}$$

subject to:

$$40x_1 +27x_2 +72x_3 +62x_4 +4x_5 +97x_6 +34x_7 +56x_8 +12x_9 -25x_{10} \leq 322$$

$$40x_1 +24x_2 +12x_3 +94x_4 +29x_5 +35x_6 +23x_7 +59x_8 + 0x_9 +23x_{10} \leq 254$$

$$-15x_1 +36x_2 +62x_3 +53x_4 +62x_5 +93x_6 +78x_7 +62x_8 +69x_9 +34x_{10} \leq 423$$

$$86x_2 +26x_3 +97x_4 +79x_5 +79x_6 +0x_8 - 0x_8 +50x_9 +52x_{10} \leq 352$$

$$0 \leq x_i \leq 1,\ i=1,\ldots 10 \text{ Integer}$$

In simulating the decision-maker's responses on the computer, a quadratic utility function with equal weights on each of the objectives is assumed. The utility function is quadratic about the idea point obtained by maximizing each objective separately. The problem has been solved by ZWA, GLT and BTI. Figures 1, 2 and 3 show the solution tree enumerated by these algorithms, respectively. All three algorithms obtained the optimal solution $(97,415)^t$ to this problem. Table I summarizes the characteristics of the three algorithms.

It can be seen from the above table that GLT requires fewer questions than both ZWA and BTI. This result has also been observed in the detailed computational study on the algorithms. However, solution time does not appear to be critical for solving bicriteria problems, although the solution time in BTI is considerably higher than those of GLT and ZWA. The optimal solution to the linear relaxation problem at the top node is $(118.54,415)^t$ and this required just one question. The initial Incumbent obtained by rounding this solution is $(97,415)^t$ which is also the optimal solution to the integer programming problem. The global cones that remained after solving the problem by each algorithm are shown in Figure 4. The solution tree of the algorithms provide insights on the nature of the algorithms, and their interpretations are given below.

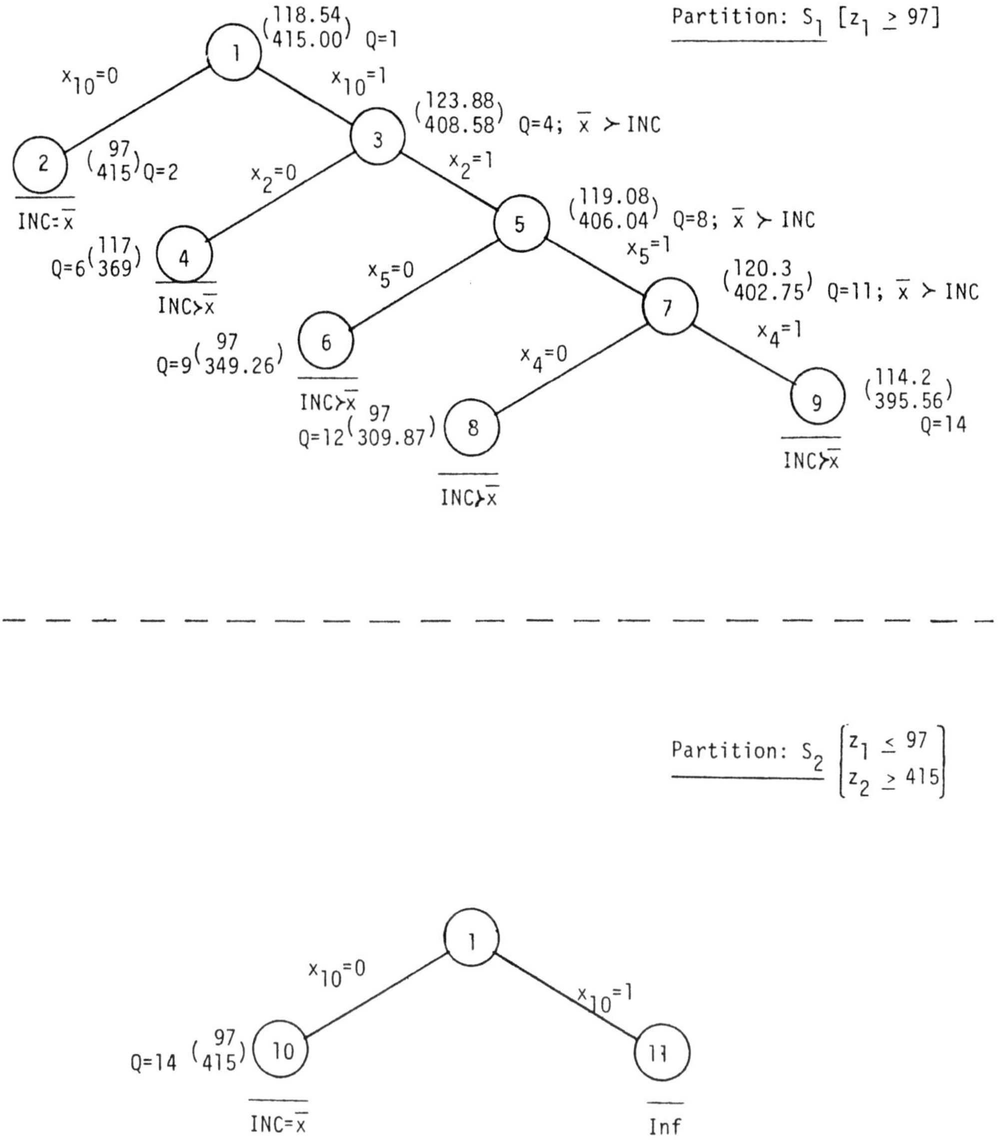

FIGURE 1. SOLUTION TREE ENUMERATION BY ALGORITHM ZWA

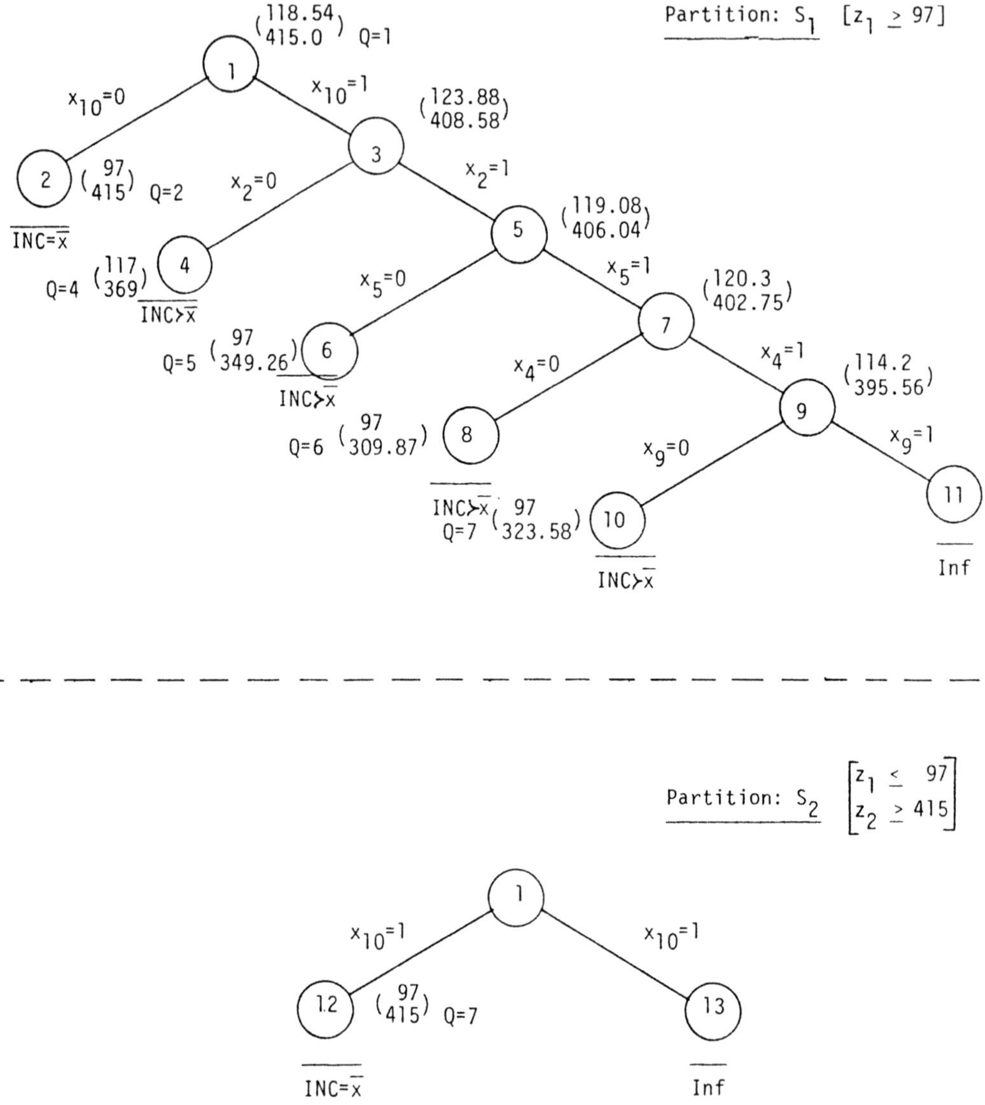

FIGURE 2. SOLUTION TREE ENUMERATION BY ALGORITHM GLT

FIGURE 3. SOLUTION TREE ENUMERATION BY ALGORITHM BTI

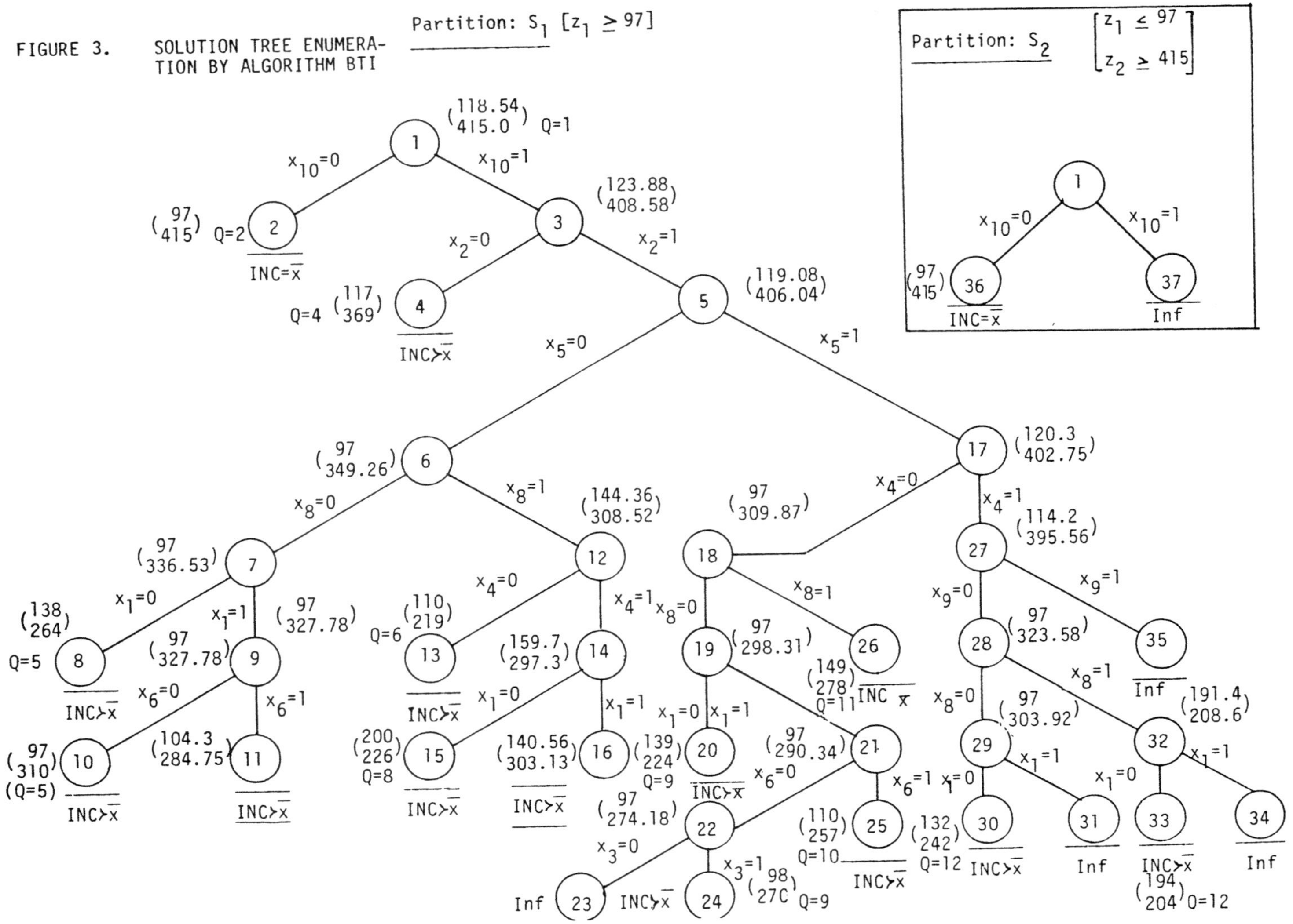

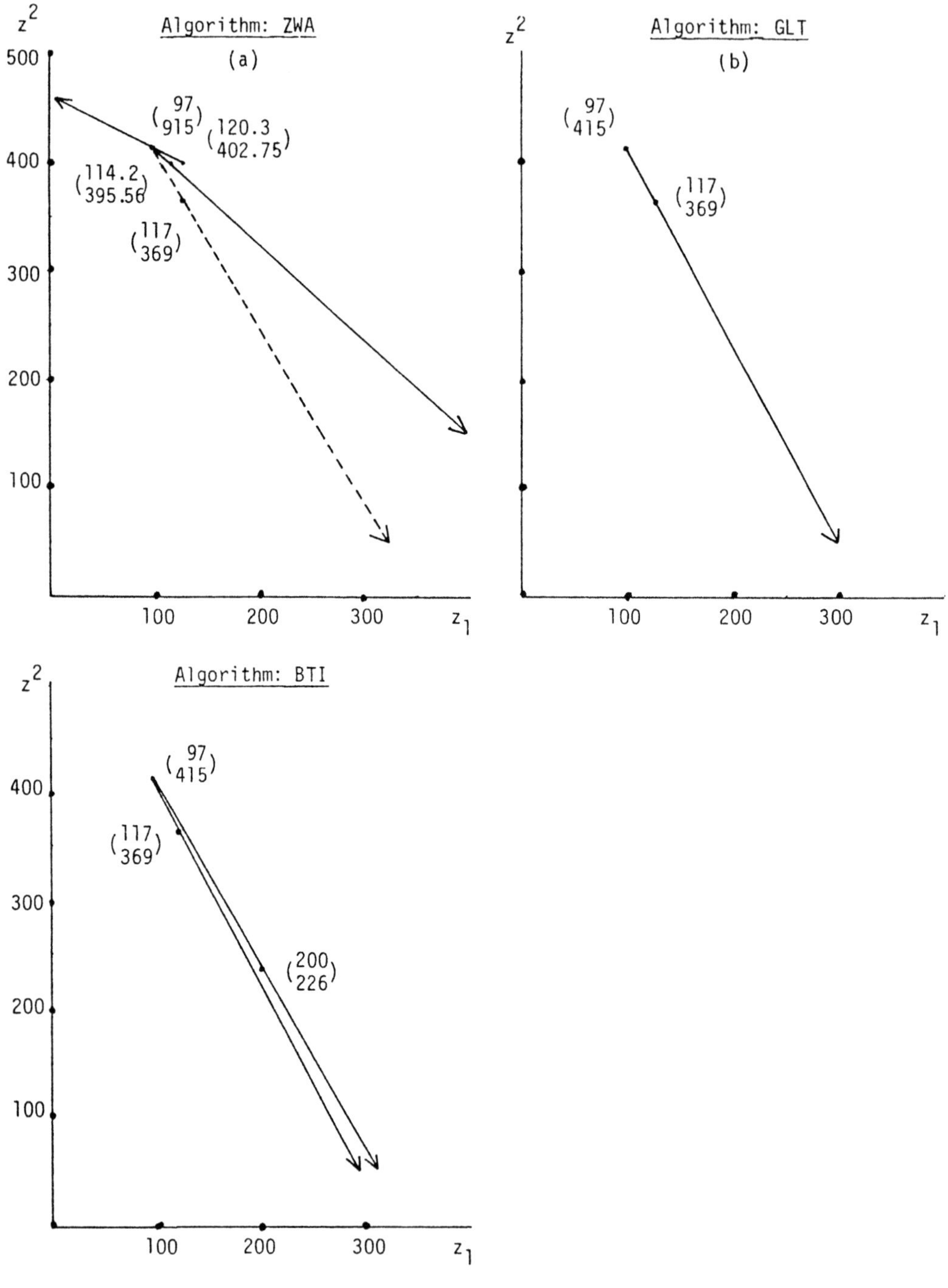

FIGURE 4. GLOBAL CONES GENERATED IN DIFFERENT ALGORITHMS

Table 1: SUMMARY OF SOLUTION CHARACTERISTICS

No.	Characteristic	ZWA	GLT	BTI
1.	Total number of questions asked	14	7	12
2.	CPU time for solution	0.262 cps	0.264 cps	0.476 cps
3.	Total number of nodes	11	13	37
4.	Total number of questions asked in solving problem problem (BCLP)	1	1	1

<u>ZWA-tree</u>

The following observations can be made from the tree in Figure 1.

1. By branching from node 1, the intiial Incumbent is again obtained at node 2.
2. At nodes 4, 6, 8 and 9, the Incumbent is preferred to the solution obtained from the ZW-procedure. Therefore, these nodes are fathomed.
3. At nodes 3, 5 and 7, the solution from the ZW-procedure is preferred to the Incumbent. Therefore, branching is continued from these nodes. Note that these solutions are continuous.
4. The solution tree in partition S_2 is fathomed after exploring only two nodes. This tree is the same in GLT as well as TII.
5. The global cone generated in GLT is shown in Figure 4. The cone $c[97,415)^t \succ (117,369)^t]$ is generated at node 4 as in ZWA. However, no further global cones are generated since the solutions obtained after performing the ZW-procedure at nodes 6, 8 and 10 are simply dominated by the Incumbent. The ZW-procedure is not performed at nodes 3, 5, 7 and 9 since the solutions obtained by branching to these nodes are not

dominated by global cones. As this illustration shows, it pays to branch from these nodes and perform the ZW-procedure as soon as the solutions obtained from branching are dominated by the global cones.

BTI-tree

The following observations can be made from the solution tree in Figure 3.

1. The solutions at nodes 1, 2, 3, 4 and 5 are the same as in the previous algorithms. The nodes 6 and 17 are obtained by branching on variable x_5 from node 6. It can be seen from Figures 1 and 2 that node 6 is fathomed in ZWA and GLT after performing the ZW-procedure with one additional question. However, in BTI, a subtree of node 6 is generated and fathomed after asking three additional questions. Node 7 in Figures 1 and 2 corresponds to node 17 in Figure 3. The subtree from this node is fathomed after five additonal questions. Thus, the enumeration scheme in BTI also results in higher solution times than the other two algorithms.
2. The global cones generated in BTI are shown in Figure 4. The cone $c[97,415)^t \succ (117,369)^t]$ is generated at node 4 as in ZWA and GLT. Another lower leaf $c[(97,415)^t \succ (200,226)^t]$ is obtained after fathoming node 15. The nodes 8, 10, 13, 20, 24, 25, 26, 30 and 33 are fathomed after determining that the Incumbent is preferred to the solution from the ZW-procedure through simulation. At node 15, the preference of the Incumbent over the solution from the ZW-procedure is determined after asking a question.

This problem illustrates the definite advantages of GLT over ZWA and BTI in terms of number of questions. The computational results on the solution strategies are given in the following section.

4. Computational Results

The solution framework has been programmed in FORTRAN V AND IMPLEMENTED ON CDC-Cyber/730 as well as IBM 3081/GX at the State University of New York at

Buffalo. The framework has been extensively tested with randomly generated problems. The decision-maker's responses have been simulated on the computer using different pseudoconcave utility functions. The overall performance has been measured in terms of number of questions asked and the computation time. A randomized block design consisting of six levels of the number of variables (10, 20, 30, 40, 60, 80) and four levels of the number of constraints (10, 20, 30, 40) has been used in this study. Each cell in this design consisted of ten random problems, and we empirically evaluated the performance of the three solution strategies (BTI, ZWA and GLT) as well as other mechanisms incorporated in the solution procedure in this study. Algorithm GLT resulted in 35% to 42.5% fewer questions than with algorithm ZWA, and 41% to 49% fewer questions than with algorithm BTI. Similarly, algorithm GLT resulted in 5.5% to 12% less computation time than that of algorithm ZWA, and 50.5% to 71.5% less time than that of algorithm BTI. This confirmed our intuitive expectations on the performance of GLT, and showed that it is possible to restrict both the number of questions and the computation time by selectively exploring the decision-maker's preference structure as in algorithm GLT. The total number of questions asked using algorithm GLT ranged between 8.9 to 28.9 and the computation time ranged between 1.97 CPU seconds to 129.44 CPU seconds in the class of problems tested.

5. Conclusions

In this research, we have developed an efficient interactive solution framework for bicriteria integer programming. Three search strategies have been developed using this framework, and their performance in terms of number of questions and computation time have been empirically evaluated. In this paper, we have presented the rationale underlying these strategies, illustrated them with an example, and summarized our results. Our experience shows that algorithm GLT is a viable approach for solving practical problems of moderate size. Possible areas of application are numerous. These include project selection, warehouse location, selection of sampling plans in quality control, strategic planning and policy analysis and many more.

BIBLIOGRAPHY

1. Chalmet, L. G., Lemonidis, L. and Elzinga, D. J., "An Efficient Algorithm for the Bi-Criterion Integer Programming Problem," Research Report No. 82-9, Department of Industrial and Systems Engineering, University of Florida, Gainesville (1983).

2. Evans, G. W., "An Overview of Techniques for Solving Multiobjective Mathematical Programs," Management Science, Vol. 30, No. 11 (1984).

3. Farquhar, P. H., "Utility Assessment methods: State of the Art," Management Science, vol. 30, No. 11 (1984).

4. Gabbani, D. and Magazine, M. J., "An Interactive Heuristic Approach for Multi-objective Integer Programming Problems," Journal of the Operational Research Society, Vol. 37, pp. 285-291 (1986).

5. Geoffrion, A. M., Dyer, J. S. and Feinberg, A., "An Interactive Approach for Multicriterion Optimization with an Application to the Operations of an Academic Department," Management Science, Vol. 19 (1972).

6. Keeney, R. L. and Raiffa, H., Decisions with Multiple Objectives: Preferences and Value Tradeoffs, John Wiley and Sons, New York (1976).

7. Klein, D. and Hannan, E., "An Algorithm for the Multiple Objective Integer Linear Programming Problem," European Journal of Operations Research, Vol. 9 (1982).

8. Korhonen, P., Wallenius, J. and Zionts, S., "Solving the Discrete Multiple Criteria Problem Using Convex Cones," Management Science, Vol. 30, No. 11 (1984).

9. Pasternak, H. and Passy, U., "Bicriterion Mathematical Programs with Boolean Variables," in Cochrane, J. L. and Zeleny, M. (Editors), Multiple Criteria Decision Making, University of South Carolina Press (1973).

10. Ramesh, R., "Multicriteria Integer Programming," Unpublished Doctoral Dissertation, Department of Industrial Engineering, State University of New York at Buffalo (1985).

11. Ramesh, R., Karwan, M. H. and Zionts, S., "A Class of Practical Interactive Branch and Bound Algorithms for Multicriteria Integer Programming," European Journal of Operations Research, Vol. 26, No. 1 (1986).

12. Ramesh, R., Karwan, M. H. and Zionts, S., "Preference Structure Representation Using Convex Cones in Multicriteria Integer Programming," Working Paper No. 646, School of Management, State University of New York at Buffalo (1985).

13. Villarreal, B. and Karwan, M. H., "Multicriteria Integer Programming: A (Hybrid) Dynamic Programming Recursive Approach," Mathematical Programming, Vol. 21 (1981).

14. Walker, J., "An Interactive Method as an Aid in Solving Bicriterion Mathematical Programming Problems," Journal of Operational Research Society, Vol. 29, pp. 915-922 (1978).

15. Zionts, S., "A Survey of Multiple Criteria Integer Programming Methods," Annals of Discrete Mathematics, Vol. 5 (1979).

16. Zionts, S. and Wallenius, J., "An Interactive Programming Method for Solving the Multiple Criteria problem," Management Science, Vol. 22, No. 6 (1976).

17. Zionts, S and Wallenius, J., "An Interactive Multiple Objective Linear Programming Method for a Class of Underlying Nonlinear Utility Functions," Management Science, Vol. 29, No. 5 (1983).

Multiple Criteria Trajectory Decision Support Using Polyscreen Workstations: A Production Planning Illustration

Ralph E. Steuer
Department of Management Science & Information Technology
Brooks Hall
University of Georgia
Athens, Georgia 30602

Lorraine R. Gardiner
Department of Management
322 Thach Hall
Auburn University
Auburn, Alabama 36849

John J. Bernardo
College of Business & Economics
University of Kentucky
Lexington, Kentucky 40506

Abstract

This paper demonstrates how the Combined Tchebycheff/Reference Point Procedure [8] can be installed in a decision support system for solving multiple criteria trajectory optimization problems. With such problems often presenting difficult modeling and solution procedure challenges, the Combined Procedure is employed because it gives good direction while providing substantial user control. The approach of this paper is futuristic in that it explores the use of polyscreen workstations (up to six screens) in order to facilitate man/machine communications at each iteration. A multiple criteria multi-time period production planning problem is used to illustrate the decision making approach.

1. Introduction

This paper presents a strategy for decision support with regard to the solution of multiple criteria <u>multi-observation point</u> (e.g., multi-time period or multi-surveillance point) optimization problems. Installed in a decision support system (DSS), the approach of this paper uses the Combined Tchebycheff/Reference Point Procedure of Steuer, Whisman and Silverman (1989) along with <u>polyscreen</u>

workstations (up to six screens) for communications and the display of output.

Multiple criteria multi-time period problems are present in production planning and manpower planning (Silverman, Steuer and Whisman (1988)). They also occur in economic planning (Grauer and Zalai (1982)), forest management (Kallio, Propoi and Seppälä (1986)), and energy planning (Grauer, Lewandowski and Schrattenholzer (1982)). Multiple criteria multi-surveillance point problems occur in river basin management (Steuer and Wood (1986)). Because each criterion is to be monitored in each time period (or at each surveillance point), the values of each given criterion can be imagined to form a "path" over the multiple observation points. Adopting the terminology of Wierzbicki (1982), we will call such multi-observation point problems multiple criteria trajectory optimization problems.

With regard to the four-component DSS paradigm of Figure 1, this paper is directed at the model, solution procedure, and user-interface components. As the illustrative vehicle of this paper, the model component is represented by the multiple criteria multi-time period production planning problem of Section 2. The solution procedure component is occupied by the Combined Tchebycheff/Reference Point Procedure, while the user-interface component is characterized by the use of polyscreen workstations.

Database Component	Model Component	Solution Procedure Component
User-Interface Component		

Figure 1. Four-Component DSS Paradigm

This paper addresses the three major user control challenges in solving multiple criteria trajectory optimization problems: (1) control over the way the model is defined, (2) control over the solution procedure search for a final solution, and (3) control over the way output is displayed.

Control over the way the model is defined is required because of the large-scale and semi-structured nature of most multiple criteria trajectory optimization problems. Given that the model is an abstraction, the modeling features requiring closest attention are often a function of our criterion aspirations and where in the feasible region we are searching. Since we will be searching different regions of the feasible region with our aspirations perhaps undergoing modifications, we thus have a need for being able to make model adjustments during the interactive process.

With regard to control over the solution procedure search for a final solution, it is likely that, during the early iterations, a DM would be interested in being guided by the solution procedure as to where to look for improved solutions. However, during the later iterations, after the DM has acquired a better feel for the problem, it is likely that a DM would want to "steer" the solution procedure toward particular neighborhoods of search.

As for control over the way output is displayed, let us consider the 8 objective, 6 time-period production planning problem of Section 2. Since each objective is to be monitored in each time period, 50 pieces of information are required to characterize the criterion value performance of each solution. Furthermore, given the necessity to cross-reference different solutions and iteratively communicate model adjustments to the DSS, we can see the need for multiple screens and flexibility over what is displayed on them.

2. Production Planning Problem

With 2 plants, 3 products, 2 levels of production, and 6 time periods, the illustrative multiple criteria multi-time period production planning problem (with 8 objectives, 214 constraints, and 277 variables) is previewed in Figures 2 and Tables 1 & 2. Note that objective (3) deals specifically with minimizing product 2 backorders and that objectives (6), (7) and (8) deal with production smoothing.

Figure 2. Production Planning MOLP

i product 1, 2, 3 ℓ level 1 (finished goods), 2 (bottle)
j plant 1, 2 t time 0, 1, 2, 3, 4, 5, 6, 7

max	(1) total net revenue
max	(2) minimum period net revenue
min	(3) maximum period product 2 backorders
min	(4) inventory carrying costs
min	(5) overtime costs
min	(6) maximum Δ period standard hours -- plant 1
min	(7) maximum Δ period standard hours -- plant 2
min	(8) workforce size variation costs

s.t.	(a) net revenue constraints	6
	(b) inventory identity constraints	36
	(c) Δ backorder balance constraints	18
	(d) requirements balance constraints	18
	(e) minimax backorder constraints	18
	(f) maximum net revenue constraints	6
	(g) regular time capacity constraints	24
	(h) maximum bottleneck capacity constraints	12
	(i) maximum Δ inventory constraints	36
	(j) minimax Δ standard hours constraints	26
	(k) Δ regular time capacity constaints	14
		214

Table 1

Variable Definitions

$B_{i\cdot 1t}$	21	backorders outstanding for product i at end of period t ϵ {1,2,...,6}.
$d^{+}_{\cdot j \cdot t}$	14	amount by which regular time hours of capacity at plant j during time period t exceed those of the period (t ϵ {1,2,...,7}).
$d^{-}_{\cdot j \cdot t}$	14	amount by which regular time hours of capacity at plant j during time period t are exceeded by those of previous period (t ϵ {1,2,...,7}).
I_{ij1t}	42	units of product i in inventory of plant j at beginning of time period t ϵ {1,2,...,7}.
$L_{\cdot j \cdot\cdot}$	2	minimum period standard hours at plant j.
$N_{\cdot\cdot 1t}$	6	(unrestricted) net profit (loss) during time period t ϵ {1,2,...,6}.
$S_{\cdot j \ell t}$	32	regular time hours of capacity at plant j for level ℓ during time period t ϵ {1,2,...,7}.
$U_{\cdot j \cdot\cdot}$	2	maximum period standard hours at plant j.
$V_{\cdot j \ell t}$	24	overtime hours at plant j for level ℓ during time period t ϵ {1,2,...,6}.
X_{ij1t}	36	units of product i produced at plant j during time period t ϵ {1,2,...,6}.
Y_{ij1t}	36	(unrestricted) backorder components of product i at plant j at end of period t ϵ {1,2,...,6}.
Z_{ij1t}	36	requirements components of product i at plant j during time period t ϵ {1,2,...,6}.
α_i	3	minimax variable for maximum product i period backlog.
α_4	1	(unrestricted) maximum period net revenue variable.
α_{j+4}	2	minimax variable for maximum Δ standard hours for plant j.
$A_{\cdot\cdot 1t}$	6	fixed costs for period t ϵ {1,2,...,6}.
	277	

Table 2

Coefficient Definitions

$a_{i\cdot\ell\cdot}$	standard hours at level ℓ per unit of finished product i
$C^B_{i\cdot 1\cdot}$	unit period backorder costs of product i
$C^H_{\cdot\cdot\cdot\cdot}$	hiring cost
$C^I_{ij1\cdot}$	unit period inventory carrying costs for product i at plant j
$C^L_{\cdot\cdot\cdot\cdot}$	layoff cost
$C^V_{\cdot j\ell\cdot}$	total hourly overtime costs at plant j for level ℓ
$C^X_{\cdot j\ell\cdot}$	regular time hourly costs at plant j for level ℓ
$F_{i\cdot 1t}$	delivery requirements for product i during time period $t \in \{1,2,\ldots,6)$
$m_{\cdot j2\cdot}$	maximum period bottleneck hours of capacity at plant j
$T_{i\cdot 1\cdot}$	unit transfer price for product i

In Figure 2, the (a) net revenue constraints are given by

$$\sum_{q=1}^{6} (a.q) + N_{\cdot\cdot 1t} = (a.7) \qquad t \in \{1, \ldots, 6\}$$

where:

(a.1) $\sum_j (\sum_\ell C^X_{\cdot j\ell\cdot}\ a_{i\cdot\ell\cdot})X_{ij1t}$ — standard production costs

(a.2) $\sum_{j\ell} (C^V_{\cdot j\ell\cdot} - C^X_{\cdot j\ell\cdot})V_{\cdot j\ell t}$ — supplemental overtime costs

(a.3) $\sum_{ij} C^I_{\cdot j1\cdot}\ I_{ij1(t+1)}$ — inventory carrying costs

(a.4) $\frac{1}{336} \sum_j C^H_{\cdot\cdot\cdot\cdot}(d^+_{\cdot j\cdot t} + d^+_{\cdot j\cdot(t+1)})$ — hiring costs (a.5) $\frac{1}{336} \sum_j C^L_{....}(d^-_{\cdot j \cdot t} + d^-_{\cdot j \cdot (t+1)})$ layoff costs

(a.6) $\sum_i C^B_{i \cdot 1 \cdot} B_{i \cdot 1t} + A_{\cdot \cdot 1t}$ backorder + fixed costs

(a.7) $\sum_i T_{i \cdot 1 \cdot} (F_{i \cdot 1t} + (B_{i \cdot 1(t-1)} - B_{i \cdot 1t}))$ sales revenue

The remaining categories of constraints are given by:

(b) $I_{ij1t} + X_{ij1t} + Y_{ij1t} = Z_{ij1t} + I_{ij1(t+1)}$ $\forall_{ij}$ $t \in \{1, \ldots, 6\}$

(c) $Y_{i11t} + Y_{i21t} = B_{i \cdot 1t} - B_{i \cdot 1(t-1)}$ $\forall_i$ $t \in \{1, \ldots, 6\}$

(d) $Z_{i11t} + Z_{i21t} = F_{i \cdot 1t}$ $\forall_i$ $t \in \{1, \ldots, 6\}$

(e) $B_{i \cdot 1t} \leq \alpha_i$ $\forall_i$ $t \in \{1, \ldots, 6\}$

(f) $N_{\cdot \cdot 1t} \geq \alpha_4$ $t \in \{1, \ldots, 6\}$

(g) $\sum_i a_{i \cdot \ell \cdot} X_{ij1t} - V_{\cdot j \ell t} \leq S_{\cdot j \ell t}$ $\forall_{j\ell}$ $t \in \{1, \ldots, 6\}$

(h) $\sum_i a_{i \cdot 2 \cdot} X_{ij1t} \leq m_{\cdot j 2 \cdot}$ $\forall_j$ $t \in \{1, \ldots, 6\}$

(i) $I_{ij1(t+1)} - I_{ij1t} - X_{ij1t} \leq 0$ $\forall_{ij}$ $t \in \{1, \ldots, 6\}$

(j) $\sum_{i\ell} a_{i \cdot \ell \cdot} X_{ij1t} - U_{\cdot j \cdot \cdot} \leq 0$ $\forall_j$ $t \in \{1, \ldots, 6\}$

$\sum_{i\ell} a_{i \cdot \ell \cdot} X_{ij1t} - L_{\cdot j \cdot \cdot} \geq 0$ $\forall_j$ $t \in \{1, \ldots, 6\}$

(k) $U_{\cdot j \cdot \cdot} - L_{\cdot j \cdot \cdot} - \alpha_{j+4} \leq 0$ $\forall_j$

The objectives are given by

(1) $\sum_t N_{\cdot \cdot 1t} = z_1$

(2) $\alpha_4 = z_2$

(3) $$\alpha_2 = z_3$$

(4) $$\sum_t (a.3) = z_4$$

(5) $$\sum_t (a.2) = z_5$$

(6) $$\alpha_5 = z_6$$

(7) $$\alpha_6 = z_7$$

(8) $$\sum_t (a.4) + \sum_t (a.5) = z_8$$

3. Combined Procedure

The Combined Procedure of [8] represents the merging of the Tchebycheff Method of [7, Chapter 14] and the Reference Point Method propounded by Wierzbicki (1982 and 1986) whose description is spelled out in Kallio, Lewandowski and Orchard-Hays (1980), and Lewandowski and Grauer (1982).

The Tchebycheff Method is a rather passive procedure that provides good direction as to where to search for improved solutions. By presenting a group of candidate solutions to the DM at each iteration, the Tchebycheff method enables a user to rapidly (a) learn about what is and what is not possible in the feasible region and (b) understand the tradeoffs inherent in the problem. The Tchebycheff method is usually most useful during the early iterations. The Reference Point Method, on the other hand, facilitates much more active involvement on the part of the user. The Reference Point Method is good at enabling a user to "steer" the solution procedure. In particular, it is good at pinpointing solutions that most closely correspond to the DM's specification of an <u>aspiration criterion vector</u>. Thus, the Reference Point Method is often very useful during the later iterations.

4. Polyscreen Workstations

Although workstations with more than two screens are not available as standard equipment, they are certainly feasible technologically. Let us assume that a workstation with six screens has been assembled for us to use on the production planning problem. Control over what is displayed might be directed via a panel that could be windowed onto any screen. In the panel, the user would specify which of the following displays are to be shown on which screens.

Display

(a) z-Value Ranges over the Nondominated Set & Tabular Display of Current Iteration's Group of Candidate Solutions. The z-values are the z-values generated by objectives (1) through (8) in Section 2. The ranges provide global information about the problem, and the tabular display provides overview information about the current iteration's group of solutions.

(b) Data Requirements. This display contains the coefficient values used to parameterize the model.

(c) Bounds. By means of this display a user can fix the value of, or alter the bounds on, any decision variable in the model.

(d) Criterion Value Trajectories of a Given Solution. This display presents line graphs to describe the "paths" of criterion values for the different objectives over the multi-time periods.

(e) Product Trajectories of a Given Solution. This display presents line graphs to describe the amount of each product produced, the backorder situation for each product, and the amount of each product in inventory during each time period.

(f) Plant Trajectories of a Given Solution. This display presents line graphs to describe the standard hours of production for each product at each level over the multi-time periods.

(g) Algorithm Parameter Specification. This display is a merger of Screens 1, 3 and 4 in [8]. In this display, the user specifies how many solutions are to be generated at each iteration, the rate at which the solution procedure is to converge, and aspiration criterion vectors (when using the Reference Point capabilities of the Combined Procedure).

5. Employing the DSS to Solve the Production Planning Problem

A way to solve the production planning problem is to run the Combined Procedure for 4 iterations, requesting between 10 or 16 solutions per iteration, as follows:

The first iteration begins from the (g) Display in which the Combined Procedure is asked to generate, as in the Tchebycheff method, 10 dispersed solutions from the (entire) nondominated set. Then, the user-interface unit is asked to show Displays (a), (b), and (c) along with three (d) Displays for cross-referencing solutions. After the user is done examining the 10 solutions, one of the (d) Displays is replaced by the (g) Display in which the user specifies:

(i) the most preferred of the 10 first-iteration solutions, designating it $z^{(1)}$
(ii) a neighborhood reduction factor of .40
(iii) that 10 dispersed solutions be generated, as in the Tchebycheff Method, from the second-iteration reduced neighborhood centered about $z^{(1)}$.

After making any model adjustments via the (b) and (c) Displays, the second iteration begins. The second iteration is carried out in the same way as the first iteration except that in the (g) Display at the end, the user specifies:

(i) the most preferred of the 10 second-iteration solutions, designating it $z^{(2)}$
(ii) a neighborhood reduction factor of .30 (bringing the cumulative effect to .12)
(iii) that 8 dispersed solutions be generated, as in the Tchebycheff Method, from the third-iteration reduced neighborhood centered about $z^{(2)}$
(iv) an aspiration criterion vector $y^{(3)}$
(v) that 8 dispersed solutions be generated from a reduced neighborhood of size .12 centered about the nondominated criterion vector that is computed to be closest to $y^{(3)}$ according to the Reference Point Method.

After making any further model adjustments via the (b) and (c) Displays, the third iteration begins. After working with Displays (a), (b), (c) and (d) as in the first two iterations, suppose that the user wishes to examine particular pairs of solutions in greater detail. In this case, the user might request that the six current displays be replaced by two (d) Displays, two (e) Displays, and two (f) Displays so that decision variable value as well as criterion value information can be studied. When done with the examinations, the (g) Display is called in which the user specifies:

(i) the most preferred of the 16 third-iteration solutions, designating it $z^{(3)}$

(ii) a neighborhood reduction factor of .20 (bringing the cumulative effect to .024)

(iii) that 8 dispersed solutions be generated, as in the Tchebycheff Method, from the third-iteration reduced neighborhood centered about $z^{(3)}$

(iv) an aspiration criterion vector $y^{(4)}$

(v) that 8 dispersed solutions be generated from a reduced neighborhood of size .024 centered about the nondominated criterion vector that is computed to be closest to $y^{(4)}$ according to the Reference Point Method.

With the fourth iteration proceeding in the same way as the third iteration, a $z^{(4)}$ is selected as the final solution.

References

[1] Grauer, M., A. Lewandowski, and L. Schrattenholzer (1982). "Use of the Reference Level Approach for the Generation of Efficient Energy Supply Strategies," in M. Grauer, A. Lewandowski, and A. P. Wierzbicki (eds.), Multiobjective and Stochastic Optimization, Laxenburg, Austria: International Institute for Applied Systems Analysis, pp. 425-444.

[2] Grauer, M. and E. Zalai (1982). "A Reference Point Approach of Nonlinear Macroeconomic Multiobjective Models," WP-82-134, International Institute for Applied Systems Analysis, Laxenburg, Austria.

[3] Kallio, M., A. Lewandowski, and W. Orchard-Hayes (1980), "An Implementation of the Reference Point Approach for Multiobjective Optimization," WP-80-35, International Institute for Applied Systems Analysis, Laxenburg, Austria.

[4] Kallio, M., A. Propoi and R. Seppälä (1986). "A Model for the Forest Sector," TIMS Studies in the Management Sciences, Vol. 21, pp. 89-112.

[5] Lewandowski, A. and M. Grauer (1982), "The Reference Point Optimization Approach - Methods of Efficient Implementation," in Grauer, M., A. Lewandowski, and A. P. Wierzbicki (eds.), Multiobjective and Stochastic Optimization, Laxenburg, Austria: Internatioal Institute for Applied Systems Analysis, pp. 353-376.

[6] Silverman, J., R. E. Steuer and A. W. Whisman (1988), "A Multiperiod, Multiple Criteria Optimization System for Manpower Planning," European Journal of Operational Research, Vol. 34, No. 2, pp. 160-170.

[7] Steuer, R. E. (1986), Multiple Criteria Optimization: Theory, Computation, and Application, John Wiley & Sons.

[8] Steuer, R. E., A. W. Whisman, and J. Silverman (1989). "A Combined Tchebycheff/Reference Point Interactive Multiobjective Programming Procedure," College of Business Administration, University of Georgia.

[9] Steuer, R. E. and E. F. Wood (1986). "On the 0-1 Implementation of the Tchebycheff Solution Approach: A Water Quality Illustration," Large Scale Systems, Vol. 10, No. 3, pp. 243-256.

[10] Wierzbicki, A. P. (1982). "Multiobjective Trajectory Optimization and Model Semiregularization." In M. Grauer, A. Lewandowski, and A. P. Wierzbicki (eds), Multiobjective and Stochastic Optimization, Laxenburg, Austria: International Institute for Applied Systems Analysis, pp. 3-38.

[11] Wierzbicki, A. P. (1986). "On the Completeness and Constructiveness of Parametric Characterizations to Vector Optimization Problems," OR Spektrum, Vol. 8, No. 2, pp. 73-88.

MULTIOBJECTIVE DECISION SUPPORT FOR SIMULATED GAMING

Andrzej P. Wierzbicki
Institute for Automatic Control,
Warsaw University of Technology
Nowowiejska 15/19, 00-665 Warsaw, Poland

Abstract

This paper presents a review of various approaches to decision support, distinguishes a methodological approach based on reference point optimization and reviews various possible levels of decision support in simulated gaming, bargaining and negotiations as well as some mathematical methods that might be used for this purpose.

1. Introduction.

The concept of a decision support system (DSS), though widely used and developed both in research and in practical applications through more than last ten years, is not yet quite precisely defined. This paper starts with an attempt at a broad definition of this concept by enumerating various classes of DSS and proceeds to possible modes and functions of decision support for multiperson decision situations where a cooperative behavior of participants or decision makers cannot be taken for granted - that is, for situation of gaming, bargaining and negotiations possibly supported by simulated gaming if a model of substantive aspects of a decision situation is used. A specific class of decision support systems is described - those that use the principle of reference point optimization for generating and evaluating decision alternatives with help of an analytical model of a decision situation. Finally, some mathematical tools of decision support in simulated gaming are described.

2. Concepts and definitions of decision support systems.

There are many proposed definition of a DSS in the current literature - see, e.g., Keen and Scott-Morton (1978), Sage (1981), Parker and Al-Utabi (1986), Gray (1986), Jarke (1986) and others. However, most of them do not take into account the fact that three main classes of decision support systems have been practically developed in applications and research. These are (see Lewandowski and Wierzbicki, 1987, Lewandowski and al,. 1987):

A) Simple tools for managerial decision support (that might be used as building blocks of more sophisticated DSS) such as modern data bases, electronic spreadsheet systems, etc. as well as more complex but pragmatically designed decision support systems somposed of such tools;

B) Decision support systems based on logical models and logical inference whose main function are to help in recognizing a logical pattern in a decision situation; these systems typically involve the use of logical programming languages, expert systems style programming, knowledge bases, other tools of artificial intelligence;

C) Decision support systems based on analytical models, multiobjective optimization and choice whose main functions concentrate on the process of analysis of various decision alternatives either specified a priori or generated with help of the system. Such systems typically include a computerized model of a decision situation formulated in analytical terms as well as some tools related to multiobjective optimization and evaluation of alternatives.

All these three classes can be further subdived according to various metodological principles. However, there are certain features that are common to all DSS. Observe that the systems of classes B and C contain explicitly models of the decision situation, although of different types. The same can be said, in fact, about the systems of the class A: when preparing a simple decision support tool such as a date base or a spreadsheet to support a definite decision process, one must assume (even if implicitly) a kind of a model of the decision situation. Thus, we can state that all decision support systems contain models of decision situation.

All DSS can be subdivided into two large classes: those that are designed to serve essentially one user or decision maker versus those that are explicitely designed to serve many users or multiple decision makers. The latter class can be further subdivided into two esentially different subclasses: those that serve cooperative group decision making or collegial decisions versus those that are designed to help in truly game-like situations that might involve conflict escalation through noncooperative decisions and thus serve bargaining and negotiations (through they might and should try to help reaching cooperative decisions, such systems do not take cooperative behavior of users for granted, see Wierzbicki, 1983 a,b). We concentrate in the further part of this paper precisely on this latter case, specifically on decision support for simulated gaming. However, for all multiuser decision support systems another universal feature becomes apparent: all decision support systems should be designed not to serve reaching a single decision, but to help in organizing a decision process.

This essential feature of DSS was noted by many authors - see, e.g., Parker and Al-Utabi (1986). An early characterization of a decision process was given by Simon (1958); it consists of the following three steps: intelligence - searching the environment for opportunities calling for a decision, design - defining the decision situation, inventing, developing and analysing possible courses of action, finally choice - selecting a particular course of action from those available. However, the experience in analysing decision process might contain more essential elements. Cooke and Slack (1984) combine the decision making with problem solving process and define its phases as observation, a formal recognition of a problem, interpretation and diagnosis, the definition of a decision problem, the determination of options or alternatives, an evaluation of options, their selection or choice, finally decision implementation and monitoring.

The results of Dreyfus (1984) indicate that an essential distinction should be made between familiar (even comlex) decision situations and novel decision situations as well as between various levels of expertize of the decision maker in a given field. A master expert in a decision field is able to treat most of the decision situations as familiar ones, recognize them immediately and select a decision which quality exceeds considerably the quality of decision achieved by any computerized system; we stil do not have adequate models and interpretations of the paraller processing of information performed in human mind. However, even a master expert recognizes (through certain feeling of uneasiness) situations that are novel and deliberates about them. Again, the process of such deliberation is not understood fully and is certainly not as ordered and linear as the models of decision processes described above; it ends in a sudden recognition of a decision pattern or in a deeper understanding of the decision problem. An expert of a lower level or a novice in a decision field treats more decision situations as novel and thus needs more logical or analytical decision support.

When seen from this perspective, every decision process is a part of a longer process of learning in order to become a master expert. Thus, a decision process in all novel situations is not necessarily linearly ordered, can have many recourses to earlier stages, while as a decision situation becomes more familiar for a given decision maker, the decision process becomes shorter and finally looses its distinctive phases. This is similar to an adaptive treatment of uncertainty, to the old concept of Feldbaum (1962) of the dual role of control - this of control and that of of learning. This also indicates several concepts of dynamics in a decision process. One is related to the fact that decisions are concerned with future events and have dynamic consequences - even if we do use sometimes static models of their consequences in more simple cases. The second reflects the fact that even reaching a single decision is a process, possibly with many phases and recourses. The third reflects the fact that separate decision processes are imbedded in a longer learning process of the decision maker to become a master expert, with its much more complicated dynamics. We can conclude also that the requirement of consistency of a decision maker, essential to many classical approaches to decision analysis, has a fundamental drawback: a learning decision maker can often gain by being inconsistent.

All this indicates that DSS can have multiple functions in a decision process. Most important are three general functions: (1) organizing a decision process procedurally (which actually assumes a definite but often implicit model of the decision situation), (2) helping the decisions maker to learn about the decision situation (to familiarize it by playing with the proxy of reality provided by the DSS) and (3) filling in details to the outlines of a decision suggested by the decision maker. This also suggests that the emphasis on the phase of decision choice, typical for more classical approaches to decision processes and decision support, is actually misplaced: if adequately supported, humans can make (until now, and probably for a long time to come) much better decisions than most advanced computerized systems - and the problem is not how to replace, but how to support human decision making. There certainly are decision problems of repetitive type that might and shuld be automated - because of the necessary speed of decisions, because of their tediousness for humans, because of the reliability of automata that do not grow tired and do not have the human right to change their minds - but this becomes then the field of decision automation or automatic control, not of decision analysis and support.

3. Functions and modes of decision support in simulated gaming.

A frequent mode of analysing complex multiperson decision situations in modern systems analysis consists in building simulation models and performing gaming experiments with them. Thus, the first function of a decision support system should be to help in model formalization, edition, simulation, parametric analysis etc. Naturally, models used in decision support can be of various types - very simple or more complex, of logical or analytical nature, etc. - and contemporary DSS cannot work with all possible classes of models, are necessarily specialized. Nevertheless, good DSS should contain a model edition and simulation interface and a directory of models together with a data base of the results of experiments with these models.

Organizing such directories and data bases must take into account the specific character of simulated gaming experiments. Typically, such experiments can consist of (I) repetitive static games, where the same simple game is played repetitively for the purpose of learning, and (I) dynamic development games, which are composed of a sequence of simpler static games with conditions depending on the outcomes of proceeding decisions. The data base of results of such experiments should obviously contain, in both cases, the history of gaming experiments; however, a DSS for such experiments could contain much more elements and functions.

While such simulated gaming experiments have consiserable educational and analytical advantages, not much has been achieved yet in equipping gaming and simulation models with tools of decision analysis and support. Without such tools, however, simulated gaming experiments model quite simplified decision situations since analytical and computerized decision support is typically used in real life decisions if they are complex and important enough.

Thus, in a first mode of decision support in simulated gaming we should include at least means for a tentative game evaluation, that is, the possibility given to each player to assume tentatively the decisions of other players and his own, to analyse the tentative results with help of the model simulated in the system and thus to select his actual gaming decision with some elementary support.

More advanced modes of decision support in simulated gaming might involve various levels of applications of multiobjective optimization and game theory. We shall list here some possible modes (Wierzbicki, 1987):

A second mode of decision support in simulated gaming should include a basic game analysis module that would at least compute current and minimal conflict coefficients as compared to some status quo outcomes (see section 5), for example, established in the first round of the game. This should provide the players with the information whether they escalate or deescalate conflicts in the course of the repetitive game.

A third mode of decision support in simulated gaming might include tentative arbitration module that would compute, beside conflict coefficients, also one-shot cooperative solution of the game (for example, of Raiffa (1953) or Nash (1953) or Kalai-Smorodinsky (1975) type) and propose it to the players. They would not be obliged to accept such arbitrated solution, but they might communicate between themselves whether to accept it jointly; the knowledge of such solution might have a considerable impact on their behavior.

A fourth mode of decision support in simulated gaming might include multiobjective iteractive decision support systems for individual players. While assuming some moves of other player could select interactively his moves of other players, each player could select interactively his moves that best serve his multiple objectives (for the games of the dynamic development character, these objectives can also have the the form of trajectories, see Lewandowski et. al. 1987).

A fifth mode of decision support in simulated gaming would include, beside individual DSS, also an interactive fair mediation module based e.g. on the single text procedure (see Raiffa, 1982) augmented by the principle of limited confidence (see Fandel and Wierzbicki, 1985, Bronisz et. al., 1988). This mode should also include all functions of the previous modes to prepare data for the mediation module. The players must be asked for their acceptance of mediation, since they must disclose to the mediation module their selected objectives and the aspiration values or weighting coefficients for these objectives. The players should be also free to accept a mediated solution or not.

There already exists theoretical background and even tested software components for organizing a DSS that would include all modes described until now. A general structure of such a system is shown in Fig.1.

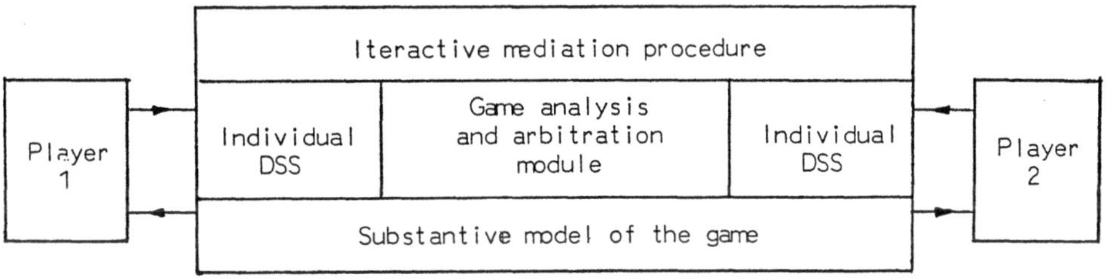

Fig. 1. The structure of a decision and mediation support systems for simulated gaming.

Much less advanced is the theory and software for a next, sixth mode of decision support for simulated gaming, designed for supporting noncooperative equilibria in the game. In a multiobjective game, the number of noncooperative equilibria might be quite large, what typically leads to conflict escalation (see Wierzbicki, 1983a); even if the players cannot - for some reason - agree on a cooperative solution, they might stil need a mediation support to deescalate conflict and to reach a mutually acceptable selection between nonunique noncooperative equilibria (which, once achieved, need not be stabilized by any formal agreement, but might be quite difficult to achieve without mediation support).

It is practically impossible to identify players' utility functions - that might be changing due to a learning process - in DSS performing repetitively such complicated functions. Moreover, if we insist on the souvereignity of a human decision maker and consider the system as supporting mostly learning and filling in details, then each decision choice proposed by the system must be only tentative and the user must have convenient means of influencing this choice. Various methods of multiobjective decision analysis can be used for this purpose; however, the use of the principle of reference point optimization described later in some detail might be most appropriate for DSS supporting simulated gaming.

The insistence on the users souvereignity is a relatively new feature of DSS developed in the last decade together with "high tech - high touch" trend in modern societies (see Naisbitt, 1982). Older approach to DSS, while stressing that such a system should help decision makers in reaching decisions, was not quite consistent with this assumption. Typically, such systems communicate the following message to the user in the phase of choice: "if your answers to my questions have been consistent, your best decision is as follows". This often helps the user, but not sufficiently: he does not know which of his answers are responsible for this particular choice, nor to change general instructions to the system in order to influence the final decision if he does not like it for some reason.

4. The principle of reference point optimization in DSS.

While leaving a more detailed review of various frameworks of rationality to another paper (see Lewandowski et al., 1987), we stress here firstly some essential facts related such review.

Any mathematical formalization of a rationality framework is typically converned with two preorderings of the spaces of decision outcomes (attributes, objectives) and decision variables (alternative decisions):

- a partial preordering in the space of outcomes that is usually implied by the decision problem and usually has some obvious interpretation, such as maximization of profit competing with the maximization of market share, etc.; a standard assumption is that this preordering is transitive and can be expressed by a positive cone D.
- a complete preordering in the spaces of outcomes and decisions or, at least, in the set of attainable outcomes and decision alternatives, which is usually not given in any precise mathematical form, but is contained in the mind of the decision maker, such as how actually the preferences betweem the maximization of profit and the maximization of market share should be distributed in a market analysis case.

The main differences between various frameworks of rationality that lead to diverse approaches to interactive decision support result from assumptions about this complete preordering and the way of its utilization in the DSS. This issue is also closely related with the way in which the DSS interacts with the decision maker; some variants of DSS require that the user answers enough questions for an adequate estimation of this complete proerdering by a utility function, some other variants need only general assumptions about the preordering, still other variants admit a broad interpretation of this preordering, still other variants admit a broad interpretation of this preordering and diverse frameworks of rationality that might be followed by the user.

The most strongly established rationality framework is based on the assumption of maximization of a value function or an utility function. Under rather general assumptions, the complete preordering that represents the preferences of the decision maker can be represented by an utility function such that by maximizing this function over admissible decisions we can select the decision which is most preferable to the decision maker; the publications related to this framework are very numerous, but for a constructive review see, for example, Keeney and Raiffa (1976).

There are many fundamental and technical difficulties related to the identification of such utility function. Leaving aside various technical difficulties, we should stress the fundamental ones. Firstly, the utility maximization framework - although it reprerents the behavior of many human decision makers - is not the universal case of human rationality - see, for example, Rappoport (1984). Secondly, while the utility maximization framework might be a good predictor of mass economic phenomena, it has many drawbacks as a predictor of individual behavior - see, for example, Fisher (1979), Erlandson (1981), Horsky (1984). Thirdly - and most importantly for applications in

DSS - an experimental identification and estimation of an utility function requires many questions and anwers in the interaction with the decision maker. Users of DSS are typically not prepared to answer that many questions, for several reasons. They do not like to disclose their preferences in too much detail because they intuitively perceive that DSS should support them in learning about the decision situation and thus they should preserve the right to change their minds and preferences. Therefore, if any approximation of an utility function is used in a DSS, it should be nonstationary in time in order to account for the learning and adaptive nature of the decision making process.

Another rationality framework, called satisficing decision making, was formulated by Simon (1958) and further extended by many researchers, see for example Erlandson (1981) for a formalization and review of this approach. Originally, this approach assumed that human decision makers do not optimize, because of the difficulty of optimization operations, because of uncertainty of typical decision environment, and because of complexity of the decision situations in large organizations. Terefore, this approach was sometimes termed bounded rationality, that is, somewhat less than perfect rationality. However. there are many indications that this approach represents not bounded, but culturally different rationality: it is related to the culture of complex organizations, while the utility framework represents the culture of an entrepreneur facing an infinite market.

While the first two reasons for not optimizing have lost today their validity (both in the calculative sense, with the development of computer technology and optimization techniques, including issues of incertainty, and in the deliberative sense - expert decision makers can intuitively optimize in quite complex situations), the third reason remains valid and has been reinforced by the results of various studies. For example, the studies of human behavior in situations of social traps or games with paradoxical outcomes - see Rappoport (1984), Axelrod (1985) - indicate that evolutionary experience forces humans to accept certain rules of ethical character that stop maximizing bahavior. Any intelligent man learns after some quarrels with his wife that maximization is not always the best norm of behavior; children learn from conflicts among themselves that cooperative behavior is also individually advantageous for a longer perspective. All these observations and studies might motivate in the future the development of a new framework of evolutionary rationkity but they certainly reinforce the conclusions of the satisficing framework that there are rational reasons to stop maximizing in complex situations.

A very important contribution of the satisficing framework is the observation that decision makers often use aspiration levels for various outcomes of decisions; in classical interpretations of the satisficing framework, these aspiration levels indicate when to stop optimizing. While more modern interpretations might prefer other rules for stoping optimization, the concept of aspiration levels is extremely useful for aggregating the results of learning by the decision maker: aspiration levels represent values of decision outcomes that can be accepted as reasonable or satisfactory by the decision maker and thus are aggregated, adaptable parameters that are sufficient for a simple representation of his accumulated experience.

There might be also other frameworks of rationality, such as the framework of goal and program oriented planning, see Glushkov (1972). Pospelov and Irikov (1976), Wierzbicki (1985), that corresponds to the culture of planning organizations. This framework has some similarities, but also some differences to the utility maximization framework, the satisficing framework and to the principle of reference point optimization - see Wierzbicki (1980) - in multiobjective optimization and decision support.

In order first to include the principle of reference point optimization into the framework of satisficing decisions and then to develop a broader framework that would be useful for decision support for decision makers representing various perspectives of rationality, Wierzbicki (1982, 1984b, 1985, 1986) proposed the following proncíples of quasisatisficing decision making.

A quasisatisficing decision situation consists of (one or several) decision makers or users that might represent any perspective of rationality and have the right of changing their mind and of stopping optimization as well as of a DSS that might be either fully computerized or include also human experts, analysts, advisors. It is assumed that:

- The user evaluates possible decisions on the basis of a set (or vector) of attributes or objective outcomes. These factors can be expressed in numerical scale (quantitatively) or in verbal scale (qualitatively). Each factor can be additionally constrained by specifying special requirements on it that must be satisfied. Beside this, objective outcomes can be characterized by their type: maximized, minimized, stabilized - that is, kept close to a given level (which corresponds to foregoing optimization).

- One of the basic means of communication of the user with the desicion support system is his specification of aspiration levels for each objective outcome. In more complex situations, the user can specify to levels for each objective outcome - an aspiration level interpreted as a reasonable value and a reservation level interpreted as the lowest acceptable calue for the given objective outcome.

- Given the information specified by the user - i.e., the specification of objective outcomes and their types, together with aspiration and possibly reservation levels - a DSS following the quasisatisficing principle should use this guiding information totether with other information contained in the system, in order to propose to the user one or several alternative decisions that are best attuned to this information. When preparing (generating or selecting) such alternative decisions, the DSS should not impose on the user the optimizing or the satisficing or any other behavior, but should follow the behavior that is indicated by the types of objective outcomes. This means that the DSS should optimize when at least one objective outcome is specified as minimized or maximized and should satisfice (stop optimizing upon reaching aspiration levels) when all objective outcomes are specified as stabilized.

The later case corresponds actually to the technique of goal programming, see e.g Ignizio (1978), hence the quasisatisficing decision support can be also considered as a generalization of this technique. By using aspiration or resevation levels for some objective outcomes as constraints, also the goal - and program oriented behavior can be supported by a quasisatisficing DSS.

In order to illustrate possible responses of a quasisatisficing DSS to the guiding information given by the user, let us assume that all specified objective outcomes are supposed to be maximized and have specified aspiration levels or references points. In this original formulation of the principle of reference point optimization we can distinguish the following cases:

Case 1: the user overestimated the possibilities implied by available resources and there is no admissible decision such that the the values of all objective outcomes are exactly equal to their aspiration levels. In this case, the DSS should tentatively propose a decision for which the values of objective outcomes are as close as possible (while using some uniform scaling, for example implied by the aspiration and reservation levels) to their aspiration levels.

Case 2: the user underestimated the possibilities implied by admissible decisions and there exists decisions which result in the values of objective outcomes exactly equal to or better than the specified aspiration levels. In this case, the DSS should tentatively propose a decision which improves all objective outcomes uniformly as much as possible.

Case 3: the user, by a chance or as a result of a learning process, has specified aspiration levels there are uniquely attainable by an admissible decision. The DSS should inform the user about this case and specify the details of the decision that results the attainment of aspiration levels.

In the process of quasisatisficing decision support, all aspiration levels and the corresponding decisions proposed by the system have tentative character. If a decision proposed by the system is not satisfactory to the user, he can modify the aspiration levels and obtain new proposed decisions, or even modify the specification of objective outcomes or constraints; the process is repeated until the user learns enough to make the actual decision himself or to accept a decision proposed by the system.

The process of quasisatisficing decision making can be formalized mathematically - see, e.g., Wierzbicki (1986) - and the mathematical formalization can be interpreted in various wasy. Let us consider an interpretation that corresponds to the framework of utility maximization. We assume that the user has a nonstationary utility function that changes in time due to his learning about a given decision situation. At each time instant, however, he can intuitively and tentatively (possibly with errors concerning various aspects of the decision situation) maximize his utility; let this tentative maximization determine his aspiration levels.

When he communicares these aspiration levels to the DSS, the system should use this information, together with the specification of the decision situation, in order to construct an approximation of his utility function that should be relatively simple and easily adaptable to the changes of aspiration levels, treated as parameters of this approximation. By maximizing such an approximative utility function while using more precise information about the attainability of alternative decisions and other aspects of the decision situation - for example, expressed by a model of the decision situation incorporated by expert advice into the DSS - a tentative decision can be proposed to the user. The user might accept this decision or change aspiration levels.

The question of convergence of a decision support process based on interactive changes of aspiration levels by a decision maker and on quasisatisficing responses of a DSS to a maximum of a "true" utility function of the decision maker (who has right to err and to change his mind provided that his utility function settles in time in a stochastic sense) has been investigated by Michalevich (1986): a stochastic quasigradient procedure is sufficient to guarantee convergence even in such general case.

A tentative approximation of the user's utility function, constructed in the DSS only in order to propose a tentative decision to the learinig decision maker, is called order-consistent achievement function or simply achievement function. It should be stressed that the concept of achievement function has been also used in the context of goal programming, but without the requirement of order consistency (achievement functions in goal programming are equivalent to norms and thus satisfy the requirements of Cases 1 and 3 listed above but fail to satisfy the requirements of Case 2).

There are many other interpretations of an order-consistent achievement function (see Wierzbicki, 1986): penalty function related to aspirations treated as soft constrains, a utility function not of the decision maker, but of the DSS interpreted as an ideal staff trying to follow instructions given by it's boss, a device for automatically switching from norm minimization to maximization in generalized goal programming upon crossing the boundary of attainable outcomes, a mathematical tool for closely approximating the positive cone D in the space of outcomes, an extention of the concept of membership function in a fuzzy set approach to multiobjective optimization, etc.

The general idea of reference point optimization has been independently developed or further used and extended by many researchers - Steuer and Cho (1983), Nakayama and Savaragi (1935), Korhonen and Laakso (1986). The more specific use of order - consistent achievement functions in decision support has been developed in many papers of IIASA - see Lewandowski et al. (1987). For this paper, however, more important are some mathematical aspects related to reference point optimization that might be useful in DSS for simulated gaming.

5. Matematical tools for decision support in simulated gaming.

The first concept that is very useful in DSS for simulated gaming is that of a conflict coefficient proposed first by Kręglewski (see Wierzbicki, 1984b). Consider first a situation where each of n players has only one objective (payoff or utility function). Let d denote a disagreement point or a status quo point in the objective (payoff) space and S the set of attainable outcomes $q = f(x)$ in the same space, while x are elements of decision soace and f is an outcome mapping; suppose d int S. Then the utopia point components relative to the status quo point are defined by:

$$(1) \qquad U_i(d) = \max_{q \in S_d} q_i; \quad S_d = \{q \in S: \; q_j \geq d_j, \quad j = 1,\ldots n\}$$

provided S_d is compact (actually, the optimisation in (1) is performed in the decision space, but we confine the presentation in this paper to the outcome space only). The coflict coefficient at an outcome q relative to the status quo d is defined by:

$$(2) \qquad C(d,q) = D(d,q)(2+(D(d,q)-1)n(n-2)/(n-1))$$

where:

$$(3) \qquad D(d,q) = \max_{i \in N} ((U_i(d) - q_i)/(U_i(d) - d_i)); \quad N = \{1,\ldots n\}$$

denotes the Chebyshev distance of the outcome q from the utopia point U(d) while scaled by the distances between the status quo point d and the utopia point; the transformation (2) is used in order to make the conflict coefficient independent of the number of players for the case of outcomes belonging to a linear manifold (corresponding to a constant sum game) and obtain the value of the conflict coefficient in such a case equal to 1. Since $C(d,U(d)) = 0$ and $C(d,d) = 2$, values of the conflict coefficient greater than 2 indicate that the current outcome of the game is worse than the status quo point, probably as a result of a conflict escalation process. A DSS could also compute, for the information of players, the minimal value of the conflict coefficient:

$$(4) \qquad C_{min}(d) = \min_{q \in S_d} C(d,q)$$

For convex S, this minimal value is between 0 and 1 and is attained at the Kalai-Smorodinsky cooperative solution (the value 0 indicates that the utopia point is attainable and thus is no need of any conflict, while the value 1 indicates that the Pareto boundary of S has linear manifold character). For nonconvex S, this minimal value is attained at a generalization of Kalai-Smorodinsky solution and is always greater than 1, while values close to 2 indicate strong nonconvexity and essential conflict character of the game.

The next concept useful for DSS simulated gaming is the principle of α-limited confidence (Fandel and Wierzbicki, 1985). This principle states that in a single round of a mediation procedure aimed at attaining a cooperative solution and based on a single text proposals (that is, consecutive proposals of agreements prepared by the mediator and taking into account interests of all players) no player can accept a proposal that gives to other players improvements of their payoffs greater than a specified part α of maximal possible improvements of these payoffs (in S_d, without deterioration of other payoffs):

$$(5) \qquad d^+ - d \leq \alpha(U(d) - d)$$

where d^+ denotes a new status quo point that can be reached due to a round of mediation. For n players, a typical value of the confidence coefficient would be $\alpha \leq 1/n$; see Bronisz et.al. (1988) for the relations between the confidence coefficient and minimal conflict coefficient. The concept of a confidence coefficient can be used also interactively, when each player communicates to the mediator his confidence coefficient and the minimal one is used in preparing a single text mediating proposal. It the number of rounds of negotiations would be infinite, each player would be motivated to propose very small confidence coefficients; but with a limited number of rounds, a reasonable confidence coefficient approaches 1/n. An interactive mediation process of such type has been characterized axiomatically in Bronisz et.al., op.cit.

If each player has many objectives, the situation is more complicated because each player could potentially use dummy objectives in order to increase their number for manipulating the mediation process in his favor. For example, the use of a classically defined (1) utopia point in the product space of multiple payoffs, although admissible when defining an approximative conflict coefficient, does not give symmetry properties to possible generalisations of the cooperative solution of Kalai-Smorodinsky type. Therefore, it is necessary to define a special utopia point relative to aspirations; before doing so, let us first consider issues of interactive decision support for a player.

Suppose the i-th player has n^i objectives (all to be maximized), $N^i = \{1,...n^i\}$, and denote the vector of his objectives by q^i. Suppose he communicates with the DSS by specifying the aspiration levels for these objectives; denote the vector of these aspirations by q^{ai}. There are two or three possible cases of the analysis of his possible decisions:

a) he can investigate his own decisions under assumed moves of other players;

b) he can investigate cooperative decisions of all players while assuming, however, that no other player should be worse off than the status quo solution. For doing this, he must either (b1) assume the objectives of other players or (b2) use a DSS that can automatically impose the constraints of status quo solution without informing him about the objectives of other players.

The functions of a DSS in all these cases are similar; let us consider the case (b2). Given any spiration point q^{ai}, the DSS can comute the corresponding efficient solution (decisions of all players in this case) and outcome $\hat{q}^i$ by maximizing over $q \in S_d$ an order-consistent achievement function, for example in the form:

$$S(q^i,q^{ai}) = \min_{j \in N^i} (q_j - q^{ai}_j)/(U^i_j(d)-q^{ai}_j) + \varepsilon \sum_{j \in N^i} (q^i_j - q^{a}{}^i_j)/(U^i_j(d)-q^{ai}_j) \tag{6}$$

where $U^i_j(d)$ is a component of classically defined utopia point and $\varepsilon > 0$ is a coefficient; any maximal point of this function over q^i such that $q \in S_d$ is properly efficient with bound ε, see Wierzbicki (1986). Suppose the player changes his aspirations q^{ai} until he finds a preferred outcome $\hat{q}^i$ and repeat such a process for each player. Then a vector $RU(d,q^a)$ composed of such $\hat{q}^i$ for all players $i \in N$ is typically not an attainable outcome and is called utopia point relative to aspirations.

If we now use $RU(d,q^a)$ instead of $U(d)$ in the definition (3) of the Chebyshev distance, modify accordingly the definition (2) of the conflict coefficient and select a cooperative decision that minimizes this modified conflict coefficient over $q \in S_d$, then this decision and its outcome determine a generalization of the Kalai-Smorodinski solution that is independent of the number of objectives of each player and has symmetry properties, see Bronisz et.al. (1988). Similarly, we can use $RU(d,q^a)$ instead of $U(d)$ to construct a fair mediation process based on the limited confidence principle as in (5).

These shortly reviewed mathematical aspects of decision support for simulated gaming make it possible to construct experimental versions of DSS for such purpose; one of such systems has been constructed by Bronisz, Kruś and Łopuch, see Bronisz at.al.op.cit. However, much remains yet to be done in the development of DSS for simulated gaming, and the interest in such DSS is rapidly increasing.

6. Conclusions.

While combining various concepts and tools from multiobjective optimization and decision theory, bargaining and cooperative game theory, it is possible to give much more decision support for gaming experiments than it is typically done in simulated gaming. This possibility opens, however, many new research questions.

7. References.

Axelrod, R. (1985). The Evolution of Cooperation. Basic Books, New York.

Bronisz, P., L. Kruś and B. Łopuch (1987). An experimental system supporting multiobjective bargaining problems. A methodological quide. In Theory, Software and Testing Examples for Decision Support Systems, ed. A. Lewandowski and A.P. Wierzbicki, IIASA WP-87-26, Laxenburg, Austria.

Bronisz, P., L.Kruś and A. P. Wierzbicki (1988). Towards interactive solutions in bargaining problems. RP.I.02.1.4, report of the Institute of Automatic Control, Warsaw University of Technology.

Cooke, S. and N. Slack (1984). Making Management Decisions. Prentice-Hall, Englewood Cliffs.

Dreyfus, R.E. (1984). Beyond rationality. In: M. Grauer, M. Thompson, A.P. Wierzbicki, Eds: Plural Rationality and Interactive Decision Processes. Proceedings Sopron, Hungary, 1984. Lecture Notes in Economics and Mathematical Systems, Vol. 248. Springer-Verlag, Berlin.

Fandel, G. and A.P. Wierzbicki (1985). A procedural selection of equilibria for supergames (private communication).

Erlandson, F.E. (1981). The satisficing process: A new look. IEEE Trans. on Systems, Man and Cybernetics, Vol. SMC-11, No. 11, November 1981.

Feldbaum, A.A. (1962). Foundations of the theory of optimal control systems (in Russian: Osnovy teorii optimalnych avtomaticheskikh sistem). Nauka, Moscow.

Fisher, W.F. (1979). Utility models for multiple objective decisions: Do they accurately represent human preferences? Decision Sciences, Vol. 10, pp. 451-477.

Glushkov, V.M. (1972). Basic principles of automation in organizational management systems (in Russian). Upravlayushcheye Sistemy i Mashiny, 1.

Gray, P. (1986). Group Decision Support Systems. In: E. McLean and H.G. Sol. Eds: Decision Support Systems: A Decade in Perspective, Proceedings of the IFIP WG 8.3 Working Conference on Decision Support Systems, Noordwijkerhout, The Netherlands.

Horsky, D. (1984). Estimation of attribute weights from preference comparisons. Management Science, Vol. 30, No 7.

Ignizio, J.P. (1978). Goal programming - a tool for multiobjective analysis Journal for Operational Research, 29, pp. 1109-1119.

Jarke, M. (1986). Group Decision Support through Office Systems: Developments in Distributed DSS Technology. In: E. McLean and H.G.Sol. Eds: Decision Support Systems: A Decade in Perspective, Proceedings of the IFIP WG 8.3 Working Conference on Decision Support Systems, Noordwijerhout, The Netherlands.

Kalai, E. and M. Smorodinsky (1975). Other solutions to Nash's bargaining problem. Econometrica, Vol. 43, pp 513-518.

Keen, P.G.W. and M.S. Scott Morton (1978). Decision Support Systems - An Organizational Perspective. Addison-Wesley Series on Decision Support.

Keeney, R.L. and H. Raiffa (1976). Decisions with Multiple Objectives: Preferences and Value Tradeoffs, Willey, New York, 1976.

Korhonen, P., and J. Laakso (1986). Solving a generalized goal programming problem using a visual interactive approach. European Journal of Operational Research 26, pp. 355-363.

Lewandowski, A., S. Johnson, and A.P.Wierzbicki (1986). A prototype selection committee decision analysis and support system, SCDAS: theoretical background and computer implementation. Working Paper WP-86-27, International Institute for Applied Systems Analysis, Laxenburg, Austria.

Lewandowski, A., T. Kręglewski, T. Rogowski and A.P. Wierzbicki (1987). Decision support systems of DIDAS family. In A. Lewandowski, A.P. Wierzbicki, Eds: Theory, Software and Testing Examples for Decision Support Systems. Working Paper WP-87-26, International Institute for Applied Systems Analysis, Laxenburg, Austria.

Lewandowski, A. and A.P. Wierzbicki (1987). Interactive decision support systems - the case of discrete alternatives for committee decision making. IIASA WP-87-38. Laxenburg, Austria.

Michalevich, M.V. (1986). Stochastic approaches to interactive multicriteria optimization problems. Working Paper WP-86-10, International Institute for Applied Systems

Analysis, Laxenburg, Austria.
Naisbitt, J. (1982). Megatrends. Warner Books, New York.
Nakayama, H., and Y. Sawaragi (1983). Satisficing trade-off method for multiobjective programming. In M. Grauer and A.P. Wierzbicki, eds: Interactive Decision Analysis. Springer-Verlag, Berlin-Heidelberg.
Nash, J.F. (1953). Two person cooperative games. Econometrica, Vol. 21, pp.129-140.
Parker, B.J. and G.A. Al-Utabi (1986). Decision support systems: The reality that seems to be hard to accept? OMEGA Int. Journal of Management Science, Vol. 14, No. 2, 1986.
Pospelov, G.S. and V.A. Irikov (1976). Program- and Goal Oriented Planning and Management (in Russian). Sovietskoye Radio, Moscow.
Rapoport, A. (1984). The uses of experimental games. In: M. Grauer, M. Thompson, A.p. Wierzbicki, eds: Plural Rationality and Interactive Decision Processes, Proceedings Sopron, Hungary, 1984. Lecture Notes in Economics and Mathematical Systems, Vol. 248. Springer-Verlag, Berlin.
Raiffa, H. (1953) Arbitration schemes for generalized two-person games. Annals of Mathematical Studies, No. 28, pp. 361-387, Princeton.
Raiffa, H. (1982). The Art and Science of Negotiations. Harvard University Press, Cambridge, Massachussets.
Sage, A.P. (1981). Behavioral and organizational considerations in the design of information systems and processes for planning and decision support. IEEE Trans. on Man, Systems and Cybernetics, Vol. SMC-11, No. 9, September 1981.
Simon, H. (1958). Administrative Behavior, McMillan, New York.
Steuer, R., and E.V.Cho. (1983). An interactive weighted Chebyshev procedure for multiple objective programming. Mathematical Programming 26., pp. 326-344.
Tversky, A., D. Kaheman and P. Slovic (1983). Judgment Under Uncertainty: Heuristics and Biases, Cambridge University Press.
Vlacic, Lj., B. Matic and A. P. Wierzbicki (1986). Aggregation Procedures for Hierarchically Grouped Decision Attributes with Application to Control System Performance Evaluation. International Conference on Vector Optimization, Darmstadt,1986.
Wierzbicki, A.P. (1980). The use of reference objectives in multiobjective optimization. In G. Fandel and T. Gal, Eds: Multiple Criteria Decision Making, Theory and Applications. Springer Verlag, Heidelberg .
Wierzbicki, A.P. (1982). A mathematical basis for satisfying decision making. Math. Modelling, Vol. 3, pp. 391-405.
Wierzbicki, A.P. (1983a). Negotiation and mediation in conflicts: The role of mathematical approaches and methods. In H. Chestnut et al., Eds: Supplemental Ways to Increase International Stability. Pergamon Press, Oxford 1983.
Wierzbicki, A.P. (1983b). Critical essay on the methodology of multiobjective analysis. Regional Science and Urban Economics, Vol. 13. pp. 5-29.
Wierzbicki, A.P. (1984a). Models and Sensitivity of Control Systems. Elsevier, Amsterdam 1984.
Wierzbicki, A.P. (1985). Negotiation and mediation in conflicts: Plural rationality and interactive decision processes. In: M. Grauer, M.Thompson, A.P. Wierzbicki, Eds: Plural Rationality and Interactive Decision Processes, Proceedings Sopron, Hungary, 1984. Lecture Notes in Economics and Mathematical Systems, Vol. 248. Springer Verlag, Berlin.
Wierzbicki, A.P. (1986). On the completeness and constructiveness of parametric characterizations to vector optimization problems. OR-Spektrum. Vol. 8, pp.73-87.
Wierzbicki, A.P. (1987). Towards interactive procedures in simulation and gaming: implications for multiperson decision support. Seminar Days: New Advances in Decision Support Systems, Albena (Bulgaria), October 1987.

RESEARCH ON DECISION-AIDING: HOW TO BE A DSS

Ken Bowen
Royal Holloway and Bedford New College
University of London
Egham, Surrey, TW20 OEX

Introduction

A researcher in the field of decision-aiding must work on real problems in real environments. Ideas about what may help managers can come from logical argument, from hunches, and from past experience. However, whatever theories or near-theories are set up, they are no more than hypotheses, however compelling, until they are tested in action and their power to influence is understood. It is a difficult task and progress is inevitably slow.

From research activities, with clients, a number of issues have been illuminated;

how does one get such research started?

will the techniques be used in the environments in which they were developed?

if not, what has been the main impact of the researcher's involvement, and can there be confidence in the techniques for use elsewhere?

Above all, there is the issue of the involvement of decision-makers themselves, both in the research process and in later applications of whatever techniques emerge.

The author has been involved in debating and researching the OR-process, as a sub-process of the decision-process, for the last 20 years. Before that he was concerned with managing projects in defence OR. Over the years, his emphasis has shifted, moving steadily from mathematical and statistical modelling to the structuring of problems and their environments, including his own problem of how to be a Decision-Support System.

It is perhaps being such a system, both for his own team and for his client, that is the main criterion of a good operational researcher, and makes him or her more than a well-informed technician.

Some emphasis in the paper is given to recent work: for MOD(Navy), where the real decision-maker was only available by proxy; for British Airways, where a study initially concerned with stock-control became something quite different; and for British Railways, where a study primarily concerned with statistics of potential hardware deficiences has been transformed into a study of human factors and communication.

Theory and Practice

It is all too easy, in principle, to conceive ideas to which people ought to adhere, in a perfect world, when making choices. However, the world is not perfect, nor are the people in it, including the analysts. Multiple criteria for choice are invariably part of any problem-situation, the situation is seldom clear cut, and the desired end is equally vague. Research has to encompass much more than the structuring of a new decision-aid. Even modification of what has been done before in similar situations needs careful research. Decision aids need to be custom-built: they are not just equipment and analytic process, but also require rules as to when these should or should not be used.

Increasingly, methodologies to guide the client-analyst interaction are being developed: the end-product should provide a methodology for the client in his decision-making role. With theories, the circumstances to which they apply are essential: with methodologies, similar but less prescriptive statements are required. In any case, in most problem-situations, decision aids have to be modified to suit changing circumstances: this modification, whether by analyst or client, is more likely to be based on experience than on explicit analysis (where the circumstances are completely new, analysis may be crude and informed guessing the best that can be done).

The analyst is a part of any decision aid during its creation and early use: eventually the clients who share this role take it over entirely. Importantly, at the even earlier stage of choosing what sort of decision aid is required, the analyst may be the primary decision support system to help the clients to choose the form of aid that they seek.

Getting Research Going

Academic research, which looks at existing OR processes, invents new ones, and studies the compatibility (and pros and cons) of all these, is a useful activity. From this and from experience in managing OR in complex problem-situations, gaps in understanding may be identified and ideas for filling them produced. Yet if these are to be more than just ideas they must be tested and further developed in practical contexts.

First find your client! There must be many ways in which the search can be carried out; I will talk only of the three of which I have experience.

1. The client comes to you, because you have what he sees as a suitable reputation as a researcher and as a practitioner in the area of his interest.
2. You advertise your wares to company research groups whose natural clients have, or are likely to have, the types of problem related to your research programme, and for which you could be a useful consultant.
3. You offer your services (or those of your staff or students) as a quid pro quo for being able to observe specific operating environments for your purposes: in this way you get a laboratory, trading off the managerial time given to you against the potential value to them of what you come up with.

Basically, these are

offers of a specific task-oriented environment;
the search for a research-oriented environment with particular potential applications;
negotiation for an experimental environment.

They have different possibilities but, broadly, money for research is the greatest with 1 and least with 3, while the matching of opportunities to a planned research programme is greatest with 3 and least with 1 (although this last always depends on what opportunities arise).

Offers of Research Environments

The two largest studies undertaken by me at Royal Holloway and Bedford New College have come as a result of requests to propose long term research studies into difficult problem areas. In both cases, an important feature was the negotiation of terms of reference to establish the work as research rather than as a project study.

The first study was of the information requirements and decision processes of a Naval Task Force Commander in the periods leading up to and during a major operation. It was specified that the aim would be to move towards the definition, and perhaps provision, of suitable decision-aids. At the same time, other research units were examining more tightly defined issues at lower command level. Because I had long-experience of Naval studies in my earlier MOD employment, no period of 'proof of competence' was required. One full-time Research Assistant was employed, working under joint supervision by me and by an academically qualified Commander, R.N., who was the MOD contract officer.

The second study was for British Rail. In this case there was a requirement for a six-month "trial" period, although their research officer who requested my involvement did so because he knew at first hand of my interest in complex problems. However, the Board, because of the sensitivity of the problem, wanted to be sure of our competence. This raised two issues, one being the employment in an uncertain situation of a competent researcher, the other being the funding of the feasibility study. Fortunately, the Research Assistant previously employed was still available and willing to accept the challenge, the College agreed to fund two-thirds of the money required and BR agreed to fund the remainder subject to progress. The end product was a 3-year research contract, to develop a suitable data-analysis system, and to suggest the form of a management control system, to cope with the understanding and management of a particular area of rail safety.

Both studies had one thing in common. There was a constant pull by the client group to move faster towards an implementation stage than the knowledge gained from research properly allowed, and also to want a concentration upon parts of the study which could only be seen in perspective within a total systems approach. These features are not unfamiliar to researchers. It is part of the political battle before and during the research to create and maintain a satisfactory research environment, without losing the clients' confidence. It is a negotiation process in which give and take is needed on both sides.

Trawling for interested clients

Research which takes one too far away from base can give many problems. It is desirable to find reasonably near-by groups whose work is likely to be relevant to one's own research area and who offer, potentially, a practical or a problem environment of value. There are, for an academic researcher, two somewhat different reasons for keeping in touch with relevant people. The first is to extend one's own abilities as a consultant, and so extend one's methodological research and test earlier ideas on new applications. The second is to find homes for doctoral students to test ideas developed in their embryo theses.

It seems that the first of these requires patience, cultivation over a considerable period before the right moment comes to reap any harvest. It is not a question of marketing for financial gain (the normal pressure on a consultant) but of marketing to make people want your ideas, and their development, as part of their research programme. Two such involvements have recently come about: some indication of their relevance to my research programme is given later.

The second has been quite an enlightening process. Environments have been found for two students, one concerned with research into the management learning process, and one with conflict management and conflict resolution. In each case, the students were accepted to look into specific problem-situations for their own research purposes: the time and expertise of managers was made available to them without any promise of specific returns, but assurances that whatever was unearthed would be their property. No financial support was asked or provided. The students went in with a good knowledge of the methodologies they were going to use and to develop. In both cases, they rapidly achieved respect and effective (unpaid) consultant status : this was undoubtedly due in part to the coherence and professionalism of their approach, but also, of course, to their personalities and abilities to communicate and not be seen as a nuisance (one has to choose the right people to do any research, and those looking at OR methodologies should have the qualities to interface with clients satisfactorily).

Research Outcome

The initial study, for the Navy, produced a total methodology[1] for inquiring into data needs, decision-making styles, and the process of decision, for senior executives. The first two stages of the methodology, system-based interviewing[2] and expert games[3] were utilised with senior naval officers, although it proved impossible in the time available to get the involvement of Rear-Admirals. Several suggestions relevant to the later construction and use of decision aids were made in the full report[1]. This report also includes an account of the third stage of the methodology, data collection-in-action: it was used in a study for the marketing division of an international manufacturing company. The time and environment of a group of managers was "borrowed" by the researcher, who was able to gain the interest of a senior member of the company.

It is worth noting that parts of other people's methodologies were used, primarily the Checkland Formal Systems Model[4] in the development of the system-based interviewing technique. Also used were Checkland's root-definition and conceptual models[4] and the cognitive mapping process developed by Eden et al[5].

Just prior to all this, I had produced a Problem-Formulation Methodologyy[6,7], but, apart from work with "experimental" clients[8], a group of teachers at Aston University, it had not been tested. The opportunity came with a study of models for management learning, since the student involved chose to use my methodology as most appropriate for him and, potentially, for his clients.
His application work, carried out for British Airways (B.A.) at Heathrow, not only 'proved' the methodology but also added usefully to the diagrammatic notation used for communication. The learning model that most influenced the managers was the OR process as the student used it. A second model was a simple micro-computer aid. Managers, who had originally believed their major problem to be an inadequate stock-control system, became acutely aware of their need for aide-memoires for the whole range of their tasks. Software was devised on the basis that only they could devise the basic data format and provide, update and modify data input as necessary over time. A summary report[9] given at the 1986 EURO Conference in Lisbon is an appendix to the full account of the work in the student's thesis[10].

In parallel with the Naval and B.A. studies, a third student was researching on conflict in organisations. His thesis[11] is so far the only published document. His work looks at the use of existing Process of OR developments and various non-OR methodologies for examining the management and resolution of conflict. It was the sort of study that might have been done by my first student and research assistant, Dr. Moynihan, had not the Navy study intervened: however, contacts he had with the West Midland Police on the subject of conflict, led to a useful environment for testing the other's ideas in a real situation. This situation, briefly, was that of growing differences in a committee composed of a large number of interested parties (police, council, ethnic groups and voluntary bodies) examining how to deal with crime in their community; it was indeed conflict within a would-be conflict management group. Interest was shown in this work when an informal presentation was given of some of the key ideas at the 1987 Annual OR Conference at Edinburgh. The West Midlands group are still in touch with the student whose report to them (annex to the thesis[11]) proved valuable.

Through the contacts with the managers, with whom and for whom my students were working, I developed an interest in what the precursors might be to the satisfactory production of expert systems[12]. I now hope to develop these ideas further in a large knowledge elicitation study for which I may become a consultant (and the supervisor of a mature PhD student) to one of the groups in a consortium. At the time of writing, I do not know exactly how we shall proceed, but I now have good research contacts and they are already briefed on the methodologies and techniques which might be employed and developed.

The second of my consultancy roles is in supporting a firm of consultants in a study that they are engaged in for an overseas client. The purpose is to look at the preemption or control of potential threats to the stability and well-being of the country concerned. It has various attractions to me. It is a large and complex problem; there are problem formulation problems; there are conflict analyses to be carried out; and there are nice judgements to be made as to what is appropriate to the needs and style of the client. In the team management area, I have also gained experience in the use of Synectics[13]. The work is still in progress.

Finally, there is the current study for British Rail. We are concerned with all aspects of signals being passed at danger. Dr. Moynihan is the prime researcher and his expertise as a psychologist (his first degree) is invaluable since there are naturally many problems such as stress and vigilance to be taken into account. Methodologically, we are using the system-based interviewing technique to study managerial needs and styles, and using the problem-formulation diagrammatic notation to communicate with individuals (and later, as planned, to intercommunicate in groups). This and other studies of data and needs for data up and down the hierarchy of a most complex organisation are essential to decide what statistical and other analyses are required for an ultimate management data and control system. Conflict issues arise, but as far as conflict study is concerned this is a question of our managing to keep research studies going in areas of some sensitivity. There are at present no published papers.

Concluding Remarks

This has been an account, necessarily brief, of how I have managed research on my concerns about aiding the decision process. There is no doubt that the task of getting involved with organisations and carrying out research in their environments has gone very well. It has not necessarily been the case that techniques developed have been carried through and used in their original environments. That they have been useful to progress in understanding of the clients in those environments seems a reasonable and modest claim. That they can be used in similar ways in quite different environments has been established in some cases. At a minimum, I can say that I and my researchers have been well-received in new areas and that we have had confidence in handling our interactions with clients and other researchers., It will however take many more years before I could state what has been most valuable and how and when it is applicable.

I believe that, working and thinking within a structured process of inquiry, the most important decision-support system that operational researchers provide is themselves. In the end, it is the clients who have to decide how to organise their affairs. One may provide technical systems that help, but these are of little use unless they embody a style and a data content which the clients have come to recognise and understand. Accordingly, my central interest has been how we can assist this recognition and understanding, to help surface it and make it explicit. Once that is done, the managers themselves can encase it in black boxes which will not be opaque to them.

Acknowledgements

I wish to acknowledge, with gratitude, the support of Dr. Peter Moynihan, Dr. Sabah R. Dakhel and Dr 'Yemi Adegoke. Without their enthusiasm, their ability and their capacity for hard work, progress would have been very limited and slow.

References

1. Moynihan P (1987) A study of senior executive decision-making in high level military and commercial decision-making situations. PhD thesis. London University.

2. Moynihan P, Bowen, K (1987) System based interviewing. In: Rand G (ed) Operational Research '87. North-Holland, Amsterdam New York Oxford Tokyo, 582-593.

3. Moynihan P (1987) Expert gaming: a means to investigate the executive decision-process. J.Opl Res.Soc. 38, 215-231.

4. Checkland P (1981) Systems thinking, systems practice. John Wiley & Sons, Chichester New York Brisbane Toronto.

5. Eden C, Jones S, Sims D (1983) Messing about in problems: an informal structured approach to their identification and management. Peragmon, Oxford New York Toronto Sydney Paris Frankfurt.

6. Bowen K (1981) A conflict approach to the modelling of problems of and in organisations. In: Brans J P (ed) Operational research '81, North-Holland. Amsterdam New York Oxford.

7. Bowen K (1983) A methodology for problem-formulation. PhD thesis. London Unversity.

8. Bowen K (1983) An experiment in problem-formulation. J. Opl Res. Soc. 34, 685-694.

9. Dakhel S R, Bowen K (1986) Exploring managers' needs for decision aids. A paper presented to the Eighth European Conference on Operational Research (EURO VIII), Lisbon, Portugal.

10. Dakhel S R (1987) The use of interactive models to formulate managers' problems and to identify decision aids, PhD thesis. London University.

11. Adegoke E O (1988) Integration system: a problem-solving framework for the examination of conflict in complex situations, PhD thesis. London University.

12. Bowen K (1985) The expertise of extracting expertise for expert systems. International Federation for Information Processing. 8 (Information Systems) Newletter/3, December, 1985.

13. Prince G M (1970) The practice of creativity: a manual for dynamic group problem solving. Harper and Row, London.

EXPERT KNOWLEDGE ACQUISITION SYSTEM

O.I.Larichev, A.I.Mechitov, E.M.Moshkovich E.M.Furems
All-Union Research Institute for Systems Studies
Moscow, Pr.60 let Octiabria,9 117312 USSR

One of the main element of expert systems and decision support systems is the system's knowledge base (KB). In the process formation it is necessary to obtain a large volume of expert information. Usually this information contains the expert's estimation of a number of different situations (states) in the problem field under consideration [1]. The process of expert's knowledge elicitation is labour intensive as far as knowledge engineers and experts are concerned. That is why the development of effective methods for KB construction allowing to decrease the number of situations estimated by experts on the base of certain assumptions about the structure of a problem field is an urgent task.

The task of diagnostic system KB construction can be presented as follows. Let there be an object which can be in different states. Object's state can be described by M attributes. For each attribute m ($m=\overline{1,M}$) a set Q_m of its possible values is given: $Q_m=\left\{q_{m1},q_{m2},\dots,q_{mn_m}\right\}$, where q_{mi} is the i-th value of attribute m. $A=Q_1xQ_2x\dots xQ_M$ defines the set of all hypothetically possible object's states: $a_i \in A$ is the i-th object's state, $a_i=(a_{i1}, a_{i2},\dots,a_{iM})$, where $a_{im} \in Q_m$, $m=\overline{1,M}$. Let us mark a set of properties which an object is able to possess as $P=\left\{P_1,P_2,\dots,P_L\right\}$. It is necessary on the base of expert's knowledge to build a classification of all possible object's states $A=\bigcup_{l=0}^{L} K_l$ in such a way: state $a_i \in A$ belongs to the class K_l if an object in this state according to the expert's opinion possesses the property P_l. Object's states in which the object does not possess any of the properties under consideration belong to the class K_0.

The hypothesis about different degree of inherence of attribute values to each property is in the basis of the proposed approach [2]. It is supposed that an expert is able to order values of each attribute scale according to the decreasing of values' inherence for each property and that this order does not depend on the values upon other attributes. This hypothesis allows to introduce antireflexive and transitive binary relations of inherence r_l^m for each property l on the set of possible values for attribute m ($l=\overline{1,L}$, $m=\overline{1,M}$): $(q_{ms},q_{mt})\in r_m^l$, if the value q_{ms} is not less inherent for the l-th property than the value q_{mt}. Let us form an antireflexive and transitive binary relation of dominance upon inherence for the l-th property

$$R^l=\left\{(a_s,a_t)\in A\times A \mid \forall m=\overline{1,M} \quad (a_{tm},a_{sm})\bar{\in}r_m^l \quad \text{and} \quad \exists m_0 \quad 1\le m_0\le m \right.$$
$$\left. \text{so that} \quad (a_{sm_0},a_{tm_0})\in r_m^l\right\} ; \qquad l=\overline{1,L} .$$

It is natural to suppose that if an expert marks the presence of some property in some object's state, than the state being described by a set of attributes' values not less inherent for the same property also possesses this property. That is may be put down formally in the following way:

$$\text{if} \quad a_s\in K_l \quad \text{and} \quad (a_t,a_s)\in R^l \quad \text{than} \quad a_t\in K_l \tag{1}$$

That is why if some object's state is not inherent for some property, then the state less inherent for this property can not possess this property, that is:

$$\text{if} \quad a_s\bar{\in}K_l \quad \text{and} \quad (a_s,a_t)\in R^l \quad \text{than} \quad a_t\bar{\in}K_l \tag{2}$$

This assumption about possibility to make a conclusion for a number of states without their direct estimation by an expert allows to construct a rational procedure of expert's interview to classify all possible object's states.

In dependence with peculiarities of the task under consideration the requirements to the type of reflexion of a set of object's states into a set of classes may be formulated differently. In accordance with this requirements there may be different presentations of the expert classification task. At first let us consider the task of nominal classification, assuming that on the set of properties there is no any relation. It is nesessary to build up a covering of the initial set of states when in each state an object can possess several properties.

In the process of expert's interview the expert is presented with some object's state and the expert gives a conclusion about classes of inherence for this state (properties which an object possesses in this state). This way the expert classifying the state a_i marks two sets of classes: those of inherence for this state K_i^+ (corresponding to the properties it possesses) and those that this state does not possess K_i^- (corresponding to the properties that the object does not possess).

This information allows to decrease the uncertainty about classes of inherence for object's states in the process of the following two-stage procedure. At the first step for each class of the first set all states dominating the state a_i upon the relation of inherence are determined. For each of them this class is considered to be defined (that is for $K_j \in K_i^+$ if $a_s R^j a_i$ then $a_s \in K_j$). At the second stage for each class of the second set it is possible to determine less inherent states than the estimated one. For these states it is possible to exclude this class from acceptable ones for them (that is for each $K_j \in K_i^-$ if $a_i R^j a_s$ then $a_s \bar{\in} K_j$).

This way classification by an expert of one state allows in an indirect way to define classes of inherence for some other states on the base of introduced binary relations.

Let us consider the task of ordinal classification of object's states. In this task there are antireflexive and transitive binary relations on the set of classes which define order on classes (from the best K^n to the worst K^s). for each attribute it is defined only one binary relation of inherence R_l and it is assumed that object's state a_i, dominating upon inherence the state a_j can not be put into a less inherent class:

$$\text{if} \quad a_i R_l a_j \quad \text{and} \quad a_j \in K^s, \quad \text{then} \quad a_j \bar{\in} K^n, \quad n<s \tag{3}$$

The interpretation of this task presentation may be the following. An object in different state defined by a set of attributes' values is able to possess some property. The degree of its inherence to this property is able to be different in differing object's states. For each attribute values are ordered upon the degree of inherence for the property under consideration. In the desired classification an object in a less inherent for this property state than the other one must not possess this property in a more inherent degree. The use of binary relations allows and in this task of expert classification to decrease uncertainty about other states on the base of the classification of one state [2].

It is possible to formulate a more general task of nominal and ordinal clàssification. Within such task an object in each of its states is able to possess some of L properties in different degree of inherence. It is necessary on the base of expert's knowledge to build up such a classification of object's states that passes the requirements of nominal classification task and requirements of an ordinal classification task.

The solution of nominal and ordinal classification task can be obtained by consequent implementation of methods for solution of nominal classification task and ordinal classification task. At first the nominal classification task is being solved. Within this stage an expert while classifying the state a_i points out not only corresponding properties but also their degree of inherence in this state. These stages is over when all object's states are classified and so it is known L set of states possessing one and the same property. At the second stage L tasks of ordinal classification are being solved. Within each of them for all states belonging to the l-th class the degree of inherence for the l-th property is defined $(l=\overline{1,L})$. For object's states being classified directly by an expert in the cause of nominal classification task solution degrees of inherence are known. For all other states these degrees are to be determined in the cause of ordinal classification task solution.

In any procedure of expert's interview it is necessary to take into account the possibility of erroneous answers [3,4]. There can be random errors and errors connected with nonconsistency of answers. Errors in an expert classification conclude in violence of conditions (1) and (2) and in violence (3) for ordinal classification task.

The proposed procedure for processing of expert information allows to detect errors in an expert's answers. In fact if an expert puts some object's state into class which has been excluded for it from possible ones on the base of an expert's prevous answers this means that the last answer contradicts with some from the previous ones. To eliminate this contradiction both corresponding answers are presented to the expert for reclassification of corresponding states.

The approach to knowledge base construction described has been used in development of a medical diagnostic system for cardia and some other diseases. Knowledge base was constructed on the base of the described procedure of expert's interview carried out as computerized interactive complex. Interface with an expert is user-friendly: a situation displayed for an expert is analogous to some patient's history. Answers expected from an expert are formulated in a traditional form of diagnostic conclusions. Besides this an expert evaluates the degree of suspision on the disease (in terms "high", "middle", "low"). Also it is possible to answer "situation is contradictory" in the case when the data of attributes under consideration are not able to be presented in one situation.

The developed interactive system for expert knowledge elicitation allows to construct a complete knowledge base for structured problem field. The implementation of binary relation of inherence makes it possible to decrease essentially the duration of expert's interrogatory and also to detect and eliminate errors in expert's answers. According to our opinion the described system for expert knowledge elicitation is sufficiently universal and may be used while constructing knowledge base in different tasks (management, financial, technical) of expert classification.

REFERENCES

1. Winston P. (1977). Artificial Intelligence. Addison-Wesley.
2. Larichev O., Mechitov A., Moshkovich E., Furems E. (1987). Expert knowledge acquisition systems. Technicheskaya Kibernetica, N.2, (in Russian).
3. Larichev O. (1979). The science and art of decision making. Moscow: Nauka (in Russian).
4. Larichev O., Moshkovich E. (1988). Limits to Decision-Making Ability in Direct Multiattribute Alternative Evaluation. Organizational Behavior and Human Decision Processes, 42.

PROBLEM STRUCTURING TOOLS FOR PRE-MCDM STAGE

D. P. Solomatin

Institute for Systems Studies (VNIISI),
Prospekt 60-Let Oktyabria, 9
117312 Moscow USSR

The phase of problem analysis (clear formulation, problem decomposition, maintaining the links between problem's parts or elements) is always present as an early stage of any decision making process. Usually structural models are used on this stage. Examples of their usage are given in [1,2,3,4]. Generally speaking structural modeling deals with the relationships between the various elements of the system. Under "system" we understand here or a problem situation facing the decision maker.

Very often SMs are described by the binary relations (rarely by relations of higher order) or by directed graphs (weighted or not) - they are widely and effectively used in many disciplines.

SMs (graph models) are used also in the process of multi-criteria decision making itself being a tool for representing the connections between alternatives, criteria, objectives, experts, decision outputs (decision trees, objectives trees, preference graphs, decision matrices etc.).

Besides the task of constructing SM and its analysis there is another typical job in a problem situation study - maintaining the elements relative significance (weight) according to some or other criterion (criteria) or scale.

The problem of eliciting and processing experts' judgements is actually a part of an MCDM process itself. (The term "experts" will equally relate to decision makers also, if not specified otherwise). Nevertheless the early stages of MCDM (the problem analysis being one of them) need more of less simple and robust procedure of this type. We'll consider an approach which we believe to be adequate for a pre-decision stage of the analysis.

So, here we shall focus on the two problems of structural

modeling that may probably evolve during the problem analysis (and decision making):

1) to structurize the problem situation under study: identify its elements and maintain the links between them, i.e. build its SM (we restrict ourselves by the class of digraphs);

2) to maintain a comparative significance of the system's elements according to the criterion (criteria) and scale chosen by experts.

Constructing graph models: main issues and the proposed method

There is a widespread and often effective method of constructing graph SM - step-by-step decomposition used in PATTERN-like techniques for objectives trees construction. Unfortunately this method isn't free of drawbacks [4]. The most significant is the impossibility to construct SM on the basis of a given set of elements. Furthermore, generating of elements and their structurization set different requirements to the intellectual activity of the expert group. One of the ways out is to separate in time the phases of generating and structuring in the process of SM construction.

There are many effective expert techniques to generate a list of the analysed system's main elements, Very often, however, it is difficult or even impossible to form SM manually and a computer can be useful when: a) the number of elements is large and problems of scale arise; b) elements are related in a complex way and it is difficult to maintain and analyze their interconnections; c) structuring is performed by a multidisciplinary team of experts and a systematic procedure for organizing their work is necessary.

It is generally accepted that the human ability to perceive, remember and process large amounts of data is limited. For this reason one of the most reliable means of eliciting expert judgements is to request them to excercise a systematic pairwise comparison of objects [5].

Consider the problem of synthesis (reconstruction) of the binary relation R (i.e. graph G(V,R)) when this relation is transitive. In a variety of applications this is the case (the majority of relations of support, impact, preference, cause - effect, time precedence and others are transitive).

The algorithm of interactive synthesis of the graph G(V,R) (the relation R is supposed to be transitive and reflexive, i.e. the partial quasi-order) is actually the sequential (cyclic) realization of the following steps:

1) choose a pair of elements to be asked about (the question on their connection);

2) get the experts' answer (the arc in the graph or its absence);

3) define all the logical inferences of the answer, and put them into the graph.

If the binary relation is transitive it becomes possible to optimize the process of pairwise comparisons. (The idea of reducing the number of comparisons, made possible by the transitivity, is implicitly used for reconstructing the relations of linear quasi-order - in algorithms of quick sorting).

For reconstructing the partial quasi-order we developed [7] the following method of getting logical inferences. It is based on the fact that the transitivity of R leads not only to the inferences

(a) iRk & kRj --> iRj (let's call them 1-inferences), but also to:

(b) kRi & NOT kRj --> NOT iRj and

(c) jRk & NOT iRk --> NOT iRj (let's call them 0-inferences).

In constructing relation R two types of element pairs emerge: 1) those, that iRj; 2) those, that NOT iRj ; 3) those, whose connections we yet know nothing about. Assume, that the subset of pairs of the type 2 generates the binary relation Q. So, the current state of the graph is described by two relations: R and Q.

1-infereneces of the type (a) are derived from the answers of the type iRj by ordinary transitive closure $R^{\wedge}$. Given the information about the connections between elements (iRj) and the absence of such connections (NOT iRj) then all possible 0-inferences of the types (a) and (b) can be defined with the following operation:

$$Q \longleftarrow R^{\wedge T} \quad Q \quad R^{\wedge T}.$$

The choice of a question (two elements to be compared) is performed to maximize the average number of inferences (1-inferences go with the higher priority).

The graph (relation R) construction goes on until R U Q = VxV. The technique based on these ideas (algorothm for successive d-transitive closure, or d-ASTC) is behind the STRUM system along with the algorithms for SM automatic analysis and mapping.

Simulation testing and the experience of d-ASTC application indicate that the total number of questions it suggest is 40 to 70% lower if compared with the number of all possible pairwise comparisons.

Use of STRUM in the problem structuring process

Following the selection of the elements and relation their verbal representations are entered to the computer. The construction of a structural model is then an interactive procedure, which takes the following form.

STRUM first puts questions to the user as "Indicate the direction of the relation R between elements A and B" (where R, A and B are verbal representations). (By "user" we mean one or more experts working on a specific SM; currently STRUM doesn't allow to take into account their different opinions). On obtaining the answer STRUM automatically introduces transitive inferences into the SM (if relation R is assumed to be transitive). Then STRUM by special optimizing rules chooses the next pair of elements to be considered, asks the next question and the process is repeated until the SM is complete.

When the model is complete it must be analysed and corrected. Visual analysis is aided by a graphical facility which displays the graph of the SM. Automatic analysis involves identification of levels, cycles, subgraphs on subsets of elements etc. Individual links between the elements may be added or deleted at any stage of construction process.

Maintaining the comparative significance of elements: COMPA system

The other important task of problem analysis is to obtain experts' knowledge on the elements comparative significance. COMPA system helps an expert to obtain the weights of elements (definition of "weight" may vary on the problem under study: importance, volume, etc.). If there are more than 5 elements then the problem becomes non-trivial for most applications. This occurs due to the considerable peculiarities of the human mind which is not able to hold and process correctly more than 5-7 elements.

Following STRUM system COMPA prompts the user to compare only two elements simultaneously at every step of the interactive procedure. In a number of such steps the vector of elements weights is obtained.

Elements may be compared on the basis of various criteria (the result of each procedure is the vector of weights). Then the criteria may be compared too and the elements absolute weights easily calculated. This last task is solved with the help of the COMPA-Binder subsystem.

There are quite various methods for finding the vector of elements' weights [5]. In the method described in Saati's publications [6] the user should compare the elements using a verbal equi-distant scale. The vector of weights is calculated as a normalized eigenvector of the pairwise comparisons matrix.

The drawbacks of the scale used by Saati are analysed in [8] and the more adequate non-linear scale is suggested. Besides method of logarithmic regression with this scale is presented. It is especially effective when the opinions in the experts group vary.

It is important that these methods don't force the user to be transitive in comparing various elements, i.e. he may say that A > B, B > C, but C > A. COMPA then will ask him to compare other elements with a view to decreasing "intransitivity index". The COMPA user may use any of three available scales.

At every step of the procedure the user can directly enter his opinion about the elements' relative importance, make corrections to the opinions entered earlier, change the used scale, to print list of elements, vector of current weights, etc.

Case study: structuring the problem "management of construction sector in the USSR"

This case study was a part of studies of national management undertaken by the USSR Academy of Sciences. It should be mentioned that this study was undertaken in 1985. Owing to the latest improvements in the economic management some of the problems listed in this paper are not so urgent now.

The problematique of construction management on the national level is notable for high complexity and entangled structure. In the very beginning of the study the expert team decided that it would be difficult to effectively communicate further and develop the common decision without the detailed analysis of the problem situation. STRUM and COMPA systems turned to be useful tools for this purpose.

On the first stage the expert group decided to construct and analyse the graph of inter-dependencies between the problems. A list of problems was prepared by brainstorming and consisted at first of 18 to be later extended to 30. The relation of structurization was formulated as "problem A sufficiently impacts problem B". The full list follows.

1. Inadequate balance between financial resources of the customer and the productive capacity of construction units (CU)

2. Planning too high output for CUs

3. Allocating resources to numerous construction projects

4. Too protracted construction cycle

5. Far from perfect criteria of CUs effectiveness

6. Mismatching economic interests of CUs and organizations participating in the construction process

7. CUs are sometimes not interested in effective resource utilization

8. No managerial body on the national level to be fully responsible for different construction projects

9. Large number of small ineffective CUs

10. Irrational loading of CUs

11. Complexities of pursuing a unified scientific and technological policy in construction

12. Decentralized management of the CUs in a region

13. Multi-layer hierarchy of the sectoral construction management

14. Difficulties in improving the organazational and production structure of CUs

15. Complexity of material resources supply

16. Non-coordinated realization of various stages of the construction process

17. Poor relationship between labour results and payments

18. Unsatisfactory oranizational and technological support of construction

19. Poor quatity of construction

20. Construction industry is poorly equipped with modern

machinery and appliances

21. Inefficient machinery utilization
22. Low productivity
23. High labour turnover
24. Hard labour conditions in construction
25. Shortage of skilled workers, technicians and engineers
26. Standards and norms of construction quality are not full and correct
27. Unsatisfactory quality control in construction
28. Non-coordinated interests of engineering personnel and self-supporting workers brigades
29. Absence of actual CUs self-supporting
30. Underestimation of social factors in management

Experts answered to 192 STRUM questions questions like "show which of the two listed problems strongly impact the other". Theoretical number of questions (in the case of comparing all pairs) is 0.5 n (n-1) = 435 (here n = 30) i.e. there is a 55 % decrease.

The graph of the resulting structural model of the problem situation is presented in Fig. 1 (direction of links is up).

The STRUM system enabled the expert group to analyse in detail the problems' pairwise inter-relationships, stimulated constructive discussion and helped to clarify the structure of the problem situation in general.

A lot of time was devoted to the analysis of the obtained graph of problems' inter-dependencies. The main idea here was to identify various chains of elements which are actually the possible ways of solving these problems. Non-trivial chains were chosen for the additional home work to be analysed in detail:

8 - 11 - 26 - 27 - 19

(17, 29) - 7 - 3 - 21 - 22 - 4

30 - 24 - 23 - 25 - 18 - 4

8 - 12 - 14 - 9 - 4.

The other task was to identify the problems critical for accomplishing the main objective of the system under study.

The general objective of the national construction sector was formulated as "full satisfaction of the national economy requirements in effective construction activities".

The general objective was decomposed and the objective tree constructed. We'll consider only 3 partial objectives from it (short formulations are used): 1) reduce construction cycle; 2) improve quality of construction; 3) increase productivity.

Following the additional analysis of the graph the expert group decided that the most critical problems in this context were problems 1, 5, 6, 8, 11, 17, 29, 30. They are located at the lower layers of the graph.

The 8 problems were considered then in the context of their contribution to reaching these objectives, i.e. it is reasonable to consider these 3 objectives as criteria for assessing the "impact" of solving the particular problem on reaching the objectives.

So the expert group used COMPA system to obtain 3 vectors of weights (relative importance of problems' impact for each criterion). The COMPA-Binder subsystem was used to make an integral assessment of the problems. The group assigned preference weights (within the range 1 ... 100) to 3 criteria which reflected group's relative valuation of their importance. Pairwise comparisons were not used here because of insufficient number of criteria to be compared. The vector follows.

40	1. Reduce construction cycle
75	2. Improve quality of construction
30	3. Increase productivity

These weights were entered to COMPA-Binder and on the basis of additive integral criterion it calculated the integral weights of the 8 problems.

Weights accord. 1	2	3	Integ. weight	Problems to criter. no.
.093	.070	.199	.10	1. Inadequate balance between financial resources of the customer and the productive capacity of construction business units (CBU)
.134	.092	.177	.12	5. Far from perfect criteria of CBUs effectiveness
.220	.158	.155	.17	6. Mismatching economic interests of CBUs and organizations participating in the construction process
.113	.116	.088	.11	8. No managerial body on the national level to be fully responsible for different construction projects
.113	.116	.088	.13	11. Complexities of pursuing a unified scientific and technological

				policy in construction
.085	.161	.118	.16	17. Poor relationship between labour results and payments
.142	.189	.107	.09	29. Absence of actual CBUs self-supporting
.106	.128	.073	.11	30. Underestimation of social factors in management

Hence, problem 6 is the most important, followed by problems 17, 11, 5, 8, 30, 1, 7.

We understand the drawbacks of the additive integral criterion but we think it may be applied at the early stages of decision making under ill-structured situations.

The multi-context evaluation helped set priorities of the problems during the preparation of the final report with suggestions on improving the construction management.

The considered task of improvement of construction management didn't require the precise problems' assessment, but generally speaking the analysis of this kind may be of help at the early stage of an MCDM process of a wide variety.

Such integral evaluation of each problem is not the only means of analysis provided by COMPA. The current version provides a possibility for constructing Pareto-optimal set and Pareto layers of elements in the criteria space. The user may choose any 2 criteria and layers will be presented on screen in a 2-criteria space. In the nearest future the user will be able to enter criteria weights in graphical and verbal form, to indicate ranges of weights instead of their exact values, to construct fuzzy Pareto-layers. In this case it will be possible to use COMPA not only at the problem analysis stage but in the MCDM process itself.

Conclusion

The tools for complex problem structuring described in this paper proved to be useful in various decision making tasks, particularly for preparing managerial decisions in a national economy branch (construction). STRUM and COMPA systems helped:

1) get a deeper insight into the problem situation owing to the necessity of analysing the pairwise dependencies of the elements;

2) identify the chains of impacts between problems; this made it possible to suggest possible ways to problem solution;

3) identify the "critical" problems affecting most of the other problems and the decision making results in general;

4) obtain the information on the comparative significance of problems - it may help at the subsequent stages of MCDM.

The both systems were realized on an IBM PC, they are friendly, menu-driven, have graphics capabilities.

The suggested approach to the problem structuring and analysis was used earlier also in preparing management decisions, multi-criteria resource allocation tasks and technology assessment tasks and proved effective. We think that the described problem-structuring systems may be adjusted to be used as automated knowledge-acquisition tools for constructing knowledge bases for DSSs and expert systems.

We know that methods of computer analysis very often do not lead to good decisions... but other methods generate even worse decisions. We beleive that the described participative structural problem analysis may assist in getting at least acceptable decisions.

References

1. A.Vari, J.Vecsenyi, Z.Paprika. Supporting problem structuring in high level decisions. - In: B.Brehmer, H.Jungermann, P.Lourens, G.Sevon (eds), New directions in research on decision making. Amsterdam: North Holland, 1986.

2. J.H.Kim, J.Pearl. CONVINCE: A conversational inference consolidation engine. IEEE Trans. on Systems, Man and Cybernetics, vol. SMC-17, No. 2, 1987, pp. 120-132.

3. K.Niwa. A knowledge-based human-computer cooperative system for ill-structured management domains. IEEE Trans. on Systems, Man and Cybernetics, vol. SMC-16, No. 3, 1986, pp. 335-342.

4. I.A.Ganin, D.P.Solomatin. Structural models in intersectoral management: typology and problems of construction. - In: Intersectoral complexes: strategies for development and management. Moscow: VNIISI, 1983, pp. 72-81 (in Russian).

5. David H.A. The method of paired comparisons. - London: Griffin, 1963.

6. Saaty T.L. The Analytic Hierarchy Process: Planning, Priority Setting, Resource Allocation. - N.Y.: McGraw-Hill, 1980.

7. I.A.Ganin, D.P.Solomatin. STRUM - An Interactive Computer System for Modeling Binary Relations. IIASA Collaborative Paper CP-84-52. Laxenburg: IIASA, 1984.

8. F.A.Lootsma, J.Meisner, F.Schellemans. Multi-criteria Decision Analysis as an Aid to the Strategic Planning of Energy R & D. - European Journal of Operational Research, 25 (1986), pp. 216-234.

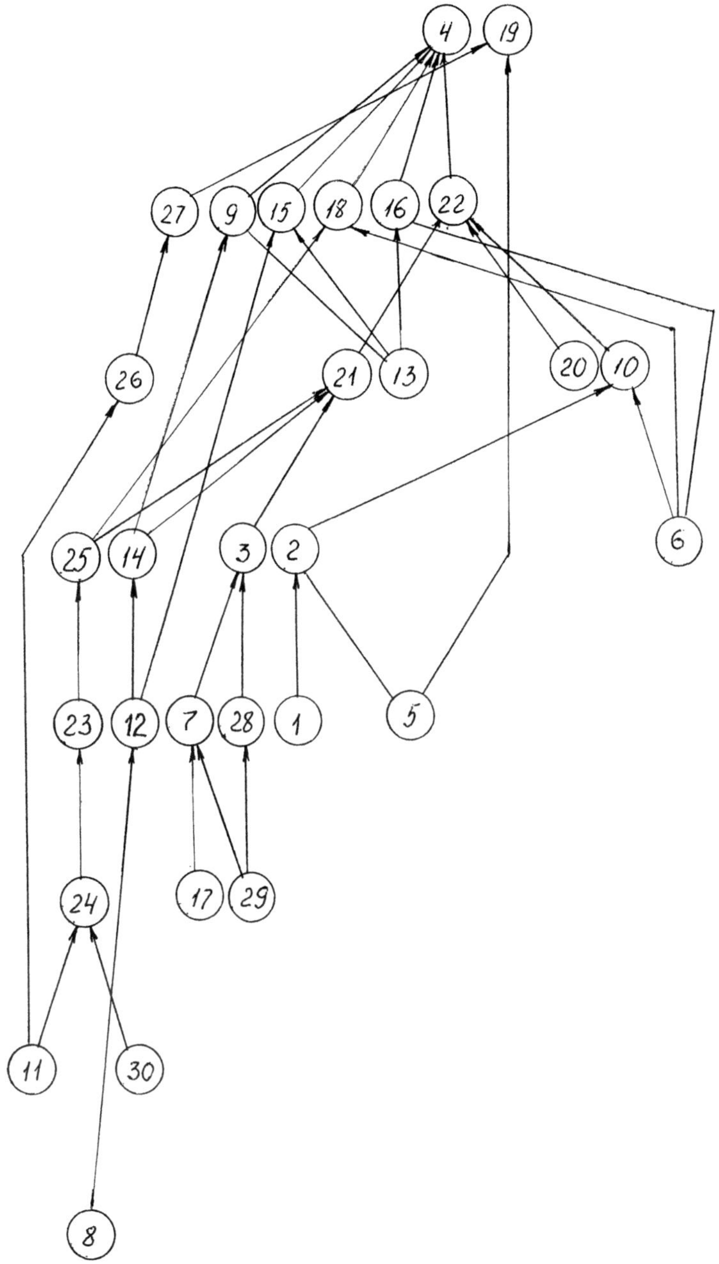

Figure 1: Graph of Structural Model Developed Using STRUM

DEFINITE
A SUPPORT SYSTEM FOR DECISIONS ON A FINITE SET OF ALTERNATIVES

M. van HERWIJNEN and R. JANSSEN
INSTITUTE FOR ENVIRONMENTAL STUDIES
FREE UNIVERSITY, P.O. BOX 7161
1007 MC AMSTERDAM, THE NETHERLANDS

1. INTRODUCTION

The use of decision support systems in public planning is slowly increasing, following developments in private enterprise. Evaluation methods such as cost benefit analysis, multicriteria and multiobjective analysis are replaced by or included in interactive computer programmes. Bij bridging the gap between methods and decision maker, Decision Support System induce a shift in the use of the methods from the analyst to the decision maker (see fig. 1).

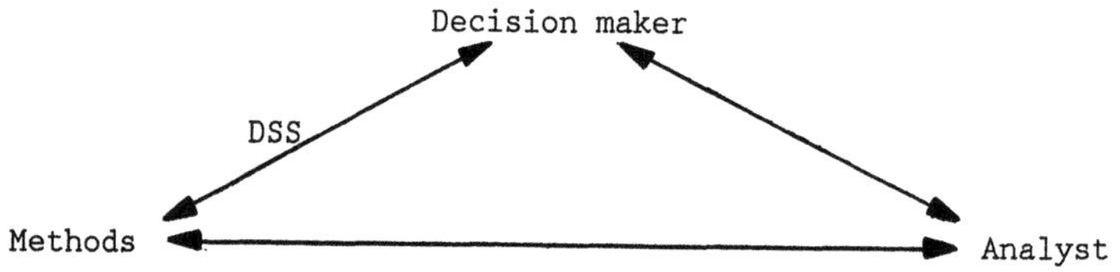

Figure 1. The role of decision support systems.

Integration of evaluation methods and models in a computer system also allows for a wider scope of decision support: from evaluation to all phases of a decision procedure. This makes it possible to use the results of the analysis for generating new or modified policy alternatives or to add new evaluation criteria.
A decision support system can be described in terms of the types of decisions it can handle, the types of users and the type of results it generates. DEFINITE, the decision support system described in this paper, supports DEcisions with a FINITE set of alternatives in relation to a finite number of criteria. DEFINITE is commissioned by the policy analysis department of the Dutch Ministry of Finance; intended users are people involved in the preparation of public planning decisions. Results range from a

complete ranking of alternatives to a graphic presentation of all relevant information. DEFINITE is to be used in all sectors of public planning, ranging from stop-go decisions on infrastructure projects to the selection of a policy scheme to abate vandalism.

The next section contains some remarks on decision support theory. Structure and aims of the DEFINITE system are described in section 3. In section 4 we describe various procedures of DEFINITE through some examples. Section 5 we try to bridge the gap between the theory of section 2 and practice of section 4.

2. DEFINITION AND THEORY OF DECISION SUPPORT SYSTEMS

Definitions on Decision Support Systems range from (Sprague and Watson, 1986):

- An interactive computermodel that helps decision makers utilize data and models to solve unstructured problems, to:
- Any system that makes some contribution to decision making.

The second definition is so broad that is almost a tautology. In this paper we will adopt the first definition. Note that this definition is still sufficiently broad to include software such as expert systems and information systems.

Important in the first definition is the concept of unstructured problems. If a decision problem is completely structured, that is if problem content and priorities are agreed on by all participants, interaction with the decision maker is not required. If at the other extreme no structure can be brought into the problem content, decision support is impossible. It is only between these extremes that DSS's are relevant (see also Ginzberg and Stohr, 1982). Bringing structure into unstructured problems is, in our view, one of the most important aspects of DSS.

The absence of a clear definition is an example of the lack of a firm theoretical foundation of decision support. The various aspects of decision support reflect contributions from a variety of disciplines such as computer science, phychology, economics and planning.

Literature in computer science describes decision support in terms of information processing and by describing the various functional elements of the system (Andriole, 1986; Spraque and Watson, 1986). A DSS consists of a Modelbase, a Database, a Modelbase and Data base Management System and finally of a Dialog Management and Generation System usually referred to as the user interface. The quality of the user interface is crucial to the acceptance of a DSS by its users. In DEFINITE, models consist of methods for generating and comparing alternatives. The database structure can contain all user input and results. The database content is specific for each application.

From Behavioural Psychology (Newell and Simon, 1972) DSS literature adopted the following phases of decision making:

- Intelligence: Searching the environment for conditions calling for decisions;
- Design: Inventing, developing and analyzing possible courses of action;
- Choice: Selecting a particular course of action from those available.

DSS accomodates all these phases and, more importantly, for the integration of these phases. A central issue in DSS is the user-system interaction: consistency of system decision rules with user decision rules and perception of results. All these issues are in the domain of behavioural psychology (Newell and Simon, 1972; Janis and Mann, 1977; Winterfeldt and Edwards, 1986; Tufte, 1985 on perception of graphs).

Economics emphasises the more normative aspects of decision making. This is in contrast to the descriptive emphasis of psychology. Given a certain problem and a certain preference structure, the optimal alternative is defined according to predefined principles of rational behaviour (Keeney and Raiffa, 1976; Rietveld, 1980). DSS literature shows a certain ambivalence between normative decision making reflected in the use of formal evaluation methods such as multicriteria methods and cost-benefit analysis, and descriptive decision making reflected in the emphasis on learning by doing and on presentation techniques.

Planning literature provides also a theory on decision procedures. Various phases of a decision procedure are distinguished. These phases can be used to structure a DSS. Procedural decision procedures, as advocated in planning literature, are usually not included in decision support systems.

Management Science distinguishes between strategic, tactical and operational decisions. DSS can be applied to all three types of decisions. Systems such as DEFINITE with emphasis on valuation and ranking are mostly applied to strategic decisions. Expert systems and information systems are more commonly applied to tactical and operational decisions.

In conclusion it can be said that although DSS as a whole lacks theoretical foundation, the various aspects do not. A priori it is not guaranteed, however, that this amalgamation of theory results in a consistent whole. The present practice of making something that works and appeals might result in a bias towards the operational and marketing aspects of a DSS at the expense of its theoretical foundation.

3. DEFINITE STRUCTURE

DEFINITE supports decisions on problems with a finite set of alternatives and a finite set of criteria. The system is divided into five modules, which represent different levels in the process of decision making.

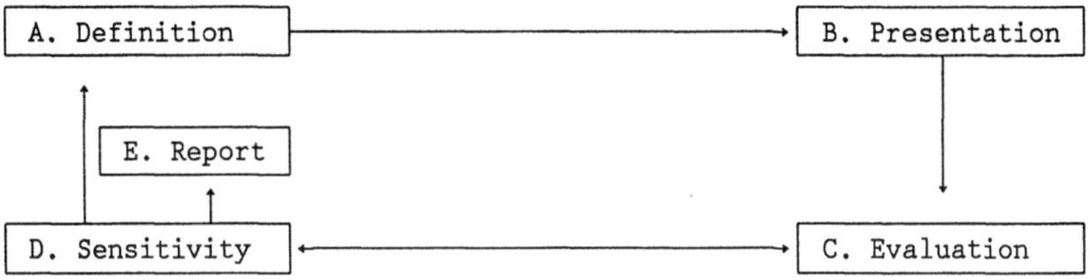

Figure 2. Modules of DEFINITE.

The procedures performing the various functions within each module are listed in the next section. In describing a selection of these procedures emphasis will be given to the following issues:

1. The completeness of the impact matrix. An important difficulty in problem definition is to reach a complete impact matrix. Two procedures are included to generate a set of alternatives which is as complete as possible: a procedure to combine aspects of a problem into alternatives and a procedure to aggregate elements into alternatives (see section 4). A set of impact criteria can be obtained through applying a hierarchy of objectives, goals and criteria.
2. Allow low information level. Impact scores are not always available as hard numbers. To be able to include all information DEFINITE accepts low information scales such as the ordinal, binary and nominal scale.
3. Creativity in problem formulation. The user is stimulated and supported to create new alternatives or new combinations of alternatives. Since procedures are available specifically to reduce the number of alternatives at a later stage, the user is not restrained from inventing non feasible alternatives.
4. Presentation versus aggregation. Presentation can replace aggregation in problems with a small number of alternatives or criteria. In these cases the overall impression provided by the presentation, replaces the scores calculated through decision rules.
5. Negotiations. In module D the sensitivity of all of the chosen values and methods can be examined. Robustness intervals can be calculated for weights and scores; within a robustness interval the ranking of alternatives is not sensitive to

changes in scores or weights. This is useful in negotiations if there are different opinions on scores and priorities (see also Rietveld, 1988).

4. DEFINITE DESCRIPTION AND EXAMPLES

DEFINITE comprises five modules, including a wide variety of procedures. Examples of procedures related to the special issues described in the previous section are included below.

The purpose of Module A, Problem Definition, is to obtain an impact matrix which represents the decision problem at hand.

Completeness of the set of alternatives may be achieved by using a combination procedure. The user is requested to identify a number of aspects of the problem at hand and to specify per aspect options for the solution of the problem. The user specifies further which combinations of options are not feasable. The procedure now generates all possible alternatives.

Example

Problem: Two cities Vlissingen and Breskens are situated on opposite sides of an estuary. The type of connection between both cities has to be selected.

Table 1. Input to the combination procedure.

Aspects	Option A	Option B	Option C
Means of transport	Train	Car	
Type of connection	Ferry-boat	Tunnel	Bridge
Capacity (persons/day)	10.000	100.000	

The user specifies that the combination train and ferry-boat and the combination of ferry-boat and a capacity of 100.000 persons/day are not feasible. Options as listed in table 1 and the restriction as specified by the user, results in 10 alternatives (table 2).

Table 2. Output of the combination procedure.

Alt 1	Alt 2	Alt 3	''''	Alt 10
Train	Train	Train		Car
Tunnel	Tunnel	Bridge		Bridge
10.000	100.000	10.000		100.000

Completeness of the set of alternatives can be achieved also by aggregating all possible combinations of elements into alternatives. The user specifies elements and combination rules. The procedure generates all combinations of elements that comply with these combination rules.

Example

Problem: Generation of alternatives for water extraction. The elements are four wells with a capacity of 3.000, 5.000, 2.000 and 6.000 m^3/year (the elements).

Table 3. Input to the aggregation procedure.

	Element 1	Element 2	Element 3	Element 4
Capacity m^3/y	3.000	5.000	2.000	6.000
Land use m^2	200	400	200	700
Groundwater level cm	-5	-10	-4	-15

Combination rule: each alternative must produce at least 10.000 m^3/y and no more than 12.000 m^3/y.

This results in the following alternatives:

Alternative 1 = Element1 + Element2 + Element3

Alternative 2 = Element1 + Element3 + element4

Alternative 3 = Element2 + Element4

The resulting impact matrix as shown in table 4 is derived by summing for each alternative the element scores (the columns of table 3) included in the alternatives.

Table 4. Output of the aggregation procedure.

	Alt.1	Alt.2	Alt.3
Capacity m^3/y	10.000	11.000	11.000
Land use m^2	800	1.100	1.100
Groundwater level cm	-19	-24	-25

Impact scores with a low information level are shown in table 5. In this example two impacts are scored on a low information qualitative scale.

Example

Problem: Traffic congestions occurs on a road between two cities. There are three possible alternatives to solve this problem.

The number of lanes of the road can be raised (Alt.1), a new road with two lanes route can be constructed (Alt.2) or a new highway can be constructed (Alt.3).

The user specifies the names of the alternatives and impact categories, the units of the categories and the scores.

Table 5. An impact matrix.

Impact matrix	Units	Alternatives		
		Alt.1	Alt.2	Alt.3
Costs	mln guilders	-49	-78	-107
Travel time	qualitative	++	+	+++
Landscape	ordinal	3	1	2
Accidents	injured/yr	-14	-20	-10

The number of alternatives can be reduced by eliminating inefficient alternatives.

The purpose of Module B, Problem Presentation, is to present the impact matrix in a form that allows for the ranking of the alternatives without applying a formal decision rule. A graphic presentation of a small impact matrix is shown in figure 3. After ordering the impacts in descending priority the alternatives can be ranked visually (the system can also sort the columns to provide an initial ranking).

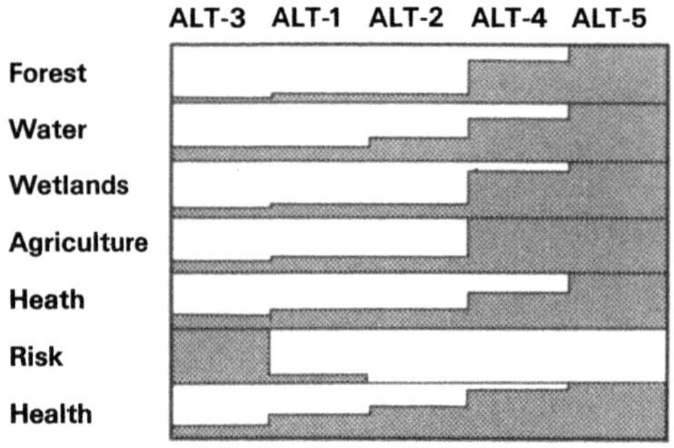

Figure 3. Graphic presentation of an impact matrix.

If an impact matrix is to large, impacts can be combined into impact categories; for example the number of accidents in table 5 may be a weighted sum of the number of accidents with material damage, the number of accidents with injuries and the number of fatal accidents.

In Module C, Problem Evaluation, alternatives are ranked. This is achieved by aggregating the impact matrix, using decision rules. Aggregation can take place via five different multicritera methods and by cost benefit analyses. Several procedures are available to transform information on priorities to weights as needed to perform multicriteria analysis. Since different methods may produce different results it is advisable to apply several methods. In this section we will show results of one method for transforming priority information into weights and results of two multicriteria analyses.

Example

The user is asked to order the impacts in descending priority order e.g. costs are more important than accidents; accidents are more important than travel time etc.. In this example this information is transformed to weights by the expected value method (Rietveld, 1980). By using one of the available multicriteria methods a ranking of the alternatives can be produced. In table 6 the weights resulting from the expected value method and the rankins resulting from the weighted summation and the Electre 2 method (Voogd, 1983) are shown.

Table 6. Results of the weighted summation and the Electre 2 method.

Weights		Weighted Summation		Electre 2	
Ranking	Score	Ranking	Score	Ranking	Score
1. Costs	0.521	1. Alt1	0.80	1. Alt1	3.00
2. Accidents	0.271	2. Alt3	0.45	2. Alt2	1.50
3. Travel time	0.146	3. Alt2	0.42	Alt3	1.50
4. Landscape	0.063				

Module D, Sensitivity Analyses, is included to analyse the sensitivity of the rankings obtained in module C to changes in evaluation methods, impact scores and weights. In this section we will show two procedures for calculating robustness intervals for weights and scores; within a robustness interval the order of two alternatives is not sensitive to changes in scores or weights (cf. Rietveld, 1988). This is usefull in negotiations if there are different opinions on scores and priorities.

Example

A procedure of module D calculates robustness intervals for impact scores. The user selects the impact score and the pair of alternatives to be tested. In table 7 the intervals of the impact score of the number of accidents resulting from Alternative 1 are calculate for the ranking of Alternative 1 and 2 and the ranking of Alternative 1 and 3. The robustness intervals presented in table 7 show that a different score for the number of accidents in Alt1 only affects the ranking of Alt1 and Alt2 if this score exceeds 60, whereas the ranking of Alt1 and Alt3 never changes.

Table 7. Example of an robustness interval for impact scores.

Ranking	Robustness interval of accidents in Alt1
Alt1 > Alt2	(0 , 60)
Alt1 < Alt2	(60 , -->)
Alt1 > Alt3	(0 , -->)
Alt1 < Alt3	empty

Example

Another procedure of module D calculates robustness intervals for weights. The user specify the impact category whose weight is to be tested and two alternatives to compare. The results are presented in table 8, which shows that the order of alternative 1 and 3 is constant between 0 and 0.61 and is reversed in the interval 0.61 - 1.00.

Table 8. Example of an robustness interval for weights.

Ranking	Weight robustness interval of Accidents
Alt1 > Alt3	(0.00 , 0.61)
Alt1 < Alt3	(0.61 , 1.00)

Note: Ratios of all other weights are held constant; weights are standardized so that weights add up to one.

5. CONCLUSIONS

DEFINITE has a bias towards an economic, normative way of decision making. With the exception of graphic presentation all rankings are established through some sort of rational decision rule.

DEFINITE will not be used by decision makers seeking a system that supports other, perhaps more subjective decision procedures. More emphasis should be put on the behavioural aspects of decision making to accomodate for these users.

The possibilities in DEFINITE to move backwards through the modules is in line with notions in Planning theory of the cyclical planning process.

DEFINITE complies with the computer sciences approach to decision support. Since the system contains no standard data it can be described as a DSS shell.

The use of DEFINITE in public planning has clear advantages. The most important are:

- It enables the decision maker to incorporate all relevant information in the decision. It counteracts the unsatisfactory practice of basing a decision only on the number of issues that can be handled by a single individual.
- The flexibility of the system, and specifically the feedback from results to problem content, stimulates "learning by doing". This will increase the understanding of the problem and stimulate the creativity of the decision maker. This reduces the risk of omitting relevant alternatives or criteria.
- DEFINITE can be used to analyse the influence of changes in data or assumptions on the relative performance of alternatives. This is specifically usefull in group decision making under conflict, where there is usually no agreement on problem definition. The use of DEFINITE can delimit the debate to crucial issues (see also Rietveld, 1988; Janssen and Hafkamp, 1986).

There is no doubt that, through an increase in the number of DSS applications, the use of DSS in public planning will increase. This increase will improve flexibility and efficiency of public decision making.

LITERATURE

Andriole, S.J. (ed.) (1986). Microcomputer decision support systems: design, implementation and evaluation. North Holland, QED Information Sciences, Inc.

Ginzberg, M.J., E.A. Stohr (1982). Decision Support Systems: Issues and Perspectives. In: Ginzberg, M.J. et al. (eds.), Decision Support Systems, North Holland, Amsterdam.

Janis, L., L. Mann (1977). Decision Making. The Free Press, New York.

Janssen, R., W. Hafkamp (1986). A Decision Support System for Conflict Analysis. The annals of regional science, Vol. 20, no. 3, pp. 67-85.

Keeny, R., H. Raiffa (1976). Decision with Multiple Objectives. John Wileys, New York.

McGrew, A.G., M.J. Wilson (ed.) (1982). Decision making; approaches and analysis. Manchester University Press.

Newell, A., H.A. Simon (1972). Human problem solving. Prentice Hall, Englewood Cliffs.

Rietveld, P. (1980). Multiple Objective Decision Methods in Regional Planning. North Holland, Amsterdam.

Rietveld, P. (1988). Beslissings Ondersteunend Systeem voor Discrete Alternatieven (BOSDA). Gevoeligheidsanalyses bij multicriteria beslissingsmethoden. Instituut voor Milieuvraagstukken, Vrije Universiteit, Amsterdam.

Sprague, R.H., E.D. Carlson (1982). Building Effective Decision Support Systems. Prentice Hall, Englewood Cliffs.

Sprague, R.H., H.J. Watson (eds.) (1986). Decision Support Systems: Putting theory into practice. Prentice Hall, New Yersey.

Tufte, E.R. (1985). The visual display of quantitative information. Graphics Press, Cheshire, Connecticut.

Voogd, J.H. (1983). Multi Criteria Evaluation for Urban and Regional Planning. Pion, London.

Winterfeldt, D. von, W. Edwards (1986). Decision Analysis and behavioural research. Cambridge University Press, Cambridge.

Wright, G. (1984). Behavioural decision theory. Sage and Harmondsworth, Beverly Hills.

AND/OR GRAPHS, SCENARIOS AND MULTIPLE CRITERIA: PROTOTYPING IN DECISION SUPPORT

G. E. Kersten and G. R. Mallory

Decision Analysis Laboratory
School of Business, Carleton University
Ottawa, Ont., K1S 5B6, Canada

Abstract

Formal tools may be accepted in mental experiments if they are easy to use, responsive and flexible. Assuming that the structure may be easily modified and manipulated we may reduce the simplifications and reductions caused by the modelling. We propose to support ill-structured decision making using rule-based formalism and prototyping-like approach. The user builds a model which does not necessarily encompass the whole decision problem. This model describes one or more parts of the problem. Then, the user develops other models together with possible links between them. At any point of the development/prototyping process a partial problem structure is created. This structure may be verified against user expectations and/or against environmental data, and scenarios and advices may be created.

1. Structuring ill-structured decisions

Complex and ill-structured decision-making is difficult to model and support. A decision problem is ill-structured when it has not been encountered before and when it is a process type decision with a set of actions and dynamic factors that has its beginning and the end (Mintzberg *et al.*, 1976). The dynamic factors are often not known from the outset, the decision-maker's rationality may be context-dependent, the goals are complex and/or ill-defined, and information received during the process influences the problem structure (Simon *et al.*, 1987). Strategic decisions, i.e., these with outcomes, resources and/or people involved being important to the decision-maker, are often complex and ill-structured.

An ill-structured problem has no unique representation nor is its best representation known. Decision-makers may try to decompose the complex problem and obtain subproblems which are manageable so that each subproblem is attacked with one approach (Newell and Simon, 1972). Thus, the decision-maker applies a basic logic to the unstructured problem and, arbitrarily often, he/she reduces it to interrelated subproblems which can be structured.

Decomposition may cause an oversimplification of the original problem because important relations may be lost. In such cases a two stage approach is often applied: first the problem is decomposed and the well defined subproblems are analyzed and solved. Then, a scenario is created that links all the subproblems. A scenario is the result of the imposition of a structure on the problem. A structure which is coherent and fits the available information, and which takes into account effects of the missing information, and the intuition, common sense and preferences of the scenario constructor.

If we look at the decision-making from the process point of view we see that it is an inferencing process in which "we observe the available bits and pieces [of information] and try to figure out what they imply about what we do not observe" (von Winterfeldt and Edwards, 1986; p. 165). The decision process is then a process of analysis of the recognized subproblems, scenario creation, assessment of its merit and value in light of data, and if required, the creation of new scenarios.

The task of constructing a scenario to fit data is the most important kind of inference making but, as von Winterfeldt and Edwards point out, we know almost nothing about it. Lindley (1986) says that decision analysis parallels other measurement problems, and that underlying the analysis are probabilities and utilities that are "true" values for the decision-maker. Applying measurement theory, and utility machinery in particular, to strategic decision-making does not produce expected results. As to do this we must deal with both a structured and well defined problem, and with an individual who does not violate rationality axioms and whose decisions are not affected by such "noises" as intuition and common sense. In the case of strategic decisions the "true" values are not only subjective but also time and context dependent, by creating a scenario and making a decision analysis we may change the true values.

We propose to use rule-based formalism to represent basic properties of a decision problem. This corresponds to the assumption that the decision is decomposable (Simon, 1973) and that problem structures (there may be many possible or feasible structures for an ill-structured problem) is of the form of inference network which is a graphical representation of antecedent-consequent rules (Winston, 1984).

An inference network may be transformed into an AND/OR tree, possibly with repetitive nodes. The variable (consequent) at the root is the (sub)problem's *goal*, and if the decision-maker has more than one goal, the root is an artificial goal. The problem's goal is the consequent of lower level goals, which in turn may be consequents of the third level goal: all the variables at the intermediary nodes, i.e., these which have their antecedents are called goals. Variables (antecedents) at the leaves of the tree are called *facts*.

The AND/OR tree is equivalent to a set of rules $S'_i,\ i \in I$

$$S'_i = \{\alpha_j \Leftarrow p_j(\alpha_k)\ k \in K_j,\ j \in J_i\}, \qquad (1)$$

where: α_j is a variable that describes a goal, J_i is the set of indices of goals in the problem S'_i, α_k describe antecedents of α_j; K_j is the set of indices of all the antecedents of goal α_j. Function p_j is a transformation of antecedents into a consequent and the permissible operators in p_j are: and, or and not. $\Leftarrow$ indicates that α_i is explained by α_k, i.e., the value of α_j is inferred from the values of α_k ($k \in K_j$). We assume that circular relations are not permitted.

2. Alwyn and his decision problem

The case used to illustrate the use of rule-based formalism in generating scenarios is drawn from the data base of the Bradford Studies of decision Making (Hickson *et al.*, 1986) who use this case to open their discussion on strategic decision making. The case is discussed in more detail by Wilson (1982) and the approach is more fully described by Kersten and Mallory (1989a, 1989b).

2.1 Overture

Before 1960, the company, Toxicem (a pseudonym), a medium sized but significant producer of intermediate chemicals, used steam in various phases of its process. This was essentially between 80 and 150 p.s.i. Demand for higher pressure steam was projected to increase and in 1962 the company bought a boiler capable of operating at 400 p.s.i. However, the projected demand did not materialize and the boiler was satisfying requirements at 80 p.s.i. To accomplish that the boiler had to be fuelled up to produce at 400 p.s.i. and the pressure subsequently reduced.

This stimulated the production director, Alwyn, and his engineers to think of ways to minimize the loss by putting this "spare" pressure to productive use. One suggestion was to use the high pressure steam to drive a turbo alternator, to produce electricity which in theory would make Toxicem independent of the state owned and supplied government network.

2.2 First movement

Following these initial thoughts some preliminary investigations confirmed the feasibility of such a project. Despite the realization that Toxicem could not afford the £150,000 necessary for the equipment, the engineers, supported by Alwyn, continued to develop this as a possible capital project. They invited the National Industrial Fuel Efficiency Service (NIFE) to conduct a general survey at the plant. This was an activity that anyway took place every few years. This time, however, NIFE was specifically asked to comment on the economics of electricity generation.

NIFE estimated that savings of around £39,000 per annum would be possible, and Alwyn put together and circulated a paper summarizing the situation. Concurrent with these moves and despite the tight capital situation it had become evident that the demand for steam was now at the point when another boiler would have to be purchased. The Board of Toxicem asked for a detailed proposal.

Alwyn and his supporters saw this as an opportunity to get a high pressure boiler in place so that they could then restate their case for the long term savings from the generation project. However, Giles the purchasing director who was also investigating potential boilers had dismissed the generation idea as a waste of time and effort. He argued that Toxicem was really in the business of chemical production and not electricity generation. Giles favored a low pressure boiler which was £15,000 cheaper than Alwyn's preferred option. Although admitting that there may be possible long term benefits, given Toxicem's current situation, it was obvious that Giles favored short term savings.

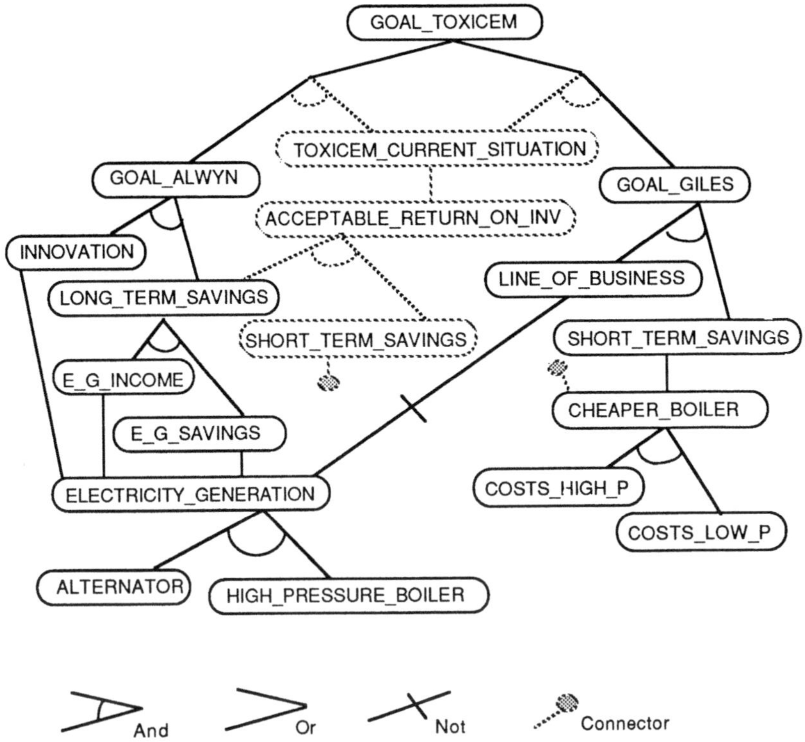

Figure 1. AND/OR graph: Alwyn's problem representation

The situation described above is presented in Fig. 1 as an AND/OR graph (to avoid repetition of nodes we do not transform this graph into an AND/OR tree). The part of this graph marked by solid lines is a graphical representation of the set of *empirical* rules which are as follows:

GOAL_TOXICEM ⇐ ALWYN_GOAL or GOAL_GILES,
ALWYN_GOAL ⇐ LONG_TERM_SAVINGS and INNOVATION,
GOAL_GILES ⇐ LINE_OF_BUSINESS and CHEAPER_BOILER,
LONG_TERM_SAVINGS(DIFFERENCE(POSITIVE)) ⇐ E_G_INCOME(VALUE) and E_G_COSTS(VALUE),
INNOVATION ⇐ ELECTRICITY_GENERATION,
LINE_OF_BUSINESS ⇐ not ELECTRICITY_GENERATION, (2)
E_G_COSTS ⇐ ELECTRICITY_GENERATION,
E_G_SAVINGS ⇐ ELECTRICITY_GENERATION,
SHORT_TERM_SAVINGS ⇐ CHEAPER_BOILER,
SHORT_TERM_SAVINGS(DIFFERENCE(POSITIVE)) ⇐ COST_HIGH_P(VALUE) and COST_LOW_P(VALUE),
ELECTRICITY_GENERATION ⇐ HIGH_PRESSURE_BOILER and ALTERNATOR.

We expanded the notation given in (1) by adding variables to goals and facts. The overall goal of the problem is GOAL_TOXICEM and it is decomposed into two goals, one describing goal of Alwyn (GOAL_ALWYN) and another describing goal of Giles (GOAL_GILES). The overall goal is true if GOAL_ALWYN or GOAL_GILES is true. Alwyn's goal is true if the project brings long term savings and is innovative. Goal LONG_TERM_SAVINGS is decomposed into two goals: savings from electricity generation (E_G_INCOME) and its costs (E_G_COSTS). However, this goal is true if there is a difference between the income value and the cost value and it is positive. This situation is indicated by the parameters DIFFERENCE and POSITIVE. Calculation of the difference is possible if we obtain values of income and costs. Therefore, we added parameter VALUE to goals E_G_INCOME and E_G_COSTS. These values can be calculated if the antecedents of these two goals are true, i.e., if ELECTRICITY_GENERATION is true. Other rules have similar interpretation.

Let us now assume that Alwyn prepared the problem representation of his mental models: one representing his decision problem, and one representing his perception of Giles' mental model. Then, he used rules to build a knowledge base (in fact two knowledge bases: his and Giles'), and used an inferencing engine to operate on this base. He may ask a question: What has to be done to satisfy both Toxicem's and my goals? The answer, obtained through backward chaining is that the purchase of alternator and high pressure boiler is required. Alwyn's next question might be: What does Giles expect? and the answer is, that because of the difference in price of low and high pressure boilers, Giles prefers the low pressure boiler.

Alwyn knows that his scenario, i.e. one that leads to his goal achievement, is not possible. The reason is that "given Toxicem's current situation" short term savings override long term savings. This information, however, has not been introduced to the support model which may suggest that achieving Alwyn's goal is sufficient to achieve Toxicem's goal. Through analysis of the scenario which based on purchase of alternator and high pressure boiler, Alwyn realizes that the support model has to be verified and new rules added to the knowledge base (dotted lines in Fig. 1).

GOAL_TOXICEM ⇐ (ALWYN_GOAL and TOXICEM_CURRENT_SITUATION) or
(GOAL_GILES and TOXICEM_CURRENT_SITUATION),
TOXICEM_CURRENT_SITUATION ⇐ ACCEPTABLE_RETURN_ON_INVESTMENT, (3)
ACCEPTABLE_RETURN_ON_INVESTMENT(VALUE(CLOSE_TO_TWO)) ⇐
SHORT_TERM_SAVINGS(VALUE) and LONG_TERM_SAVINGS(VALUE).

The rules given in (3) make Alwyn's proposal infeasible; the company's current situation (TOXICEM_CURRENT_SITUATION) requires that the return on investment is around to two years (parameter VALUE(CLOSE_TO_TWO)) and Alwyn's estimated return on investment is almost four years (£150,000/£39,000).

2.3 Interlude

Alwyn therefore cannot pursue his project of electricity generation and the immediate reason for this is Toxicem's current situation. However, Alwyn may expect this to change in the future, and use the support model for identifying points of both his and Giles' vulnerability. To do this Alwyn has to introduce other, than empirical, rules. To create scenarios which may guide him he needs inferential rules which do not model the world as empirical rules do (e.g., (2) and (3)), but which produce new empirical rules or control the inferencing.

Alwyn may ask for advice of what should/may be changed in his problem and/or the environment to make electricity generation possible. This requires analysis of the knowledge base through *inferencing* rules like the following:

not HIGH_PRESSURE_BOILER ⇐ SHORT_TERM_SAVINGS and AVAILABLE_DATA, (4)
ALTERNATOR ⇐ HIGH_PRESSURE_BOILER .

The first rule says that because of the goal SHORT_TERM_SAVINGS and the available data on the costs of boilers (new goal AVAILABLE_DATA), the high pressure boiler cannot be chosen. The second rule says that you cannot buy an alternator if you did not buy the high pressure boiler. Using the two rules in conjunction with the knowledge base, Alwyn may obtain the the following scenario:

You concentrate on purchasing a high pressure boiler as a prerequisite of electricity generation. This attempt will be successful if you can prove that there are short term savings and/or that the return on investment is close to two years. This will not lead to a resistance from Giles based on the reasoning that electricity generation is not Toxicem's line of business as long as you do not try to purchase an alternator. To attempt the latter you have to neutralize Giles' subgoal about the line of business.

Purely by coincidence a National Coal Board circular arrived on Giles desk. This stated that they had large stocks of pulverized coal at local pits and could supply it at a very low cost for at least the next 10 years. Giles, never one to pass up a bargain, a trait that had contributed to his reputation as an effective procurer both within and outside Toxicem, decided to investigate the possibilities.

This coal, while cheap, had a very low calorific value and would not be able to raise steam on low pressure boilers. However, it could fire a water tube boiler which provided superheated steam power at 600 p.s.i. The economics were so overwhelmingly in Toxicem's favor that a high pressure boiler was ordered and installed.

2.4 Variation

The boiler operating at 400 p.s.i. later reduced to 80 for the plant reached its capacity about a year later. Alwyn now made his move to have the whole electricity generation debate reopened by suggesting that the boiler be raised to full pressure and the surplus used to drive an alternator. The debate continued for several months until Giles and Tom (the development director) realized that if Toxicem did produce its own electricity there would have to be an agreement negotiated with the electricity supply utility to provide standby facilities if any breakdown should occur. This, they suggested, would be very expensive. Alwyn sought and obtained corroboration of these charges. The capital control commission advised of this, rejected the project as, Giles predicted.

The managing director of Toxicem decided to retire. Within Toxicem there were two candidates for this job, Alwyn and Giles. From Alwyn's point of view Giles having won the battle on the electricity generation project has enhanced his image and was the prime candidate. What could he do to either look good himself or somehow downplay Giles's success?

Due to the above events, the problem representation changed. First, the subproblem of purchasing a high pressure boiler is solved. This require changes in the problem representation which may be done with *restructure* rules (Kersten *et al.*, 1988) which are a type of inferencing rules. The required restructure rules are (::= means "is"):

HIGH_PRESSURE_BOILER ::= *true* $\Rightarrow$ ELECTRICITY_GENERATION $\Leftarrow$ ALTERNATOR,

(5)

HIGH_PRESSURE_BOILER ::= *true* $\Rightarrow$ GOAL_GILES $\Leftarrow$ LINE_OF_BUSINESS and E_G_COSTS.

The first restructure rule says that if the high pressure boiler is burchased then electricity generation is true when an alternator is purchased. This rule may be inferred from a *regulation schema* (Holland *et al.*, 1987) which in this case says that whenever a goal (fact) becomes true the problem's structure has to be changed so that both this goal (fact) and all its antecedents which are not antecedents of other goals are removed from the graph. Applying the regulation schema to GOAL_GILES we obtain that its antecedent is only LINE_OF_BUSINESS. However, the changes in the environment (i.e., Giles's problem representation) introduce new goal E_G_COSTS and this goal is the consequent of the rule:

E_G_COSTS $\Leftarrow$ STANDBY_COSTS and HIGH_GRADE_COAL_COSTS and MAINTENANCE_COSTS. (6)

Second, the top levels goals have changed since the electricity generation project became the mean of achieving the goal of ALWYNS_CAREER and not the means of innovation and long term savings. Alwyn, to become the managing director, has to prove that Giles's goals are against Toxicem's goals. This requires focusing on Giles's goals and not on the whole project. This requirement may be inferred from another type of rules which are similar to *law of large numbers heurestics* (Holland *et al.*, 1987), e.g.:

POLITICAL_PROBLEM $\Leftarrow$ FOCUS_ON_OPPONENTS_CASE and BOTTOM_UP_APPROACH and
CHOICE_OF_EASY_TO_UNDERMINE_GOALS, (7)
ENGINEERING_PROBLEM $\Leftarrow$ FOCUS_ON_YOUR_CASE.

Now, if the system obtains information on the retirement of the managing director it can generate the following advice:

You concentrate on Giles's problem and choose these lower level of goals Giles which you can prove to be false. Then, through forward chaining you try to prove that the overall Giles's perspective is incorrect and, by implication, your perspective to be the correct one.

If Alwyn asks for explanation of the advice, the system responses that because of the new goal (ALWYNS_CAREER) the problem became political, and therefore Alwyn should focus on Giles's case. Moreover, he should analyze Giles's lower level goals (BOTTOM_UP_APPROACH) and choose those which are easy to undermine. The system "knows" that if the value of a lower level goal is changed, then values of its consequents may change. This may lead to proving that Giles's top level goal is false.

Alwyn decided to see if there was anything in the electricity project that he could use. He began to scrutinize all the project documentation and the subsequent debate had generated. It occurred to him and his engineers that the cost of standby supply was unreasonably high and may in fact be an error.

He then did two things. He wrote to the Confederation of British Industry asking for their comments on the charges. He also gave notice to the Capital Control Committee that he wishes to reopen the debate. The reply from the Confederation confirmed that the charge had been inflated and gave a more reasonable basis for standby supply, i.e., Alwyn showed that the goal STANDBY_COSTS has different value than Giles assumed. Alwyn also stated that the maintenance costs have to be higher than those given by Giles, thus he undermined the goal MAINTENANCE_COSTS. Following this he was able to undermine the costs of electricity generation, as given in (6).

3. Conclusions

We have proposed the use of an expert system technology to develop assistant-like decision support. This uses rule-based formalism because of its flexibility, modularity, ability to describe both context-dependent problems and multiple perspective representation within one formalism. These features together with available powerful inference engines and user-friendly interfaces make it possible to use prototyping in system development. It means that the decision-maker can develop the system during the decision process. The system's knowledge-base grows and is modified with the incoming information. Moreover, the system may have the ability to use available quantitative models whenever this is required.

The system-assistant may be intelligent in a sense that it has built in problem-independent knowledge bases. An example of this is given in (7), these rules did not have to be input by Alwyn, but they can be in the knowledge base which is provided with other prototyping tools. Such a knowledge may be derrived from the research and expertise of other decision-makers and customized to the user's requirements. The case which we used to present the approach, is one of 150 analyzed by one of the authors (Hickson *et al.*, 1986). Using some of their conclusions, we may add such rules as:

DISTURB_BALANCE_OF_POWER ⇐ BRING_IN_OUTSIDERS,

PROCESS_ISSUE_THROUGH_ROUTINE_OF_YOUR_CHOICE ⇐ NEW_ISSUE.

The first rule advises that to change the balance of power one should get expertise from outside. Both Alwyn and Giles followed this rule. Following the second rule, Alwyn choose the capital control commision as the "battleground" because he could predict what will happen.

The architecture of the system-assistant was not defined, and neither were all the steps which may lead to problem resolution. We concentrated on how the problem may be structured and restructured due to changes in the environment and/or changes in the user perception, preferences, etc. Examples of simple scenarios, which may be expanded if the knowledge bases are more complex and if the language processor and justifier are well developed were given. These knowledge bases correspond to different levels of quasi-morfism and they may generate more accurate advice and prediction on the future state of the user's world.

We used an example which, although fairly simple, was a strategic decision. We showed how the problem can be structured and restructured, how knowledge which is external to the problem can be included to generate scenarios, how the modelling process can be integrated with the changing information about the environment, and how the decision-maker can progressively develop problem representations. The proposed approach can be integrated with other (e.g., quantitative) approaches and it also can be expanded with other qualitative models proposed within the strategic planning domain.

References

Hickson, D.J., R.J. Butler, D. Cray, G.R. Mallory and D.C. Wilson (1986) *Top Decisions: Strategic Decision Making in Organizations.* Oxford: Blackwell.

Holland, J.H., K.J. Holyoak, R.E. Nisbet and P.R. Thagart (1986) *Induction: Process of Inference, Learning and Discovery.* Cambridge, MA: MIT Press.

Kersten, G.E. and G.R. Mallory (1989a) "Quantitative Models and Scenarion Generation in Strategic Decision-Making", WPS 89-04, School of Business, Carleton University, Ottawa, K1S 5B6, Canada.

Kersten, G.E. and G.R. Mallory (1989b) "Supporting Problem Representations in Decisions with Strategic Interactions", WPS 89-10, School of Business, Carleton University, Ottawa, K1S 5B6, Canada.

Kersten, G.E., W. Michalowski, S. Matwin and S. Szpakowicz (1988) "Representing the Negotiation Process with a Rule-based Formalism", *Theory and Decision.* vol. 25, 225-257.

Lindley, D.V. (1986) "The Reconcilation of Decision Analyses", *Operations Research*, Vol. 34, No. 2, 289-295.

Mintzberg, H. *et al.* (1976) "The Structure of Unstructure Decision Processes", *Administrative Science Quarterly*, vol. 21, 246-275.

Newell, A. and H. Simon (1972) *Human Problem Solving*, Engelwood Clifs, NJ: Prentice-Hall.

Simon, H.A. (1973) "The Structure of Ill Structured Problems", *Artificial Intelligence*, Vol. 4, No. 3&4.

Simon, H.A. *et al.* (1987) "Decision making and Problem Solving, *Interfaces*, Vol. 17, No. 5, 11-31.

von Winterfeldt, D. and W. Edwards (1986) *Decision Analysis and Behavioral Research*, Cambridge, MA: Cambridge Univ. Press.

Wilson, D.C. (1982) Electricity and Resistance: A Case Study of Innovation and Politics, *Organization Studies*, Vol. 3, No. 2, 119-139.

Winston, P.H. (1984) *Artificial Intelligence*, Addison Wesley.

A FRAMEWORK FOR BUILDING AN EXPERT SYSTEM FOR MCDM MODELS SELECTION

M. Tawfik Jelassi
European Institute of Business Administration
77305 Fontainebleau Cedex, France

Vladimir M. Ozernoy
California State University, Hayward
Hayward, California 94542, U.S.A.

ABSTRACT

This paper discusses the purpose, structure, and software capabilities of a rule-based expert system for multiple criteria decision making (MCDM). The expert system combines information about available discrete alternative MCDM methods with procedures for guiding the user through the analysis of the decision situation and supporting him/her in selecting the most appropriate method.

1.0 INTRODUCTION

Recent advances in microcomputer hardware and software have stimulated the implementation of some of numerous MCDM methods using a decision support system framework as the one suggested by Sprague and Carlson (1982). Typically, these systems, referred to as multicriteria decision support systems (or MCDSS), assist interactively the decision maker in structuring the problem at hand and in evaluating/comparing decision alternatives. [For more detail, see, for example, Bui (1984) and Jelassi (1987)].

In this paper, the focus is on discrete alternative MCDM methods. For solving a particular decision problem, several MCDM methods may apply. However, they often represent radically different approaches to decision making, and the choice among them depends on a number of factors including the specific problem under consideration, the decision maker's MCDM knowledge, etc. The selection of the most appropriate MCDM method is an MCDM problem in itself.

This paper reports on an ongoing research effort intended to develop an expert system that advises decision makers in the MCDM method selection. This work involves three tasks: (1) identification and acquisition of knowledge about discrete alternative MCDM methods, (2) representation of the acquired MCDM knowledge, and (3) development and testing of a prototype MCDM expert system.

Section 2 of this paper briefly reviews the evolution of MCDSS with special focus on fifth generation systems which are knowledge-based tools. Section 3 deals with the design of an MCDM expert system. It addresses technical issues dealing with MCDM knowledge acquisition and representation and the reasoning strategies that can be employed. Section 4 outlines some implementation factors related to the building of a system prototype. Section 5 ends the paper with some concluding remarks.

2.0 THE FIFTH GENERATION MCDSS

Jelassi (1987) investigated the evolution of MCDM method implementations from "stand-alone" tools (referred to as First Generation MCDSS) to knowledge-based decision support systems (referred to as Fifth Generation MCDSS). A First Generation MCDSS is characterized by the use of only one MCDM method, a traditional file management system, and a rudimentary user interface. In this category, each MCDSS has components that operate independently of each other.

In the Second Generation MCDSS, data files are integrated and handled by a database management system (DBMS). The DBMS, in addition to performing basic data-handling operations, offers sophisticated capabilities for query processing, view generation, and data exchange and conversion [see Jelassi (1985) and Jelassi et al. (1985a).

Third Generation MCDSS contain an enhanced user interface. This component, with its associated dialogue generation and management software, provides menu handling, screen composition, and graphic facilities (Jarke et al., 1984).

Fourth Generation MCDSS move one step further toward a full integration of the system components by putting all the MCDM methods together in a model base. The model, data, and dialogue components provide the basis for building MCDSS that offer MCDM model builders and users many new important features. For example, a library of MCDM methods is available for decision makers; it provides for each method the required inputs, the output data, the technique(s) used, etc. Stored models can interact with each other and share information such as subroutines and statements. [A complete discussion of Fourth Generation MCDSS features can be found in Jelassi et al. (1985b)].

Fifth Generation MCDSS are knowledge-based expert systems. In addition to the three basic components (MCDM model bank, data base, and user interface), their structure contains a knowledge base, an inference engine, and an explanation system. Briefly put, the knowledge base contains knowledge about the MCDM domain. The inference engine is a general-purpose problem solver. It uses the knowledge base to make inferences and arrive at possible conclusions. The explanation system provides the motivation, justification, or rationale for the selection of a particular MCDM method. [For more detail, see Jelassi (1987)].

3.0 DESIGN OF AN MCDM EXPERT SYSTEM

It is the authors' belief that building an MCDM expert system should be based on methodologically-sound principles. To achieve this goal, three major tasks need to be undertaken:

Task 1. Identification and acquisition of MCDM knowledge.

Task 2. Representation of the acquired knowledge.

Task 3. Development and testing of an MCDM expert system prototype.

3.1 MCDM Knowledge Identification and Acquisition

The purpose of Task 1 is to determine what kind of MCDM knowledge is required to design and develop an MCDM expert system that will guide the user through the analysis of the decision situation making it possible for him/her to select the most appropriate MCDM method. Here MCDM knowledge, accumulated by researchers and practitioners in the field, needs to be identified and structured. This includes, but is not limited to, a list of available MCDM methods, assumptions that these methods make, their description, information requirements, findings of descriptive research, and applications. In addition, attributes for the quantification of MCDM methods and decision situations need to be developed.

There exists a large number of books, papers, and technical reports dealing with various aspects of MCDM. Based on a review of the relevant literature and discussions with leading decision analysts and MCDM researchers, a comprehensive list of available discrete alternative MCDM methods needs to be developed. This list serves as a basis for answering the following questions: (1) How to develop a taxonomy of MCDM methods that can be used as the basis for defining the content of the MCDM knowledge base in terms of facts and heuristic rules? (2) How can MCDM knowledge be extracted from MCDM literature and MCDM experts? (3) What are the objectives to be achieved by various methods? (4) What are the characteristics for the evaluation of decision situations? For a preliminary list of discrete alternative MCDM methods, see Ozernoy (1988).

The development of a comprehensive MCDM taxonomy could be based upon the work of Ozernoy and Gaft (1975), where the complexity of decision maker's prference information is used as a uniform classifying yardstick. The classification scheme allows one to identify a sequence of steps for obtaining the necessary preference information. Moreover, through a review of MCDM methods and discussions with MCDM experts, information presented to and supplied by decision makers for all available methods should be identified. This information could be identified in the form of specific questions posed to and by decision makers as well in a more generalized form. An example of a specific question submitted to the decision maker is (Keeney, 1977): "Consider an option leading to 700 fatalities and 2000 acres of land permanently unusable; both attributes are at their worst levels. Would you rather move up to 100 fatalities or 0 acres of land?" An example of a more generalized form of preference information is as follows: "Make a choice between two alternatives, A and B. Alternative A has the first attribute at its best level and the second attribute at its worst level. Alternative B has the first attribute at its worst level and the second attribute at its best level. All other attributes are at their best levels".

As part of Task 1, the assumptions of various MCDM methods also need to be determined. While for some of the available methods the underlying assumptions are explicitly formulated, for many others such assumptions are rather difficult to specify. Therefore, the user of a particular method is sometimes not aware about its rationale. For example, while Saaty's AHP method (1980) has been widely used for a number of years, the assumptions of the method were discussed only resently in Harker and Vargas (1987).

In the process of translating MCDM expertise to a form suitable for developing an expert system, two standard methods for extracting the knowledge from the human expert could be adopted: (1) using the present MCDM expert system investigators as both knowledge engineers and experts relying on introspection to articulate the requisite knowledge, and (2) conducting intensive interviews with the MCDM domain experts. The acquired MCDM knowledge will be encoded in Task 2 using a proper knowledge representation technique.

The overall objective of an MCDM method is to provide both a framework and techniques to order the set of feasible alternatives in accordance with the decision maker's preferences. The specific objectives to be achieved by various MCDM methods need to be identified and the attributes to measure the degree of attainment of each of these objectives arranged in a hierarchical structure. The three highest-level objectives as formulated in Ozernoy (1987) are: (1) mutual correspondence between the MCDM method and the decision situation, (2) mutual correspondence between the MCDM method, assumptions about the decision maker's preferences, and preference information required by the method, and (3) resource demand of the MCDM method. Each of these major objectives should be divided into several lower-level objectives that can be considered as criteria for the selection of the most appropriate method for a given decision situation. These criteria provide a set of screening and evaluation criteria based on the professional judgment of the user and/or the MCDM expert involved in the development of the MCDM expert system. Screening criteria allow the elimination of MCDM methods not suitable for a particular decision situation. Evaluation criteria are used to compare suitable MCDM methods.

This task also involves the development of a hierarchy of characteristics for the evaluation of decision situations. This hierarchy contains attributes for matching discrete alternative MCDM methods with decision situations. These situations could be characterized with three dimensions related to (1) the problem at hand, (2) the decision maker, and (3) resource constraints (Ozernoy, 1987). Obviously, specific attributes under these three general dimensions as well as their scales need to be developed.

3.2 MCDM Knowledge Representation

Task 2 of the proposed research strategy is intended to determine which of the existing knowledge representation techniques is most suitable for the development of an MCDM expert system. The representational framework adopted can play an important role in the stages of information elicitation, interpretation, and integration into the framework.

There exist several knowledge representation techniques, any of which can be possibly used to build an MCDM expert system. At present, the most common type of knowledge representation technique is rule-based, it cosists of factual knowledge and a set of if-then rules. Dhar and Pople (1987) argued that the basic advantage of the rule-based representation scheme is the modularity that rules provide and the simple uniform interpretive procedure that is often sufficient in rule-based systems.

A comparison of the rule-based representation of domain-specific MCDM knowledge with other representation techniques such as semantic networks, frames, and object-oriented form (Waterman, 1986) would shed some light on the appropriateness of each technique. In selecting the most suitable knowledge representation for the development of an MCDM expert system, a primary consideration should be the degree of certainty that significant MCDM knowledge will not be lost. Another important factor to take into account is the availability of inexpensive microcomputer-based expert system shells suitable for the implementation of MCDM expert systems.

This task also involves the evaluation of available inference strategies that could be used in the MCDM expert system. Possible choices are modus ponens, reasoning about uncertainty, and resolution (Harmon and King, 1985). Modus ponens is a common rule for deriving new facts from rules and known facts. Reasoning about uncertainty allows for handling incomplete information. Resolution is a way to discover whether a new fact is valid given a set of logical statements. When selecting the most suitable inference strategy, factors such as content, structure, and volume of the MCDM knowledge (acquired in Task 1) should be taken into account.

3.3 Reasonong Strategies for the MCDM Expert System

Task 3 is intended to determine the appropriate control mechanism for the MCDM expert system. Possible choices include backward vs. forward chaining, depth-first vs. breadth first search, and monotonic vs. nonmonotonic reasoning (Hayes-Roth et al., 1983; Harmon and King, 1983). In order to determine the best control strategy, a set of both user- and computer-related criteria needs to be identified and applied.

A specific activity in this context is to compare backward and forward chaining to demonstrate which one of these strategies is best suited to the needs of the MCDM expert system. Using both strategies in the system should not be ruled out as a possible alternative.

Backward chaining is an inference method where the system starts with what it wants to prove. In an MCDM expert system, backward chaining can be used to check whether the principles involved in selecting a particular MCDM method match those the decision maker thinks he/she applied in the selection process. With this approach, the system starts with the hypothesis that a particular MCDM method is the most appropriate one for a given decision situation and reasons backwards, looking for supporting facts and rules to support it, switching to a second hypothesis when the first one fails.

Forward chaining is an inference method where the if-portion of rules is matched against facts to establish new facts. In an MCDM expert system, a forward chaining strategy can be used to assist the user in selecting the most appropriate MCDM method. With each step, the system draws closer to a conclusion.

4.0 COMPUTER IMPLEMENTATION OF A SYSTEM PROTOTYPE

The development of an MCDM expert system is guided in this research by an extensive methodological analysis of the selection problem carried out in the preceding two tasks. The results obtained in these tasks are used to develop a prototype MCDM expert system. This involves the following steps: (1) identification of the scope of the system and its capabilities, (2) selection of an appropriate commercially available expert system shell, (3) development of a prototype MCDM expert system, (4) identification of methodological and technical difficulties that may arise during the computer implementation phase, (5) evaluation and enhancement of the system, and (6) testing and validation of the system.

The MCDM expert system prototype is intended to guide the user through the analysis of the decision situation. The analysis activity involves several steps. First, screening criteria specified in Task 1 are used to eliminate non-suitable methods for the decision situation at hand. For example, if only incomplete information can be obtained from the decision maker, multiattribute value functions may prove to be infeasible.

If there are no feasible methods, the expert system will explicitly state so. In addition, the system can recommend to change the representation of the decision situation. For example, if no method is available for rank-ordering a set of feasible alternatives, the problem can sometimes be reformulated and an MCDM method allowing for partial ordering may prove to be appropriate.

If some methods are feasible for the given decision situation, the attribute levels specified by the user are compared with the attribute levels of feasible methods. The output of the comparison will be a subset of methods that match or exceed the decision situation requirements in terms of all attribute levels. This subset may contain no method, or one or several methods.

If the subset contains only one method, this method is the one considered most preferred by the MCDM expert system for the decision situation at hand. If the subset contains several methods, the final choice of the method will be made by the user or by the system based on additional criteria. As an example of such criteria, the user may choose the method that he/she has the most faith in, or is the most familiar with, etc. Alternatively, an appropriate MCDM method can be used to evaluate and compare feasible methods. The testing of the expert system prototype will demonstrate the usefulness of such a tool in effectively matching the characteristics of the decision situation with those of an appropriate MCDM method.

5.0 CONCLUDING REMARKS

The overall objective of the research framework described in this paper is the advance of systematic operational methodologies for selecting the most appropriate MCDM method for a given decision problem. The basic result will consist of the development of an MCDM expert system. The latter combines information about available discrete alternative MCDM methods and a particular decision situation with tools for drawing inferences and reaching conclusions about that information.

A very important feature of the system is that it is a decision aid designed and implemented with close attention to the user's behavioral constraints. That is, the system explicitly takes into account the user's ability to provide particular types of preference information (e.g., the user's ability to answer specific questions). The system therefore selects the MCDM method that is both operational and consistent with the user's capabilities.

The MCDM expert system will address some of the major difficulties in MCDM and thus it will be useful in helping to improve the decision-making process. In particular, this system will enable the development of Fifth Generation MCDSS mentioned in Section 2. Also of significant importance will be a better understanding of the usefulness and limitations of a wide range of MCDM methods.

The MCDM expert system will perform some of the important functions of a decision analyst helping people make better informed decisions and giving them greater confidence in their decisions. The system will also provide a training facility for unexperienced users. In addition, the use of the system may give behavioral scientists new ideas and new possibilities. For example, the system will extend the possibilities of experimental investigation of the relation between decision justification and MCDM methods. It will also facilitate the determination of whether people's behavior in selecting the most appropriate MCDM method complies with the assumptions of normative theories. Thus, the MCDM expert system may find broad application not only in normative decision making, but also in descriptive decision research.

REFERENCES

Bui, T. "Building Effective Multiple Criteria Decision Models: A Decision Support System Approach", Systems, Objectives, Solutions,

Vol. 4, 1984, pp. 3-16.

Dhar, V. and H.E. Pople. "Rule-Based versus Structure-Based Models for Explaining and Generating Expert Behaviour", Communications of the ACM, Vol. 30, No. 6, 1987, pp. 542-555.

Harker, P.T. and L.G. Vargas. "The Theory of Ratio Scale Estimation: Saaty's Analytic Hierarchy Process", Management Science, Vol. 33, No. 11, 1987, pp. 1383-1403.

Harmon, P. and D. King. Artificial Intelligence in Business. New York, New York, Wiley, 1985.

Hayes-Roth, R., D.A. Waterman, and D.B. Lenat (Eds.), Building Expert Systems. Reading, Massachusetts, Addison-Wesley, 1983.

Jarke, M., M.T. Jelassi, and E.A. Stohr. "A Data-Driven User Interface for a Generalized Multiple Criteria Decision Support System", Proceedings of the IEEE Workshop on Languages for Automation, New Orleans, Louisiana, 1984, pp. 127-133.

Jelassi, M.T. "An Extended Relational Database for Generalized Multiple Criteria Decision Support Systems". Ph.D. Dissertation, Graduate School of Business Administration, New York University, 1985.

Jelassi, M.T., M. Jarke, and A. Checroun. "A Database Approach for Multiple Criteria Decision Support Systems (MCDSS)", in G. Fandel and J. Spronk (Eds.), Multiple Criteria Decision Methods and Applications. New York, New York, Springer-Verlag, 1985a, pp. 227-244.

Jelassi, M.T., M. Jarke, and E.A. Stohr. "Designing a Generalized Multiple Criteria Decision Support System", Journal of Management Information Systems, Vol. 1, No. 4, 1985b, pp. 24-43.

Jelassi, M.T. "MCDM: From 'Stand-Alone' Methods to Integrated and Intelligent DSS", in Y. Sawaragi, K. Inoue, and H. Nakayama (Eds.), Toward Interactive and Intelligent Decision Support Systems, Vol. 2, New York, New York, Springer-Verlag, 1987, pp. 90-99.

Keeney, R.L., "The Art of Assessing Multi-Attribute Utility Functions", Organizational Behavior and Human Performance, Vol. 19, 1977, pp. 267-310.

Ozernoy V.M. "A Framework for Choosing the Most Appropriate Discrete Alternative Multiple Criteria Decision Making Method in Decision Support Systems and Expert Systems", in Y. Sawaragi, K. Inoue, and H. Nkayama (Eds.), Toward Interactive and Intelligent Decision Support Systems, Vol. 2, New York, New York, Springer-Verlag, 1987, pp. 56-64.

Ozernoy, V.M. "Some Issues in Designing an Expert System for Multiple Criteria Decision Making", Acta Psychologica, Vol. 68, 1-3, 1988, pp. 237-253.

Ozernoy, V.M. and M.G. Gaft, "Multi-Criterion Decision Problems", in D.E. Bell, R.L. Keeney, and H. Raiffa (Eds.), Conflicting Objectives in Decisions, New York, New York, Wiley, 1975.

Saaty, T. The Analytic Hierarchy Process: Planning, Priority Setting, Resource Allocation. New York, New York, McGraw-Hill, 1980.

Sprague, R.H. Jr. and E.D. Carlson. Building Effective Decision Support Systems. Englewood Cliffs, New Jersey, Prentice-Hall, 1982.

Waterman, D.A. A Guide to Expert Systems. Reading, Massachusetts, Addison-Wesley, 1986.

An Expert System for Assigning Investment Quality Ratings Using the AHP

Venkat Srinivasan and Paul Bolster
Finance and Insurance Group
College of Business Administration
Northeastern University
Boston, MA 02115

Investment Quality Ratings: An Introduction

A vast majority of security issues in financial markets are evaluated by rating agencies and assigned quality ratings. These ratings provide investors with an indication of default risk associated with the securities. Ratings are assigned to numerous types of financial assets including commercial paper, bonds, sovereign government debt, common and preferred stock, and securities offered on euromarkets such as eurobonds and eurocommercial paper.

Investors rely heavily on the ratings provided by rating agencies such as S&P, Moody's, and Fitch to assess the creditworthiness of the security issuing firms. By obtaining a rating from a major agency, a firm vastly broadens the market for its securities, as institutional investors (at least in the U.S.) typically have a legal obligation to purchase only rated issues. Another reason to strive for a favorable rating is the associated reduction in interest expenses. Top rated bond issues often pay interest at rates 1%-3% below those required for bonds with similar maturities but inferior ratings.

Firms must pay a fee to obtain a rating. Ratings are extant until reviewed by the rating agency or withdrawn by the firm. Formal reviews of existing ratings are typically triggered by a perceived change in the fundamental rating factors. These reviews are infrequent, perhaps twice yearly, and often lag perceived changes in investment quality evidenced by changes in the market price for rated securities.

The crucial implications of these ratings for investors and issuers in determining the interest level of interest payment both parties will agree to has led many financial researchers to examine perceived determinants of bond ratings. These studies focus on the identification of the objective, or "hard" accounting information that apparently underlies much of the rating process. Such studies typically rely on multivariate statistical methodologies ranging from the parametric multiple discriminant analysis to nonparametric partitioning methods. These studies develop predictive models that presumably replicate the rating process.

Classification accuracy of these studies ranges from 60%-80% in holdout examples (Srinivasan et. al., 1986).

While rating prediction models based on objective information are quite useful, two sets of deficiencies can be identified with this line of research. First, they ignore the explicit claim of rating agencies that qualitative data also plays an important role in rating determination. This explains the nontrivial misclassification rate produced by models using only quantitative variables. A complete representation of the rating process must incorporate these qualitative factors that obviously influence the result. Second, the rating prediction models assume that ratings are assigned consistently over time or across issuing firms. They provide no normative framework for the rating agencies.

The value of an expert bond rating system should be considerable to all interested parties: regulators, investors, issuing firms, and rating agencies. An expert system can provide a "real time", or spontaneous rating instead of the discrete review process currently employed. Sensitivity analysis is also facilitated by an expert system. The response of ratings to changes in various factors can be modeled and thus anticipated. While the subjective judgements inherent in the production of such ratings are not eliminated by an expert system, the role of such judgements can be captured and formalized. This will improve the consistency of the rating process by eliminating potential bias or error when rating a group of firms at a single point in time or a single firm at several points in time.

In this paper we present a formal judgmental model of the bond rating process using the Analytic Hierarchy Process (AHP) relying on publicly available descriptions of the rating process (S&P, 1986). We describe an AHP-based expert support system prototype for bond rating assignment under development. The next section contains the AHP model of the industrial bond rating process adopted by Standard & Poor's Corporation.

A more detailed version of this paper is available from the authors.

The Rating Process as an AHP Hierarchy

AHP is a multiattribute modeling approach that has been applied to many varied problems including planning for a national waterway (Saaty and Vargas, 1980) and marketing decisions (Wind and Saaty, 1980). Srinivasan and Kim (1986, 1987) illustrate the applicability of AHP to many financial decisions. Recently, a major criticism of the AHP has been remedied by Saaty (1986) by provision of an axiomatic treatment of the methodology. Like other multiple criteria decision models (MCDM), AHP also attempts to resolve conflicts and to analyze judgements by determining the relative importance of a set of activities, players, or criteria.

An overall hierarchical representation of the rating process used by S&P for rating industrial bonds is presented in Exhibit 1. The hierarchy has five levels. At the top, the major goal of the rating process is to assess the debt servicing capability of the issuer. Debt servicing capability is defined as the firm's ability to make timely payments of principal and interest. Three major factors form the second level of the hierarchy: Industry Analysis, Management Evaluation, and Financial Analysis. Each of these factors is further decomposed into nine specific criteria and form level three. In turn, each of the nine criteria is assessed in the fourth level using a detailed set of subcriteria. The fifth and final level consists of the possible rating categories. For convenience, and without loss of generality, we consider only seven categories. From best to worst, these categories are: AAA, AA, A, BBB, BB, B, and CCC. Additional detail on categories is available from S&P (1986). Appendix 1 contains a complete list of factors, criteria, and subcriteria. A brief summary used to evaluate each of the major factors follows.

I. Industry Analysis

Industry Analysis is broken down into three criteria: *Industry Risk*, *Market Position*, and *Operating Efficiency*. *Industry Risk* is further divided into ten subcriteria. For example, these include importance in the economic cycle (a reference to the industry's leadership in the entire economy) and the closely related subcriteria of business cyclicality (consideration of the volatility and predictability of both revenues and earnings). Analysis of *Industry Risk* also involves supply factors. Specifically, how secure are sources of raw materials and labor? Demand considerations such as sufficient or excess plant capacity are also relevant.

The *Market Position* criterion is influenced by subcriteria that measure the individual firm's ability to retain or improve market share. Analysis of this potential requires analysis of both historical and projected future sales trends. Positive and negative aspects of the firm's distribution network are examined at the regional, national, and international levels, as are support and service functions. Other subcriteria evaluate the firm's product lines, marketing flexibility, and relationship with customers.

Operating Efficiency within the industry reflects the control it has over prices and costs. Hence, subcriteria examine such attributes as the ability to maintain or improve operating margins and the age and efficiency of plant and equipment. Cost of production inputs, raw material and labor are considered here. Relationship with employee unions or other organizations is captured here as well.

To summarize, firms will be rated favorably in the Industry Analysis segment of the rating process if they are growing, efficient leaders in a healthy industry. Ability to accurately forecast future earnings for both the industry and the issuer is a necessary although not sufficient condition for a good rating.

II. Management Evaluation

Management determines to a large extent the future prospects of the firm. Managers make decisions that affect the firm's future earnings stream, and in doing so, they influence the creditworthiness of the firm. Certainly, a highly talented management team can't extract above average earnings growth over an extended period of time from a firm in a dying industry. But, managerial effectiveness is still of fundamental importance in the rating process.

This evaluation proceeds through examination of past performance (operating and financial results). Attention is given to earnings growth both in absolute terms and relative to industry and non-industry peers. The future direction of the firm is crucial to its debt servicing capability. Therefore, the planning function is examined from both financial and strategic perspectives. The rating agency is concerned with how realistic management's stated goals are. In addition, management's commitment to past goals and plans, its consistency, and its credibility are also relevant qualitative assessments required in this segment of the rating process.

III. Financial Analysis

The Financial Analysis factor addresses the financial health of the firm explicitly. Five criteria are examined: *Accounting Quality, Earnings Protection, Financial Leverage and Asset Protection, Cash Flow Adequacy*, and *Financial Flexibility*. Briefly, these criteria, also broken down into subcriteria, analyze the more quantitative aspects of creditworthiness. *Accounting Quality* captures the consistency of accounting and auditing procedures used by the firm. *Earnings Protection* comprises analysis of the firm's historical rates of return on capital, profit margins, and earnings by business segments. Sources of future earnings growth and effects of inflation are also considered here.

Financial Leverage and Asset Protection and *Cash Flow Adequacy* also consider the firm's current level of debt, its maturity structure, and the cash needed to service it. These criteria examine the nature of the firm's assets (are they marketable? intangible?) and total cash flows that they provide and consume.

Financial Flexibility entails an assessment of the firms ability to alter financial plans. For example, a growing firm that must raise external funds frequently has little flexibility, whereas a firm that is able to finance a significant amount of its growth internally has considerable flexibility.

BONDRATE: An Expert System Prototype for Assigning Industrial Bond Ratings

BONDRATE is an experimental expert system prototype for assignment of ratings to industrial bond issues. The purpose of BONDRATE is to provide intelligent processing and analytical support to analysts. The system enables the rating agency to use expert weights consistently across ass issues to evaluate the numerous criteria outlined in the rating hierarchy. This section describes the design, capabilities, and features of the prototype version of BONDRATE.

The initial design of BONDRATE was accomplished in 4 steps:

1. Study the ratings decision process.
2. Choose the initial design for model base and data base.
3. Extract expert knowledge.
4. Implement initial design.

For the initial prototype design, the rating process was primarily studied through publicly available descriptions of the rating process (S&P, 1986; FERF, 1983). It appears that within the major rating agencies, the rating process is managed through a hierarchy of rating personnel with different groups of analysts assigned to specific industries. Each group then is responsible for rating all security issues that belong to the respective industry.

Many expert systems operate in a query mode where the system prompts the user with a series of questions in its (most efficient) search for a solution. As the user responds to each question, the system tracks the appropriate rule and returns with the next question until it can reach a conclusion with the available information. Other expert systems (e.g., Srinivasan and Ruparel, 1988) are heavily data oriented. In these systems, data is input on standard input layouts and stored in a data base. At the time of processing, the system searches the data base for available information. For many financial expert systems, this data base mode is more relevant since financial analysis is often data intensive.

BONDRATE fits the data base mode. The initial model base design for BONDRATE consists of the AHP and simple spreadsheet type models to facilitate financial analysis. Data base design is relational in nature. The data base structure is divided into 3 major compartments: Industry data base, Company specific data base, and Comparison data base. The industry data base, common to all firms in a given industry, contains linked files for financial and nonfinancial data for that industry. Financial measures can be input directly using Standard Industrial Classification (SIC) code or computed by aggregating over firms that are affiliated with a specific range of SIC codes. The company specific data base contains three linked data files: Financial, Marketing, Management. Finally, the comparison data base contains the pairwise comparisons for elements at various levels of the hierarchy.

Knowledge acquisition for BONDRATE is being accomplished in two phases. First, the prototype is being designed to analyze the judgmental comparison matrices that are input by analysts. Along with the AHP derivation of criteria weights from these comparisons, the system also provides high-end analytical support by generating intelligent analytical narratives on the issuer's financial condition, quality of management, and overall industry environment. The first phase knowledge base constitutes the system's core knowledge. In the second phase, knowledge acquisition will focus on designing rules to replicate the analyst's analytical processes. Successful execution of this phase will eliminate the need for the analyst to input comparisons. The system will generate comparison matrices.

BONDRATE is being implemented both in IBM and Macintosh environments using 4th generation data base software: Clarion for the IBM and 4th Dimension for the Macintosh. The user-system interface is entirely menu driven providing for easy and effective operation. Menus and options are briefly described below:

The Main Menu:

Data: To perform data management function including adding, editing.

	deleting, and printing data in the three major data bases.
Analyze:	To analyze available firm data and display results on the screen.
Report:	To analyze available firm data and print the results using a connected line or laser printer.
What If:	To perform sensitivity analysis on conclusions and to assess the effects on rating recommendations.
Misc.:	To perform miscellaneous housekeeping functions including calculation of industry averages, designing and mailing of form letters, etc.
Tickler:	To track and remind analyst of actions due on each rated issue.

The Data Sub-Menu:

This sub-menu comprises BONDRATE's main data handling capabilities. It is further divided into 3 options to allow access to the 3 different types of data bases: industry, company specific, and comparison. The user has the option to add, edit, delete, or print various sets of data.

The Analyze Sub-Menu:

This menu comprises BONDRATE's analytical capabilities. The user is provided with 4 options:

Rating:	To analyze all available data on a security issue and provide a rating recommendation along with a narrative summary of rating rationale.
Rating Only:	To analyze data and suggest ratings for a group of firms without the detailed narrative given by the option above.
Financial:	To analyze the financial data relating to a firm, provide a listing of relevant accounting based performance measures, and narrative.
Management:	To analyze the background and management of a firm including a detailed narrative on management performance and quality.

The Report Sub-Menu:

This menu comprises BONDRATE's report writing capabilities. The user is provided with the same 4 options shown in the Analyze menu.

The What If Sub-Menu:

This allows the user to conduct sensitivity analysis using several options:

Industry:	Perform "what if" analysis on industry membership of a firm.
Global Weights:	Perform sensitivity analysis on global AHP weights at various levels of the hierarchy. This option is password protected to restrict access to analysts of requisite authority.

Comparison: Perform sensitivity analysis on comparison data for ratings level of the hierarchy.

Financial Project firm's financials and produce rating based on projections.

The Misc. Sub-Menu:

This menu allows the user to edit and print form letters to issuing firms. For example, common requests for additional information could be printed from this menu. Another option allows the user to update industry norms based on firm specific data stored in the data base.

The Tickler Sub-Menu:

This option provides a calendar of upcoming deadlines concerning rating review dates, etc. Other options include a notepad from which memos can be composed.

An illustration of the rating recommendation from BONDRATE for Chrysler Corporation is enclosed as Exhibit 2.

Summary

Investment quality ratings are important to issuers and investors alike. To date, research into the nature of the rating process has been focussed on the more quantitative inputs. We provide a more complete representation of the ratings process by incorporating the qualitative aspects as well. BONDRATE is a prototype expert system that generates bond ratings for industrial firms using an AHP model of the bond rating process.

References

Financial Executives Research Foundation (1983) Rating Industrial Bonds, NY: Financial Executives Research Foundation (FERF).

T.L. Saaty (1980) The Analytic Hierarchy Process, New York: McGraw-Hill.

T.L. Saaty (1986) "Axiomatic Foundations of the Analytic Hierarchy Process", Management Science, July, 841-855.

T.L. Saaty and L.G. Vargas (1982) The Logic of Priorities, Boston: Kluwer-Nihoff.

V. Srinivasan and B. Ruparel (1988) "CGX: An Expert Shell for Designing Credit Granting Systems", EURO TIMS, Paris, July.

V. Srinivasan and P.J. Bolster (1987) "Predicting Commercial Paper Ratings of Industrial Corporations", Financial Management Association Meetings, Las Vegas, October.

V. Srinivasan and Y.H. Kim (1986) "Financial Applications of the Analytic Hierarchy Process", TIMS International Meetings, Gold Coast, Australia, July.

V. Srinivasan and Y.H. Kim (1987) "Credit Granting: A Comparative Analysis of Classification Procedures", Journal of Finance, July, 661-680.

Standard and Poor's Corporation (1986) CreditOverview: Corporate and International Ratings, New York: Standard and Poor's Corporation.

Exhibit 1
The Ratings Process As An AHP Hierarchy

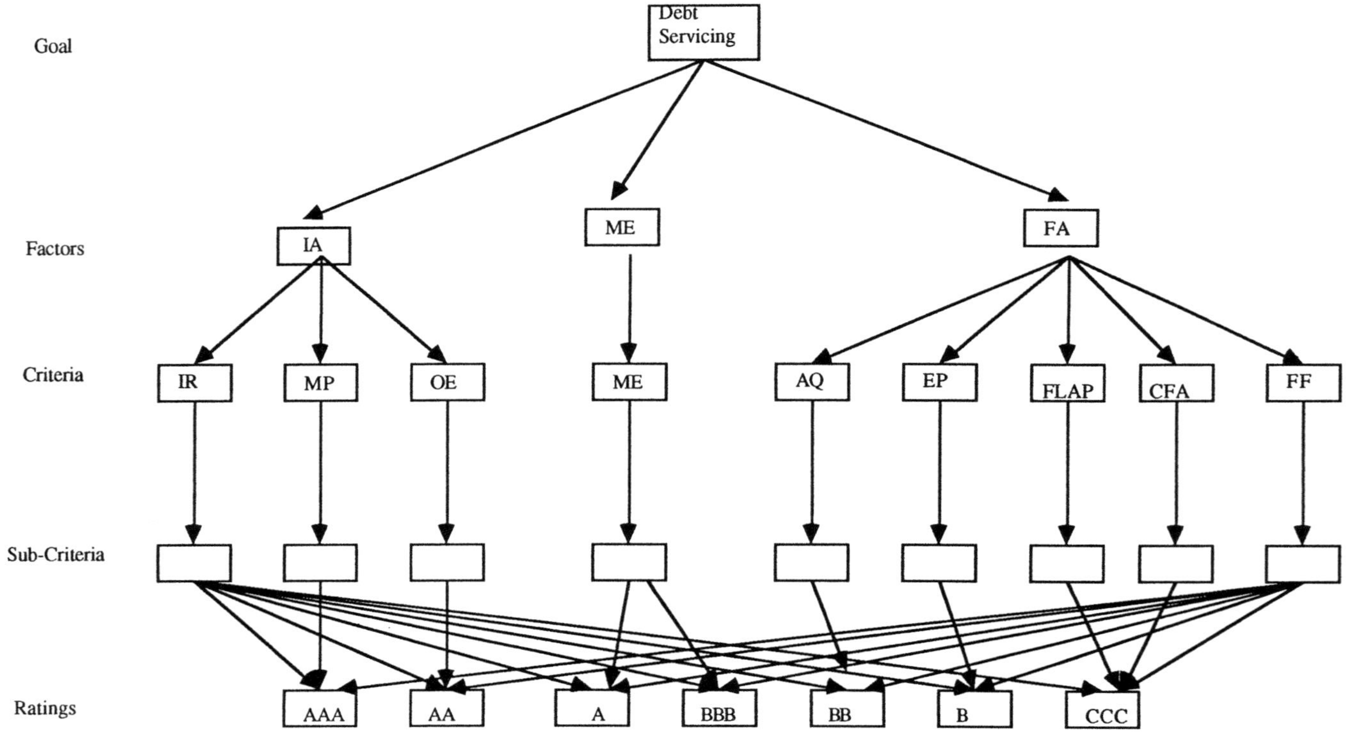

Not all arcs have been drawn to retain visual clarity

Exhibit 2
Sample Prototype Output

BONDRATE

Analysis for: CHRYSLER CORPORATION

Description of Issue:

DEBENTURE
12s
DUE 2015
$800M

Suggested Rating: A

Brief Rationale:

Industry Analysis - Automotive industry is reasonably strong. However, it is vulnerable to deterioration of economic conditions. Chrysler has an acceptable share of the domestic market and is in a better position to expand this share than other domestic producers. Growing consumer acceptance of competing foreign products is a major concern for all domestic producers.

Management Evaluation - Chrysler management especially strong among domestic peers. The firm has captured signficant market share from its competitors. Aggressive spending on new plant and equipment provides a major production cost advantage.

Financial Analysis - Profit margins and returns have improved or remained stable during recent years. Continued heavy use of debt financing. Enormous capital expense to upgrade facilities leaves firm with little flexibility. The firm's cash position is good, but highly sensitive to a change in economic conditions.

Appendix I: Factors, Criteria and Sub-criteria

I. Industry Analysis

A. Industry Risk (IA)
Importance in the economic cycle (EC)
Business cyclicality/Earnings volatility (BC/EV)
Economic forces impacts (EFI)
Demand factors (DF)
Basic financial characteristics (FC)
Supply factors (SF)
Federal, State, Foreign regulation (R)
Fragmented or concentrated business (FCB)
Barriers to entry (BE)

B. Market Position (MP)
Ability to generate sales (AGS)
Market share (MS)
Marketing/Distribution requirements (M/DR)
R&D reliance (R&D)
Support/Service organization (SS)
Dependance on major customers (MC)
Long term sales contracts (LTSC)
Product diversity (PD)

C. Operating Efficiency (OE)
Ability to maintain or improve margins (MIM)
Pricing leadership (PL)
Integration of manufacturing operations (IMO)
Plant and equipment (PE)
Supply of raw material (SRM)
Level of capital/employee productivity (CEP)
Labor supply, cost, relations (L)
Pollution control requirements (PC)
Energy costs (EC)

II. Management Evaluation
Record of achievement (RA)
Planning (P)
Controls (C)
Financing policies and practices (FPP)
Commitment, consistency, and credibility (CCC)
Overall quality of management (OQ)
Merger and acquisition considerations (MA)
Performance versus peers (PP)

III. Financial Analysis

A. Accounting Quality (AQ)
Auditor's qualifications (A)
LIFO versus FIFO inventory method (LF)
Goodwill and intangible assets (G)
Recording of revenues (RR)
Depreciation policies (DP)
Nonconsolidated subsidiaries (NS)
Accounting for pension fund liabilities (AFP)
Undervalued assets (UA)

B. Earnings Protection (EP)
Return on capital (ROC)
Pretax coverage ratio (PC)
Profit margins (PM)
Earnings on asset/business segments (EOA)
Sources of future earnings growth (SFG)
Pension service coverage (PSC)
Ability to finance growth internally (AFI)
Inflation adjusted earnings capacity (IEC)

C. Financial Leverage and Asset Protection (FLAP)
Long term debt, total debt to capital (LTD)
Total Liabilities to stockholders' equity (TLE)
Preferred stock to capitalization (PSE)
Leverage implicit in off balance sheet financing (OBSF)
Nature of assets (NA)
Working capital management (WCM)
Level, nature, value of intangibles (LNI)
Off balance sheet assets (OBSA)

D. Cash flow Adequacy (CFA)
Size/scope of total capital requirements (CRSF)
Variability of future cash flows (VFCF)
Cash flow to fixed and working capital (CFFWC)
Cash flow to debt (CFD)
Free cash flow to ST debt and total debt (FCFD)

E. Financial Flexibility (FF)
Relative financial needs (FN)
Projected financial plans (PFP)
Financing alternatives under stress (FAS)
Capital spending flexibility (CSF)
Asset redeployment potential (AR)
Off balance sheet assets/liabilities (OBSA)
High level of ST or floating rate debt (HSTD)
High debt service schedule (HDS)
High preferred stock to capitalization (HPS)
Assessment of near term sources of funds (NSU)
Ownership/affiliation (OA)

Preparing an intelligent interface for the use of multicriteria outranking methods.

Gabriella BALESTRA Alexis TSOUKIAS

DIPARTIMENTO AUTOMATICA INFORMATICA
POLITECNICO DI TORINO
Corso Duca degli Abruzzi 24 - 10129 TORINO - ITALIA

1 INTRODUCTION

In the last decade, interactive tools, as supports for decision making processes, have become a main argument both for academic and applicative research [Mc Lean and Sol, 1986], [Roy, 1987]. In this mainstream our attention has been focused in experimenting the possibility of building an intelligent interface for the use of multicriteria outranking methods.

An increasing demand for supporting creative decision making and ill structured decision situations, for flexible tools in organizational decision processes, helped to create and improve new methods in decision aid and in particular within the multicriteria approach.

Multicriteria outranking methods are now broadly implemented and used in different application domains and a considerable amount of experience is available. Moreover, experience in decision aid (methodologies, different methods, organizational approaches etc.) is now analyzed and discussed (see for example [Norese and Ostanello, 1984], [Ostanello and Hirsh, 1986] and [Ostanello, 1987]).

What seems to stand out is that key factor for a "successful decision aid" is *knowledge*, both methodological, about the tools to be used, and specific on the application domain. Decision aid cannot be considered without taking in account the individuals, the organizations and their relations involved in the decision process. In this context the analyst capacity to elicite significant information about the problem is relevant and implies use of knowledge and intelligence [Balestra et al., 1987].

On the other hand experience in building and using D.S.S. (cfr. [Landry et al., 1985], [Keen, 1986]), focuses the attention on qualitative dimensions such as analyzing and understanding problems, problem formulation, use of knowledge etc..

Our goal has been to build an intelligent interface for the use of the multicriteria outranking methods by representing the knowledge that a multicriteria expert uses in his interaction with a client. Artificial intelligence tools were used to build an *'extended D.S.S.'* [Keen, 1986] able to provide a consultant rather than an assistant.

In the second section a brief presentation of the domain and a presentation of the knowledge acquisition process are given.. Moreover the experience acquired by building a prototype under both the knowledge acquisition and representation point of view is presented. In the third section the main features of the knowledge are discussed and in the forth section a matching between the knowledge features and the problems we faced is provided and discussed.

2 THE DOMAIN AND OUR EXPERIENCE

Assuming the concept frame of the works of B. ROY [Roy, 1985] a multicriteria problem can be formulated as follows (quoting from [Ostanello, 1985]): "A finite set of potential (feasible) actions, A, is considered; each of them is supposed perfectly identified but not necessarily exactly and completely known in all its quantitative and qualitative consequences. It is thus admitted that those consequences can be analyzed by means of a 'consistent family' of criteria, g, where g(a) will characterize the evaluation made of an action a on a criterion. The technical problem consists of assessing, by an aggregation model, one of the three problems: to select one (or some) actions(s) of A - problem (α) -; to accept actions that appear 'good' and reject those appearing 'bad', with a coplementary analysis for the others - problem (β) -; to rank all the actions according to a complete weak order - problem(γ). An outranking relation is conceived so as to represent, in the aggregation model, the particular case where two actions are incomparable... Thus, when accepting incomparability situations, willingness to clarify situations of preference and indifference, an outranking relation refers to preferences modeled only for those cases where the analyst is able to establish them with an objectivity and security that he judges satisfactory." Different kinds of outranking relations are thus defined depending on the real nature of the problem and its modelization.

Knowledge on this domain has been elicited from three different sources:

1. formal knowledge, included in textbooks (mainly [Roy, 1985]), where axiomatic definitions, theorems and other formalized concepts are provided;
2. experience, acquired by analyzing several cases of interventions in real decision situations where multicriteria outranking methods were used. The intervention process has been reconstructed focusing the attention on the way followed to formalize the information about the problem in multicriteria methods concepts;
3. experience, elicited interviewing a multicriteria expert about how she uses these methods facing a client problem. Precisely the range of the analyst intervention has been discussed, how uncertain information is treated, how client incoherences or misunderstandings are elicited, how real problem complexity is reduced.

In a previous work [Balestra et al., 1987] a first model of the analyst - client interaction is presented (from the knowledge representation point of view) and five different levels of intervention are outlined:

1. acquiring data;
2. advising on using a method;
3. improve client model;
4. provide tools for acquiring and formalize significant information;
5. help the client in redefining his problem,

In the cited paper is concluded that only the first three item could realistically be supported by an interactive computer based client-oriented tool.

A first design of the prototype has been produced, based on this model of analyst-client interaction. The system is configured as an interface between a user of multicriteria outranking methods and the methods (intended as computer implemented tools). This interface acquires data, advises and assists the user in applying a method proposed by the interface itself and helps him in improving his model to better suite to the method used; selects a method from the library and transfers data under formats required by the method itself.

The system has been divided in four modules:

1. the first one defines the outranking relation asking the user about the criteria he uses, the alternatives he tries to evaluate, the kind of problem (selecting (α), sorting (β), ranking (γ), describing (δ), conceptualization (ϵ)); assists the user in using correctly the data and gives a first help in improving the user model;
2. the second one acquires and proves different sets of weights about the criteria and different sets of concordance and discordance indexes;
3. the third one selects a method suitable with the information provided by the user, transfers the data and uses the method;
4. the forth discusses the results with the user, register objections and insatisfactions, provides explanations about the results and the reasoning of the system itself and goes on modifying the data.

Our experience derives mainly from the implementation of the first module because problems arose since the work begun. Problems and considerations presented are divided in the knowledge acquisition section and the knowledge representation section. However, there have not been distinct phases, because acquisition and representation have been mutually influenced.

2.1 Knowledge acquisition phase

In the first step, after we had elicited a first main list of concepts used in defining an outranking relation, we tried to identify how these concepts were used in real decision aid processes. What clearly emerges is the non linearity of this process because the client changes continuously his representations as the process goes on. There are four main characteristics in this process:

- the process is pointed on the contradictions, incoherences and incongruences between different moments of the modelization. A set of alternatives is not always available, a set of criteria cannot be always coherently defined, preferences are not clear enough to be immediately formalized in a multicriteria evaluation table etc.;
- the decision aid process cannot be generalized. It is not possible to identify a single pattern as well as a finite number of patterns a priori defined, representing the concept use in each application. The process begins from any element, the one the user feels sure to represent and every time a different path is defined constructing the other elements of the method;
- the relations between the different concepts of the method are complex and we defined them as functional. By this term we mean that to each concept a set of possible values

is attached and that the precise value of an instantiation of a concept depends on the values reached by other concepts. A concept acquires a sense and is used depending on how other concepts are used. No causal relations can be identified between the concepts (as direct inheritance, true or false conditions for true or false conclusions etc.). When a concept is instantiated with a meaning or transformed, a propagation of meanings on the other concepts begins guided from the functional links.

- finally, during the decision aid process at least three domains are present and interrelated. The expert knowledge about the methods having a formal representation, the knowledge exchanged about the decision process that both the expert and the client construct during the process which is used by the methodology, the knowledge about the specific problem domain where the decision process is held, which has to be understood by the expert during the process. It is not easy to distinguish the relations among these domains. By now this is a limit in our representations.

2.2 Knowledge representation phase

Knowledge representation begun testing different representation techniques and different tools for the implementation. Very soon become clear that the usual techniques based on production rules were poor because of the knowledge complexity. Moreover appeared that this approach created limits in testing reliability of the system. We thus oriented our efforts to a frame-based knowledge representation.

On the other hand we recognized that the links between the concepts could be viewed as a semantic net where the arcs are not the classic inheritance or causal relations but should represent the functional relations previously described. We defined a semantic net of frames (where each frame is a concept), presented schematically in fig.1, and tried to implement it. We recognized the following characteristics:

- an adequate tool for the implementation had to be chosen between commercial shells and declarative programming languages. After several tests we have decided to use a prolog environment enabling a rapid development of prototypes including production, testing and evaluation. Problems arose with shells because they are very limited from their intrinsic inference mechanism, often incompatible or very distant from the knowledge characteristics.

- each concept is instantiated to a variable number of objects in each application. Because of the functional relations each instantiated object causes the instantiation of other objects of another concept and this propagation is different in each application. The system has to determine itself the objects, evaluate them and reason about them. Changing the application domain is not only a modification of evaluation but a new configuration of the system with a different number of criteria, alternatives etc. and different information about them. Moreover the relations between the concepts generate loops where groups of evaluation rules have to be executed a number of times depending on the evaluation made in another concept, depending on its time from another loop of executions and all of this depending on the specific application domain;

- the system has to acquire information (client representations, preferences etc.). This procedure cannot be represented by a pattern but depends on the problem itself. The system has to be instantiable from any concept with the available information and try to complete data choosing a way suitable with the client problem;

- to explain the reasoning is not an option to help the user but is a fundamental characteristic of the system. Traditional explanation mechanisms, as displaying the trace of the reasoning or the last assertions, are not sufficient. What is requested are deep explanations of the reasons producing a conclusion where the key factors are presented. This possibility has to be always available, enabling to change a sequence of reasoning shifting to another;
- finally we can assert that this knowledge cannot be represented as a tree. We deal with a complex graph where exploration depends both on evaluation on the nodes and on the nature of the arcs depending the latters from the application domain. Both representation paradigms based either on reasoning or search algorithms seem not to be adequated to these problems [Simon, 1983].

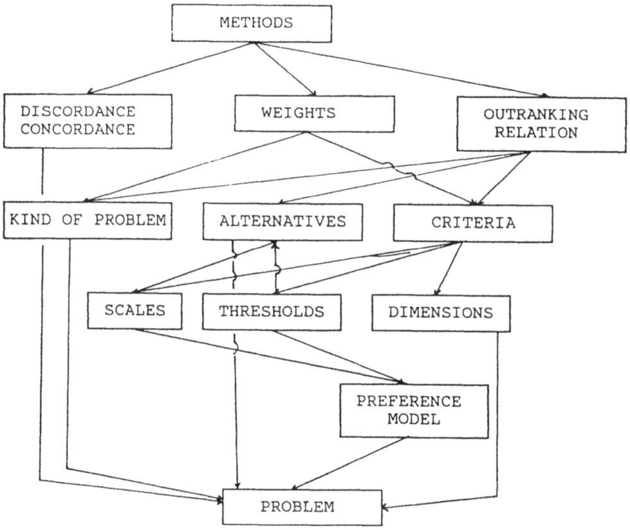

FIG. 1: SCHEMATIC REPRESENTATION OF THE KNOWLEDGE BASE

3 KNOWLEDGE CHARACTERISTICS

In our attempt to build the prototype we had to better define the characteristics of the domain to understand how it is really used, which problems in formalization could generate and which differences exist from other kinds of knowledge. We distinguish three main features of the knowledge: methodological, self-referential, a learning process.

3.1 Methodological knowledge

Methodology is *"axiomatic assumptions and a formalized behaviour about ways linking concepts defined in the assumptions"*. Different paths can be defined among the concepts; we call these paths *methods*. Hence a methodology can be viewed as reasoning or as a discourse about methods. Therefore a methodology provides uses of the methods and more important attributes different meanings at these uses. A methodology is independent from a specific application domain.

The knowledge we are coupling with, is based on the multicriteria approach in decision making using outranking methods where these are suitable. If our assumption of methodology is accepted we can assert that this is a methodological knowledge, that leads to a reasoning system independent from the specific problem domain.

On the other hand we know that this knowledge is essential for a successful decision aid activity. Modelization is significant only if it has a precise sense for both the client and the analyst and this is possible only through the use of a methodology.

Anyway, what is interesting from the knowledge representation point of view is the difference from the usual factual knowledge represented in most expert systems applications [Waterman, 1986]. Our goal is to construct a knowledge base representing elements of the multicriteria methodology and the expert knowledge about them.

3.2 Self-referential knowledge

The concept of self-referentiality (or self-organization) is borrowed from general systems theory [Varela, 1975] and indicates a system characterized from the maintenance of its internal coherence with respect on its organization, where organization indicates the necessary relations which define the system. A self-referential system is self-described, has eigenbehaviours and changes (adaptation on the environment) modifying its structure (the actual relations in the system) without violating organizational constraints (see also [Varela, 1984]). This characterization is given by an observer and underlines the observation of a self-referential behaviour rather than the existence of a self-referential system.

From our point of view we used this concept to summarize some observations on the knowledge. What we assert now is that the knowledge we handle can be observed having a self-referential behaviour. Indeed analyst reasoning is mainly guided from the search of coherence with the assumptions of the multicriteria approach. Assuming that these assumptions are the necessary relations defining the methodology, this reasoning is used to maintain the internal coherence while the methods (the structure) are adapted on the client problem (environment).

This interpretation satisfies two other observations about the knowledge. The specific multicriteria approach we observed does not make reference in external value systems, whereas it tries to represent the client preferences (a projection of the client value system) coherently with the methodological principles of the approach. Of course, extrapolating, any methodology can be viewed as self-referential, but we distinguish the existence of a behaviour (in this knowledge) where coherence is searched and not external systems are assumed.

On the other hand the analyst uses his knowledge to elicit significant information about the problem where significant is referred to requirements dictated from the methodology itself. Information provided by the client is characterized significant, organized, encoded by the methodology for the methodology.

Problems arose since we decided to represent this feature considering it as essential, at least with respect on the complexity of the problems where experience has been achieved. From the knowledge representation point of view such a knowledge base cannot be explored by heuristics or classical reasoning algorithms because the application of the principle of maintenance of coherence depends on the specific domain and the way to explore the knowledge base is defined during its consultation.

3.3 Knowledge as a Learning Process

The interaction between the analyst and his client is a reciprocal relation of learning. The analyst has to learn about the client representations of the problem, his value systems, his preferences, his information sources. The client has to understand how his problem is modelized, how his preferences are formalized and has to recognize his representations in the analyst work. This exchange of information and knowledge is finalized at the construction [Roy, 1987] of a consensual (between analyst and client) representation of the problem.

This process has an interactive nature where the client verifies different models provided by the analyst until consensus is achieved. Knowledge used by the analyst is both sensible in considering client objections or consensus and flexible to adequate itself while learns about the clients preferences, values etc.. This interaction could be represented as two models trying to fit one another by mutual adjustment. This process implies acceptation of a common language and a learning procedure between the two actors.

From the knowledge representation point of view we underline that this is an inductive reasoning guided by learning, whereas usual expert systems use deductive reasoning who do not learns [Sutherland, 1986]. Moreover the system we try to construct has to achieve consensus with the user about its performance.

4 DISCUSSION

The formal characterization of the knowledge we were working with, induced to try to understand the links between the problems we faced and the knowledge characteristics, so as to justify the use of the tools we selected, to understand the bounds of the knowledge base, to individuate new techniques and/or tools more suitable.

To summarize we can say that we defined a conceptualization [Genesereth and Nillson, 1987] of the knowledge to represent where:

- the universe of discourse is the set of concepts of the multicriteria methodology;
- the functional set and the relational set are the links between the concepts defining the internal coherence of the methodology.

We choose a declarative representation of the knowledge for the following reasons:

- the system has to be used in many different ways that cannot be defined a priori;
- the system has to explain its reasoning and has to represent itself to the user.

The choice of a frame-based representation seem to be confirmed as the only one able to couple with the knowledge base complexity. What seems not to be very suitable are the inference mechanisms based on prepositional logic and predicate calculus. On the other hand we decided to use what already exists and not to create ad-hoc tools for our knowledge representation.

In the next paragraphs we focus our attention on the links between the knowledge characteristics, the knowledge representation and the problems we faced.

4.1 The "methodological" consequences

We can attribute to the methodological characteristics of the knowledge the following problems:

- a methodology, as stated before, provides a formal behaviour. That is, each time we use a methodology we can generate a new path between the concepts, the one suitable to the problem it has to be solved. Of course the continuous use of a methodology can generate recurrent patterns when the "same kind" of problem is presented. We can assert that there are infinite paths between the concepts and that the methodology defines one when applied on a problem. Try to define a system able to represent the generation of the infinite paths is impossible. We decided to individuate and represent the elements of the methodology that enable the system to recognize the correspondence (pattern matching) between the user problem and some formalized patterns going under the name of *outranking methods*;

- the links between concepts inside a methodology are of "it depends on" type. These relations, we called functional, are evaluated after the instantiation of a concept in objects depending on the application domain. So, the concept A (in our case for instance: "outranking relation") depends on the values of the objects instantiated by the concept B (in our case for example: "criterion") and so on. Both the objects and the relations between the concepts have to be evaluated. More precisely these evaluations are generally *constructed* rather than *elicited.*

  At least one concept has to be evaluated by the user himself, the one he feels sure about, so as to generate a propagation of evaluations. Generally in a methodology should be possible to have infinite different evaluations. The pattern matching approach limits the search to finite sets of values reducing thus the complexity of the knowledge base;

- for the same reasons as above inside a methodological knowledge is not predefined how the evaluation has to be executed. The knowledge base has to use metarules or procedural algorithms to guide the use of the evaluation rules after the consultation begun, bounded to the possible states the pattern matching admits;

- the network of the conceptualization is a graph. As previously said a methodology transforms an initial problem of a specific domain in a formal representation instantiating the concepts in objects representing the information provided by the user on the problem. In our knowledge base the system tries to match the problem with any of the patterns available. The problem consists on the fact that the search goes on a graph and not on a logical tree. This is the reason for which the traditional reasoning or search algorithms are not adequate and generate problems reducing the trasparency of the system.

4.2 The "self-referential" consequences

We can attribute to the self-referential characteristics of the knowledge the following problems:

- as previously defined a fundamental characteristic of the methodology is the existence of a behaviour guided by the principle of internal coherence maintenance. On the other hand we recognize that the decision aid process is described by its incoherences, misunderstandings, informative lacks etc.. The methodology tries to modify the client representation of the problem in a new representation coherent with the methodology assumptions. In our case the system has to verify the matching between the user problem representation and the patterns, individuating incoherences and suggesting modifications;

- the different evaluations of the links between the concepts can be viewed as the adaptation of the existing relations on the problem or the generation of new ones, without violating the methodological assumptions as the self-referential behaviour requires.

From this point of view the system has a set of relations coherent with the methodology trying to apply them to the information provided by the user. If there is a way to match the user problem with a method the system finds it;

- a self-referential system is selfdescribed. The problem modelization evolves with the client-analyst interaction achieving reciprocal consensus through successive representations. Consensus is possible when the client recognizes in the formalization, imposed by the methodology, his problem. This identification is achieved explaining to the user the reasoning applied by means of the methodology to match his problem with the concepts. The system has to describe its behaviour related to the user problem presenting both the reasoning and the methodological principles used to generate the precise reasoning besides explaining the information that caused this kind of reasoning.

4.3 The "learning process" consequences

The learning dimension of the knowledge seem to be the more untreatable of the problems we faced. The knowledge used in the decision aid process enables both the analyst to understand the client problem and the client to recognize the incoherences of his representations. Moreover this knowledge increases dynamically during the decision process as its three domains generate experience, induce new reasoning, create new patterns.

The system we are working with cannot face this kind of problems as the learning mechanisms are just in an experimental phase among the Artificial Intelligence community. We limit our work in generating a system able to acquire the information provided by the user in a pattern matching guided way so that the information is verified, evaluated, discussed with the user, and explanations should be available.

The graph representation of the knowledge base generates some problems from this point of view as the information is not propagated in a monotonic way but circulates on the graph.

5 CONCLUSIONS

The problems and the first results in implementing an intelligence interface for the use of multicriteria outranking methods have been presented. A knowledge base system representing the knowledge used by a multicriteria expert in his interaction with a client is actually under construction in this direction.

The main problems are analyzed and a formal characterization of them is provided defining three features of the knowledge represented: methodological, self-referential, learning process.

A discussion is provided from the knowledge representation point of view where some solutions, as a pattern matching approach, a frames based representation etc., are individuated to face the complexity of the knowledge base.

References

[Balestra et al., 1987] G. BALESTRA, A. OSTANELLO, A. TSOUKIAS "Rappresentazione della conoscenza di un esperto di analisi multicriteri con techniche di Intelligenza Artificiale" (in Italian), proceedings of 1987 AIRO meeting, Franco Angeli, Milano, 379 - 394, 1987.

[Genesereth and Nillson, 1987] M.R. GENESERETH, N.J. NILLSON *Logical foundations of Artificial Intelligence*, Morgan Kaufmann Publishers, Los Altos, California, 1987.

[Keen, 1986] P.G.W. KEEN "Decision Support Systems: the next decade", in *D.S.S.: a decade in perspective*, E. Mc LEAN, H.G. SOL (eds.), North Holland, Amsterdam, 221 - 237, 1986.

[Landry et al., 1985] M. LANDRY, D. PASCOT, D. BRIOLAT "Can DSS evolve without changing our view of the concept of problem?", *Decision Support Systems*, vol. 1, 25 - 36, 1985.

[Mc Lean and Sol, 1986] E. Mc LEAN, H.G. SOL, (eds.) *Decison Support Systems: a decade in perspective*, North Holland, Amsterdam, 1986.

[Norese and Ostanello, 1984] M.F. NORESE A. OSTANELLO "Planning processes and technician interventions: an integrated approach" *Sistemi Urbani*, n.2, 247-259, 1984.

[Ostanello, 1985] A. OSTANELLO "Outranking methods", in *Multiple Criteria Decision Methods and Applications*, G. FANDEL, J. SPRONK (eds.), Springer Verlag, Berlin, 41 - 60, 1985.

[Ostanello and Hirsh, 1986] A. OSTANELLO, G. HIRSH "Ou se situent les méthodes multicritères dans le processus au sein des organisations", *23th Meeting of the EURO Group on MCDM*, Rotterdam, 1986.

[Ostanello, 1987] A. OSTANELLO "Comparaison d'approches pour la définition de poids des critères", *25th Meeting of the EURO Group on MCDM*, Bruxelles, 1987.

[Roy, 1985] B. ROY *Mèthodologie multicritère d'aide à la décision*, Economica, Paris, 1985.

[Roy, 1987] B. ROY "Meanining and validity of interactive procedures as tools for decision making", *European Journal of Operational Research*, vol 31, 297 - 303, 1987.

[Simon, 1983] H.A. SIMON "Search and reasoning in Problem Solving", *Artificial Intelligence*, vol. 21, 7 - 29, 1983.

[Sutherland, 1986] J.W. SUTHERLAND "Assessing the artificial intelligence contribution to decision technology", *IEEE Transaction on Systems, Man and Cybernetics*, 16, 3-20, 1986.

[Varela, 1975] F.J. VARELA "A calculus for Self-reference", *International Journal of General Systems*, vol. 2, 5 - 24, 1975.

[Varela, 1984] F.J. VARELA "Two princiles of self-organization", in *Self-organization and management of Social Systems*, H. ULRICH, G.J.B. PROBST (eds.), Springer Verlag, Berlin, 25 - 32, 1984.

[Waterman, 1986] D.A. WATERMAN *A guide to Expert Systems*, Addison Wesley, Amsterdam, 1986.

Abstracts of papers presented at the conference but not included in the proceedings

N.ASSIMAKOPOULOS*

An MCDM Approach for Distributed Decision Making

Distributed decision making regarding agricultural loans by evaluating different criteria that affecting the procedure paths and regulating the operation is examined in this paper. Decision making nodes function as multitasking operators in a complex cybernetic system by executing command decomposition for the concurrent implementation of various loan procedures. A hierarchical organisation of N decision makers is examined where each one deals with a finite number of objects. An illustrative example for the concurrent operation is given.

M. CIURLA, J. PLOCINSKI*

Organisational Alternatives to Improve Decision Making in the Industrial Company

The paper presents results of investigations aimed at improving the organisational structure of industrial companies operating at the region of Lower in Poland. The basic goal of these investigations was to improve decision making process, circulation of management information and documents and thus to reduce a number of indirect links responsible for preparing and taking decisions. The investigations were carried on in the two stages:

- identification of the existent organisational state
- elaboration of projects of organisational alternatives to
- achieve the previously mentioned goal.

The main tool used was the SPIN method which allowed for identification of all significant routine managerial activities performed by all organisational cells of the company, classification of these activities and evaluation of their performance laboriousness on the basis of predetermined labour standards. In the elaborated organisational projects included were legal regulations compulsory in Poland and some recommendations developed by theory of organisation and management.

So the proposed structural actions for the company postulated considerable changes in its departments and organisational cells, creation of new cells and considerable reduction of administrative posts. The projects are also intended to increase the company management system efficiency.

J. CLIMACO, C. ANTUNES*, A. MARTINS, J. MARTA, A. ALMEIDA

A Novel Approach to Power Generation Expansion Planning Using TRIMAP

The main purpose of this paper is to discuss the application of a tricriteria interactive algorithm to an outline study of the expansion policy for a power generation system. The three objective functions quantify the total system cost, the risk of supply and the environmental impact. Special emphasis has been placed on interactive graphical capabilities that facilitate the decision maker tasks.

A first study with this type of approach was put forward by the authors previously. Some suggestions from the scientific community have contributed for the new version of the model now being presented.

Y. DUBOV*

Interactive Design and Interactive Decision Support in Urban Planning

The models developed in the Institute for Systems Studies for estimating alternative urban planning strategies may be considered as a set of tools providing new opportunities for urban planners and objective estimates of decisions outcomes. The natural continuation of these studies consists in direct implementation of the models into urban planning practice leading to the development of new urban planning and management technology based on the system of mathematical models. Realisation of this stage depends on a number of problems - from reconsidering the approach to models building to certain changes of existing organisational structure of urban management.

The most attractive way of implementing the existing models is their inclusion into interactive urban planning systems. This paper deals with the basic principles for developing such systems and the experience of their use in urban planning practice.

T. GAL, K. WOLF*

The Impact of Variations of the Domination Structure on the Efficient Solutions in a Decision Problem Under Several Criteria.

Sensitivity or stability investigations in the area of multiple criteria decision making (MCDM) are mostly restricted to the examination of the impact of changes of the objective functions coefficients, the "A"-matrix or the right hand side on the set of all efficient solutions. The effect of changing the domination structure which represents the preference attitude of the decision maker(s) is scarcely considered up to now. Given a Pareto efficient point the question is how much can be the preference cone changed (enlarged, reduced) such that this point remains strongly or at least weakly efficient. Because of formal reasons this research is divided into two parts: The impact of 1) enlarging the preference cone and 2) reducing the preference cone on the Pareto efficient set. This viewpoint can be enlarged such that enlarging or reducing the preference cone is feasible as long as the underlying set is at least weakly efficient. The corresponding formalism and an algorithm to determine the "new feasible preference cones" is briefly represented. A possibility to apply such kind of sensitivity analysis to interactive methods is indicated.

A. GONCALVES*

Market Equilibrium and Internal Equilibrium Solutions to MCDM

A multicriteria linear programming problem is used to represent the typical situation where K firms plan to collaborate exploring a common technology. Each firm contributes with its own resources, according to its individual structure, and have different objectives, which may be conflicting. For this problem, two subsets of efficient solutions are identified. The "Internal

Equilibrium Solution (IES)", with the following property : for any IES, the return for each firm is equal to the marginal value of its resource contribution. And the "Market Equilibrium Solutions (MES)", which maximise the total marginal value of the resources, when a MES is adopted as a plan to explore the common technology.

Procedures to find both internal and market equilibrium solutions are developed, and its characteristics emphasised.

B. GRONQUIST*

Why Use Logic Programming to Support Judgement Modelling

The development of expert systems in the field of management seems to be promising. It will provide better decision support, especially in areas where problems are ill structured and/or not fully specified. In those cases, choice depends narrowly on the judgement of the decision maker. The ability of the decision support system (DSS) to assist the manager in considering different options, criteria and evolution of solutions is essential.

Actually expert systems work well in specific domains. Their use for judgement modelling however requires some particular adaptations in order to take into account the nature of human decision-making. Specifically, the user should be able to express his own preferences about goals and constraints.

DSS) and expert system (ES) techniques, two approaches are possible. The first one is to add some DSS-features to existing expert systems. The second one is to add some ES-features to existing decision support systems. This is the approach we are considering in this paper. The aim is to show how existing methods can be enhanced by the use of Prolog for integrating facilities.

J. GUERARD, G. McCABE, K. and S. LAWRENCE, B. STONE

MCDM and Firm-Financing : Further Results

Recent financial and economic studies have found that the R & D, Investment, dividend, and new debt decisions are interdependent and the firm should use a goal programming model to allocate corporate resources to maximise its stock price. This study takes a closer look at the costs and benefits of estimating pre-emptive goal weights as opposed to the use of a naive equal-weighting scheme for goal weights.

T. HEMMING*

A Multi-Criterion Decision Support System for Strategic Planning

The purpose of this paper is to discuss how computer supported games and interactive multicriterion optimisation methods can be used to improve corporate strategic planning.

It is conjectured that the relations between financial criteria and decision variables are more readily modelled mathematically than other relationships of relevance to strategic planning and that financial models also are more 4 easily accepted by managers. Estimates of data used to input to financial models have to be obtained as a part of the planning process. A game approach is suggested as a complement to a regular planning process in terms of traditional studies. The idea of the game approach is to create an integrated vision of the company and its mission and also to provide data to the multi-criterion financial model. The approach suggested here can be seen as an attempt to integrate soft and hard approaches to decision making in a productive way.

One way to proceed in strategic planning is to plan subject to conditional certainty ;and to consider just a discrete number of alternative ways to implement different strategies subject to different scenarios. But in spite of such simplifcation the number of alternatives easily increase beyond what is manageable when traditional planning methods are used. One function of the financial model is its use as a tool for eliminating inferior strategies as early as possible in the planing process.

As discussed here strategic planning calls for the use of models of the multi-criterion type not to solve overall problems but to make sure that the strategies being evaluated are feasible and to restrict the set of alternatives which are being evaluated according to intuition, feeling and other not formalised procedures outside the financial model.

M. HENIG*

Modelling and Consistency in Interactive Programming

In recent years serious effort has been directed to listing and comparing the various interactive methods for solving multicriteria problems in organisations. In addition to other yardsticks, we suggest that the methods should also be judged by:

1. their ability to present the alternatives and the criteria explicitly and accurately as a mathematical model.

2. their ability to select an alternative which is consistent with the organisation's revealed preferences.

On the first point, since an interactive method, at least in its first phase,is a process of simultaneous modelling and solving the model, the characteristics of the problem are evinced while solving it. On the second point, we argue that selecting a preferred alternative in a multicriteria environment is not a one-time event, and that the method should, therefore, be designed as a long-term instrument with the ability to yield alternatives which are consistent with preferences as revealed through time.

I. KALISZEWSKI*

Substantially Efficient Solutions of Vector Optimization Problems

The notion of proper efficiency can be used to impose a hierarchy over the efficient solutions of a vector optimisation problem. We strengthen this notion and introduce so called substantially efficient solutions. Necessary and sufficient conditions for a solution to be substantially efficient are given. A method to verify substantial efficiency of solutions is proposed. The importance of substantially efficient solutions in multiobjective optimisation is discussed.

M.M. KOKSALAN*, O.V. TANER

Experiments and a Method for Solving the Discrete Alternative Multiple Criteria Problem

A number of interactive methods that use convex cones for solving the discrete alternative multiple criteria problem have appeared in the literature. Convex cones are used to identify inferior alternatives based on the previous responses of the decision maker. The aim is to identify as many inferior alternatives as possible using the cones so that the decision maker will be required to make as few pairwise comparisons as possible.

In this study we make experiments to see the effects of number of cone generators, the selection procedure of cone generators, and the order in which pairwise comparisons are made, on the total number of pairwise comparisons. We also develop an algorithm based on the results of the experiments and report our computational experience.

E. KONARZEWSKA-GUBALA*

PC-Implementation of the Multicriteria Decision Analysis with Bipolar Reference System

The paper presents the variants of the decision analysis executed by the decision support PC-program EVA. It has been elaborated as a computer implementation of the multiple criteria decision aid model with bipolar reference objectives. The proposed methodology imitates the human reference objectives. The proposed methodology imitates the human behaviour in some decision situations and forces the decision make to realise what is for him/her desirable and what is not acceptable. The "position" of the given decision alternative is defined by two indicators: the degree of reaching the success and the degree of avoiding the failure. In the case of incomparability of the decision alternatives to be bipolar reference system we propose to use the PC-dialog for the purpose of the initial problem re-analysis.

M. LE LOUARN*

Introducing the Values and Value Systems of Actors in the Decision Process

Unlike the natural sciences, the social sciences cannot evade the questions of meaningfulness of conduct. In particular, decision-aid is very much concerned with the nature of human values, since these are supposed to preside over human action. But a close review of the research perspectives shows that the term "values" has been assigned very different meanings, usually suited to the accompanying epistemological premises. These positions can be classified using three paradigms labelled "concept-values", "proposition-values" and "system-values", which will be described and analysed. The system-values paradigm is recommended as best fit to today's epistemological ideas, but its carrying out presents practical inconvenience, as will be lastly shown.

A. LEWANDOWSKI*

SCDAS - Decision Support System for Group Decision Making : Information Processing Issues

Most of research in the field of computerised Group Decision Support Systems is devoted to analysis and support the quantitative phase of decision processes using various methods of multiple-criteria analysis. The experience shows that the soft side of the decision process needs also certain support. This relates mostly to distribution of textual information which augments the quantitative side of decision process and to providing the linkage between such information and numerical data. This aspect is especially important when decision support system is implemented in distributed computing environment. In the paper possible forms of information processed within the SCDAS system are anlaysed as well as the framework for implementation the software providing such processing functions is presented.

D. LI, Y. HAIMES

Multiobjective Dynamic Programming : The State of the Art

Multiple objectives and dynamic control characterise many real-world sequential decision making problems. For example, an adaption of mulitobjective dynamic optimisation has been used in the design of advanced aircraft (Tabak et al 2979; Sridhar and Salzwedel 1982; Kreisselmeier and Steinhauser 1983). The past decade has seen an increasing concern to integrate system's multiple objectives with dynamic programming, leading to the emergence of a new research field known as multiobjective dynamic programming (MODP). It is the concern of this paper to review the major concepts used in MODP and to examine current progress in the development of such theory and methodology.

M. LUPTACIK*, F. TURNOVEC

Lexicographic Geometric Programming

The paper presents the Kuhn-Tucker conditions and the duality theory for the lexicographic minimization problem:

$$\text{lex min } \{F(x) \mid x \in R_n, \quad G(x) \leq 0_m\}$$

where F, G are vector valued functions. The lexicographic duality is then applied for problems with linear objectives and linear constraints and for problems with polynomials in the objectives and in the constraints. The duality theory for lexicographic geometric programming is developed and illustrated by a numerical example.

B. MATARAZZO*

Some Considerations for a More Effective Application of the MAPPAC and PRAGMA Methods

This article introduces a new concept with respect to the MAPPAC and PRAGMA methods, that of indifference areas. Moreover, some preference structures are considered and particular discordance indices are taken into consideration in order to obtain relationships of vetopreference and of incomparability.

A. SKULIMOWSKI*

MMCDM Problems in Control of Discrete Event Systems

The goal of this paper is to demonstrate that the paper description of conflicts and multiple functions of discrete production systems leads to a formulation of a multiple objective optimisation problem. Then we will present a multicriteria problem statement combining quantitative criteria with a qualitative preference structure defined on the set of final states of the system. We will propose a solution method for feedback systems based on a generalisation of the multicriteria shortest-path method for variable-structure networks as well as on the reference points and reference trajectory approaches. Finally, we will analyse a real-life example of control for several interconnected industrial devices.

G. SLAVOV, B. DANEV, B. METEV, G. BOSHNAKOV*

Interactive Decision Support system "MicroMULCRE"

There are many areas in the practice, where man makes a choice between different alternatives from finite sets with respect to many criteria. Many management decisions, some cases of resource allocation and project financing are of this type. For example, let's have some projects and each of them needs some funds. Giving less than the announced need leads to the realisation of some parts of the project only. And there are no reasons to give them more funds. When the total existing amount is less than the sum of needed funds, the solution is to choose a subset only of the projects under consideration and usually with respect to many criteria. The choice of a good subset can be made following different methods and techniques, which render an account of different types of initial and additional information.

This paper contains a description of the interactive decision support system (IDSS) "microMULCRE V2.0", which is intended for IBM PC XT (or compatible). IDSS "microMULCRE V2.0", has program realisations of several normative methods for multicriterial estimation and choice from a finite set of elements. The first program of MULCRE program series is for main frame computers and is used for applications since 1983. The first version for 16-bit personal computers is used for applications in the last two years.

T. SZAPIRO*

Differential Modeling of Dynamics of Negotiation

The majory of multiobjective decision problems is investigated in the frameworks of optimisation or vector-optimisation approaches. While axioms of the first one are violated by some experiemental evidence, the second one leads not to unique solution of decision problems and requires its restructuralisation. An attempt to resolve this difficulty is proposed in the so-called interactive procedures aiming as well in creation of successive improvements of considered alternatives, as in elicitation of own preferences of decision maker. These procedures are frequently based on use in the models of some reference objects such as ideal or nadir points or trajectories, externally imposed reference opinions of independent advisers, compromise solutions or others. In the paper there is developed a model or negotiation process with dynamics described by some simplified system of ordinary differential equations. The machinery of theory of such equations/existence, stability of solutions, control of the system/is interpreted in terms of decision analysis of multicriteria problem. The model enables consideration of some internally defined reference points, which can be used in the support of decision makers in negotiations.

E. TAKEDA*, P.L. YU*

Eigen Weight Vectors from Incomplete Revealed Weight Ratios

In the AHP, the eigenvector method has been widely used for estimating the relative weights from a reciprocal matrix. The reciprocal matrix is constructed by all pairwise comparison judgments of the DM. As the judgments in pairwise comparisons are dependent, to a large extent, on personal experience, learnings, situations and state of mind, the degree of easiness or confidence to make the judgments on the ratios can be different. If one is more familiar with the comparison of i against j, then he/she can be more confident in assessing the ratio analysis. Usually, some elements of the reciprocal matrix may be obtained with greater confidence than others if the DM is forced to answer all pairwise comparisons. Furthermore, when the number of pairwise comparisons is large, it could be a great burden to require the DM to express all pairwise comparisons.

The present paper proposes a procedure for eliciting the relative weights using "incomplete" reciprocal matrix. It does not require the DM to answer all the pairwise comparisons. The DM needs to answer only those pairs which he/she feels comfortable or confident to do so.

J. TEMESI*, F. FORGO

Computer Aided Multiattribute Decision Making Method and Its Industrial Applications

To select the best alternative or to rank some products, plans, alternatives etc., in the real world the decision maker is required to pursue many objectives simultaneously.

To help the decision maker in this ranking process we developed a computer-aided system based on some theoretical aspects of multicriteria decision making. This is put in an interactive, very flexible framework allowing for maximum freedom on the part of the decision maker. Rather than giving a well-defined algorithm which produces the "best" alternative from the input data without asking for the active participation of the decision maker in the selection, we wanted to model the process allowing the user to choose his/her preferences as to certain parameters and methods.

Basically, our model requires two kinds of input data.

- parameters (technical, economic, numerical or non-numerical)
- subjective comparisons of alternatives according to the criteria taken one-by-one and assessment of relative importance of criteria.

We grouped the criteria in three subsets (modules) and the alternatives are first ranked within the modules. Then the three modules are combined and the "best" alternative is selected. A certain kind of sensitivity analysis is an inherent part of the system.

L.B. VLACIC*, S. ZIMONJIC

Planning at the Strategic Level Aided by the Rules of Quasisatisficing Behaviour

With the decision making situations aimed at the differentiation of support to technologies, (namely their development, introduction and implementation), - which are to be selected from the entire set of basic and generative technologies - it is usual to establish the Committee to represent diverse constituencies, from technology advisers (experts and analysts) through business managers to various authorities at the social planning level. Thus,

each actor of this situation tries to select a decision he thinks would be preferable for his constituency, and proposes it to the Committee as his preferred alternative. Hence, the actual choice is the result of a special process, which has a distinctive (mostly procedural) rationality, arising from the experience in making the social decisions.

In such cases, when it is necessary to choose the satisficing outcome from an admissible decision set, it is reasonable to introduce the rules of behaviour.
The rules of behaviour create the conditions for making the decisions which implicate the qualitative and structural changes not only within the environment they refer to, but within the entire society as well.

Various approaches may be employed for the elaboration of the rules of behaviour of Committee decision makers. The subject approach resulted into a procedural behaviour model aided by quasisatisficing decision making.

H. WALLENIUS*, T.M. LESCHINE, A. VERDINI

Multiple Criteria Decision Methods in Formulating Marine Pollution Policy : A Comparative Investigation

Contamination of the New York Bight has been a growing concern of the Environmental Protection Agency, and the neighbouring municipalities in recent years. In this paper we perform a set of experiments to compare, contrast and evaluate the use of Multiple Objective Linear Programming (Pareto Race) and Multiple Objective Decision Analysis (MATS) approaches to solving the problem of sludge disposal in the Bight. Three different disposal sites are considered. Both Marine Policy students and experts were tested. The design and the results of the experiments are discussed.

G. WRIGHT*, P. AYTON

Psychological Aspects of Forecasting With Statistical Methods

Judgment is implicated in many ways with a variety of forecasting methods. This paper briefly outlines the major methods (pure judgment, extrapolation and econometric modelling) and focus on the role of judgment in forecasting with statistical models by considering anecdotal reports of the use of these models. We then evaluate studies of the quality of human judgment, most of which have been considered by psychologists using paper-and-pencil tasks with undergraduate students as experimental subjects. Finally we evaluate studies of expert judgment and discuss the implications of these for the quality of judgmental "adjustments" to statistical models in work-a-day situations.

P. YU*, Y.R. LEE, Y. SHI

Construct Optimal Contingency Plans for Linear Optimal Designing Problems

In optimal selection problems we try to select a subset from the opportunity set for commitment for a period of time. As the external parameters can vary with time, optimal contingency plans must be constructed to cope with the external conditions. We describe the formulation and methods for locating potentially good solutions and their corresponding optimal contingency plans using MC - simplex method to solve the optimal designing problems.

M. ZELENY*

Knowledge-Based Management Systems : MCDM Perspective

Knowledge has become the most important productive force. It has upstaged the medieval dependency on land, labor, technology and capital and has become the crucial input and condition of a globally-competitive enterprise. Country's store of knowledge is its principal asset and the greatest source of wealth. Other forms of capital simply "follow", flowing from the knowledge-poor to knowledge-rich areas of the world.

Humanity-degrading categories of economics, like "labour-intensive" or "capital-intensive", are becoming irrelevant in the "knowledge-intensive" world. Having all the land, resources, labor, technology and money is necessary, but not sufficient, if one does not know how and why to use them.

Knowledge is the primary form of capital.
We speak of "knowledge industries", "knowledge systems", "knowledge workers" or even "knowledge engineering". Productive systems of respectable performance and viability are characterised by their scope, management and expansion of knowledge, not by their labor, technology or information processing.

In lagging or declining economies, knowledge has become increasingly specialised, atomised and splintered, leading to overwhelming ineffectiveness of labor-oriented processes, task specialisation and division of labor. Accelerating economies are characterised by dominant emphasis on task, labor and knowledge reintegration, multifunctionality, flexibility and integrative education..

How do we manage in a knowledge-intensive environment?

S. ZIONTS*, M. KARWAN

Toward Group Decision Support

An approach that is intended to lead a group decision support system is developed. Group decisions of various types are considered, and a paradigm is explored. Though several different decision situations are considered, one called an aggregate approach is most thoroughly developed. It begins with an individual and helps him to explore his preferences, and then helps him to determine what is most desirable from his perspective. Then, members that comprise a party (of individuals) used the system to determine the most preferred outcome for the party. Finally, the parties to the decision work as a group using the system to determine the group decision. The objective of our work is to develop software that can be used in a computer-based decision room, which can be used to help groups make decisions. In this paper we present the plan for our research.

K.G. ZOGRAFOS*, S. SAMARA

A Multiobjective Location Routing Model for Improving the Safety of Hazardous Wastes Transportation and Disposal

This paper introduces a Multiobjective Location Routing Model (MLR) for hazardous wastes transportation and disposal. The model determines the location of hazardous waste disposal sites and the flow of hazardous wastes to them by considering the following objectives: 1) No population center is located closer than a prespecified minimum distance from its nearest disposal facility, 2) the sum of the distances from population centers to disposal facilities is maximum, 3) the transportation risk is minimum, and 4) the total travel time is minimum. A hypothetical example is used to display the application of the model.

* See Conference Participants for list of addresses

CONFERENCE PARTICIPANTS

C. Antunes
University of Coimbra
Dept. of Electrical Engineering
3000 COIMBRA
Portugal

N. Assimakopoulos
Agricultural Bank of Greece
54 Eratosthenous Street
GR-173 43
ATHENS
Greece

K. Backwell
PA Computers and Telecommunications
Rochester House
33 Greycoat Street
LONDON SW1P 2QF
United Kingdom

V. Belton
University of Strathclyde
Dept. of Management Science
GLASGOW G1 1XQ
United Kingdom

P. Bolster
Northeastern University
College of Business Administration
413 Hayden Hall
BOSTON
Massachusetts 02115
U.S.A

G. Boshnakov
Bulgarian Industrial Association
134, Rakovski Str.
1000 SOFIA
Bulgaria

D. Bouyssou
Universite de Paris-Dauphine
LAMSADE
Place du Marechal de Lattre de Tassigny
75775 PARIS CEDEX 16
France

K. Bowen
Royal Holloway and Bedford New College
Department of Mathematics
Egham Hill
EGHAM
Surrey TW20 0EX
United Kingdom

ENG U. Choo
Simon Fraser University
Faculty of Business Administration
BURNABY
British Columbia V5A 1S6
Canada

R.J. Clinton
Northeast Louisiana University
College of Business Administration
MONROE
Louisiana 71209
U.S.A.

J. Cohon
John Hopkins University
276 Garland Hall
BALTIMORE MD 21218
U.S.A.

B. Cox
ICI Pharmaceutical Division
Bioscience II Department
Alderley Park
MACCLESFIELD SK10 4TG
United Kingdom

D. Cray
Carleton University
School of Business
Decision Support Syst.Res.Lab.
OTTAWA
Ontario K1S 5B6
Canada

H.G. Daellenbach
University of Canterbury
Operations Research
CHRISTCHURCH
New Zealand

J.P. Dauer
University of Nebraska-Lincoln
Dept. of Mathematics and Statistics
810 Oldfather Hall
LINCOLN
Nebraska 68588
U.S.A.

C.F. Dinis das Neves
Technical University of Lisbon
IST/CESUR
Av. Rovisco Pais
1000 LISBOA
Portugal

Yu.A. Dubov
Institute for Systems Studies
Prospect 60 Let Oktyabrja 9
117312 MOSCOW
U.S.S.R.

L.S. Franz
University of Missouri-Columbia
School of Business
214 Middlebush
COLUMBIA
Missouri 65211
U.S.A.

S. French
University of Leeds
School of Computer Studies
LEEDS LS2 9JT
United Kingdom

L.R. Gardiner
Auburn University
Department of Management
322 Thach Hall
AUBURN
Alabama 36849-5241
U.S.A.

T. Gear
Decision Dynamics Limited
Parsons Walk
P.O. Box 142
WIGAN
Lancashire WN1 1RW
United Kingdom

A.S. Goncalves
Universidade Tecnicade Lisboa
Instituto Superior de Economia
Rua Miguel Lupi 20
GAB.604
1200 Lisboa
Portugal

B. Gronquist
Univ.des Sciences Sociales de Grenoble
CRISS
Economie Gestion n 90
38400 ST-MARTIN-D'HERES
France

M. Halme
Helsinki School of Economics
Runeberginkatu 14-16
00100 HELSINKI
Finland

T. Hemming
University of Stockholm
Dept. of Business Administration
106 91 STOCKHOLM
Sweden

M. Henig
Tel-Aviv University
Faculty of Management
Campus Ramat-Aviv
TEL-AVIV 69978
Israel

B. Hetherington
Manchester Business School
University of Manchester
Booth Street West
MANCHESTER M15 6PB
United Kingdom

S. Huang
Jiaozuo Mining Institute
JIAOZUO CITY
Henan Province
P.R. China

G. Islei
Manchester Business School
University of Manchester
Booth Street West
MANCHESTER M15 6PB
United Kingdom

I. Kaliszewski
Polish Academy of Science
Systems Research Institute
ul. Newelska 6
01-447 WARSZAWA
Poland

B. Karpak
Youngstown State University
Williamson School of Business Admin.
Department of Management
YOUNGSTOWN
Ohio 44555
U.S.A.

G. Kersten
Carleton University
School of Business
Decision Support Syst. Res. Lab.
OTTAWA
Ontario K1S 5B6
Canada

R. Klimberg
Boston University
School of Management
704 Commonwealth Avenue
BOSTON
Massachusetts 02215
U.S.A.

M. Koksalan
Middle East Technical University
Dept. of Industrial Engineering
Inoenue Bulvari
ANKARA
Turkey

E. Konarzewska-Gubala
Oskar Lange Academy of Economics
Institute of Cybernetics
Komandorska 118/120
53-345 WROCLAW
Poland

P. Korhonen
Helsinki School of Economics
Runeberginkatu 14-16
00100 HELSINKI
Finland

J. Kornbluth
Hebrew University
Jerusalem School of Business Admin.
JERUSALEM
Israel

M. Le Louarn
Universite Paris IX Dauphine
LAMSADE
Pl. Ml de Lattre de Tassigny
75775 PARIS CEDEX 16
France

A. Lewandowski
I.I.A.S.A.
System and Decision Science Program
2361 LAXENBURG
Austria

A.G. Lockett
Manchester Business School
University of Manchester
Booth Street West
MANCHESTER M15 6PB
United Kingdom

V. Lotfi
University of Buffalo
School of Management
310D Jacobs Management Centre
BUFFALO
New York 14269
U.S.A.

M. Luptacik
Technische Universitaet Wien
Inst. fuer Oekonometrie und O.R.
Argentinierstrasse 8/119
1040 WIEN
Austria

G.R. Mallory
Carleton University
Decision Analysis Laboratory
School of Business
OTTAWA
Ontario K1S 5B6
Canada

B. Matarazzo
Facolta di Economia e Commercio
Istituto di Matematica
Corso Italia
55-95129 CATANIA
Italy

A.I. Mechitov
Institute for Systems Studies
VNIISI
Prospect 60 let Octiabrya, 9
117312 MOSCOW
U.S.S.R.

W. Michalowski
Carleton University
School of Business
OTTAWA
Ontario K1S 5B6
Canada

V. Miranda
INESC (Norte)
FEUP-DEEC
Rua dos Bragas
4099 PORTO CODEX
Portugal

A. Money
Henley Management College
HENLEY-ON-THAMES
RG9 3AU
United Kingdom

T. Nakagawa
System Sougou Kaihatsu Co. Ltd.
Hongou 4-8-17
Bunkyo-Ku
TOKYO 113
Japan

H. Nakayama
Konan University
Dept. of Applied Mathematics
Higashinada
KOBE 658
Japan

S.C. Narula
Virginia Commonwealth University
School of Business
RICHMOND
Virginia 23284-0001
U.S.A.

P. Naude
University of Manchester
Manchester Business School
Booth Street West
MANCHESTER M15 6PB
United Kingdom

M.F. Norese
Politechnico di Torino
Dip. Automatica e Informatica
C.so Duca degli Abruzzi 24
10129 TORINO
Italy

W. Ogryczak
Warsaw University
Institute of Informatics
PKiN VIII p.850
00-901 WARSAW
Poland

S. Opricovic
Faculty of Civil Engineering
Bul. Revolucije 73
11000 BELGRADE
Yugoslavia

R. Ostermark
Abo Academy
Dept. of Business Administration
Henriksgatan
20500 ABO 50
Finland

V.M. Ozernoy
California State University
School of Business and Economics
HAYWARD
California 94542
U.S.A.

P. Parizek
Institute for Applied Cybernetics
Hanulova 5/a
844 16 BRATISLAVA
Czechoslovakia

H. Pastijn
Royal Military Academy
Centre for Informatics
30 Renaissance Avenue
1040 BRUSSELS
Belgium

A.D. Pearman
University of Leeds
School of Economic Studies
LEEDS LS2 9JT
United Kingdom

J. Plocinski
Sc. Soc. for Organ. and Management
Wroclaw Branch
Rynek 33
50-102 WROCLAW
Poland

R. Ramesh
State University of New York
School of Management
325 Jacobs Management Center
BUFFALO
New York 14260
U.S.A.

M. Read
Decision Dynamics Ltd
Parsons Walk
P.O. Box 142
WIGAN
Lancashire WN1 1RW
United Kingdom

D. Rios Insua
University of Leeds
School of Computer Studies
LEEDS, LS2 9JT
United Kingdom

Y. Rommel
Universite de Paris Dauphine
LAMSADE
Place du Marechal de Lattre de Tassigny
75775 PARIS CEDEX 16
France

Y. Sawaragi
The Japan Inst. of Systems Research
Yoshida Ushinomiya-cho 4
Sakyo-ku
KYOTO 606
Japan

B. Schoner
Simon Fraser University
Faculty of Business Admin.
BURNABY
British Columbia V5A 1S6
Canada

F. Seo
Kyoto University
Kyoto Inst. of Economic Research
KYOTO 606
Japan

A. Sissouras
University of Patras
RION PATRAS
Greece

A.M. Skulimowski
Eidgenossische Techn. Hochschule Zurich
Inst. fur Automatik und Ind. Elektronik
ETH-Zentrum/ETL
8092 ZURICH
Switzerland

D. Smith
Ministry of Acriculture,Fisheries & Food
Great Westminster House
Horseferry Road
LONDON SW1P 2AE
United Kingdom

D. Solomatin
Institute for Systems Studies
USSR Academy of Science
Prospekt 60-Let Oktyabria,9
117312 MOSCOW
U.S.S.R.

D.I. Soloveitchik
Ministry of Energy and Infrastructure
Policy and Planning Division
JERUSALEM
Israel

J. Spronk
Erasmus University Rotterdam
Postbus 1738
3000 DR ROTTERDAM
Netherlands

A. Stam
University of Georgia
Department of Management Sciences
Brooks Hall
ATHENS
Georgia 30602
U.S.A.

R.E. Steuer
University of Georgia
College of Business Administration
Brooks Hall
ATHENS
Georgia 30602
U.S.A.

M.J.W. Stratford
ICI Corporate Management Services
Brunner Business Centre
Brunner House, P.O. Box 7
NORTHWICH
Cheshire CW5 4DJ
United Kingdom

T. Szapiro
Central School of Planning and Stats.
Institute of Econometrics
Al. Niepodleglosci 164
02-544 WARSZAWA
Poland

W. Szymanowski
Agricultural University of Warsaw
ul. Gierrymskiego 11 m.9
00-772 WARSAW
Poland

M.T. Tabucanon
Asian Institute of Technology
Division of Ind.Engin. & Management
P.O. Box 2754
BANKOK 10501
Thailand

E. Takeda
Ashiya University
Dept. of Industrial Education
Rokurokuso
ASHIYA 659
Japan

J. Temesi
University of Economics
Inst.of Maths. and Comp.Science
Dimitrov ter 8
BUDAPESTH-1093
Hungary

D. Timmermans
University of Gróningen
Department of Psychology
Kraneweg 4
9718 JP GRONINGEN
Netherlands

A. Tsoukias
Politecnico di Torino
Dip. Automatica e Informatica
c.so Duca degli Abruzzi 24
10129 TORINO
Italy

E. Van Dyk
University of Sussex
Mathematics & Physics Building
Falmer
BRIGHTON BN1 9QH
United Kingdom

M. Van Herwijnen
Free University
Institute for Environmental Studies
PO Box 7161
1007 MC AMSTERDAM
Netherlands

D. Vanderpooten
Universite de Paris-Dauphine
LAMSADE
Place du Marechal de Lattre de Tassigny
75775 PARIS CEDEX 16
France

S.P. Vickers
Heriot-Watt University
Dept. of Accountancy and Finance
EDINBURGH
EH1 1HX
United Kingdom

L.B. Vlacic
Inst. of Control and Computer Science
IRCA
P.O. Box 158
71000 SARAJEVO
Yugoslavia

C. Vlek
University of Groningen
Department of Psychology
Kraneweg 4
9718 JP GRONINGEN
Netherlands

H. Wallenius
University of Jyvaeskylae
Seminaarinkatu 15
40100 JYVAESKYLAE 10
Finland

J. Wallenius
Helsinki School of Economics
Runeberginkatu 14-16
00100 HELSINKI
Finland

T. Wegner
University of Cape Town
Dept. of Business Science
Private Bag
RONDEBOSCH 7700
South Africa

H.R. Weistroffer
Virginia Commonwealth University
School of Business
1015 Floyd Avenue
RICHMOND
Virginia 23284
U.S.A.

A.P. Wierzbicki
Warsaw University of Technology
Inst. of Automatic Control
Nowowiejska 15/19
00-665 WARSAW
Poland

K. Wolf
Fernuniversitaet Hagen
Fachbereich Wirtschaftswissenschaft
Lehrgebiet O.R.
5800 HAGEN
Postfach 940
W.Germany

G. Wright
Bristol Polytechnic
Coldharbour Lane
Frenchay
BRISTOL BS16 1QY
United Kingdom

H. Yano
Iwate University
Dept.of Computer Science
Faculty of Engineering
MORIOKA 020
Japan

P.L. Yu
University of Kansas
The School of Business
350E Summerfield Hall
LAWRENCE
Kansas 66045
U.S.A.

M. Zeleny
Fordham University
Grad. School of Business Admin.
113 West 60th Street
NEW YORK
U.S.A.

S. Zionts
State University of New York
BUFFALO
New York 14260
U.S.A.

K.G. Zografos
University of Miami
College of Engineering
P.O. Box 248294
CORAL GABLES
Florida 33124
U.S.A.

K. Zorychta
Warsaw University
Institute of Informatics
PKiN VIII p.850
00-901 WARSAW
Poland

ORGANISATION

European Organising Committee

Chairman : Professor A.G. Lockett	**University of Manchester (UK)**
Dr.V. Belton	**University of Strathclyde (UK)**
C.C. French	**PA Computers and Telecommunications (UK)**
Prof. T. Gal	**Fernuniversität Hagen (FRG)**
Dr. T. Gear	**Royal Ordnance (UK)**
B. Hetherington	**University of Manchester (UK)**
Dr. G. Islei	**University of Manchester (UK)**
Dr. D. Smith	**Ministry of Agriculture, Fisheries and Food (UK)**
Prof. J. Spronk	**Erasmus Universiteit Rotterdam (NL)**

Session Chairmen:

Prof. K. Bowen (UK)
Dr. T. Gear (UK)
Prof. S. French (UK)
Prof. P. Korhonen (SF)
Dr. A. Lewandowski (A)
Prof. B. Matarazzo (I)
Prof. S.C. Narula (USA)
Dr. A.D. Pearman (UK)
Prof. Y. Sawaragi (J)
Dr. D. Smith (UK)
Prof. J. Spronk (NL)
Prof. R.E. Steuer (USA)
Prof. J. Wallenius (SF)
Prof. A.P. Wierzbicki (PL)
Dr. G. Wright (UK)
Prof. P.L. Yu (USA)
Prof. S. Zionts (USA)

LIST OF SPONSORS

Abbey National Building Society

Bank of England

The BOC Group plc

British Gas plc

British Nuclear Fuels plc

N Brown Group plc

Cooperative Retail Service Ltd.

Cooperative Wholesale Society Ltd.

Courtaulds plc

Courtaulds Spinning

Halifax Building Society

Imperical Chemical Industries plc
(Colours and Fine Chemicals)

Imperial Chemical Industries plc
(Pharmaceuticals)

International Computers Ltd.

Journal of Management Studies

Kellogg Company of Great Britain Ltd.

Manchester Business School

Ocean Transport & Trading plc

Operational Research Society

Provincial Insurance

H & J Quick Group plc

Royal Insurance (U.K.) Ltd.

Scapa Group plc

Silentnight Holding plc

Springer Verlag

Unilever plc

Vickers Shipbuiling & Engineering Ltd

The Wellcome Foundation Ltd.

Whitecroft plc

Vol. 236: G. Gandolfo, P.C. Padoan, A Disequilibrium Model of Real and Financial Accumulation in an Open Economy. VI, 172 pages. 1984.

Vol. 237: Misspecification Analysis. Proceedings, 1983. Edited by T.K. Dijkstra. V, 129 pages. 1984.

Vol. 238: W. Domschke, A. Drexl, Location and Layout Planning. IV, 134 pages. 1985.

Vol. 239: Microeconomic Models of Housing Markets. Edited by K. Stahl. VII, 197 pages. 1985.

Vol. 240: Contributions to Operations Research. Proceedings, 1984. Edited by K. Neumann and D. Pallaschke. V, 190 pages. 1985.

Vol. 241: U. Wittmann, Das Konzept rationaler Preiserwartungen. XI, 310 Seiten. 1985.

Vol. 242: Decision Making with Multiple Objectives. Proceedings, 1984. Edited by Y.Y. Haimes and V. Chankong. XI, 571 pages. 1985.

Vol. 243: Integer Programming and Related Areas. A Classified Bibliography 1981–1984. Edited by R. von Randow. XX, 386 pages. 1985.

Vol. 244: Advances in Equilibrium Theory. Proceedings, 1984. Edited by C.D. Aliprantis, O. Burkinshaw and N.J. Rothman. II, 235 pages. 1985.

Vol. 245: J.E.M. Wilhelm, Arbitrage Theory. VII, 114 pages. 1985.

Vol. 246: P.W. Otter, Dynamic Feature Space Modelling, Filtering and Self-Tuning Control of Stochastic Systems. XIV, 177 pages. 1985.

Vol. 247: Optimization and Discrete Choice in Urban Systems. Proceedings, 1983. Edited by B.G. Hutchinson, P. Nijkamp and M. Batty. VI, 371 pages. 1985.

Vol. 248: Plural Rationality and Interactive Decision Processes. Proceedings, 1984. Edited by M. Grauer, M. Thompson and A.P. Wierzbicki. VI, 354 pages. 1985.

Vol. 249: Spatial Price Equilibrium: Advances in Theory, Computation and Application. Proceedings, 1984. Edited by P. T. Harker. VII, 277 pages. 1985.

Vol. 250: M. Roubens, Ph. Vincke, Preference Modelling. VIII, 94 pages. 1985.

Vol. 251: Input-Output Modeling. Proceedings, 1984. Edited by A. Smyshlyaev. VI, 261 pages. 1985.

Vol. 252: A. Birolini, On the Use of Stochastic Processes in Modeling Reliability Problems. VI, 105 pages. 1985.

Vol. 253: C. Withagen, Economic Theory and International Trade in Natural Exhaustible Resources. VI, 172 pages. 1985.

Vol. 254: S. Müller, Arbitrage Pricing of Contingent Claims. VIII, 151 pages. 1985.

Vol. 255: Nondifferentiable Optimization: Motivations and Applications. Proceedings, 1984. Edited by V.F. Demyanov and D. Pallaschke. VI, 350 pages. 1985.

Vol. 256: Convexity and Duality in Optimization. Proceedings, 1984. Edited by J. Ponstein. V, 142 pages. 1985.

Vol. 257: Dynamics of Macrosystems. Proceedings, 1984. Edited by J.-P. Aubin, D. Saari and K. Sigmund. VI, 280 pages. 1985.

Vol. 258: H. Funke, Eine allgemeine Theorie der Polypol- und Oligopolpreisbildung. III, 237 pages. 1985.

Vol. 259: Infinite Programming. Proceedings, 1984. Edited by E.J. Anderson and A.B. Philpott. XIV, 244 pages. 1985.

Vol. 260: H.-J. Kruse, Degeneracy Graphs and the Neighbourhood Problem. VIII, 128 pages. 1986.

Vol. 261: Th.R. Gulledge, Jr., N.K. Womer, The Economics of Made-to-Order Production. VI, 134 pages. 1986.

Vol. 262: H.U. Buhl, A Neo-Classical Theory of Distribution and Wealth. V, 146 pages. 1986.

Vol. 263: M. Schäfer, Resource Extraction and Market Structure. XI, 154 pages. 1986.

Vol. 264: Models of Economic Dynamics. Proceedings, 1983. Edited by H.F. Sonnenschein. VII, 212 pages. 1986.

Vol. 265: Dynamic Games and Applications in Economics. Edited by T. Başar. IX, 288 pages. 1986.

Vol. 266: Multi-Stage Production Planning and Inventory Control. Edited by S. Axsäter, Ch. Schneeweiss and E. Silver. V, 264 pages. 1986.

Vol. 267: R. Bemelmans, The Capacity Aspect of Inventories. IX, 165 pages. 1986.

Vol. 268: V. Firchau, Information Evaluation in Capital Markets. VII, 103 pages. 1986.

Vol. 269: A. Borglin, H. Keiding, Optimality in Infinite Horizon Economies. VI, 180 pages. 1986.

Vol. 270: Technological Change, Employment and Spatial Dynamics. Proceedings 1985. Edited by P. Nijkamp. VII, 466 pages. 1986.

Vol. 271: C. Hildreth, The Cowles Commission in Chicago, 1939–1955. V, 176 pages. 1986.

Vol. 272: G. Clemenz, Credit Markets with Asymmetric Information. VIII, 212 pages. 1986.

Vol. 273: Large-Scale Modelling and Interactive Decision Analysis. Proceedings, 1985. Edited by G. Fandel, M. Grauer, A. Kurzhanski and A.P. Wierzbicki. VII, 363 pages. 1986.

Vol. 274: W.K. Klein Haneveld, Duality in Stochastic Linear and Dynamic Programming. VII, 295 pages. 1986.

Vol. 275: Competition, Instability, and Nonlinear Cycles. Proceedings, 1985. Edited by W. Semmler. XII, 340 pages. 1986.

Vol. 276: M.R. Baye, D.A. Black, Consumer Behavior, Cost of Living Measures, and the Income Tax. VII, 119 pages. 1986.

Vol. 277: Studies in Austrian Capital Theory, Investment and Time. Edited by M. Faber. VI, 317 pages. 1986.

Vol. 278: W.E. Diewert, The Measurement of the Economic Benefits of Infrastructure Services. V, 202 pages. 1986.

Vol. 279: H.-J. Büttler, G. Frei and B. Schips, Estimation of Disequilibrium Models. VI, 114 pages. 1986.

Vol. 280: H.T. Lau, Combinatorial Heuristic Algorithms with FORTRAN. VII, 126 pages. 1986.

Vol. 281: Ch.-L. Hwang, M.-J. Lin, Group Decision Making under Multiple Criteria. XI, 400 pages. 1987.

Vol. 282: K. Schittkowski, More Test Examples for Nonlinear Programming Codes. V, 261 pages. 1987.

Vol. 283: G. Gabisch, H.-W. Lorenz, Business Cycle Theory. VII, 229 pages. 1987.

Vol. 284: H. Lütkepohl, Forecasting Aggregated Vector ARMA Processes. X, 323 pages. 1987.

Vol. 285: Toward Interactive and Intelligent Decision Support Systems. Volume 1. Proceedings, 1986. Edited by Y. Sawaragi, K. Inoue and H. Nakayama. XII, 445 pages. 1987.

Vol. 286: Toward Interactive and Intelligent Decision Support Systems. Volume 2. Proceedings, 1986. Edited by Y. Sawaragi, K. Inoue and H. Nakayama. XII, 450 pages. 1987.

Vol. 287: Dynamical Systems. Proceedings, 1985. Edited by A.B. Kurzhanski and K. Sigmund. VI, 215 pages. 1987.

Vol. 288: G.D. Rudebusch, The Estimation of Macroeconomic Disequilibrium Models with Regime Classification Information. VII, 128 pages. 1987.

Vol. 289: B.R. Meijboom, Planning in Decentralized Firms. X, 168 pages. 1987.

Vol. 290: D.A. Carlson, A. Haurie, Infinite Horizon Optimal Control. XI, 254 pages. 1987.

Vol. 291: N. Takahashi, Design of Adaptive Organizations. VI, 140 pages. 1987.

Vol. 292: I. Tchijov, L. Tomaszewicz (Eds.), Input-Output Modeling. Proceedings, 1985. VI, 195 pages. 1987.

Vol. 293: D. Batten, J. Casti, B. Johansson (Eds.), Economic Evolution and Structural Adjustment. Proceedings, 1985. VI, 382 pages. 1987.

Vol. 294: J. Jahn, W. Krabs (Eds.), Recent Advances and Historical Development of Vector Optimization. VII, 405 pages. 1987.

Vol. 295: H. Meister, The Purification Problem for Constrained Games with Incomplete Information. X, 127 pages. 1987.

Vol. 296: A. Börsch-Supan, Econometric Analysis of Discrete Choice. VIII, 211 pages. 1987.

Vol. 297: V. Fedorov, H. Läuter (Eds.), Model-Oriented Data Analysis. Proceedings, 1987. VI, 239 pages. 1988.

Vol. 298: S.H. Chew, Q. Zheng, Integral Global Optimization. VII, 179 pages. 1988.

Vol. 299: K. Marti, Descent Directions and Efficient Solutions in Discretely Distributed Stochastic Programs. XIV, 178 pages. 1988.

Vol. 300: U. Derigs, Programming in Networks and Graphs. XI, 315 pages. 1988.

Vol. 301: J. Kacprzyk, M. Roubens (Eds.), Non-Conventional Preference Relations in Decision Making. VII, 155 pages. 1988.

Vol. 302: H.A. Eiselt, G. Pederzoli (Eds.), Advances in Optimization and Control. Proceedings, 1986. VIII, 372 pages. 1988.

Vol. 303: F.X. Diebold, Empirical Modeling of Exchange Rate Dynamics. VII, 143 pages. 1988.

Vol. 304: A. Kurzhanski, K. Neumann, D. Pallaschke (Eds.), Optimization, Parallel Processing and Applications. Proceedings, 1987. VI, 292 pages. 1988.

Vol. 305: G.-J.C.Th. van Schijndel, Dynamic Firm and Investor Behaviour under Progressive Personal Taxation. X, 215 pages. 1988.

Vol. 306: Ch. Klein, A Static Microeconomic Model of Pure Competition. VIII, 139 pages. 1988.

Vol. 307: T.K. Dijkstra (Ed.), On Model Uncertainty and its Statistical Implications. VII, 138 pages. 1988.

Vol. 308: J.R. Daduna, A. Wren (Eds.), Computer-Aided Transit Scheduling. VIII, 339 pages. 1988.

Vol. 309: G. Ricci, K. Velupillai (Eds.), Growth Cycles and Multisectoral Economics: the Goodwin Tradition. III, 126 pages. 1988.

Vol. 310: J. Kacprzyk, M. Fedrizzi (Eds.), Combining Fuzzy Imprecision with Probabilistic Uncertainty in Decision Making. IX, 399 pages. 1988.

Vol. 311: R. Färe, Fundamentals of Production Theory. IX, 163 pages. 1988.

Vol. 312: J. Krishnakumar, Estimation of Simultaneous Equation Models with Error Components Structure. X, 357 pages. 1988.

Vol. 313: W. Jammernegg, Sequential Binary Investment Decisions. VI, 156 pages. 1988.

Vol. 314: R. Tietz, W. Albers, R. Selten (Eds.), Bounded Rational Behavior in Experimental Games and Markets. VI, 368 pages. 1988.

Vol. 315: I. Orishimo, G.J.D. Hewings, P. Nijkamp (Eds.), Information Technology: Social and Spatial Perspectives. Proceedings, 1986. VI, 268 pages. 1988.

Vol. 316: R.L. Basmann, D.J. Slottje, K. Hayes, J.D. Johnson, D.J. Molina, The Generalized Fechner-Thurstone Direct Utility Function and Some of its Uses. VIII, 159 pages. 1988.

Vol. 317: L. Bianco, A. La Bella (Eds.), Freight Transport Planning and Logistics. Proceedings, 1987. X, 568 pages. 1988.

Vol. 318: T. Doup, Simplicial Algorithms on the Simplotope. VIII, 262 pages. 1988.

Vol. 319: D.T. Luc, Theory of Vector Optimization. VIII, 173 pages. 1989.

Vol. 320: D. van der Wijst, Financial Structure in Small Business. VII, 181 pages. 1989.

Vol. 321: M. Di Matteo, R.M. Goodwin, A. Vercelli (Eds.), Technological and Social Factors in Long Term Fluctuations. Proceedings. IX, 442 pages. 1989.

Vol. 322: T. Kollintzas (Ed.), The Rational Expectations Equilibrium Inventory Model. XI, 269 pages. 1989.

Vol. 323: M.B.M. de Koster, Capacity Oriented Analysis and Design of Production Systems. XII, 245 pages. 1989.

Vol. 324: I.M. Bomze, B.M. Pötscher, Game Theoretical Foundations of Evolutionary Stability. VI, 145 pages. 1989.

Vol. 325: P. Ferri, E. Greenberg, The Labor Market and Business Cycle Theories. X, 183 pages. 1989.

Vol. 326: Ch. Sauer, Alternative Theories of Output, Unemployment, and Inflation in Germany: 1960–1985. XIII, 206 pages. 1989.

Vol. 327: M. Tawada, Production Structure and International Trade. V, 132 pages. 1989.

Vol. 328: W. Güth, B. Kalkofen, Unique Solutions for Strategic Games. VII, 200 pages. 1989.

Vol. 329: G. Tillmann, Equity, Incentives, and Taxation. VI, 132 pages. 1989.

Vol. 330: P.M. Kort, Optimal Dynamic Investment Policies of a Value Maximizing Firm. VII, 185 pages. 1989.

Vol. 331: A. Lewandowski, A.P. Wierzbicki (Eds.), Aspiration Based Decision Support Systems. X, 400 pages. 1989.

Vol. 332: T.R. Gulledge, Jr., L.A. Litteral (Eds.), Cost Analysis Applications of Economics and Operations Research. Proceedings. VII, 422 pages. 1989.

Vol. 333: N. Dellaert, Production to Order. VII, 158 pages. 1989.

Vol. 334: H.-W. Lorenz, Nonlinear Dynamical Economics and Chaotic Motion. XI, 248 pages. 1989.

Vol. 335: A. G. Lockett, G. Islei (Eds.), Improving Decision Making in Organisations. Proceedings. IX, 606 pages. 1989.

GPSR Compliance

The European Union's (EU) General Product Safety Regulation (GPSR) is a set of rules that requires consumer products to be safe and our obligations to ensure this.

If you have any concerns about our products, you can contact us on ProductSafety@springernature.com

In case Publisher is established outside the EU, the EU authorized representative is:

Springer Nature Customer Service Center GmbH
Europaplatz 3
69115 Heidelberg, Germany

Batch number: 10838062

Printed by Printforce, the Netherlands